Lecture Notes in Computer Science 11942

More information about this series at http://www.springer.com/series/7412

Bhabesh Deka · Pradipta Maji ·
Sushmita Mitra · Dhruba Kumar Bhattacharyya ·
Prabin Kumar Bora · Sankar Kumar Pal (Eds.)

Pattern Recognition and Machine Intelligence

8th International Conference, PReMI 2019
Tezpur, India, December 17–20, 2019
Proceedings, Part II

 Springer

Editors
Bhabesh Deka
Tezpur University
Tezpur, India

Sushmita Mitra
Indian Statistical Institute
Kolkata, India

Prabin Kumar Bora
Indian Institute of Technology Guwahati
Guwahati, India

Pradipta Maji
Indian Statistical Institute
Kolkata, India

Dhruba Kumar Bhattacharyya
Tezpur University
Tezpur, India

Sankar Kumar Pal
Indian Statistical Institute
Kolkata, India

ISSN 0302-9743   ISSN 1611-3349 (electronic)
Lecture Notes in Computer Science
ISBN 978-3-030-34871-7   ISBN 978-3-030-34872-4 (eBook)
https://doi.org/10.1007/978-3-030-34872-4

LNCS Sublibrary: SL6 – Image Processing, Computer Vision, Pattern Recognition, and Graphics

This Springer imprint is published by the registered company Springer Nature Switzerland AG
The registered company address is: Gewerbestrasse 11, 6330 Cham, Switzerland

Foreword

Welcome to PReMI 2019, the 8th International Conference on Pattern Recognition and Machine Intelligence. We were glad to have you among us at this prestigious event to share exciting results of your pattern recognition and machine intelligence research. This year we were fortunate to have the conference located in a historic city of Northeastern India–Tezpur. We hope that you enjoyed this idyllic place on the bank of the mighty Brahmaputra and that you took the opportunity to enjoy the beauty and uniqueness of this cultural hub of Assam. The Technical Program Committee worked very hard to put together an outstanding program of 90 full-length oral papers and 41 short oral-cum-poster presentations that aptly reflect the recent accomplishments in machine intelligence and pattern recognition research from across the globe. This represents a significant growth in the number of presentations compared to previous years. Although, we had planned to adopt the single-session format so that everyone could attend all oral presentations and learn from them, due to the large number of high-quality papers from multiple sub-themes, we decided to have parallel sessions.

The conference began with two parallel tutorial sessions followed by another, led by three eminent academic experts. The first session was on 'Compressed Sensing' by Dr. Ajit Rajwade from IIT Bombay. The second session was led by Prof. Chiranjib Bhattacharyya from IISc Bangalore on 'Machine Learning and its Application in Industrial Problem Solving.' The third session was delivered by Dr. M. Tanveer from IIT Indore on 'Machine Algorithms for the Diagnosis of Alzheimer's Disease.' The main conference began the next day led by two plenary talks, five invited talks, and two industry talks, followed by the oral presentations.

The first plenary talk was delivered by Prof. Witold Pedrycz of University of Alberta, Canada on 'Granular Artificial Intelligence' to highlight its applications in modeling environments and pattern recognition. Prof. Jayaram Udupa of University of Pennsylvania, Philadelphia, USA delivered the next plenary talk on 'Biomedical Imaging.' The technical sessions were organized in parallel oral presentations under ten major themes: Bioinformatics, Biomedical Signal Processing, Deep Learning, Soft Computing, Image and Video processing, Information Retrieval, Machine Learning and Pattern Recognition, Remote Sensing and Signal Processing, and Smart and Intelligent Sensors. Each of these sessions began with an invited academic or industry talk. There were five invited academic talks and two industry talks. Prof. Pushpak Bhattacharyya of IIT Patna delivered an invited talk on 'Imparting Sentiment and Politeness on Computers.' Prof. C. V. Jawahar of IIIT Hyderabad delivered his invited talk on 'Beyond Text Detection and Recognition: Emerging Opportunities in Scene Understanding.' Another invited talk on 'Cognitive Analysis using Physiological Sensing' was delivered by Prof. S. K. Saha of Jadavpur University. The fourth invited talk focused on 'The Rise of Hate Content in Social Media,' and was presented by Dr. Animesh Mukherjee of IIT Kharagpur. Another invited talk on wireless networks was delivered by Prof. Sudip Misra of IIT Kharagpur.

PReMI 2019 was also be an excellent platform for facilitating industry-academic collaboration. As in previous years, this year's event included industry talks to better expose academic researchers to real-life problems and recent application-oriented developments. The first industry talk was delivered by Dr. Praneeth Netrapalli from Microsoft Research India. His talk focused on 'How to Escape Saddle Point Efficiently.' The second industry talk was on 'Machine Learning and its Applications in Remote Sensing Data Classification,' was delivered by Dr. Anil Kumar from the Department of Photogrammetry and Remote Sensing, Indian Institute of Remote Sensing, Indian Space Research Organization (ISRO), Dehradun, India.

Another attractive and distinguishing feature of PReMI 2019 was the inclusion of a Doctoral Symposium in the name of late Prof. C. A. Murthy, which was held on the first day of the conference. This forum was useful for PhD students to showcase their research outcomes and seek advice and mentorship from prominent scientists and engineers. Although we received a good number of submissions for this symposium, we were only able to select 12 presentations.

Finally, we feel privileged to acknowledge and appreciate the expertise and hard work of the various committees, sub-committees, and individuals of PReMI 2019, who were deeply involved in making this prestigious event an outstanding success. It was truly a memorable experience for us to work with such a professional group. Our sincere and heartfelt thanks to all.

Please enjoy the proceedings!

December 2019   Sushmita Mitra
                                         Dhruba Kumar Bhattacharyya
                                                 Prabin Kumar Bora

Preface

Recent technological advancements have played a prominent role in directing Information Technology research towards making 'intelligent' machines. Traditionally, intelligence has been commonly associated with humans as an intellectual characteristic, by virtue of which they demonstrate the ability to transcend trivial computations or decisions. Intelligent computations or decisions act as a driving force to deal with problems from a wide range of domains. Intelligent machines have the ability to acquire crucial knowledge from the environment, which enables them to learn and draw significant inferences based on evidences. Since these machines are knowledge-oriented, they possess the ability to generalize which makes them quite reliable. This volume covers all these aspects of machine intelligence, as an outcome of PReMI 2019, an international conference on Pattern Recognition and Machine Intelligence, held at Tezpur University, India, during December 17–20, 2019. It includes 90 full-length and 41 short papers from across the globe. It aims to provide a comprehensive and in-depth discussion of the contemporary research trends in the domain of pattern recognition and machine intelligence. The conference began with two plenary talks followed by five invited talks and oral presentations. The first plenary talk on 'Granular Artificial Intelligence' highlighted its applications in modelling environments and pattern recognition and was delivered by Prof. Witold Pedrycz of University of Alberta, Canada. The second plenary talk by Prof. Jayaram Udupa of University of Pennsylvania, Philadelphia, USA focused on 'Biomedical Imaging.'

The technical sessions included parallel oral presentations under six major themes: Machine Learning and Deep Learning, Bioinformatics and Medical Imaging, Pattern Recognition and Remote Sensing, Intelligent Sensor and Information Retrieval, Signal, Image, and Video Processing, and Evolutionary and Soft Computing. The sessions included five academic talks and two industry talks. Prof. P. Bhattacharyya of IIT Patna delivered the first invited talk on 'Imparting Sentiment and Politeness on Computers.' Another invited talk on 'Beyond Text Detection and Recognition: Emerging Opportunities in Scene Understanding' was by Prof. C. V. Jawahar of IIIT Hyderabad. Prof. S. K. Saha of Jadavpur University focused his talk on 'Cognitive Analysis using Physiological Sensing.' The fourth invited talk, presented by Dr. Animesh Mukherjee of IIT Kharagpur, was on 'The Rise of Hate Content in Social Media.' Another invited talk on wireless networks was delivered by Prof. Sudip Misra of IIT Kharagpur. As in previous editions, PReMI 2019 intended to bridge the gap between academia and industry by adopting a collaborative approach in the relevant fields. As in previous years, PReMI 2019 included two industry talks to provide better exposure to academic researchers in real-life problems and recent application-oriented developments. Dr. Praneeth Netrapalli from Microsoft Research India deliberated on 'How to Escape Saddle Point Efficiently.' Another industry talk by Dr. Anil Kumar from the Department of Photogrammetry and Remote Sensing, Indian Institute of Remote Sensing,

Indian Space Research Organization (ISRO), Dehradun, India, focused on 'Machine Learning and its Applications in Remote Sensing Data Classification.'

Section I of this volume primarily deals with Machine Learning and Deep Learning and their applications in diverse domains. Kannadasan et al. propose an approach to predict performance indices in Computer Numerical Control (CNC) milling using regression trees. Subramanyam et al. introduce a machine learning based method to detect dyscalculia. Dammu and Surampudi explore the temporal dynamics of the brain using Variational Bayes Hidden Markov model with applications in autism. Dutta et al. present a robust dense or sparse crowd identification technique based on classifier fusion. Buckchash and Raman present a novel sampling algorithm for sustained Self-supervised pre-training for Temporal Order Verification. John et al. analyze the retraining conditions of a network after pruning layer-wise. Jain and Phophalia introduce M-ary Random Forest algorithm, where multiple features are used for splitting at a time instead of just one in the traditional Random Forest algorithm. Jain et al. propose a dynamic weighing scheme for Random Forest algorithm. Alam and Sobha present a method for instance ranking using data complexity measures for training set selection. Karthik and Katika introduce an identity independent face anti-spoofing technique based on Random scan patterns. Das et al. propose a method for automatic attribute profile construction for spectral-spatial classification of hyperspectral images. Rathi et al. propose an enhanced depression detection method from facial cues using univariate feature selection techniques. Shanmugam and Tamilselvan propose a game theoretic approach to design an efficient mechanism for identifying a trustworthy cloud service provider by classifying the providers based on how they cooperate to form a coalition. Prasad et al. propose an incremental k-means clustering method to improve the quality of clusters. This session also includes several significant contributions on Deep Learning. Challa et al. propose a multi-class deep all CNN architecture for detection of diabetic retinopathy using retinal fundus images. Maiti et al. aim to provide a solution to the problem faced in real-time vehicle detection in aerial images using skip-connected convolutional network. Repala and Dubey use Convolutional Neural Networks for unsupervised depth estimation. Basavaraju et al. introduce a Deep CNN based technique with minimum number of skip connections to derive a High Resolution image from a Low Resolution image. Mazumdar et al. propose a technique to detect image manipulations using Siamese Convolution Neural Networks. Mishra et al. present a deep learning based model comprising of causal convolutional layers for load forecasting. Trivedi et al. present a technique for facial expression recognition using multichannel CNN. Gupta et al. propose a data driven sensing approach for action recognition. Sharma et al. propose a method for gradually growing residual and self-attention based dense deep back propagation network for large scale super-resolution image. M J et al. propose a method to solve 3D object classification on point cloud data using 3D Grid Convolutional Neural Networks (GCNN). Singh et al. introduce a new stegananalysis approach to learn prominent features and avoid loss of stego signals using densely connected convolutional network. Rastogi and Gangnani outline a generalized semi-supervised learning framework for multi-category classification with generative adversarial networks.

Section II covers topics from the field of Bioinformatics and Medical Imaging. Mahapatra and Mukherjee introduce GRAphical Footprint for classifying species in large scale genomics. While Pant and Paul present an effective clustering method to produce enriched gene clusters using biological knowledge, Paul et al. report the impact of continuous evolution of Gene Ontology on similarity measures. Kakati et al. introduce DEGNet, a deep neural network to predict the up-regulating and down-regulating genes from Parkinson's and breast cancer RNA-seq datasets. Saha et al. perform a survival analysis study with the integration of RNA-seq and clinical data to identify breast cancer subtype specific genes. Barnwal et al. present a deep learning based Optical Coherence Tomography (OCT) image classifier for Intra Vitreal Anti-VEGF therapy. In another effort, Patowary and Bhattacharyya present an effective method for biomarker identification of Esophageal Squamous Cell Carcinoma using integrative analysis of Differentially Expressed genes. Jana et al. present a gene selection technique using a modified Particle Swarm Optimization approach. This session also includes several noteworthy contributions in the field of Biomedical Applications. Baruah et al. propose a model to simulate the effects of signal interference by mathematically modeling a pair of dendritic fibers using cable equation. While Dasgupta et al. cluster EEG signals of epileptic patients and normal individuals using feed forward neural networks, Dammu et al. report a new approach that uses brain dynamics in the classification of autistic and neurotypical subjects using rs-fMRI data. While Singh et al. use Stockwell transform to detect Dysrhythmia in ECG, Kumar et al. propose an energy efficient MECG reconstruction method. Chowdhury et al. investigate the changes in the brain network dynamics between alcoholic and non-alcoholic groups using electro-encephalographic signals. Das and Mahanta analyze segmentation techniques for cell identification from biopsy tissue samples of childhood medulloblastoma microscopic images based on conventional machine learning methods. Mahanta et al. introduce a method for automatic counting of platelets and white blood cells from blood smear images. Kumar and Maji present an effective method for automatic recognition of virus particles in Transmission Electron Microscopy (TEM) images. Roy et al. use Deep Convolutional Neural Network to detect Necrosis in mice liver tissue by classifying microscopic images after dividing them into small patches in the preprocessing phase. Deshpande and Bhatt employ Bayesian deep learning to register the medical images that are rendered corrupt by nonlinear geometric distortions. Das and Mahanta develop a novel method where the features of Adenocarcinoma and Squamous Cell Carcinoma of a histological image are taken from various statistical and mathematical models implemented on the coefficients of wavelet transform of an image. Banerjee et al. propose a new iterative method to remove the effects of motion artifacts from multiple non-simultaneous angiographic projection. Datta and Deka propose a Compressed Sensing based parallel MRI reconstruction method. Deka et al. introduce Single-image Super Resolution technique for diffusion-weighted and spectroscopic MR images. Konar et al. propose a Quantum-Inspired Bidirectional Self-Organizing Neural Network architecture for fully automatic segmentation of T1-weighted contrast enhanced MR images. Sasmal et al. present a two-class classification of colonoscopic polyps using multi directions and multi frequency texture analysis. Section III is focused on Pattern Recognition and Remote Sensing. Ahmed and Nath introduce a modified Conditional FP-tree, an efficient frequent pattern mining approach that uses both bottom-up and top-down approach to generate frequent patterns. Malla and Bhavani present a method to predict link weights for directed Weighted Signed Networks using features from network and it's dual. Dhar et al. present a content resemblance based Author Identification System using ensemble learning techniques. Baruah and Bharali present a comparative study of the airline networks on India with ANI based on some network parameters. Chittaragi and Koolagudi present a spectral feature-based dialect classification method from stop consonants. Saini et al. introduce neighborhood concepts to enhance the traditional Self-organizing maps based multi-label classification. This session also includes several interesting contributions in the field of Remote Sensing. Gadhiya et al. propose a multi frequency PolSAR image classification algorithm with stacked autoencoder based feature extraction and superpixel generation. Parikh et al. introduce an ensemble technique for land cover classification problems. Sarmah and Kalita present a supervised band selection method for hyperspectral images using information gain ratio and clustering. Baruah et al. present a non-sub-sampled shearlet transform based remote sensing image retrieval technique. Hire et al. propose a perception based navigation approach for autonomous ground vehicles using Convolutional Neural Networks (CNN). Shah et al. introduce a deep learning architecture using Capsule Network for automatic target recognition from Synthetic Aperture Radar images. Rohit and Mishra present an end-to-end trainable model of Generative Adversarial Networks used to hide audio data in images. Mankad et al. investigate feature reduction techniques for replay anti-spoofing in voice biometrics.

Session IV includes several interesting research outcomes in the field of Intelligent Sensor and Information Retrieval. Devi et al. propose a method where two layers of CNT-BioFET are fabricated for Creatinine detection. While Hazarika et al. present the modeling and analyzing of long-term drift observed in ISFET, Jena et al. present Significance-based Gate Level Pruning (SGLP) technique to design an approximate adder circuit for image processing application. Senapati and Sahu model mathematically a MOS-based patch electrode multilayered capacitive sensor. Nath et al. propose the design, implementation and working of a syringe based automated fluid infusion system integrated with micro-channel platforms for lab-on-a-chip application. Several significant works on Information Retrieval will also be discussed in this session. Chakrabarty et al. report a Joke Recommendation system, based primarily on Collaborative Filtering and Joke-Reader segmentation influenced by the similarity of the user's preference patterns. Vijai Kumar et al. propose TagEmbedSVD, a tag-based Cross-Domain Collaborative model aimed to enhance personalized recommendations in the cross-domain setting. Two of the many important aspects of recommendation systems are diversity and long tail item recommendation, which can be improved by an efficient method as presented by Agarwal et al. Pahal et al. introduce a context-aware reasoning framework that caters to the need and preferences of a group of users in a smart home environment by providing contextually relevant recommendations. Anil et al. The authors introduce a method to apply meta-path for network embedding in mining heterogeneous DBLP network. Kumar et al. introduce two techniques, one to automatically detect and filter null tweets in the Twitter data and another to identify sarcastic tweets using context within a tweet. Yumnam and Sharma propose a grammar-driven approach to parse Manipuri language using Earley's parsing algorithm. Gomathinayagam et al. introduce an information fusion-based approach for query expansion for news articles retrieval. Kumar and Singh propose a prioritized named entity driven LDA, a variant of topic modeling method LDA, to address the issue of overlapping topics by prioritizing named entities related to the topics. Yadav et al. perform sentiment analysis to extract sentiments from a piece of text using supervised and unsupervised approach. Kundu et al. propose a method for finding active experts for a new question in order to improve the effectiveness of a question routing process in community question answering services.

Section V covers topics from the fields of Signal, Image, and Video Processing. Saikia et al. describe a case study involving a framework to classify facies categories in a reservoir using seismic data by employing machine learning models. Sahoo and Dandapat introduce a technique to analyze the changes in speech source signals under physical exercise induced out-of-breadth condition. Mukherjee et al. present a long short-term memory-recurrent neural network for segregating musical chords from clips of short durations which can aid in automatic transcription. Kamble et al. propose a novel Teager energy based sub-band features for audio acoustic scene detection and classification. Pj et al. propose an audio replay attack detection technique using non-voiced audio segments and deep learning models like CNN to classify the audio as genuine or replay. Jyotishi and Dandapat use inverse filtering-based technique to present a novel feature to represent the amount of nasalization present in a vowel. Bhat and Shekar propose an approach for iris recognition by learning fragile bits on multi-patches using monogenic riesz signals. Baghel et al. present a shouted and normal speech classification detection mechanism using 1D CNN architecture. Qadir et al. aim to provide a quantitative analysis of electroencephalographic-based cognitive load while driving in Virtual Reality (VR) environment compared to a fixed non-VR environment. A good number of contributions on Image and Video Processing are also part of this session. Mukherjee et al. propose an unsupervised detection technique for mine regions with reclamation activity from satellite images. Meetei et al. introduces a text detection technique in natural scene and document images in Manipuri and Mizo. Mukherjee et al. propose a method to generate segmented surface of a mapping environment in modeling 3D objects from Lidar data. Koringa and Mitra propose a class similarity based orthogonal neighborhood preserving projection technique for recognition of images with face and hand-written numerals. Mondal et al. introduce a novel combinatorial algorithm for segmentation of articulated components of 3D digital objects using Curve Skeleton. Rajpal et al. propose EAI-Net, an effective and accurate iris segmentation network based on U-Net architecture. Memon et al. use ANN and XGBoost algorithms to classify data for land cover categorization in Mumbai region. Selvam and Mishra introduce a multi-scale attention aided multi-resolution feature extractor baseline network for human pose estimation. Tiwari et al. present a CNN-based method to detect splicing forgery in images using camera-specific features. Vishwakarma et al. propose a multi-focus image fusion technique using sparse representation and modified difference. Biswas and Barma focus on the quality assessment of agricultural product based on microscopic image, generated by Foldscope. Kagalkar et al. propose a learning-based pipeline with image clustering and image selection methods for 3D reconstruction of heritage sites. Sharma and Dey describe a method to analyze the texture quality of fingerprint images using Local Contrast Phase Descriptor. Adarsh et al. present a novel framework for detecting and filling missing regions in point cloud data using clustering techniques. Mullah et al. present a sparsity regularization based spatial-spectral super-resolution method for multispectral remote sensing image. Chetia et al. introduce a quantum image edge detection technique based on four directional sobel operator. Hatibaruah et al. propose a method for texture image retrieval using multiple low and high pass filters and decoded sparse local binary pattern. Devi and Borah propose a new feature extraction approach where both multi-layer and multi-model features are extracted from pre-trained CNNs for aerial scene classification. Bhowmik et al. present a novel high-order Vector of Locally Aggregated Descriptors (VLAD) with increased discriminative power for scalable image retrieval. Pal et al. perform Image-based analysis of patterns formed in drying drops of a colloidal solution. Choudhury and Sarma propose a two-stage framework for detection and segmentation of writing events in air-written Assamese characters. Purwar et al. present an innovative method to offer a promising deterrent against alcohol abuse using Augmented Reality (AR) as a tool. Roy and Bag introduce a detection mechanism for handwritten document forgery by analyzing writer's hand-writing. Prabhu et al. use it as a feature extractor for face recognition techniques. Naosekpam et al. propose a novel scalable non-parametric scene parsing system based on super-pixels correspondence. Saurav et al. introduce an approach towards automatic autonomous vision-based powerline segmentation in aerial images using convolutional neural networks. Bhunia et al. present a new correlation filter based visual object tracking method to improve accuracy and robustness of trackers. Kumar et al. introduce an efficient way to detect objects in 360° videos for robust tracking. Singh and Sharma introduce an effective hybrid change detection algorithm in real-time applications using centre-symmetric local binary patterns. Jana et al. present a novel multi-tier fusion strategy for event classification in unconstrained videos using deep neural networks.

In Session VI, several significant contributions in the field of Evolutionary and Soft Computing are reported. Dhanalakshmy et al. analyze and empirically validate the impact of mutation scale factor parameter of the Differential Evolution algorithm. Pournami et al. discuss a scheme to modify the conventional PSO algorithm and use it to present an image registration algorithm. Lipare and Edla propose an approach where a shuffling strategy is applied to PSO algorithm for improving energy efficiency in WSN. Srivatsa et al. propose a GA based solution to solve the classic Sudoku problem. Indu et al. present a critical analysis of an existing prominent graph model of evolutionary algorithms. Singh and Bhukya present an evolutionary approach based on a steady-state GA for selection of multi-point relays in Mobile Ad-Hoc networks. Shaji et al. propose a new Aggregated Rank Removal Heuristic applied to adaptive large neighborhood search to solve Work-over Rig Scheduling Problem. Saharia and Sarmah report a method for optimal design of a DC-DC converter with the goal of minimizing overall losses. Bandagar et al. propose a MapReduce based distributed/parallel approach for standalone fuzzy-rough attribute reduction algorithm. Bar et al. attempt to find an optimal rough set reduct having least number of induced equivalence classes or granules with the help of A* Search algorithm.

We are very grateful to all the esteemed reviewers who were deeply involved in the review process and helped improve the quality of the research contributions. We are privileged and delighted to acknowledge the continuous support received from various committees, sub-committees, and Springer LNCS in preparing this volume of the prestigious PReMI 2019. Thanks are also due to Sushant Kumar, Muddashir, and

Kabita for taking on the major load of secretarial jobs. Finally, we also extend our heartfelt thanks to all those who helped host the PReMI 2019 reviewing process on the EasyChair.org site. We hope that PReMI 2019 was an academically productive conference and you will find the proceedings to be a valuable source of reference for your ongoing and future research.

We hope you will enjoy the proceedings!

December 2019

Pradipta Maji
Bhabesh Deka
Sushmita Mitra
Dhruba Kumar Bhattacharyya
Prabin Kumar Bora
Sankar Kumar Pal

Message from the Honorary General Chair

I am delighted to see that the eighth edition of the biennial International Conference on Pattern Recognition and Machine Intelligence, PReMI 2019, was held, for the first time, in the north-east region of our country at Tezpur University, Assam, India, during December 17–20, 2019. Assam is the most vibrant among the eight states in the north-east, rich in natural resources, and has a great touristic charm. PReMI 2017 was held in the year that marked the 125th birthday of late Prof. Prasanta Chandra Mahalanobis, the founder of our Indian Statistical Institute (ISI). PReMI 2019, on the other hand, was organized when our ISI was preparing to celebrate the birth centenary of another doyen in statistics, namely, Prof. C. R. Rao, a living legendary. Prof. Rao has always been an inspiration to us, and was associated in different capacities with PReMI.

Since its inception in 2005, PReMI has always drawn big responses globally in terms of paper submission. This year, PReMI 2019 was no exception. It has a nice blend of plenary and invited talks, and high-quality research papers, covering different facets of pattern recognition and machine intelligence with real-life applications. Both classical and modern computing paradigms are explored. Special emphasis has been given to contemporary research areas such as big data analytics, deep learning, AI, Internet of Things, and Smart and Intelligent Sensors through both regular and special sessions. Some pre-conference tutorials were also arranged for the beginners. All this made PReMI 2019 an ideal state-of-the-art platform for researchers and practitioners to exchange ideas and enrich their knowledge.

I thank all the participants, speakers, reviewers, different chairs, and members of various committees for making this event a grand success. My thanks are due to the sponsors for their support, and Springer for publishing the PReMI proceedings under the prestigious LNCS series. Last, but not the least, I sincerely acknowledge the support of Tezpur University in hosting the event. I believe, the participants had an academically fruitful and enjoyable stay in Tezpur.

December 2019   Sankar Kumar Pal

Organization

Conference Committee

Patrons

Vinod Kumar Jain — Tezpur University, India
Sanghamitra Bandyopadhyay — Indian Statistical Institute, Kolkata, India

Honorary General Chair

Sankar Kumar Pal — Indian Statistical Institute, Kolkata, India

General Co-chairs

Sushmita Mitra — Indian Statistical Institute, Kolkata, India
Dhruba K. Bhattacharyya — Tezpur University, India
Prabin K. Bora — Indian Institute of Technology Guwahati, India

Program Co-chairs

Bhabesh Deka — Tezpur University, India
Pradipta Maji — Indian Statistical Institute, Kolkata, India

Organizing Co-chairs

Partha Pratim Sahu — Tezpur University, India
Kuntal Ghosh — Indian Statistical Institute, Kolkata, India
Prithwijit Guha — Indian Institute of Technology Guwahati, India

Joint Organizing Co-chair

Vijay Kumar Nath — Tezpur University, India

Plenary Co-chairs

Ashish Ghosh — Indian Statistical Institute, Kolkata, India
Nityananda Sarma — Tezpur University, India

Industry Liaisons

Manabendra Bhuyan — Tezpur University, India
P. L. N. Raju — NESAC (ISRO), India
Darpa Saurav Jyethi — Indian Statistical Institute, NE Centre, Tezpur, India

International Liaisons

Tianrui Li — Southwest Jiaotong University, China

| E. J. Ientilucci | RIT, Rochester, USA |
| Sergei O. Kuznetsov | HSE, Moscow, Russia |

Tutorial Co-chairs

| Malay Bhattacharyya | Indian Statistical Institute, Kolkata, India |
| Arijit Sur | Indian Institute of Technology Guwahati, India |

Publication Co-chairs

| Sarat Saharia | Tezpur University, India |
| Swati Choudhury | Indian Statistical Institute, Kolkata, India |

Publicity Co-chairs

Manas Kamal Bhuyan	Indian Institute of Technology Guwahati, India
Utpal Sharma	Tezpur University, India
Sanjit Maitra	Indian Statistical Institute, NE Centre, Tezpur, India

Webpage Chair

| Santanu Maity | Tezpur University, India |

Advisory Committee

Anil K. Jain	Michigan State University, USA
B. L. Deekshatulu	Jawaharlal Nehru Technological University, India
Andrzej Skowron	University of Warsaw, Poland
Rama Chellappa	University of Maryland, USA
Witold Pedrycz	University of Alberta, Canada
David W. Aha	Naval Research Laboratory, USA
B. Yegnanarayana	IIIT Hyderbad, India
D. K. Saikia	Tezpur University, India
Jiming Liu	Hong Kong Baptist University, China
Ronald Yager	Iona College, USA
Henryk Rybinski	Warsaw University of Technology, Poland
Jayaram Udupa	University of Pennsylvania, USA
Malay K. Kundu	Indian Statistical Institute, Kolkata, India
Sukumar Nandi	IIT Guwahati, India
S. N. Biswas	Indian Statistical Institute, Kolkata, India

Technical Program Committee

A. R. Vasudevan	NIT Calicut, India
Aditya Bagchi	ISI, Kolkata, India
Ajay Agarwal	CSIR-CEERI, Pilani, India
Alok Kanti Deb	IIT Kharagpur, India
Alwyn Roshan Pais	NIT Surathkal, India

Amit Sethi	IIT Bombay, India
Amita Barik	NIT Durgapur, India
Amrita Chaturvedi	IIT(BHU), Varanasi, India
Ananda Shankar Chowdhury	Jadavpur University, Kolkata, India
Animesh Mukherjee	IIT Kharagpur, India
Anindya Halder	NEHU, Shillong, India
Anjana Kakoti Mahanta	Gauhati University, India
Anubha Gupta	IIIT Delhi, India
Anuj Sharma	Punjab University, India
Arindam Karmakar	Tezpur University, India
Arnab Bhattacharya	IIT Kanpur, India
Arun Kumar Pujari	Central University of Rajasthan, India
Aruna Tiwari	IIT Indore, India
Ashish Anand	IIT Guwahati, India
Asif Ekbal	IIT Patna, India
Asim Banerjee	DA-IICT, Gandhinagar, India
Asit Kumar Das	IIEST, Shibpur, India
Ayesha Choudhary	JNU, Delhi, India
B. Surendiran	NIT, Puducherry, India
Bhabatosh Chanda	ISI, Kolkata, India
Bhargab Bhattacharya	ISI, Kolkata, India
Bhogeswar Borah	Tezpur University, India
Birmohan Singh	SLIET, Punjab, India
Debanjan Das	IIIT Naya Raipur, India
Debasis Chaudhuri	DIC, DRDO, Panagarh, India
Debnath Pal	IISc, Bangalore, India
Deepak Mishra	IIST, Trivandarum, India
Devanur S. Guru	University of Mysore, India
Dinabandhu Bhandari	Heritage Institute of Technology, Kolkata, India
Dinesh Bhatia	NEHU, Shillong, India
Dipti Patra	NIT Rourkela, India
Dominik Slezak	University of Warsaw, Poland
Francesco Masulli	University of Genova, Italy
Goutam Chakraborty	Iwate Prefectural University, Japan
Hari Om	IIT (ISM), Dhanbad, India
Hrishikesh Venkataraman	IIIT Sri City, India
Indira Ghosh	JNU, Delhi, India
Jagadeesh Kakarla	IIIT-D&M, Kancheepuram, India
Jainendra Shukla	IIIT Delhi, India
Jamuna Kanta Sing	Jadavpur University, India
Jayanta Mukhopadhyay	IIT Kharagpur, India
Joydeep Chandra	IIT Patna, India
Kamal Sarkar	Jadavpur University, India
Kandarpa Kumar Sarma	Gauhati University, India
K. Manglem Singh	NIT Manipur, India
Krishna P. Miyapuram	IIT Gandhinagar, India

Satish Chand	JNU, Delhi, India
Shajulin Benedict	IIIT Kottayam, India
Shanmuganathan Raman	IIT Gandhinagar, India
Sharad Sinha	IIT Goa, India
Shiv Ram Dubey	IIIT Sri City, India
Shubhra Sankar Ray	ISI, Kolkata, India
Shyamanta M. Hazarika	IIT Guwahati, India
Sivaselvan Balasubramanian	IIIT-D&M, Kancheepuram, India
Snehashish Chakraverty	NIT Rourkela, India
Snehasis Mukherjee	IIIT Sri City, India
Soma Biswas	IISc, Bangalore, India
Somnath Dey	IIT Indore, India
Sourangshu Bhattacharya	IIT Kharagpur, India
Srilatha Chebrolu	NIT Andhra Pradesh, India
Srimanta Mandal	DA-IICT, Gandhinagar, India
Srinivas Padmanabhuni	IIT Tirupati, India
Sriparna Saha	IIT Patna, India
Subhash Chandra Yadav	Central University of Jharkhand, India
Sudip Paul	NEHU, Shillong, India
Sujit Das	NIT Warangal, India
Sukanta Das	IIEST, Shibpur, India
Sukhendu Das	IIT Madras, India
Sukumar Nandi	IIT Guwahati, India
Suman Mitra	DA-IICT, Gandhinagar, India
Susmita Ghosh	Jadavpur University, India
Suyash P. Awate	IIT Bombay, India
Swagatam Das	ISI, Kolkata, India
Swanirbhar Majumder	Tripura University, India
Swapan Kumar Parui	ISI, Kolkata, India
Swarnajyoti Patra	Tezpur University, India
Tanmoy Som	IIT(BHU), Varanasi, India
Tony Thomas	IIITM-K, Kerala, India
Ujjwal Bhattacharya	ISI, Kolkata, India
Utpal Garain	ISI, Kolkata, India
V. M. Manikandan	IIIT Kottayam, India
V. Susheela Devi	IISc, Bangalore, India
Varun Bajaj	IIIT-D&M, Jabalpur, India
Vijay Bhaskar Semwal	MANIT, Bhopal, India
Vinod Pankajakshan	IIT Roorkee, India
Viswanath Pulabaigari	IIIT Sri City, India

Additional Reviewers

Abhijit Dasgupta
Abhirup Banerjee
Abhishek Bal
Abhishek Sharma
Airy Sanjeev
Ajoy Mondal
Alexy Bhowmick
Amalesh Gope
Amaresh Sahoo
Anirban Lekharu
Ankita Mandal
Anwesha Law
Aparajita Khan
Apurba Sarkar
Arindam Biswas
Arpan K. Maiti
Ashish Phophalia
Ashish Sahani
Atul Negi
Avatharam Ganivada
Avinash Chouhan
B. S. Daya Sagar
Bappaditya Chakraborty
Barnam J. Saharia
Bhabesh Nath
Bikramjit Choudhury
Binayak Dutta
Chandra Das
Debadatta Pati
Debamita Kumar
Debasis Mazumdar
Debasish Das
Debasrita Chakraborty
Debjyoti Bhattacharjee
Debojit Boro
Deepak Gupta
Deepika Hazarika
Dibyajyoti Chutia
Dipankar Kundu
Dipen Deka
Ekta Shah
Gaurav Harit
Haradhan Chel
Helal U. Mullah
Ibotombi S. Sarangthem

Indrani Kar
Jaswanth N.
Jay Prakash
Jaya Sil
Jayanta K. Pal
Jaybrata Chakraborty
K. C. Nanaiah
K. Himabindu
Kannan Karthik
Kaushik Das Sharma
Kaushik Deva Sarma
Kishor Upla
Kishorjit Nongmeikapam
L. N. Sharma
Lipi B. Mahanta
M. Srinivas
M. V. Satish Kumar
Manish Sharma
Manuj Kumar Hazarika
Milind Padalkar
Minakshi Gogoi
Monalisa Pal
Nabajyoti Mazumdar
Nabajyoti Medhi
Nayan Moni Kakoty
Nayandeep Deka Baruah
Nazrul Hoque
Nilanjan Chattaraj
P. V. S. S. R. Chandra
    Mouli
Pankaj Barah
Pankaj Kumar
Parag Chaudhuri
Parag K. Guha Thakurta
Paragmoni Kalita
Partha Garai
Partho S. Mukherjee
Pranav Kumar Singh
Prasun Dutta
Pratima Panigrahi
Ragini (CUJ)
Rajiv Goswami
Rahul Roy
Rajesh Saha
Rajeswari Sridhar

Rakcinpha Hatibaruah
Ram Sarkar
Reshma Rastogi
Riku Chutia
Rishika Sen
Rosy Sarmah
Rupam Bhattacharyya
Rutu Parekh
Sai Charan Addanki
Saikat Kumar Jana
Sanasam Ranbir Singh
Sanghamitra Nath
Sanjit Maitra
Sankha Subhra Nag
Sathisha Basavaraju
Shaswati Roy
Shilpi Bose
Shobhanjana Kalita
Sibaji Gaj
Siddhartha S. Satapathy
Sipra Das Bit
Smriti Kumar Sinha
Soumen Bera
Subhadip Boral
Sudeb Das
Sudip Das
Sujoy Chatterjee
Sujoy M. Roy
Suman Mahapatra
Sumant Pushp
Sumit Datta
Supratim Gupta
Sushant Kumar
Sushant Kumar Behera
Sushil Kumar
Sushmita Paul
Swalpa Kumar Roy
Swarup Chattopadhyay
Swati Banerjee
Tandra Pal
Thoudam Doren Singh
Vivek K. Mehta
Yash Agrawal

Sponsoring Organizations

Endorsed by

 International Association for Pattern Recognition

Technical Co-sponsors

 Centre for Soft Computing Research, ISI

 International Rough Set Society

 Web Intelligence Consortium

 Springer International Publishing

Financial Sponsors

Diamond Sponsors

 North Eastern Council
Government of India

North Eastern Council   Oil India Limited

Gold Sponsor

Indian Space Research Organization

Silver Sponsor

Council of Scientific & Industrial Research

Abstracts of Invited Talks

Granular Artificial Intelligence: A New Avenue of Artificial Intelligence for Modeling Environment and Pattern Recognition

Witold Pedrycz[iD]

Department of Electrical and Computer Engineering, University of Alberta,
Edmonton, Canada
wpedrycz@ualberta.ca

Recent advancements in Artificial Intelligence fall under the umbrella of industrial facets of AI (Industrial AI, for short) and explainable AI (XAI). We advocate that in the realization of these two pursuits, information granules and Granular Computing play a significant role. First, it is shown that information granularity is of paramount relevance in building meaningful linkages between real-world data and symbols commonly encountered in AI processing. Second, we stress that a suitable level of abstraction (information granularity) becomes essential to support user-oriented framework of design and functioning AI artifacts. In both cases, central to all pursuits is a process of formation of information granules and their prudent characterization. We discuss a comprehensive approach to the development of information granules by means of the principle of justifiable granularity; here various construction scenarios are discussed. In the sequel, we look at the generative and discriminative aspects of information granules supporting their further usage in the formation of granular artifacts, especially pattern classifiers. A symbolic manifestation of information granules is put forward and analyzed from the perspective of semantically sound descriptors of data and relationships among data.

Imparting Sentiment and Politeness on Computers

Pushpak Bhattacharyya[1,2]

[1] Department of Computer Science and Engineering,
Indian Institute of Technology Patna, India
[2] Department of Computer Science and Engineering,
Indian Institute of Technology Bombay, India
pb@cse.iitb.ac.in

In this talk we will describe the attempts made at making machines "more human" by giving them sentiment and politeness abilities. We will give a perspective on automatic sentiment and emotion analysis, with a description of our work in this area, touching upon the challenging problems of sarcasm arising from numbers, and multitask and multimodal sentiment and emotion analysis. Subsequently we touch upon the interesting problem of "making computers polite" and our recent work on this. We will end with noting the places of rule based, classical ML based and deep learning-based approaches in NLP.

Beyond Text Detection and Recognition: Emerging Opportunities in Scene Understanding

C. V. Jawahar

Centre for Visual Information Technology, IIIT Hyderabad, India
jawahar@iiit.ac.in

Recent years have seen major advances in the performance of reading text in natural outdoor. Methods for text detection and recognition, are reporting very high quantitative performances on popular benchmarks. In this talk, we discuss a set of opportunities in scene understanding where text plays a critical role. Many associated challenges, ongoing research and emerging opportunities for research are discussed.

Cognitive Analysis Using Physiological Sensing

Sanjoy Kumar Saha

Department of Computer Science and Engineering, Jadavpur University,
Kolkata, India
sks_ju@yahoo.co.in

Cognitive load is a measure of the processing done using working memory of the brain. The effectiveness of an activity is dependent on the amount of cognitive load experienced by an individual. Subjected to a task, assessing the cognitive load of an individual may be useful in evaluating the individual and/or task. Physiological sensing can help in cognitive analysis. EEG, GSR, PPG sensors, Eyegaze tracker can sense different aspects. Talk will mostly focus on EEG signal and eye gaze data and their processing. Their applicability in assessing the readability of text materials will also be highlighted.

The Rise of Hate Content in Social Media

Animesh Mukherjee

Department of Computer Science and Engineering,
Indian Institute of Technology, Kharagpur, India
animeshm@gmail.com

The recent online world has seen an upheaval in the fake news, misinformation, misbehavior and hate speech targeted toward communities, race and gender. This has resulted in severe consequences. Reports say, that the last US election was heavily influenced by the social media (https://www.bbc.com/news/technology-46590890). The EU referendum similarly was under social media influence (https://www. referendumanalysis.eu/eu-referendum-analysis-2016/section-7-socialmedia/impact-of-social-media-on-the-outcome-of-the-eu-referendum/). Facebook has been considered responsible for the spread of unprecedented volume of hate content resulting into Rohingya genocide. The Pittsburg Synagogue shooter was an active member of the extremist social media website GAB where he continuously posted anti-Semitic comments finally resulting into the shooting. Similar cases have been reported for the Tamil Muslim community in Sri Lanka, attacks on refugees in Germany and the Charleston church shooting incident. A concise report of the events and the damages caused thereby is present in this article from the Council on Foreign Relations (https:// www.cfr.org/backgrounder/hatespeech-social-media-global-comparisons).

Since 2017, we have started to put in focused efforts to tackle this problem using computational techniques. Note that this is not a computer science problem per se; this is a much larger socio-political problem. As a first work we show how simple opinion conflicts among social media users can lead to abusive behavior (CSCW 2018, https:// techxplore.com/news/2018-10-convolutional-neural-networkabuse-incivility.html). As a next step we investigate the GAB social network and show how hateful users are much more densely connected among each other compared to others; how the messages posted by hateful users spread far, wide and deep into the social network compared to the normal users (ACM WebSci 2019). Consequently, we propose a solution to this problem; as such suspending hateful accounts or deleting hate messages is not a very elegant solution since this curbs the freedom of speech. More speech to counter hate speech has been thought to be the best solution to fight this problem. In a recent work we characterize the properties of such counter speech and show how they vary across target communities (ICWSM 2019). Presently, we are also investigating how the hate patterns change if they are allowed to evolve in an unmoderated environment. To our surprise we observe that hatespeech is steadily increasing and new users joining are exposed to hate content at an increased and faster rate. Further the language of the whole community is driven to the language of the hate speakers. We believe that this is just the beginning. There are many challenges that need to be yet addressed. For instance, we plan to investigate how misinformation and hate content are related - do they influence each other? Can containing one of them automatically contain the other to some extent at least? In similar lines, how does hate content affect gender and cause gender/sex related discrimination/crime.

In this talk, we will try to present a summary of some of the above experiences that we have had in the last few years relating to our ventures into hate content analysis in social media.

How to Escape Saddle Points Efficiently?

Praneeth Netrapalli

Microsoft Research India
praneeth@microsoft.com

Non-convex optimization is ubiquitous in modern machine learning applications. Gradient descent based algorithms (such as gradient descent or Nesterov's accelerated gradient descent) are widely used in practice since they have been observed to perform well on these problems. From a theoretical standpoint however, this is quite surprising since these nonconvex problems are teeming with highly suboptimal saddle points, and it is well known that gradient descent (and variants) can get stuck in such saddle points. A key question that arises in resolving this apparent paradox is whether gradient descent based algorithms escape saddle points where the Hessian has negative eigenvalues and converge to points where Hessian is positive semidefinite (such points are called second-order stationary points) efficiently. We answer this question in the affirmative by showing the following: (1) Perturbed gradient descent converges to second-order stationary points in a number of iterations which depends only poly-logarithmically on dimension (i.e., it is almost "dimension-free"). The convergence rate of this procedure matches the wellknown convergence rate of gradient descent to first-order stationary points, up to log factors, and (2) A variant of Nesterov's accelerated gradient descent converges to second-order stationary points at a faster rate than perturbed gradient descent. The key technical components behind these results are (1) understanding geometry around saddle points and (2) designing a potential function for Nesterov's accelerated gradient descent for non-convex problems.

Machine Learning and Its Application in Remote Sensing Data Classification Applications

Anil Kumar

Indian Institute of Remote Sensing, Dehradun, India
anil@iirs.gov.in

From 2000 onwards sub-pixel or later called soft classification has been explored very extensively. Later learning based algorithm taken over when modified forms of learning algorithms were proposed. Artificial Neural networks (ANN) is a generic name for a large class of machine learning algorithms, most of them are trained with an algorithm called back propagation. In the late eighties, early to mid-nineties, dominating algorithm in neural nets was fully connected neural networks. These types of networks have a large number of parameters, and so do not scale well. But convolutional neural networks (CNN) are not considered to be fully connected neural nets. CNNs have convolution and pooling layers, whereas ANN have only fully connected layers, which is a key difference. Moreover, there are many other parameters which can make difference like number of layers, kernel size, learning rate etc. While applying Possibilistic c-Means (PCM) fuzzy based classifier homogeneity within class was less while observing learning based classifiers homogeneity was found more. Best class identification with respect to homogeneity within class was found in CNN soft output as shown in the figure. With this it gives a path to explore various deep leaning algorithms in various applications of earth observation data like; multi-sensor temporal data in crop/forest species identification, remote sensing time series data analysis. As learning based algorithms require large size of training data, but in remote sensing domain it's difficult to generate large training data sets. This issue also has been resolved in this research work.

Brief History of Topic Models

Chiranjib Bhattacharyya

Department of Computer Science and Automation, Indian Institute of Science,
Bangalore, India
chiru@iisc.ac.in

The success of Machine Learning is often attributed to Supervised Learning models. Comparatively, progress on learning models without supervision is limited. However, Unsupervised Learning has the potential of unlocking a whole new class of applications. In this tutorial we will discuss Topic models, an important class of Unsupervised Learning Models which have proven to be extremely successful in practice. The tutorial will discuss a self-contained introduction to learning Latent variable models (LVMs) and will discuss Topic models as a special case of LVMs. Time permitting, recent results on deriving sample complexity of Topic models will be also covered.

Introduction to Compressed Sensing

Ajit Rajwade

Department of Computer Science and Engineering,
Indian Institute of Technology Bombay, India
ajitvr@cse.iitb.ac.in

Compressed sensing is a relatively new sensing paradigm that proposes acquisition of images directly in compressed format. This is different from the more conventional sensing methods where the entire 2D image array is first measured followed by JPEG/MPEG compression after acquisition. Compressed sensing basically aims to reduce *acquisition time*. It has shown great results in speeding up MRI (magnetic resonance imaging) acquisition where time is critical, in improving frame rates of videos, and in general in improving acquisition rates in a variety of imaging modalities.

Central to compressed sensing is the solution to a seemingly under-determined system of linear equations, i.e. a system of equations where the number of unknowns (n) is greater than the number of knowns (m). Hence at first glance, there will be infinitely many solutions. However the theory of compressed sensing states that if the vector of unknowns is sparse, and the system's sensing matrix obeys certain properties, then the system is provably well-posed and unique solutions can be guaranteed. Moreover, the theory also states that the solution can be computed efficiently, and is robust to measurement noise or slight deviations from sparsity.

In this talk, I will give an introduction to the above concepts. I will also introduce a few applications, and enumerate a few research challenges/directions.

Novel Support Vector Machine Algorithms for the Diagnosis of Alzheimer's Disease

M. Tanveer

Discipline of Mathematics, Indian Institute of Technology Indore, India
mtanveer@iiti.ac.in

Support vector machine (SVM) has provided excellent performance and has been widely used in real world classification problems due to many attractive features and promising empirical performance. Different from constructing two parallel hyperplanes in SVM, recently several non-parallel hyperplane classifiers have been proposed for classification and regression problems. In this talk, we will discuss novel non-parallel SVM algorithms and their applications to Alzheimer's disease. Numerical experiments clearly show that non-parallel SVM algorithms outperform traditional SVM algorithms. This talk concludes that non-parallel SVM variants could be the viable alternative for classification problems.

Contents – Part II

Bioinformatics and Biomedical Signal Processing

Information Retrieval

Smart and Intelligent Sensors

Contents – Part I

Machine Learning

Deep Learning

Soft and Evolutionary Computing

Image Processing

l Contents – Part I

Medical Image Processing

On the Study of Childhood Medulloblastoma Auto Cell Segmentation from Histopathological Tissue Samples

Daisy Das and Lipi B. Mahanta[✉] [ID]

Central Computational and Numerical Studies Department,
Institute of Advanced Study in Science and Technology, Guwahati 781035, India
daisy.dd08@gmail.com, lbmahanta@iasst.gov.in

Abstract. Whole slide imaging in histopathology is one of the most important aspects of computational pathology. Nucleus identification and extraction can play a critical part in digital microscopic examination. This work is an extension of our previous published work on childhood medulloblastoma biopsy machine learning classification where the classifier was based on ground truth annotated data. However complete automation would entail automatic segmentation of the cells. The paper explores various segmentation techniques for cell identification from biopsy tissue samples of childhood medulloblastoma microscopic images based on conventional machine learning methods. The study is based on indigenous patient data collected from medical centers of the region. The performance of the segmentation algorithms was compared using Jaccard and Dice coefficient metric.

Keywords: Segmentation · Cell · Childhood medulloblastoma

1 Introduction

Identification of the nucleus characteristics is the most important study of a pathologist in the identification of any disease. The problem in the manual assessment of histopathology images includes inter-observer variability for challenging tissue samples with tenuous visual features. Moreover due to lack of adequate patient to doctor ratio the whole slide scanning period for an individual clinical expert is time-consuming. There are a significant number of nucleus present in a slide and inspecting them carefully is a tedious job for the clinicians where patient throughput is tremendously high. Computational pathology [1] with computer vision methods has gained momentum in the recent past to alleviate the conventional approaches [2–5]. The prime focus in digital pathology applications is to closely identify the nucleus area and its morphological characteristics. Therefore a principal component in computational pathology is segmentation [4, 6–8] of the nucleus eliminating the background from the whole slide images. Thus extracting the nucleus more accurately contributes significantly to medical software. The goal of this work is to segment nuclei relating to childhood medulloblastoma microscopic images and contribute towards computational pathology research. We have collected images from real-time patient data and

B. Deka et al. (Eds.): PReMI 2019, LNCS 11942, pp. 3–12, 2019.
https://doi.org/10.1007/978-3-030-34872-4_1

annotated the ground truth, under the guidance of a clinical expert. We studied the result of various conventional segmentation method on our data. Most of the work on medulloblastoma are texture based and no work on medulloblastoma has been carried out till date on the identification of the nuclei from the tissue samples, which is a vital part of the diagnosis. The segmentation of tissue images is a lot more challenging than cytological images due to the high presence of debris, hemorrhages and nuclei diffusion. Also, hyperchromasia and overlapped nuclei are always seen. Machine learning in the past has shown conceivable results in challenging datasets [9–11]. The serious problem in medical research is the availability of data. We did not have a benchmark medulloblastoma data set available and so we address the problem for future researchers by creating our own data set with painstakingly marked ground truth nuclei in the samples.

2 Background

Literature reveals that work on medulloblastoma is very recent and the World Health Organization [12] (W.H.O) has characterized it into grade IV malignant tumors only in 2002. Computer-aided study for such tumor has been seen only from 2012 to the best of our knowledge. All the research [13–16] works are performed by St. Jude's Children Hospital, USA. However, no work on cell segmentation has been attempted in the past years. Further, the studies carried out were based on the textural feature with the help of which it was attempted to categorize medulloblastoma into anaplastic and non-anaplastic subtypes. But, according to W.H.O, it can be classified into four different subgroups, based on its severity and cell characteristics. Prognosis and other health management depend largely on the subtype the tumour is classified as. Hence, for a cell-based study, the most vital role is that of automated cell segmentation from the tissue samples, which we plan to address in this paper. Previously classification of childhood medulloblastoma into its W.H.O. subtypes have been achieved [17] based on manually segmented cells. Complete automation would be achieved by automated cell extraction.

3 Methods and Materials

3.1 Study Region

The study region is North East India, Guwahati city in particular.

3.2 Data Collection

The tissue blocks were collected from Guwahati Medical College and Hospital (GMCH), Guwahati from the Neurosurgery Department. Next, the blocks were stained with H&E at Ayursundra Pvt. Ltd and finally images were captured under clinical supervision at Guwahati Neurological Research Center (GNRC), Dispur. The images were captured at 100× microscopic resolution and stored in jpeg format. A total of 94 images were collected from 15 slides. 18 normal cell images were also captured from the slides, where present in the sample. Ground truth, for identifying the cells in the tissue samples, were meticulously marked by us in red using MS paint and was later verified under expert supervision. For the particular study, we have marked 37 images which include both properly and poorly stained whole slides. From these images, a total of 1272 cells were identified and marked. Few images are shown in Fig. 1. The ground truth includes both normal and abnormal cells including the various subtypes. The marked ground truth was later segmented using Kmeans color segmentation method for evaluation purpose.

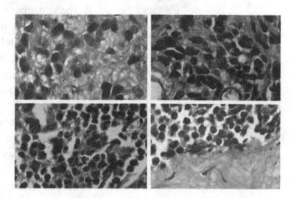

Fig. 1. Figure showing microscopic images collected at 100x magnification.

3.3 Ethical Statement

This study was a part of a joint project undertaken by the Institute of Advanced Study in Science and Technology (IASST) and GMCH. Permission for the same was granted from ethical bodies of both the institutions [IASST: Registration number ECR/248/Indt/AS/2015 of Rule 122DD, Drugs and Cosmetics Rule, 1945 of India; GMCH:MC/190/2007/pt-1/E-C/32 dated 30.5.2017].

3.4 Inclusion/Exclusion Criteria

The samples were taken from children less than the age of 15 years and only cases confirmed by the biopsy were included for the study. The cases that do not fall under MB after the clinical findings, summation of history and pathologic features were not considered in the study.

3.5 Overview of the Work

In our work, we studied 8 segmentation techniques that include color based segmentation, region-based segmentation, segmentation based on clustering technique, local and global segmentation methods. The model is depicted in Fig. 2. The algorithms for this purpose were Adaptive segmentation [18, 19], HSV color segmentation [20],

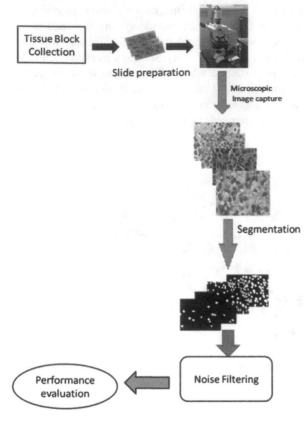

Fig. 2. Model of the work.

YCbCr color segmentation, Ostu segmentation [21], Fuzzy [22, 23] modelling based segmentation, Watershed segmentation [24, 25], and Kmeans [26] clustering and entropy [27, 28] filter segmentation. The HSV and YCbCr were color-based segmentation method where the images were first converted to the respective color channels and then global thresholding was applied on the pixel intensity values. The thresholding value for color-based segmentation was static and therefore proper smooth segmentation of the cells were not possible as different images had different color intensity histograms. We then used the adaptive and Ostu segmentation which was based on finding an iterative best threshold value for cell segmentation. This method gave a better result than our previous attempt but had a high segmentation of overlapped nuclei. We then tried the traditional watershed segmentation for minimizing the segmentation output of overlapped cells, but the watershed segmentation underperformed due to inefficiency to choose an appropriate local minima and maxima regions in the image. Finally, Kmeans and Fuzzy clustering were used for segmentation, which performed higher than the previous attempts. The Kmeans segmentation grouped the whole image into clusters of different colors based on a similarity value. The initial cluster value used was 3. Before applying the Kmeans clustering method the RGB images were first converted to their respective LUV color channel. All the segmentation methods applied is based on hard partition, where each point belong to any one cluster, while the fuzzy method is a soft partition method, where a single data point can belong to two different clusters based on its membership value and the total sum of all membership points is 1. The fuzzy clustering method gave us a better output than most of the other methods. However the most appropriate was entropy-based segmentation for our images. The segmentation module for all algorithms was followed by same morphological operations of opening using a disk structuring element of radius 10 pixels in length for noise removal. Next, the objects that had less than 80 pixels were empirically identified as noise and were removed from consideration. Later, the performance evaluation among the segmentation modules and the ground truth were done using Dice and Jaccard coefficient metric. Mathematically, it is given as

$$Dice\ coefficient = (2 * TP)/(2 * TP + FP + FN) \tag{1}$$

The Jaccard metrics is given by

$$Jaccard\ coefficient = TP/(TP + FP + FN) \tag{2}$$

where TP = True positive values where both the segmented image and the ground truth image has pixel intensity value 1, FP = False positive where the segmented image has pixel intensity value 1 but ground truth has pixel intensity value 0 for the same region, FN = False negative where segmented image has intensity value 0 but ground truth has pixel intensity value 1.

4 Results

This section presents the result of our experimentation with various segmentation techniques with the ground truth segmented data. An image of annotated and ground-truth data is displayed in Fig. 3. Next, various segmentation techniques, as detailed above, are depicted in Fig. 4. A total of 37 images were tested for all the segmentation methods and Dice (using Eq. 1) and Jaccard coefficient (using Eq. 2) for each image was calculated based on the ground truth information. The individual image scores for 25 samples out of 37 are given in Tables 1 and 2. The average

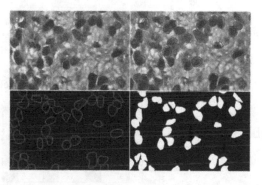

Fig. 3. Figure showing K-means segmentation of the ground truth annotation. Original Image (*top left*), Ground truth marking (*top right*), K-means boundary detection (*bottom left*), segmentation (*bottom right*)

scores of the various segmentation algorithms is shown using bar graph in Fig. 5(a) and (b). Boxplot (Fig. 6) is an efficient and simple technique to represent graphically a set of data which belong to a same variable. It displays not only the location of the data values but also variation and are especially useful for showing comparisons. The performance of the segmentation methods were in the order HSV<Watershed<YCbCr<Kmeans<Adaptive<Ostu<Fuzzy<Entropy. The highest Dice and Jaccard coefficient that we have got is 79.2% (Table 1) and 66.1% (Table 2) using entropy-based segmentation method. From Fig. 6 it is clear that Entropy, Fuzzy and Ostu have

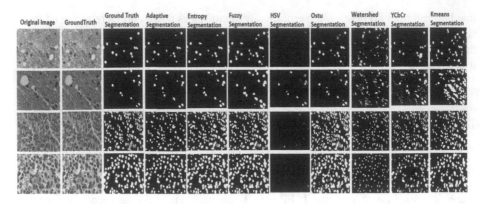

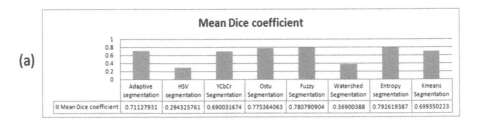

Fig. 4. Segmentation output of the different algorithms.

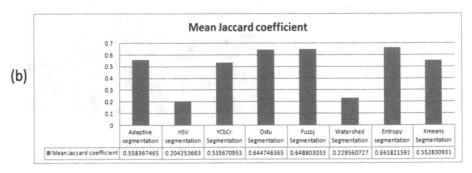

Fig. 5. Figure showing the comparative (a) Dice coefficient and (b) Jaccard coefficient of various segmentation techniques with ground truth.

the highest median, almost equal to each other. Further among these three Entropy median is located more in the middle of the 1st and 3rd quartile indicating that the scores are more evenly distributed. It also has the shortest distance between the 1st and 3rd quartile indicating higher consistency of scores and smaller whiskers indicating lower variation in the scores.

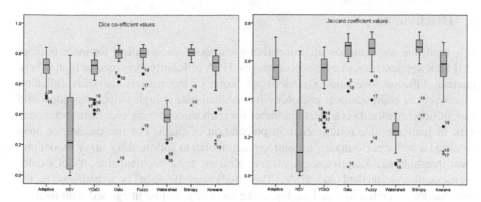

Fig. 6. Figure showing Boxplot for Dice and Jaccard coefficient values.

Table 1. Dice coefficient of various segmentation methods used.

Sl. no	Adaptive	HSV	YCbCr	Ostu	Fuzzy	Watershed	Entropy	Kmeans
1	0.669	0.747	0.736	0.761	0.79	0.285	0.793	0.738
2	0.768	0.002	0.646	0.832	0.856	0.379	0.857	0.786
3	0.696	0	0.651	0.814	0.835	0.389	0.836	0.778
4	0.714	0.271	0.765	0.769	0.788	0.366	0.788	0.765
5	0.755	0.108	0.686	0.651	0.768	0.377	0.759	0.755
6	0.84	0.182	0.764	0.811	0.856	0.452	0.856	0.823
7	0.803	0.009	0.666	0.807	0.839	0.389	0.833	0.82
8	0.69	0.209	0.756	0.771	0.787	0.432	0.787	0.739
9	0.667	0.048	0.684	0.768	0.8	0.35	0.8	0.757
10	0.759	0.577	0.79	0.827	0.836	0.375	0.849	0.803
11	0.723	0.442	0.735	0.822	0.812	0.391	0.815	0.771
12	0.732	0.303	0.686	0.825	0.774	0.376	0.829	0.707
13	0.762	0.488	0.736	0.797	0.821	0.435	0.815	0.7
14	0.761	0.289	0.712	0.798	0.816	0.492	0.806	0.685
15	0.808	0.03	0.624	0.839	0.715	0.233	0.817	0.595
16	0.504	0.51	0.495	0.089	0.302	0.124	0.459	0.207
17	0.696	0.725	0.427	0.783	0.568	0.213	0.74	0.229
18	0.537	0.575	0.468	0.613	0.665	0.114	0.56	0.41
19	0.8	0.009	0.665	0.811	0.846	0.381	0.84	0.82
20	0.685	0.047	0.695	0.769	0.817	0.366	0.817	0.767
21	0.747	0.533	0.722	0.749	0.785	0.438	0.786	0.736
22	0.753	0.307	0.687	0.758	0.776	0.479	0.768	0.685
23	0.781	0.78	0.786	0.822	0.835	0.331	0.833	0.806
24	0.784	0.79	0.795	0.824	0.813	0.366	0.813	0.798
25	0.788	0.757	0.768	0.811	0.817	0.305	0.815	0.792
Average	0.711	0.294	0.69	0.775	0.781	0.369	0.793	0.699

5 Discussions

It is seen that segmentation of histopathological tissue samples is a lot more difficult [29] then segmentation of cytological images. High cellularity, presence of high debris, nuclear diffusion, overlapping nucleus are some of the reasons for such difficulty. Moreover the computational pathology for medical data is highly data dependent and lack of benchmark data is a great hindrance for such work to carry out. Data variation is seen in particular due to difference in preparation of slides. For our dataset we have obtained a better performance for entropy segmentation followed by fuzzy modelling, Ostu thresholding, Adaptive segmentation, Kmeans color segmentation, YCbCr color segmentation, watershed and HSV. This study can be seen as a pathway to the introduction of digital MB cell analysis. The module could be integrated with medical image analysis for digital report generation and also may be extended for other tissue samples.

Table 2. Jaccard coefficient of various segmentation methods used.

Sl. no	Adaptive	HSV	YCbCr	Ostu	Fuzzy	Watershed	Entropy	Kmeans
1	0.502	0.596	0.582	0.615	0.653	0.166	0.657	0.585
2	0.624	0.001	0.477	0.713	0.748	0.234	0.749	0.647
3	0.534	0	0.483	0.686	0.717	0.242	0.718	0.636
4	0.555	0.157	0.619	0.625	0.65	0.224	0.65	0.619
5	0.607	0.057	0.522	0.482	0.623	0.232	0.611	0.607
6	0.724	0.1	0.619	0.682	0.748	0.292	0.748	0.7
7	0.671	0.004	0.499	0.676	0.722	0.242	0.713	0.695
8	0.527	0.117	0.608	0.628	0.649	0.275	0.649	0.587
9	0.5	0.025	0.52	0.623	0.667	0.212	0.667	0.609
10	0.612	0.405	0.654	0.706	0.718	0.231	0.737	0.671
11	0.566	0.284	0.581	0.697	0.684	0.243	0.687	0.627
12	0.577	0.179	0.522	0.702	0.631	0.232	0.708	0.547
13	0.616	0.323	0.582	0.662	0.697	0.278	0.688	0.538
14	0.614	0.169	0.553	0.664	0.689	0.326	0.675	0.521
15	0.678	0.015	0.454	0.723	0.557	0.132	0.69	0.423
16	0.337	0.342	0.329	0.047	0.178	0.066	0.298	0.115
17	0.534	0.569	0.272	0.644	0.396	0.119	0.587	0.129
18	0.367	0.403	0.306	0.441	0.499	0.061	0.388	0.258
19	0.667	0.004	0.498	0.681	0.733	0.235	0.724	0.694
20	0.521	0.024	0.532	0.625	0.69	0.224	0.69	0.622
21	0.596	0.363	0.565	0.598	0.646	0.281	0.648	0.582
22	0.604	0.181	0.524	0.61	0.634	0.315	0.624	0.521
23	0.641	0.639	0.648	0.697	0.716	0.199	0.713	0.675
24	0.644	0.652	0.66	0.7	0.685	0.224	0.686	0.664
25	0.651	0.61	0.623	0.682	0.691	0.18	0.688	0.656
Average	0.558	0.204	0.536	0.645	0.649	0.23	0.662	0.553

Acknowledgment. We thank IASST, GMCH, GNRC and Ayursundra Healthcare Ltd. for giving us the platform to carry our work.

Conflict of Interest
The Authors declare no conflict(s) of interest.

References

1. Gurcan, M., et al.: Histopathological image analysis: a review. IEEE Rev. Biomed. Eng. **2**, 147–171 (2009)
2. Petushi, S., et al.: Large-scale computations on histology images reveal grade-differentiating parameters for breast cancer. BMC Med. Imaging **6**(1), 1–14 (2006)
3. Naik, S., et al.: Gland segmentation and computerized Gleason grading of prostate histology by integrating low-, high-level and domain-specific information. MI- AAB Workshop, vol. 34, pp. 1–8. Springer, Australas (2007). Australas Phys. Eng. Sci. Med.
4. Lu, C., et al.: A robust automatic nuclei segmentation technique for quantitative histopathological image analysis. Anal. Quant. Cytol. Histol. **34**, 296–308 (2012)
5. Su, H., Xing, F., Yang, L.: Robust cell detection of histopathological brain tumor images using sparse reconstruction and adaptive dictionary selection. IEEE Trans. Med. Imaging **35**(6), 1575–1586 (2016)
6. Chang, H., et al.: Invariant delineation of nuclear architecture in glioblastoma multi-forme for clinical and molecular association. IEEE Trans. Med. Imaging **32**(4), 670–682 (2013)
7. Filipczuk, P., et al.: Computer-aided breast cancer diagnosis based on the analysis of cytological images of fine needle biopsies. IEEE Trans. Med. Imaging **32**(12), 2169–2178 (2013)
8. Sethi, A., et al.: Computational pathology for predicting prostate cancer recurrence. In: AACR s106th Annual Meeting, vol. 75, no. 15, pp. 18–22. American Association for Cancer Research, Philadelphia (2015). https://doi.org/10.1158/1538-7445.AM2015-LB-285
9. Szegedy, C., et al.: Going deeper with convolutions. In: IEEE Conference on Computer Vision and Pattern Recognition (CVPR), vol. 1, pp. 1–9. IEEE, Boston (2015). https://doi.org/10.1109/CVPR.2015.7298594
10. Simonyan, K., Zisserman, A.: Very deep convolutional networks for large-scale image recognition. CoRR, vol. abs/1409.1556, pp. 730–734. IEEE, Kuala Lumpur (2014)
11. Socher, R., et al.: Parsing natural scenes and natural language with recursive neural networks. In: ICML, pp. 129–136 (2011)
12. WHO classification of CNS tumours. https://radiopaedia.org/articles/who-classification-of-CNS-tumours-1. Accessed 28 Jan 2016
13. Galaro, J., et al.: A method for medulloblastoma tumor differentiation based on convolutional neural networks and transfer learning. In: SPIE 9681, 11th International Symposium on Medical Information Processing and Analysis, IEEE, Boston (2016). https://doi.org/10.1117/12.2208825
14. Lai, Y., et al.: A texture-based classifier to discriminate anaplastic from non-anaplastic medulloblastoma. In: IEEE 37th Annual Northeast on Bioengineering Conference (NEBEC), pp. 141–171. IEEE, Troy (2011)
15. Roa, A.C., et al.: An integrated texton and bag of words classifier for identifying anaplastic medulloblastomas. In: 11th International Symposium on Medical Information Processing and Analysis, pp. 3443–3446. IEEE, Cuenca (2015)

16. Tchikindas, L., et al.: Segmentation of nodular medulloblastoma using random walker and hierarchical normalized cuts. In: IEEE 37th Annual Northeast Bioengineering Conference, pp. 4–5. IEEE, Troy (2011)

17. Das, D., et al.: Study on contribution of biological interpretable and computer- aided features towards the classification of childhood medulloblastoma cells. J. Med. Syst. **42**(151), 5–12 (2018)

18. Pachai, C., et al.: Unsupervised and adaptive segmentation of multispectral 3D magnetic resonance images of human brain: a generic approach. In: Niessen, W.J., Viergever, M.A. (eds.) MICCAI 2001. LNCS, vol. 2208, pp. 1067–1074. Springer, Heidelberg (2001). https://doi.org/10.1007/3-540-45468-3_127

19. Lee, L.K., Liew, S.C., Thong, W.J.: A review of image segmentation methodologies in medical image. In: Sulaiman, H.A., Othman, M.A., Othman, M.F.I., Rahim, Y.A., Pee, N.C. (eds.) Advanced Computer and Communication Engineering Technology. LNEE, vol. 315, pp. 1069–1080. Springer, Cham (2015). https://doi.org/10.1007/978-3-319-07674-4_99

20. Priya, T., Kalavathi, P.: HSV based histogram thresholding technique for MRI brain tissue segmentation. In: Thampi, S.M., Marques, O., Krishnan, S., Li, K.-C., Ciuonzo, D., Kolekar, M.H. (eds.) SIRS 2018. CCIS, vol. 968, pp. 322–333. Springer, Singapore (2019). https://doi.org/10.1007/978-981-13-5758-9_27

21. Satapathy, C.S., et al.: Multi-level image thresholding using Otsu and chaotic bat algorithm. Neural Comput. Appl. **29**(12), 1285–1307 (2018)

22. Srinivas, B., Sasibhushana Rao, G.: Performance evaluation of fuzzy c means segmentation and support vector machine classification for MRI brain tumor. In: Bansal, J.C., Das, K.N., Nagar, A., Deep, K., Ojha, A.K. (eds.) Soft Computing for Problem Solving. AISC, vol. 817, pp. 355–367. Springer, Singapore (2019). https://doi.org/10.1007/978-981-13-1595-4_29

23. Zhou, H., Schaefer, G., Shi, C.: Fuzzy c-means techniques for medical image segmentation. In: Jin, Y., Wang, L. (eds.) Fuzzy Systems in Bioinformatics and Computational Biology. STUDFUZZ, vol. 242, pp. 257–271. Springer, Berlin (2009). https://doi.org/10.1007/978-3-540-89968-6_13

24. Lim, P.U., Lee, Y., Jung, Y., Cho, J.H., Kim, M.N.: Liver extraction in the abdominal CT image by watershed segmentation algorithm. In: Magjarevic, R., Nagel, J.H. (eds.) World Congress on Medical Physics and Biomedical Engineering 2006. IFMBE, vol. 14, pp. 2563–2566. Springer, Berlin (2007). https://doi.org/10.1007/978-3-540-36841-0_646

25. Singh, S.K., Goyal, A.: A novel approach to segment nucleus of uterine cervix pap smear cells using watershed segmentation. In: Singh, D., Raman, B., Luhach, A.K., Lingras, P. (eds.) Advanced Informatics for Computing Research. CCIS, vol. 712, pp. 164–174. Springer, Singapore (2017). https://doi.org/10.1007/978-981-10-5780-9_15

26. Chapman, C., Feit, E.M.: Segmentation: clustering and classification. R For Marketing Research and Analytics. UR, pp. 299–340. Springer, Cham (2019). https://doi.org/10.1007/978-3-030-14316-9_11

27. Melloul, M., Joskowicz, L.: Segmentation of microcalcification in X-ray mammograms using entropy thresholding. In: Lemke, H.U., Inamura, K., Doi, K., Vannier, M.W., Farman, A.G., Reiber, J.H.C. (eds.) CARS 2002 Computer Assisted Radiology and Surgery, pp. 671–676. Springer, Berlin (2002). https://doi.org/10.1007/978-3-642-56168-9_112

28. Manjunath, A.P., Rachana, C.S., Ranjini, S.: Retinal vessel segmentation using local entropy thresholding. In: Sridhar, V., Sheshadri, H., Padma, M. (eds.) Emerging Research in Electronics, Computer Science and Technology. LNEE, vol. 248, pp. 1–8. Springer, New Delhi (2014). https://doi.org/10.1007/978-81-322-1157-0_1

29. Belsare, A.D., Mushrif, M.M.: Histopathological image analysis using image processing techniques: an overview. Int. J. Signal Image Process. **3**, 23–36 (2012)

Automated Counting of Platelets and White Blood Cells from Blood Smear Images

Lipi B. Mahanta[1](✉) ⓘ, Kangkana Bora[2], Sourav Jyoti Kalita[3], and Priyangshu Yogi[3]

[1] CCNS, IASST, Guwahati 781035, India
lbmahanta@iasst.gov.in
[2] Department of Electronics and Electrical Engineering,
Indian Institute of Technology, Guwahati 781039, Assam, India
kangkana.bora89@gmail.com
[3] Department of Computer Science and Engineering,
Assam Engineering College, Guwahati 781013, India
kalitasourav12@gmail.com, priyangshuy@gmail.com

Abstract. Platelet Detection and Count are one of the major analysis of the pathological test of the blood. Conventional methods of analysis involve observation of blood smear samples under the microscope and manually identifying and counting the numbers. This process is slow and tedious. This work presents a method to automatically detect and count the number of platelets. A sample size of 270 images collected indigenously is used for carrying out the experiments with the proposed methodology, which result in an accuracy of 95.59% for platelets and 100% for WBCs respectively.

Keywords: Platelet · WBC · Segmentation · Counting · Binary thresholding · Morphological operation

1 Introduction

Platelet count is a very important pathological analysis which assists in the identification of many diseases such as malaria, dengue, yellow fever, etc. and therefore in many cases, immediate diagnosis of platelet count is required to know the state of a person's health. Platelets are tiny blood cells that help our body to form clots to stop bleeding. Normal platelet count is 150,000 to 450,000 per microliter in human beings. But due to various factors, there may be deviations in the platelet count from the normal range. Medical conditions with abnormal platelets and platelet counts are listed in Table 1.

Similarly, White Blood Cell (WBC) count is also a very important analysis in pathology to identify various diseases. White blood cells are the infection-fighting cells in the blood. They are distinctive from the red blood cells which are oxygen-carrying and are known as erythrocytes. The normal range for the white blood cell count varies between laboratories but is usually between 4,300 and 10,800 cells per cubic millimeter of blood. Medical conditions with abnormal WBCs and WBC counts are listed in Table 2.

© Springer Nature Switzerland AG 2019
B. Deka et al. (Eds.): PReMI 2019, LNCS 11942, pp. 13–20, 2019.
https://doi.org/10.1007/978-3-030-34872-4_2

Platelets and WBC counts are achieved by observing the blood smears under a microscope and manually counting the cells. This is tedious and time-consuming. Also, it suffers due to the presence of a non-standard precision as it depends on the operator's skills. The main goal of this work is to develop a system for automatic detection and count of platelets and WBCs from blood smear images using image processing techniques. It would reduce the workload and time taken for analysis as well as reduce dependency on the non-standard operator skill.

Table 1. Medical conditions associated with platelets.

Sl. No.	Medical condition	Cause	Symptoms
1	Thrombocytopenia	Bone marrow makes too few platelets or platelets are destroyed	Bleeding from cuts or nose bleed won't stop, internal bleeding, etc.
2	Thrombocythemia	Bone marrow makes too many platelets	Blood clots might form and they may block the blood supply to the head and heart. It may cause heart attacks
3	Thrombocytosis	Not caused by bone marrow abnormality but by diseases or conditions stimulating the bone marrow to make more platelets	Headache, chest pain, dizziness, weakness
4	Platelet dysfunction	Platelet dysfunction may be due to a problem in the platelets themselves or to an external factor that alters the function of normal platelets	Easy bruising or excessive bleeding after minor injuries

Table 2. Medical conditions associated with WBC.

Sl. No.	Medical condition	Cause	Symptoms
1	Leukocytosis (High WBC Count)	Infection, immunosuppression, emotional stress, injury	Weight loss, fever, stomach ache, chest pain
2	Leukopenia (Low WBC count)	Bone marrow disorders or damage, Radiation treatments for cancer, infections	Fever and chills, swelling and redness, pain or burning when urinating

2   Literature Review

A lot of work has been done in the segmentation of RBCs and WBCs but not much on platelets. On the basis of our survey, Prasannakumar et al. [1] used gray conversion, median filter, Gaussian filter, normalization in pre-processing and Fuzzy C means in segmentation to obtain an accuracy of 96%. Dela Cruz et al. [2] used HSV conversion and HSV thresholding and obtained an accuracy of 90%. Dey et al. [3] used RGB to LAB conversion and Chromaticity layer extraction, binary thresholding, and morphology and obtained an accuracy of 92.71%. Adapting an iterative structured circle detection algorithm for the segmentation and counting of WBCs and RBCs Alomari et al. [4] achieved an average accuracy of 95.3% for RBCs and 98.4% for WBCs.

3   Objective

The objective of this work is to develop a computer-aided system to detect and count the number of platelets and WBCs from a given blood smear image using image processing techniques.

4   Methodology

This method proposes a means to segment the platelets and WBCs from the blood smear images and provides a method of counting the number of platelets. We captured 270 microscopic blood smear images from 12 different blood sample slides, collected from few pathological laboratories in the city, viz. Guwahati, Assam, India. The microscope used was Leica (Leica ICC 50 HD microscope, 24-bit color depth). The images with very less number of platelets, improper stained platelets or hazy images were rejected. Thus from the 270 images in the dataset, a total of 249 images were selected for our work. The resolution of the image is 2048 × 1536 pixels. The algorithm is implemented in Python 3.6.5 using opencv 3.2.0. The ground truth labeling was done separately for platelets and WBCs using MS-Paint.

The block diagram (Fig. 1) shows the overall workflow of the method. The proposed work is implemented in 4 steps: (A) Image Preprocessing (B) Image segmentation (C) Post-processing (D) Platelet and WBC count.

A. **Image pre-processing:** In this step, Gaussian filter using a window of size (5 × 5) was applied to remove the random noise in the image. The image was converted after that to LAB color space. The L*a*b color space comprises of three different layers namely luminosity layer 'L*', chromaticity layers 'a*' and 'b*' (Fig. 2).

B. **Image segmentation:** The transformed image was binarized by the method of masking the pixels having values under a certain threshold. The threshold value was determined experimentally by hit and trial method. Morphological operation of the opening was performed on the binarized image and over that, dilution was performed using a structuring element of radius 3px to remove the noise occurred during binarization. The contours having a size less than 162 pixels (found out experimentally) are drawn using the function *drawContours()*. This image now contains all the platelets. Bitwise AND is then performed to get the image containing all the WBC.

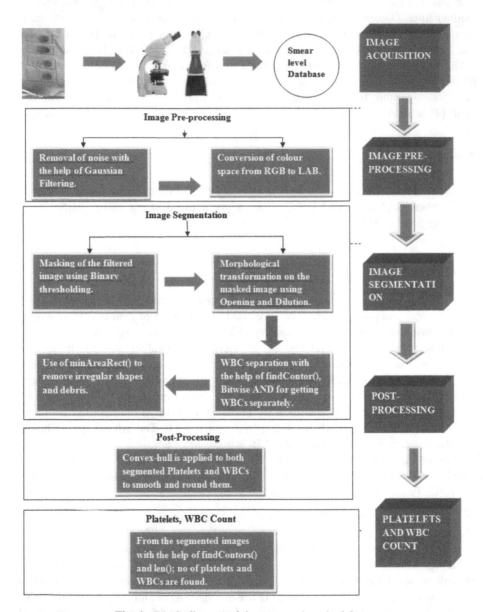

Fig. 1. Block diagram of the proposed methodology

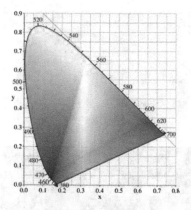 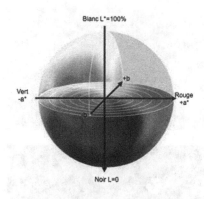

Fig. 2. Chromaticity layers of LAB color space

To remove the small debris, the same method is used to eliminate the contours having a size less than 18 pixels. To remove the objects of irregular shapes which may be stains or other debris, *minAreaRect()* function is used which draws a rectangle bounding the object. The aspect ratio, which is the ratio between the width and height of the rectangle is then measured. If the ratio is less than 0.65 or more than 1.35, the object is eliminated. These values are determined experimentally.

C. **Post-processing:** Convex hull was applied to both the platelet images and WBC images. This rounds up the irregularities and the objects become smoother and circular. These are stored as final images in our directory.

D. **Platelets and WBC count:** The number of objects in the segmented image is counted giving us the platelet count of the blood smear sample. For this, the contour detection function of opencv was *findContours()*. After storing the contours in a matrix (say contours), its length was found out using *len()* function of python. This gives us the required platelet count. The count values are tested for statistical significance using Karl-Pearson's t-test and further analysed by Box and Whisker plot, using SPSS (version 17.0).

The same method is used to count the number of WBCs.

5   Results

Following the methodology mentioned above the results of the experiment are presented in Fig. 3. Further, comparing the processed images with the ground truth labelled images, TRUE-POSITIVE (TP), TRUE-NEGATIVE (TN), FALSE-POSITIVE (FN) and FALSE-NEGATIVE (FN) values are determined. The result of ten samples is shown in Table 3 and depicted in Fig. 4. From the sample size of 270 images, we have obtained an accuracy of 95.59% for platelets and 100% for WBCs respectively (Table 4). The P value of t-test for the samples is 0.6598. Since P value is >0.5, we can conclude that the means of the two sets of data have are not significantly different.

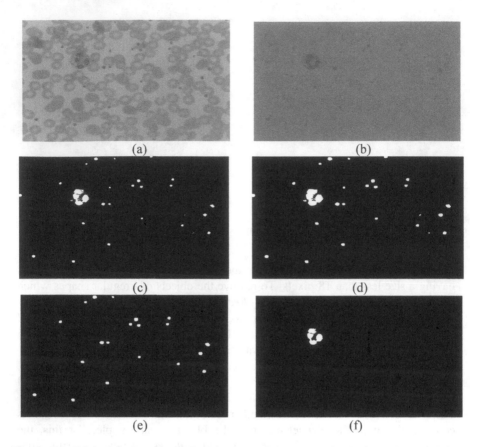

Fig. 3. (a) Original image, (b) LAB color space image, (c) Binarization, (d) Morphology, (e) Platelets, (f) WBC

Table 3. Comparison of manual count and automated count for platelets.

Sl. No.	TP	TN	FP	FN	TC(original)	TC(processed)
1	11	1	0	0	11	11
2	13	1	0	2	15	13
3	7	1	1	0	7	8
4	9	1	0	0	9	9
5	8	1	0	0	8	8
6	6	1	0	0	6	6
7	9	1	0	0	9	9
8	8	1	1	0	8	9
9	11	1	1	0	11	12
10	10	1	1	0	10	11

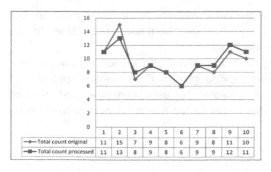

Fig. 4. Graph showing comparison of platelet count before and after segmentation.

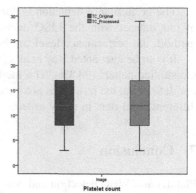

Fig. 5. Box and whisker plots of the original and estimated TC (platelet) values.

To highlight the similarity between manual count and automatic count based on the proposed methodology, box and whisker plots of the Total count values are illustrated in Fig. 5. It is seen that the median count, as well as the first and third quartile, of the proposed approach, coincide well with the manual count. Again, between these two methods, the proposed method has a slightly lower upper whisker, indicating lower variation in the counts.

Table 4. Performance evaluation for platelets and WBC over our collected dataset.

Type	TP	TN	FP	FN	Accuracy	Sensitivity	Specificity	Precision
Platlet	3159	249	50	107	95.59%	96.72%	83.27%	98.44%
WBC	106	249	0	0	100%	100%	100%	100%

6 Discussion

Though we have obtained an accuracy of around 96%, it could be better if we use cleaner blood slides. Our proposed method was applied to the LISC (Leukocyte Images for Segmentation and Classification) [4] dataset over 64 images after eliminating the bad quality images. We have obtained an accuracy of 87.56% on platelets and 100% on WBCs (Table 5).

Table 5. Performance evaluation for platelets over the LISC dataset.

Type	TP	TN	FP	FN	Accuracy	Sensitivity	Specificity	Precision
Platelet	464	64	13	62	87.56%	88.21%	83.11%	97.27%
WBC	71	64	0	0	100%	100%	100%	100%

The performance depletion is due to the color variation in the blood smear images of our dataset with the LISC dataset. By changing the thresholding values in our method, the performance level can be increased.

It is to be also noted Roy et al's [5] paper has reported the highest accuracy among the studied papers (98.8%). The methodology adopted by the study was applied over our dataset but the result was not good. The dataset they used is unknown and might be different from ours in many aspects. This might be the reason for poor performance.

7 Conclusion

Platelet and WBC detection and count is a very important pathological analysis. It can be used as an effective aid to detect various disorders and diseases. We have used 249 images from the dataset and obtained an accuracy of 95.59% for platelets and 100% for WBCs. Thus we conclude that this method can be taken forward and applied in the medical field.

References

1. Prasannakumar, S.C., Puttamadegowda, J.: White blood cell segmentation using fuzzy C means and snake. In: International Conference on Computational Systems and Information Systems for Sustainable Solutions (2016)
2. Cruz, D., et al.: Determination of blood components (WBCs, RBCs, and Platelets) count in microscopic images using image processing and analysis. In: IEEE 9th International Conference on Humanoid, Nanotechnology, Information Technology, Communication and Control, Environment and Management (HNICEM), Manila, pp. 1–7 (2017)
3. Dey, R., Roy, K., Bhattacharjee, D., Nasipuri, M., Ghosh, P.: An automated system for segmenting platelets from microscopic images of blood cells. In: Proceedings of 2015 International Symposium on Advanced Computing and Communication (ISACC), Silchar, India (2015). https://doi.org/10.1109/isacc.2015.7377347
4. Alomari, Y.M., Sheikh Abdullah, S.N., Zaharatul Azma, R., Omar, K.: Automatic detection and quantification of WBCs and RBCs using iterative structured circle detection algorithm. Comput. Math. Methods Med. **2014**, 979302 (2014). https://doi.org/10.1155/2014/979302
5. LISC: Leukocyte Images for Segmentation and Classification. http://users.cecs.anu.edu.au/~hrezatofighi/Data/Leukocyte%20Data.htm. Accessed 20 Mar 2019

An Efficient Method for Automatic Recognition of Virus Particles in TEM Images

Debamita Kumar[(⊠)] and Pradipta Maji

Biomedical Imaging and Bioinformatics Lab, Machine Intelligence Unit,
Indian Statistical Institute, Kolkata, India
{debamita_r,pmaji}@isical.ac.in

Abstract. The conventional approach for virus detection is based on analyzing the negative stain Transmission Electron Microscopy (TEM) images. In this regard, a new method is presented here based on judicious integration of the theory of rough sets and the merits of local texture descriptors. The proposed method identifies the relevant local texture descriptor for representing the intrinsic properties of each pair of virus classes. It also selects important features from each of the relevant descriptors, which can suitably describe significant characteristics of the virus particles present in TEM images. The hypercuboid equivalence partition matrix of rough sets is employed to evaluate the relevance of texture descriptors. Finally, support vector machine (SVM) is used to categorize the virus samples into one of the known virus classes. The efficiency of the proposed method, along with a comparison with related approaches, is demonstrated on publicly available Virus data set.

Keywords: Virus recognition · Transmission Electron Microscopy (TEM) images · Rough sets · Feature selection · Local texture descriptor

1 Introduction

Analysis of negative stain TEM images has long been used for the discovery and detection of virus particles. The visualization of virus structures has been possible only after the development of TEM. The study of virus structures through TEM images allows organism-specific reagents for the recognition of pathogenic agent, without any a priori knowledge of pathogens present in a sample [3]. This enables broader examination of viruses, making TEM an inevitable method for virus diagnosis. Negative staining is a preferred technique for observing small particles, such as viruses, in fluids. However, high maintenance costs, level of expertise and time required for manual inspection of the TEM images, and advancement in automated image acquisition process entail the development of automated analysis of TEM images for virus detection.

© Springer Nature Switzerland AG 2019
B. Deka et al. (Eds.): PReMI 2019, LNCS 11942, pp. 21–31, 2019.
https://doi.org/10.1007/978-3-030-34872-4_3

TEM images provide important information regarding the shape and size of the virus particles, which can be used to group the particles into various classes, for example, viruses like Adeno, Astro and Rota form icosahedral structure, Cowpox, Dengue and Lassa viruses reflect regularity in their structures, whereas Ebola, Influenza and Marburg viruses are highly irregular in shape. Early approaches for virus identification were mostly dependent on the morphology of the particles [1,2]. But, the morphological information obtained from the TEM images does not proved to be sufficient for classification of the viruses.

Apart from morphology, different viruses exhibit different surface textures when imaged using TEM. This information can be efficiently utilized using various texture analysis techniques for automatic classification of the virus particles from TEM images. Matuszewski et al. [12] have transformed the spatial information of the images into frequency domain using discrete Fourier transform and then, extracted features from the multiple spectral rings formed at the magnitude spectrum of TEM images to differentiate the four icosahedral structured viruses, namely, Adeno, Astro, Rota and Calici. In [17], a method has been proposed based on higher order spectral features, to capture the contour and texture information from the same four icosahedral virus images for the diagnosis. The radial density profile (RDP) has been used in [19] to discriminate the intensity variation between three maturation stages of human cytomegalovirus capsids present in the cell sections of TEM images. The RDP has also been computed based on Fourier magnitude spectrum (FRDP) in [9], which can be interpreted as a generalization of the spectral rings considered in [17]. Moreover, Harandi et al. [6] have mapped the original data to a high-dimensional Hilbert space and then computed the Covariance Descriptors (CovDs) in infinite-dimensional spaces, to address the virus classification problem. Finally, several Bregman divergences have been used to compute the dissimilarities between the resulting CovDs in Hilbert space, among which Jeffreys and Stein divergences based on the SVM have been found to achieve the best result. Recently, a new method, based on convolutional neural network, has been proposed in [7], which transforms a TEM image to a probabilistic map for virus particle detection.

The concepts of local binary pattern (LBP) [16] and RDP (LBP+RDP) have been used in [9] to identify different viruses from TEM images using random forest classifier. Nanni et al. [13] have presented two methods, namely, NewH and Fusion; the former focuses on extracting descriptors from the co-occurrence matrix with the goal of enhancing the performance of Haralick's descriptors [5], while the latter is an ensemble of local phase quantization variants with ternary encoding. From the results reported in literature, it can be noticed that different virus classes are well described by different textures, which hinders the overall virus detection ability. One of the main problems in virus detection is uncertainty, which may have been originated due to incompleteness in class definition and overlapping class boundaries. An effective paradigm to deal with incompleteness and uncertainty is the theory of rough sets [18]. It provides a mathematical framework to model uncertainties related to the given data.

In this regard, the paper presents a new method for automatic virus classification from TEM images, by judiciously integrating the merits of local texture descriptors and the theory of rough sets. Given a set of virus TEM images, the proposed method first identifies a subset of important features from the original feature set of the local texture descriptors, computed at a specific scale and then forms the relevant feature set corresponding to each class-pair which can appropriately differentiate the pair of virus classes. The theory of rough hypercuboid approach is employed to evaluate the relevance of a texture descriptor in categorizing virus samples into multiple classes. In the proposed approach, the important feature set is selected in such a way that it not only characterizes the virus samples present in the data set, but also the classes to which the samples belong. The set of relevant texture descriptors, selected from the important feature sets, corresponding to each the class-pair is considered to form the feature set for multiple virus classes. The SVM with linear kernel is used to evaluate the performance of the proposed method. The effectiveness of proposed method, along with a comparison with related approaches, is demonstrated on a real-life Virus image database.

2 Proposed Algorithm

The inherent texture properties of virus TEM images are quite different from each other. The inter-class structural similarities, intra-class variations in appearance, uncertainty in virus class definitions, and presence of noise further amplify the difficulty of texture identification. In this context, the proposed method introduces a new approach for classifying viruses from TEM images. It first identifies important features, which effectively capture significant characteristics of each pair of virus classes, from the relevant texture descriptors evaluated at a particular scale, and finally forms the feature set for multiple virus classes.

Suppose, $\mathbb{U} = \{y_1, \cdots, y_k, \cdots, y_n\}$ denote the set of n training images of virus particles, where each image $y_k \in \Re^m$. Let $\mathbb{C} = \{F_1, \cdots, F_j, \cdots, F_m\}$ represent the set of m features. So, a virus TEM image y_k can be represented by a set $H_k = \{H_{k1}, \cdots, H_{kj}, \cdots, H_{km}\}$, containing m feature values corresponding to y_k, where $H_{kj} = y_k(F_j)$ represents the value of feature F_j obtained for image y_k. In the proposed approach, four local texture descriptors are considered, which are LBP: local binary pattern [16], LBP^{ri}: rotation-invariant LBP [15], LBP$^{\text{riu2}}$: rotation-invariant uniform LBP [15], and CoALBP: co-occurrence of adjacent LBP [14]. So, H_k represents the normalized histogram, obtained from either of the four local descriptors, for the image y_k. Let, H_k be arranged in descending order and denoted by $\tilde{H}_k = \{\tilde{H}_{k1}, \cdots, \tilde{H}_{kj}, \cdots, \tilde{H}_{km}\}$ such that $\tilde{H}_{k1} \geq \tilde{H}_{k2} \geq \cdots \geq \tilde{H}_{km}$, and the corresponding feature indices of $\tilde{H}_k$ is preserved in the set $J_k = \{J_{k1}, \cdots, J_{kj}, \cdots, J_{km}\}$. Assume, each of the TEM images of $\mathbb{U}$ belongs to one of the c virus classes $\mathbb{U}/\mathbb{D} = \{B_1, \cdots, B_i, \cdots, B_c\}$, where $\mathbb{D}$ denotes the set containing class label information.

Primarily, significant properties of an image y_k is represented by the values of H_k, which represents the normalized histogram of y_k. However, the proposed method assumes that not all the feature values of H_k contribute uniformly in representing the characteristics of the image y_k. Indeed, only a subset of features of $\mathbb{C}$ can precisely illustrate important properties of y_k, which is defined as the important feature set of y_k and denoted by I_k, where $I_k \subseteq \mathbb{C}$. However, each image has it's own characteristics, which can be reflected by a specific set of important features.

The cumulative sum of first q features of the sorted normalized histogram $\tilde{H}_k$ is computed to select the important features of y_k and denoted by $E(y_k, q)$, which is defined as the energy function in the current study. Certainly, $E(y_k, m) = 1$ and $E(y_k, q) \in [0, 1], \forall y_k \in \mathbb{U}$. It signifies the fraction of total energy, present in $\tilde{H}_k$, which is retained by the first q features of y_k. Thus, relevant information regarding the properties of the image y_k can be suitably represented by the energy of y_k, evaluated from $\tilde{H}_k$. In order to retain a given fraction of energy E_0 of the sample y_k, the required number of important features d_k is obtained as the minimum value of q for which $E(y_k, q)$ attains E_0 value. The average number of important features $\overline{d}$ corresponding to the entire set of samples $\mathbb{U}$ is computed from the individual d_ks. As defined in [10], the set of important features I_k of the sample y_k can be obtained from the first $\overline{d}$ features of sorted histogram $\tilde{H}_k$, as follows:

$$I_k = \{F_j \mid J_{kq} = j \quad \text{and} \quad q \leq \overline{d}\}. \tag{1}$$

Thus, the important feature set $I_k \subseteq \mathbb{C}$ of the image y_k consists of only those features which can efficiently describe the inherent properties of y_k.

Now, it is most likely that the samples from a specific virus class will be represented by similar sets of important features, whereas samples from different classes will be represented by different important feature sets. Therefore, the probability of occurrence $P(F_j | B_i)$ of a feature F_j in the important sets of samples belonging to a particular class B_i is obtained and a threshold parameter ϵ is introduced to differentiate noisy features from important features. The feature set $\mathcal{C}(B_i)$ representing the class B_i is formed with only those features F_j's which have $P(F_j | B_i)$ values greater than or equal to ϵ. So, the set $\mathcal{C}(B_i)$ will only contain a feature F_j if it is found to be important for most of the samples of B_i and also reflects the relevant characteristics of the virus class B_i. Let, the significant characteristics of a pair of classes, say $\{B_i, B_r\}$, be represented by the set $\mathcal{C}(\{B_i, B_r\})$. So, the set should contain those features which can efficiently reflect the properties of both the classes B_i and B_r. Hence, $\mathcal{C}(\{B_i, B_r\})$ is formed by taking intersection of the two sets $\mathcal{C}(B_i)$ and $\mathcal{C}(B_r)$.

Let us consider a set of t modalities $\mathcal{M} = \{\mathcal{M}_1, \cdots, \mathcal{M}_p, \cdots, \mathcal{M}_t\}$, where each modality corresponds to a specific texture descriptor evaluated at a particular scale. Given the set of modalities, the proposed method computes the relevance $\Gamma_p(\{B_i, B_r\})$ of the feature set $\mathcal{C}_p(\{B_i, B_r\})$ under modality $\mathcal{M}_p$ to assess the efficacy of the set in describing important properties of the class-pair $\{B_i, B_r\}$. In order to quantify relevance, the concept of hypercuboid equivalence partition matrix of rough hypercuboid approach [11] is employed. The relevance

$\Gamma_p(\{B_i, B_r\})$ of the feature set $\mathcal{C}_p(\{B_i, B_r\})$ with respect to the pair of classes $\{B_i, B_r\}$ is obtained as follows [11]:

$$\Gamma_p(\{B_i, B_r\}) = 1 - \frac{1}{n_{ir}} \sum_{k=1}^{n_{ir}} v_k(\mathcal{C}_p(\{B_i, B_r\})) \qquad (2)$$

where, n_{ir} is the number of samples which belongs to class-pair $\{B_i, B_r\}$ and

$$v_k(\mathcal{C}_p(\{B_i, B_r\})) = \begin{cases} 1 & \text{if } h_{ik}(\mathcal{C}_p) = 1 \text{ and } h_{rk}(\mathcal{C}_p) = 1 \\ 0 & \text{otherwise,} \end{cases} \qquad (3)$$

$$\text{where } h_i(\mathcal{C}_p) = [h_{ik}(\mathcal{C}_p)]_{1 \times n_{ir}} = \bigcap_{F_j \in \mathcal{C}_p} h_i(F_j). \qquad (4)$$

Here, $h_{ik}(F_j) \in \{0, 1\}$ represents the membership of sample y_k in i-th equivalence partition induced by the feature F_j and is determined as follows:

$$h_{ik}(F_j) = \begin{cases} 1 & \text{if } L(B_i) \leq y_k(F_j) \leq U(B_i) \\ 0 & \text{otherwise.} \end{cases} \qquad (5)$$

The interval $[L(B_i), U(B_i)]$ is the value range of the feature F_j for the class B_i. So, the feature value $y_k(F_j)$ of each object y_k belonging to the class B_i corresponding to the feature $F_j \in \mathcal{C}_p(\{B_i, B_r\})$ falls within the interval $[L(B_i), U(B_i)]$, which implies that an equivalence partition is nonempty. The intersection between every two such intervals corresponding to the features of set $\mathcal{C}_p(\{B_i, B_r\})$ may form an implicit hypercuboid, as indicated by the shaded rectangle in Fig. 1. It encloses the misclassified objects that belong to more than one equivalence partitions with respect to the attribute set $\mathcal{C}_p(\{B_i, B_r\})$. The relevance $\Gamma_p(\{B_i, B_r\})$ of the set $\mathcal{C}_p(\{B_i, B_r\})$, with respect to the pair of classes $\{B_i, B_r\}$, depends on the cardinality of the implicit hypercuboid.

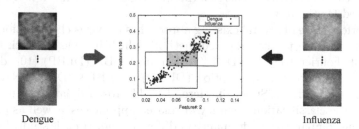

Dengue   Influenza

Fig. 1. Example of rough hypercuboids in two dimensions: two class hypercuboids corresponding to upper approximations of Dengue and Influenza virus classes, along with implicit hypercuboid (boundary region) denoted by the shaded region.

So, it can be observed from (2) that the relevance increases with decrease in the cardinality of implicit hypercuboids. If $\Gamma_p(\{B_i, B_r\}) = 1$, then no implicit hypercuboid is formed and both the classes B_i and B_r can be defined precisely using the knowledge of $\mathcal{C}_p(\{B_i, B_r\})$. On the other hand, if $\Gamma_p(\{B_i, B_r\}) = 0$, then both the classes cannot be defined using the information of $\mathcal{C}_p(\{B_i, B_r\})$. However, if $\Gamma_p(\{B_i, B_r\}) \in (0, 1)$, then B_i and B_r can be approximated using the feature set $\mathcal{C}_p(\{B_i, B_r\})$. The most relevant feature set $\tilde{C}_{ir}$ for the class-pair $\{B_i, B_r\}$ is formed by considering the relevance value of the feature set $\mathcal{C}_p(\{B_i, B_r\})$, corresponding to each of the t modalities, as follows:

$$\tilde{C}_{ir} = \underset{\mathcal{C}_p(\{B_i, B_r\})}{\arg\max} \{\Gamma_p(\{B_i, B_r\})\}. \tag{6}$$

Finally, the feature set $\mathcal{D}$, corresponding to all the virus classes, is formed as

$$\mathcal{D} = \bigcup \tilde{C}_{ir}. \tag{7}$$

The SVM is used to predict the texture patterns present in TEM images of test samples of virus particles, based on the final feature set $\mathcal{D}$ obtained using (7).

3   Performance Analysis

In order to establish the efficacy of the proposed method in classifying the virus particles present in TEM images, extensive experiment is conducted on Virus image database and the corresponding results are presented in this section.

3.1   Algorithms Compared

The proficiency of the proposed method is validated through comparison with several texture descriptors, which are LBP [16], LBP^{ri} [15], LBP$^{\mathrm{riu2}}$ [15] and CoALBP [14], obtained at different scales. Here, $\mathcal{S}_1$: scale 1, $\mathcal{S}_2$: scale 2, $\mathcal{S}_3$: scale 3, $\mathcal{S}_4$: scale 4, $\mathcal{S}_{123}$: concatenation of $\mathcal{S}_1$, $\mathcal{S}_2$ and $\mathcal{S}_3$, and $\mathcal{S}_{124}$: concatenation of $\mathcal{S}_1$, $\mathcal{S}_2$ and $\mathcal{S}_4$ are considered. In the current study, the descriptors along with the corresponding scales, are chosen arbitrarily and therefore, any other sets of descriptors are equally compatible to be used in the proposed descriptor selection method. In case of CoALBP, 4-neighborhood is considered, while 8-neighborhood is considered for the rest.

The performance of several existing approaches for virus classification is analyzed with reference to the proposed method, which include Haralick textural features [5], Fourier RDP (FRDP) [9], dominant LBP (DLBP) [10], discriminative features for texture description (DFTD) [4], LBP+RDP [9], NewH [13], Fusion [13] and Jeffreys/Stein [6]. Here, 10-fold cross-validation is performed to evaluate the classification accuracy of related approaches as well as proposed method. The comparative performance of different algorithms is studied through box-and-whisker plots, tables of means, medians, standard deviations, and p-values computed through both paired-t and Wilcoxon signed-rank tests, with 95% confidence level. In box-and-whisker plots, the central line on each box represents the median, the upper and lower quartiles are depicted by upper and lower boundaries of the box, respectively. The whiskers are drawn from mean to three standard deviations, so that extreme data points are also included. The outliers are plotted individually, denoted as '+'.

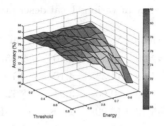

Fig. 2. Effect of energy E and threshold ϵ on classification accuracy.

3.2 Description of Data Set

The effectiveness of the existing approaches as well as proposed method is studied through evaluation on the real-life Virus data set [8]. The samples of the database are imaged using negative stain TEM. The set includes 15 different virus classes, each of which is represented by 100 TEM images. Different viruses exhibit different structural properties. However, the diameter or cross-section remains almost constant within a particular virus class. Usually, the diameter varies from 25 nm to 270 nm depending on the morphology of the viruses. The virus particles, which are presented in the data set, include Dengue, Cowpox, Ebola, Influenza, Adenovirus, Astrovirus, Rotavirus, Norovirus, Crimean-Congo Haemorrhagic Fever, Lassa, Marburg, Orf, Papilloma, Rift Valley, and WestNile.

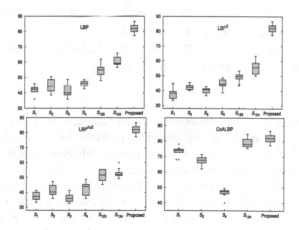

Fig. 3. Performance of different local texture descriptors and proposed method.

3.3 Optimum Values of Different Parameters

In the current study, the feature set $\mathcal{C}(B_i)$, representing the important properties of the class B_i, is obtained for a particular value of Energy E and threshold ϵ.

Table 1. Classification accuracy of various local descriptors and proposed method

Descriptors	Scales	Mean	Median	StdDev	Paired-t:p	Wilcoxon:p
LBP	S_1	42.00	42.33	2.70	2.35E−11	2.52E−03
	S_2	44.60	44.33	4.10	2.75E−10	2.50E−03
	S_3	41.20	40.00	3.91	2.72E−10	2.52E−03
	S_4	45.73	46.00	**1.78**	2.24E−11	2.50E−03
	S_{123}	55.00	55.00	4.21	1.22E−08	2.53E−03
	S_{124}	60.53	59.33	3.20	4.80E−08	2.52E−03
LBP^{ri}	S_1	37.87	38.33	3.45	5.71E−13	2.53E−03
	S_2	42.33	42.33	1.87	2.02E−11	2.52E−03
	S_3	40.00	40.67	2.04	4.12E−11	2.52E−03
	S_4	44.53	44.33	2.96	1.62E−12	2.53E−03
	S_{123}	48.87	49.67	3.19	4.90E−10	2.53E−03
	S_{124}	55.67	55.67	4.36	4.53E−08	2.52E−03
LBP^{riu2}	S_1	37.53	37.33	3.06	4.62E−11	2.53E−03
	S_2	41.13	40.33	3.61	3.86E−10	2.52E−03
	S_3	36.47	36.00	3.19	1.14E−11	2.52E−03
	S_4	42.40	44.00	4.67	2.04E−09	2.53E−03
	S_{123}	51.20	51.67	3.73	9.55E−11	2.46E−03
	S_{124}	52.93	52.00	3.03	1.40E−10	2.50E−03
CoALBP	S_1	73.20	74.00	3.14	1.82E−05	2.52E−03
	S_2	67.47	67.67	2.81	5.54E−09	2.47E−03
	S_4	46.73	47.33	2.64	2.07E−11	2.52E−03
	S_{124}	78.87	78.00	2.92	2.75E−03	7.49E−03
Proposed		**82.07**	**82.00**	2.62		

So, the values of both the parameters have an influence on the performance of the proposed method in classifying the virus particles into multiple classes.

The value of E is varied from 0.50 to 1.00 at an interval of 0.05, while the value of ϵ is varied from 0.00 to 0.70 at an interval of 0.10 and the corresponding classification accuracy of the proposed method is noted in order to obtain the optimum values of the parameters. The SVM classifier with linear kernel is applied on virus TEM images and the effect of the values of E and ϵ on 10-fold classification accuracy is reported in Fig. 2. The results depicted in Fig. 2 demonstrate that the accuracy of the proposed method increases with increase in energy E and decrease in threshold ϵ, and highest classification accuracy is obtained at $E = 0.95$ and $\epsilon = 0.30$, which validates the concept of energy and threshold defined in the current study. While low energy provides restricted representation of the important feature sets of samples, high energy signifies more information captured from the images, leading to a descriptive representation of the sets. The features, which have occurred in the important feature sets inadvertently with low probability of occurrence, are removed by incorporating the threshold parameter ϵ in the algorithm. So, for the proposed approach, the values of parameters E and ϵ are fixed at 0.95 and 0.30. Hence, a precise representation of the set $\mathcal{C}(B_i)$, reflecting the characteristics of the class B_i, is obtained with the features which preserves 95% of total energy by at least 30% samples of each B_i.

3.4   Comparison with Existing Approaches

In general, a specific descriptor, obtained at a particular scale, may be used to describe the textural properties of all the TEM images present in the Virus data set. However, the proposed method identifies class-pair specific modality to capture the inherent characteristics of each of the virus classes.

In order to validate the significance of class-pair relevant modalities over uniform modalities, extensive experiment is carried out on Virus data set, considering fifteen modalities corresponding to four local descriptors LBP, LBP^{ri}, LBP^{riu2} and CoALBP, evaluated at both individual and concatenated scales. Figure 3 and Table 1 report the classification accuracy achieved by the proposed method as well as various local texture descriptors on the samples of Virus database. The results presented in Fig. 3 and Table 1 reveal that highest mean and median values are achieved by the proposed method, irrespective of descriptors and scales considered. Also, statistical significance analysis reveals that the proposed method attains significantly lower p-values in all the 44 cases. In this regard, it is to be mentioned that the proposed method attains 82.07% accuracy on Virus database with 2327 number of features, selected from the initial feature sets of fifteen modalities corresponding to the four local descriptors, evaluated at four different scales. If the feature sets of all the modalities are concatenated, then 80% accuracy can be achieved with 4280 features. Thus, the proposed method can efficiently identify class-pair specific modalities to describe each pair of classes with lesser number of features.

Finally, the performance of several existing approaches for virus classification is analyzed with reference to the proposed method, which include Haralick textural features [5], FRDP [9], DLBP [10], DFTD [4], LBP+RDP [9], NewH [13], Fusion [13] and Jeffreys/Stein [6]. In these methods, important information is captured from the TEM images to categorize the virus samples into one of the known fifteen virus classes. The performance of the proposed approach with reference to different existing methods is analyzed in Fig. 4. It is evident from Fig. 4 that the methods corresponding to Haralick features, DLBP, DFTD and FRDP exhibit poor performance in recognizing viruses from TEM images. On the other hand, the methods like LBP+RDP, NewH, Fusion, and Jeffreys/Stein show improvement in the performance. However, the proposed method attains highest classification accuracy on Virus data set with respect to all the existing methods.

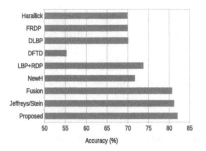

Fig. 4. Analysis of performance of the proposed approach and existing methods.

4 Conclusion

In the current study, a new method is developed for automatic recognition of virus particles from negative stain TEM images. The main contribution of the paper lies in considering relevant modality corresponding to each pair of classes present in Virus database, rather than considering uniform modalities for all the classes. The relevance of each modality is computed based on hypercuboid equivalence partition matrix of rough hypercuboid approach. A subset of important features is selected for each of the relevant modalities by reducing the impact of both noisy pixels present in a virus image as well as noisy virus images present in a particular virus class. The proficiency of the proposed approach with respect to different existing method is established on real-life Virus image data set.

Acknowledgment. This publication is an outcome of the R&D work undertaken in the project under the Visvesvaraya PhD Scheme of Ministry of Electronics and Information Technology, Government of India, being implemented by Digital India Corporation.

References

1. Almeida, J.D.: A classification of virus particles based on morphology. Can. Med. Assoc. J. **89**(16), 787–798 (1963)
2. Almeida, J.D., Waterson, A.P.: Some implications of a morphologically oriented classification of viruses. Arch. Virol. **32**(1), 66–72 (1970)
3. Goldsmith, C.S., Miller, S.E.: Modern uses of electron microscopy for detection of viruses. Clin. Microbiol. Rev. **22**(4), 552–563 (2009)
4. Guo, Y., Zhao, G., Pietikäinen, M.: Discriminative features for texture description. Pattern Recogn. **45**(10), 3834–3843 (2012)
5. Haralick, R.M., Shanmugam, K., Dinstein, I.: Textural features for image classification. IEEE Trans. Syst. Man Cybern. **3**(6), 610–621 (1973)
6. Harandi, M., Salzmann, M., Porikli, F.: Bregman divergences for infinite dimensional covariance matrices. In: Proceedings of the IEEE Conference on Computer Vision and Pattern Recognition, pp. 1003–1010 (2014)

7. Ito, E., Sato, T., Sano, D., Utagawa, E., Kato, T.: Virus particle detection by convolutional neural network in transmission electron microscopy images. Food Environ. Virol. **10**(2), 201–208 (2018)
8. Kylberg, G., Uppstrom, M., Borgefors, G., Sintorn, I.-M.: Segmentation of virus particle candidates in transmission electron microscopy images. J. Microsc. **245**, 140–147 (2012)
9. Kylberg, G., Uppström, M., Sintorn, I.-M.: Virus texture analysis using local binary patterns and radial density profiles. In: San Martin, C., Kim, S.-W. (eds.) CIARP 2011. LNCS, vol. 7042, pp. 573–580. Springer, Heidelberg (2011). https://doi.org/10.1007/978-3-642-25085-9_68
10. Liao, S., Law, M.W.K., Chung, A.C.S.: Dominant local binary patterns for texture classification. IEEE Trans. Image Process. **18**(5), 1107–1118 (2009)
11. Maji, P.: A rough hypercuboid approach for feature selection in approximation spaces. IEEE Trans. Knowl. Data Eng. **26**(1), 16–29 (2014)
12. Matuszewski, B.J., Shark, L.-K.: Hierarchical iterative Bayesian approach to automatic recognition of biological viruses in electron microscope images. In: Proceedings of the International Conference on Image Processing, pp. 347–350 (2001)
13. Nanni, L., Paci, M., Brahnam, S., Ghidoni, S., Menegatti, E.: Virus image classification using different texture descriptors. In: Proceedings of the International Conference on Bioinformatics and Computational Biology, Las Vegas, NV (2013)
14. Nosaka, R., Ohkawa, Y., Fukui, K.: Feature extraction based on co-occurrence of adjacent local binary patterns. In: Ho, Y.-S. (ed.) PSIVT 2011. LNCS, vol. 7088, pp. 82–91. Springer, Heidelberg (2011). https://doi.org/10.1007/978-3-642-25346-1_8
15. Ojala, T., Pietikainen, M., Maenpaa, T.: Multiresolution gray-scale and rotation invariant texture classification with local binary patterns. IEEE Trans. Pattern Anal. Mach. Intell. **24**(7), 971–987 (2002)
16. Ojala, T., Pietikäinen, M., Harwood, D.: A comparative study of texture measures with classification based on feature distributions. Pattern Recogn. **29**(1), 51–59 (1996)
17. Ong, H.C.L.: Virus recognition in electron microscope images using higher order spectral features. Ph.D. thesis, Queensland University of Technology (2006)
18. Pawlak, Z.: Rough Sets: Theoretical Aspects of Reasoning About Data. Kluwer Academic Publishers, Dordrecht (1991). ISBN 0792314727
19. Sintorn, I.-M., Homman-Loudiyi, M., Söderberg-Nauclér, C., Borgefors, G.: A refined circular template matching method for classification of human cytomegalovirus capsids in TEM images. Comput. Methods Programs Biomed. **76**(2), 95–102 (2004)

Detection of Necrosis in Mice Liver Tissue Using Deep Convolutional Neural Network

Nilanjana Dutta Roy[1], Arindam Biswas[1(✉)], Souvik Ghosh[2], Rajarshi Lahiri[2], Abhijit Mitra[3], and Manabendra Dutta Choudhury[3]

[1] Indian Institute of Engineering Science and Technology, Shibpur, Shibpur, India
barindam@gmail.com
[2] Department of Computer Science and Engineering,
Institute of Engineering and Management, Kolkata, India
[3] Department of Life Science and Bioinformatics,
Assam University, Silchar, India

Abstract. Acute Hepatic Necrosis is an early sign of liver dysfunction. Liver dysfunction is one of the major reasons for increasing death rate. Accurate diagnosis in less time, along with proper medication, show a ray of hope in controlling the aggravation of the situation. To overcome the unfavorable effects of harmful drugs, medicinal plant extract has become major thrust area nowadays. This research work has presented a way to show the improvements in mice liver tissue after applying the designated composition of the plant extract. And the performance is measured with our designed deep convolutional neural network (CNN) architecture along with the preprocessing techniques that has shown to be competent to classify microscopic images of mice hepatic tissues. Considering a small database of 30 images, we introduced a preprocessing stage which included the dividing of the original microscopic images to small patches. The accuracy of the classification results using the proposed CNN based classifier was 99.33%.

Keywords: Hepatic necrosis · Medicinal plant extract · Medical imaging · Deep learning · Computer vision

1 Introduction

Liver disorders correspond to the variety of diseases, infections and symptoms influencing the morphology, histology and physiological functions of the liver. Hepatic abnormalities have been reported to be the major cause of around 20,000 deaths every year [1]. The susceptibility of the liver to different fatal diseases especially jaundice, hepatitis, liver cirrhosis etc. is provoked by the excessive assimilation of various drugs as well as chemicals in the liver hepatocytes [2,12]. Phytomedicines are routinely being used for the treatment of such hepatic disorders and have been found to be much more efficient than other drugs. Hence

© Springer Nature Switzerland AG 2019
B. Deka et al. (Eds.): PReMI 2019, LNCS 11942, pp. 32–40, 2019.
https://doi.org/10.1007/978-3-030-34872-4_4

exploration and development of more and more herbal drugs from ethnomedicinally important medicinal plants for fast and effective therapy of liver diseases is the demand of time. Pteris semipinnata L. is a potential medicinal pteridophyte that is used for the treatment of hepatitis, enteritis and snake bite [5]. Paste of fronds of this plant is used externally for the treatment of carbuncle by reang tribe of Tripura [15]. This plant is widely used in curing diarrhoea as well as inflammatory disorders also [6]. The liver is an essential organ for survival and no other organ can compensate for it if the liver is damaged. Necrosis is a widely recognized form of hepatocyte damage. Microscopic imaging is an important tool in medical imaging and it provides rich and complex information about the structure of tissues and the current state of the cells. These microscopic images help pathologists study the cells and derive information from it. Due to complexity of the microscopic images, it takes time to make the diagnosis and sometimes leads to disagreement between pathologists which suggests a rising demand for stable, reliable and robust computational methods to increase the diagnosis efficiency. In the recent years, deep learning has contributed greatly to the analysis of images in the medical field. Convolutional neural networks (CNNs) has given state of the art performance on image classification [16]. CNNs are widely used for object detection and other macro landscapes. Researchers all across the world have applied CNNs to medical images like X-rays, microscopic images etc. CNN classifiers are used to segment and classify histopathological images. Classification assigns a label to the image i.e. normal/abnormal. CNNs are used in segmentation where the contours and boundaries are detected. Pre-Processing the microscopic images improve the performance of CNNs. In this experiment we present a CNN model based to classify healthy or damage liver tissues. We have combined a binary classification strategy with image splitting to calculate the amount of damage in the given sample. We also use data processing techniques and data augmentation as we lack in sufficient amount of data.

2   Related Works

CNN's are widely used in biomedical imagery applications as they overpower traditional methods which require handcoded feature extraction. A comprehensive tutorial with selected use cases that uses deep learning for digital pathology image analysis [7]. A deep CNN is used for endotracheal tube position and X-ray image classification [9]. Cell detection in pathology and microscopy images with multi-scale CNN [13]. A categorization for liver cirrhosis has been proposed using CNNs for health informatics [17]. A customized CNN uses shallow convolution layer to classify lung image patches from interstitial lung disease [10]. A deep CNN network classifies mice hepatic granuloma microscopic images [18]. There has been a requirement for applying computational approaches to get better diagnosis efficiency. A search algorithm known as Modified cuckoo segments microscopic images hippocampus [4]. Another CNN based method uses clustering and manifold learning for diabetic plantar pressure imaging dataset [11]. An optimized LPA-ICI filter denoises light microscopy image [3]. A deep CNN is used to analyse the histopathological snapshots by segmentation-based classification of epithelial and stromal regions. [8]. An arrowhead detection technique for biomedical images [14]. A deep CNN is used for segmenting and classifying epithelial and stromal regions in histopathological images [19].

3   Pre-processing

Several Pre-processing techniques were used to make our model learn features easily Fig. 1. Grayscaling is done so that the image sharpness increases also the light and dark areas are easily identifiable. Dataset Normalisation is done so that all the values of the data lie at zero centered mean and a variance of one. To increase the overall contrast and making the highlights look brighter CLAHE is used. Gamma Correction is again used for brightness control and to make sure the images don't become too dark or bleached out.

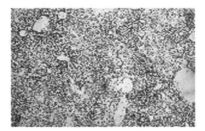

Fig. 1. Pre-processed output generated

3.1   RGB to Grayscale Conversion

The usual method for grayscale conversion is the average method which is the most simple one but it does not give a dynamic grayscale image. In this project, we have chosen to go with the original ITU-R recommendation (BT.601, specifically) which gives a more dynamic grayscale image. The formula is:

$$Gray = (B \times 0.114 + +G \times 0.587 + R \times 0.299) \tag{1}$$

where R, G, B is in the range 0–255. After this the image has one channel.

3.2   Dataset Normalisation

The images in the dataset is normalized by subtracting the mean of the dataset which serves to 'center' the data. Next it is divided by the standard deviation of the dataset. This process is done because in while training our network, we're going to be multiplying (weights) and adding to (biases) these initial inputs in order to cause activations of the neurons that we then backpropagate with the calculated gradients to train and update the model. Normalisation is done for each feature to have a similar range so that our gradients don't go out of control. The pseudo code for normalisation is:

$$(z - z.mean())/z.std()) \tag{2}$$

For standardizing the values between 0 and 1 the following was done where x represents each image in the dataset.

$$(z - z.min())/(z.max() - z.min()) \tag{3}$$

3.3   Contrast-Limited Adaptive Histogram Equalization (CLAHE)

Histogram equalization is used to give a linear trend to the cumulative probability function associated to the image and improve it's quality. This process images to adjust the contrast of an image by modifying the intensity distribution of the histogram. Histogram equalization uses the cumulative probability function (cdf). The cdf is a cumulative sum of all the probabilities lying in its domain and defined by:

$$cdf(x) = \sum_{k=-\infty}^{x} P(k) \tag{4}$$

These images are encoded with 8 bits will only have $2^8 = 256$ possible intensity values going from 0 representing black to $L-1 = 255$ representing white. The following formula gives the new pdf:

$$S_k = (L - 1)cdf(x) \tag{5}$$

In CLAHE, the image is divided into small blocks called 'tiles' (8×8 pixels). Then every tile is histogram equalized. So in a tile, the histogram would be localized to a small region. If noise is present, it will be amplified and so to avoid this, contrast limiting is used. If any histogram bin is greater than the contrast limit that is specified then those pixels are clipped and distributed uniformly to other bins before applying histogram equalization.

3.4   Gamma Correction

Gamma, which is denoted by the Greek letter γ, is described as the relationship between an input that will be the RGB intensity values of an image and the output. The equation for computing the resulting output image is as follows:

$$I' = 255 \times (\frac{I}{255})^{\gamma} \tag{6}$$

When the gamma is 1.0 the input equals the output producing a straight line. For calculating gamma correction the formula is as follows:

$$I' = 255 \times (\frac{I}{255})^{\frac{1}{\gamma}} \tag{7}$$

4 Proposed Methodology

4.1 Preparation of Test

Male Swiss albino mice were randomly classified into four groups ($n = 6$). A single water dose (25 ml/kg) was orally administered to the animals of group 1 (control), daily for 6 consecutive days. The same group of animals was also administered olive oil (10 ml/kg, i.p) on 1st and 2nd day. The animals of group 2 (CCl4) were given water (25 ml/kg) for 6 days and CCl4 (0.2%) in olive oil (10 ml/kg, i.p.) was administered on 1st and 2nd day. Group 3 animals were orally administered with the standard drug silymarin (100 ml/kg) once in a day for 4 days from 3rd to 6th day. They also received CCL4 (0.2%) in olive oil (10 ml/kg, i.p) on 1st and 2nd day. On day 7, the mice ($n = 6$ per group) were anaesthetised by chloral hydrate (350 mg/kg b.w.; I.P.), and then mice were perfused transcardially with 50 mL each of ice-cold 0.1M phosphate buffered saline (PBS; pH 7.4) and 4% w/v paraformaldehyde (in PBS). After perfusion, liver of the mice were dissected out, and stored in 4% paraformaldehyde, cryoprotected overnight in 30% w/v sucrose solution, 5 μm thick liver sections were taken on poly-L-lysine coated slides and Haematoxylin-Eosin staining was performed. This dataset has been prepared by the technicians of Biotech Hub, Assam University using foldscope.

4.2 Model

After the images were pre-processed with shape $240 \times 320 \times 1$ they were then fit to the convolution model. Our architecture is made up of repeating convolution block Fig. 2.

Fig. 2. Convolution block which was repeatedly used in our architecture

Each convolution block consists of three parts mainly first a convolution layer then a Max Pooling layer and finally Batch Normalization. The Convolution layer has a kernel size 3, stride of 1 and uses same padding. After the convolution ReLU activation was applied on it. Then Max Pooling was performed with a pool size of 2 and a stride of 2. Finally Batch Normalization was performed to stabilize the training. The first convolution block had 32 filters then the second had 64 filters and the third had 128 filters. The fourth and the fifth block had 256 filters.

The sixth and seventh consists of 512 filters after that the output was flattened to a dense layer with hidden units of 512 and activation of ReLU. Final layer consists of one hidden unit and an activation of sigmoid was used. Last layer predicts whether the image is normal or damaged Fig. 3.

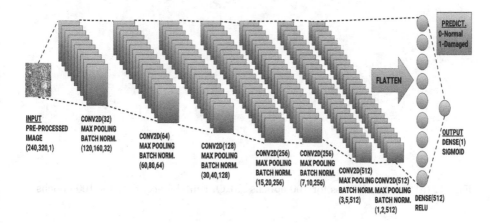

Fig. 3. Our proposed model architecture

5 Training

The training data in our hand was very low, we had 4 images for normal mice liver tissue and 8 images for damaged mice liver tissue. For validation purposes we hold out 1 image from both normal and damaged liver tissue to test our network. As deep learning model requires a lot of data for training the model, we used several techniques mentioned below to augment the data and make it feasible for our network to learn.

1. As the number of damaged images in our dataset is almost twice of the normal images, to balanced the dataset we took 500 random images from damaged tissue and 1000 random cropped images from the normal tissue.
2. The images were random cropped with a dimension of 240 × 320 which is 16 times smaller from the original image of dimension 960 × 1280.
3. If the image taken is damaged then set the label of the 1000 random cropped image as 1 else if the image taken is normal then set the label to 0.
4. The images are then converted from RGB to grayscale.
5. Then they are normalized by subtracting the mean and dividing by variance and 255.
6. After normalization Clahe Equalization was performed with a clip size of 2 and tile grid size of 8 × 8.
7. Gamma Adjustment was done with Gamma value of 1.
8. The prepared data was further augmented by taking vertical flip, horizontal flip and rotated by a value of 20.

6   Result

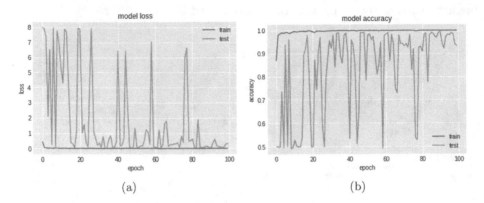

Fig. 4. (a) Our model loss for 100 epochs (b) Our model accuracy for 100 epochs

The loss used in our experiment to train the model was Binary Cross Entropy function and for optimizing our model Adam Optimizer was used. The weights of our model were saved whenever the validation loss was less than the previous validation loss. The model was trained for 100 epochs with a batch size of 150. Learning rate decay was used with a factor of 0.8 which monitored the validation loss for 10 epochs. The Graphs for losses and accuracy have been plotted Fig. 4 (Tables 1 and 2).

Table 1. Our experiments results

Loss and accuracy			
Train loss	Test loss	Train accuracy(%)	Test accuracy(%)
0.0030	0.0272	99.93	99.33

Table 2. Confusion matrix on test dataset

		Predicted	
		Damaged	Normal
Actual	Damaged	372	3
	Normal	2	373

7   Conclusion

Liver dysfunction is one of the major reasons for increasing death rate. Accurate and fast diagnosis with safe medication can control the aggravated situation. Presently, the analysis of microscopic liver images are required to find out various anomalies and using Artificial Intelligence to assist pathologists is the future of medical development. In this experiment, we propose a novel methodology to detect Necrosis in mice liver tissues using a combination of pre-processing and a deep convolutional neural network. In future, we are required to deal with the problem of less number of training images. To assist pathologists more comprehensively, there is room for detection of the damaged areas more precisely.

Acknowledgment. The authors would like to take this opportunity to thank the Department of Biotechnology, Govt. of India for funding this research.

References

1. Bhardwaj, A., Khatri, P., Soni, M.: Potent herbal hepatoprotective drugs: a review. J. Adv. Sci. Res. **2**(2), 15–20 (2011)
2. Dhiman, A., Nanda, A., Ahmad, S.: A recent update in research on the antihepatotoxic potential of medicinal plants. J. Chin. Integr. Med. **10**, 117–119 (2012)
3. Ashour, A.S., et al.: Light microscopy image de-noising using optimized LPA-ICI filter. Neural Comput. Appl. **29**(12), 1517–1533 (2018)
4. Chakraborty, S., et al.: Modified cuckoo search algorithm in microscopic image segmentation of hippocampus. Microsc. Res. Tech. **80**(10), 1051–1072 (2017)
5. Ding, H.: The Chinese Medicinal Crytogam, p. 104. Shanghai Publishing House of Science and Technology, Shanghai (1982)
6. Herz, W., Falk, H.: Progress in the Chemistry of Organic Natural Product. Springer, New York (1988). https://doi.org/10.1007/978-3-7091-6507-2
7. Janowczyk, A., Madabhushi, A.: Deep learning for digital pathology image analysis: a comprehensive tutorial with selected use cases. J. Pathol. Inform. **7**, 29 (2016)
8. Kainz, P., Pfeiffer, M., Urschler, M.: Segmentation and classification of colon glands with deep convolutional neural networks and total variation regularization. Peer J. **5**, e3874 (2017)
9. Lakhani, P.: Deep convolutional neural networks for endotracheal tube position and x-ray image classification: challenges and opportunities. J. Digit. Imaging **30**(4), 460–468 (2017)
10. Li, Q., Cai, W., Wang, X., Zhou, Y., Feng, D.D., Chen, M.: Medical image classification with convolutional neural network. In: 2014 13th International Conference on Control Automation Robotics & Vision (ICARCV), pp. 844–848. IEEE (2014)
11. Li, Z., et al.: Convolutional neural network based clustering and manifold learning method for diabetic plantar pressure imaging dataset. J. Med. Imaging Health Inform. **7**(3), 639–652 (2017)
12. Nirmala, M., Girija, K., Lakshman, K., Divya, T.: Hepatoprotective activity of musa paradisiaca on experimental animal models. Asian Pac. J. Trop. Biomed. **10**, 11–15 (2012)
13. Pan, X., et al.: Cell detection in pathology and microscopy images with multi-scale fully convolutional neural networks. World Wide Web **21**(6), 1721–1743 (2018)

14. Santosh, K., Alam, N., Roy, P.P., Wendling, L., Antani, S., Thoma, G.R.: A simple and efficient arrowhead detection technique in biomedical images. Int. J. Pattern Recogn. Artif. Intell. **30**(05), 1657002 (2016)
15. Shil, S., Dutta Choudhury, M.: Ethnomedicinal importance of pteridophytes used by reang tribe of Tripura, North East India. Ethnobot. Leaflets **13**, 634–643 (2009)
16. Shin, H.C., et al.: Deep convolutional neural networks for computer-aided detection: CNN architectures, dataset characteristics and transfer learning. IEEE Trans. Med. Imaging **35**(5), 1285–1298 (2016)
17. Suganya, R., Rajaram, S.: An efficient categorization of liver cirrhosis using convolution neural networks for health informatics. Cluster Comput. **22**, 1–10 (2017)
18. Wang, Y., et al.: Classification of mice hepatic granuloma microscopic images based on a deep convolutional neural network. Appl. Soft Comput. **74**, 40–50 (2019)
19. Xu, J., Luo, X., Wang, G., Gilmore, H., Madabhushi, A.: A deep convolutional neural network for segmenting and classifying epithelial and stromal regions in histopathological images. Neurocomputing **191**, 214–223 (2016)

Bayesian Deep Learning for Deformable Medical Image Registration

Vijay S. Deshpande(ID) and Jignesh S. Bhatt(✉)(ID)

Indian Institute of Information Technology, Vadodara, India
{201761003,jignesh.bhatt}@iiitvadodara.ac.in

Abstract. Deformable image registration is one of the challenging inverse problems in medical images due to inevitable geometric distortions during the imaging. Conventional convolutional neural networks (CNNs) are limited to the fixed weights, and require substantial examples for supervised training. In this paper, we employ Bayesian deep learning in order to register the medical images that are corrupted by nonlinear geometric distortions. Given a pair of reference and moving image, our objective is to register the moving image onto the reference image coordinates while maintaining the pixel distribution of moving image. To this end, we propose a novel Bayesian deep architecture that contains 5 downsampling and 4 upsampling layers with kernel size of 3×3 interleaved by pooling layer with *tanh* and rectified linear units (ReLU) as activation functions. The architecture estimates the displacement vector field (DVF) of the moving image. Finally, it undergoes the grid resampling to warp the moving image using estimated DVF. We demonstrate that the proposed Bayesian deep architecture learns the probability distribution over the weights and produce accurate registration. Experiments are performed on the publicly available Cardiac MRI and Lungs CT scan datasets. The proposed model achieves better performance when compared to state-of-the-art approaches.

Keywords: Bayesian deep learning · Deformable registration · Displacement vector field

1 Introduction

Image registration is the process of aligning two or more images of the same scene with respect to each other where the images are captured by different sensors, at different times and/or from different viewpoints [1]. It is an essential task for automated medical image analysis since the images are subject to unpredictable geometric distortions during the acquisition. Typically, an image is considered as the reference image and images which are to be aligned with respect to the reference image are referred to as the moving images. Image registration is broadly classified as: (i) rigid or parametric registration, and (ii) non-rigid or deformable registration [1]. In rigid registration, the algorithms typically depend

© Springer Nature Switzerland AG 2019
B. Deka et al. (Eds.): PReMI 2019, LNCS 11942, pp. 41–49, 2019.
https://doi.org/10.1007/978-3-030-34872-4_5

on a small set of parameters, i.e., few matching points between moving and reference image, from affine group of transformations. The goal of such methods is to obtain optimal parameters describing the geometric mapping between the reference and moving image by maximizing a similarity metric. On the other hand, deformable image registration algorithms aim to estimate the underlying displacement vector field (DVF) considering the distortions across all the pixel locations in the moving image [2].

Recently, deep learning based registration techniques have found performing faster and are robust when compared to the classical approaches. Studies where deep learning based methods are used for image registration include convolutional stacked autoencoder [3] which is applied to LONI and ADNI brain MRI datasets. A Multi-resolution image registration framework which learns spatial transformation at different resolution in a self supervised fully convolutional network (FCN) is proposed in [4]. In [5], spatial transformer network to estimate the transformation parameters to align the image is proposed. In [6], FCN based techniques are proposed as Deformable Image Registration network (DIRnet). Image registration is also attempted by deep generative model in [7]. More recently, *Voxelmorph* is proposed where the network is trained with Brain MRI dataset, while incorporating auxiliary information such as segmentation in order to ease the registration task [8]. Note that, all this conventional deep learning models require substantial amount of data examples for the supervised training and may also prone to underfitting/overfitting. Moreover, once such CNN is trained, it relies on its fixed set of weights to predict the output of the test image. However, there is high chance of underfitting and uncertainity involved in medical data due to scarcity of data examples and unpredictable geometric distortions, respectively. To overcome these issues, in this paper, we propose a novel Bayesian deep learning architecture for achieving deformable medical image registration. Unlike existing approaches, our architecture learns the posterior distribution over the weights with limited examples by the backpropagation. The network then learns the DVF by analyzing reference and moving image pair. In turn the DVF together with moving image is used to perform grid resampling in order to register it with the reference coordinates.

2   Proposed Architecture

In this section, we discuss the proposed Bayesian deep architecture to perform deformable image registration of medical data. Given a pair of reference and moving image our objective is to register the moving image onto the reference image coordinates while maintaining the pixel distribution of the moving image.

In the conventional convolutional neural network (CNN) set-up, given a set of training examples $\mathbf{D} = (\mathbf{x}_i, \mathbf{y}_i)_i$ where $\mathbf{x}_i$ and $\mathbf{y}_i$ are the ith input and output respectively, the goal is to learn weights $\mathbf{w}$ that shall predict the output $\mathbf{y_{test}}$ corresponding to test input $\mathbf{x_{test}}$. The training is typically done by the backpropagation, where one assume that $\log P(\mathbf{D}|\mathbf{w})$ is differentiable in $\mathbf{w}$. Note that, at the end of training, the estimated weights $\mathbf{w}$ are fixed real numbers. Hence, the network has to rely on these hard (fixed) weights (numbers) to predict the output which may lead to severe error especially in medical data.

Unlike the conventional CNNs, Bayesian neural networks estimate the posterior distribution of the weights given the training data $P(\mathbf{w}|D)$ by backpropagation [9]. The predictive distribution of output $P(\mathbf{y_{test}})$ of a test data input item $\mathbf{x_{test}}$ is then given by

$$P(\mathbf{y_{test}}|\mathbf{x_{test}}) = \mathbb{E}_{P(\mathbf{w}|D)}[P(\mathbf{y_{test}}|\mathbf{x_{test}}, \mathbf{w})], \tag{1}$$

where $\mathbb{E}_{P(\mathbf{w}|D)}$ is conditional expectation over posterior distribution of weights given the training data $\mathbf{D}$. Thus taking the conditional expectation under the posterior distribution on weights (Eq. (1)) is equivalent to using a large number of neural networks for learning the weights. We propose one such deep Bayesian architecture Fig. 1 for achieving deformable medical image registration.

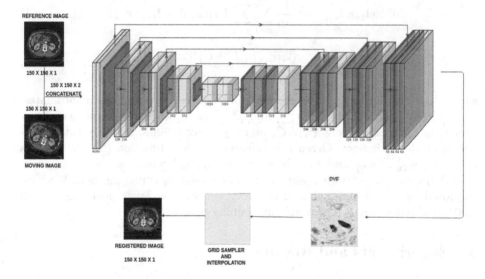

Fig. 1. Proposed deep Bayesian architecture for deformable medical image registration.

As shown in Fig. 1 a pair of reference and moving image of size 150×150 pixels concatenated into 2-channel 2D image ($150 \times 150 \times 2$) is then passed through downsampling and upsampling layers. It consists of 5 downsampling and 4 upsampling layers. Each downsampling layer consists of two convolutional layers with filters of size 3×3 where stride is 1 and pooling layer with a stride of 2. Whereas each upsampling layer consists of bilinear interpolation with scale size of 2 followed by two convolutional layers with filters of size 3×3 where stride is 1. In order to incorporate non-linearity, we have used rectified linear units (ReLU) and $tanh$ as activation functions. As shown in Fig. 1, the output of the architecture is the DVF for the moving images. It estimates the geometric displacement required by each pixel coordinate of moving image in order to get it registered with respect to the reference image. The estimated DVF along with grid resampling and bilinear interpolation of moving image gives us the final registered image as shown in Fig. 1.

2.1 Training/Test

The supervised training of proposed approach (Fig. 1) is done as follows: we use a differentiable loss function L that is a combination of structural similarity index (SSIM) [10] and mean-squared error (MSE) [11],

$$L(\mathbf{I_{ref}}, \mathbf{I_{reg}}) = SSIM(\mathbf{I_{ref}}, \mathbf{I_{reg}}) + MSE(\mathbf{I_{ref}}, \mathbf{I_{reg}}), \qquad (2)$$

where $\mathbf{I_{ref}}$ is reference and $\mathbf{I_{reg}}$ is registered image, i.e., $(\mathbf{I_{mov}} \circ \psi)$ where ψ is DVF, and

$$MSE(\mathbf{I_{ref}}, \mathbf{I_{reg}}) = \frac{1}{mn} \sum_{x=1}^{m} \sum_{y=1}^{n} [\mathbf{I_{ref}}(x,y) - \mathbf{I_{reg}}(x,y)]^2, \qquad (3)$$

$$SSIM(\mathbf{I_{ref}}, \mathbf{I_{reg}}) = \frac{(2\mu_{I_{ref}}\mu_{I_{reg}} + C_1)(2\sigma_{I_{ref}I_{reg}} + C_2)}{(\mu_{I_{ref}}^2 + \mu_{I_{reg}}^2 + C_1)(\sigma_{I_{ref}}^2 + \sigma_{I_{reg}}^2 + C_2)}, \qquad (4)$$

where $\mathbf{I_{ref}}(x,y)$ and $\mathbf{I_{reg}}(x,y)$ are reference and registered image coordinates, $\mu_{I_{ref}}$ and $\mu_{I_{reg}}$ are means and $\sigma_{I_{ref}}^2$ and $\sigma_{I_{reg}}^2$ are variances of reference and registered image, respectively. Here C_1 and C_2 are constants, m, n are total number of pixels for images. Given the differentiable loss function (2), the weights $\mathbf{w} = [w_1, w_2, ..., w_n]$ and bias $\mathbf{b} = [b_1, b_2, ..., b_m]$ where $w_i \sim \mathcal{N}(\mu_w, \sigma_w)$, $b_i \sim \mathcal{N}(\mu_b, \sigma_b)$ [12] of each layers are updated using backpropagation. We have employed batch normalization after every convolution layer, learning rate of 0.001, and batch size of 1 throughout training process.

3 Experiments and Results

In this section, we discuss the results obtained by applying the proposed Bayesian deep learning approach for the deformable medical image registration and compared with recent state-of-the-art approaches. In our experiment, we have used publicly available lungs computerized tomography (CT) scan data from the deformable image registration laboratory [13] as well as cardiac magnetic resonance imaging (MRI) data [14]. We implement the algorithm in Python and pytorch 1.0 on a machine with Intel core i5, Nvidia Geforce GTX 1050 4GB with 24 GB RAM.

3.1 Results on Lung CT Dataset

In this subsection, we show results of our approach on the lung CT images [13] and compare it with state-of-the-art approaches. The dataset consists of Thoracic 4DCT images as well as inspiratory and exipratory breath-hold CT image

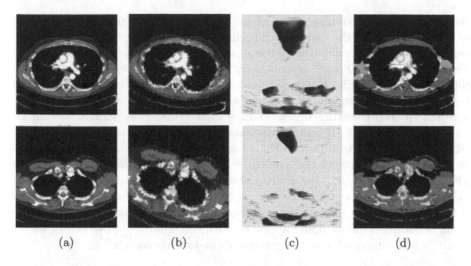

(a)   (b)   (c)   (d)

Fig. 2. A few challenging cases in lungs CT dataset [13]. (a) reference images, (b) moving images, (c) estimated DVF, and (d) corresponding registered images using proposed Bayesian deep architecture.

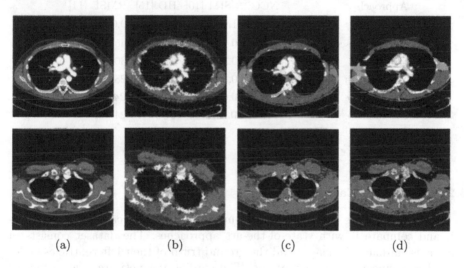

(a)   (b)   (c)   (d)

Fig. 3. Comparative results on the challenging cases in Lungs CT data [13]. (a) and (b) are reference and moving images, respectively, (c) DIRnet [6], and (d) proposed.

pairs. Entire dataset is arranged in 10 different cases. The images are of size $512 \times 512 \times 128$ voxels with a dimension of each voxel $0.97 \times 0.97 \times 2.5$ mm. In this experiment we have used 15 slices out of them, and resized it to 150×150 pixels i.e, it yields $150 \times 150 \times 15$. In order to generate moving images with different degrees of geometric distortions, we applied deformable or elastic distortion of order 0.8. We have generated 1000 distorted images for a single slice, giving a total of 15000 image pairs where the original slice (image) is considered as the reference and the corresponding distorted images are treated as its moving images. For the validation purpose, we split the dataset into training and test of 5000 and 3000 pairs, respectively. We first show the visual results of few challenging cases in Fig. 2. We now compare the reference image with estimated registered images using our approach as well as state-of-the-art approaches with different similarity metrics i.e, normalized cross correlation, structural similarity index (SSIM) [10], Hausdorff distance (HD) [15], root mean-squared error (RMSE) [11] for the test data of 3000 image pairs. We first show visual comparisons in Fig. 3. The calculated average values are as listed in the Table 1 along with results using state-of-the-art registration frameworks such as selfsupervised FCN [4], DIRnet [6], and Voxelmorph [8]. Note that, ϵ^{100} and ϵ^{1000} are the means of 100 and 1000 samples from the posterior distribution. Since we are taking the mean of samples, the proposed approach with ϵ^{100} and ϵ^{1000} have shown better results when compared to existing state-of-the-art approaches.

Table 1. Results on Lung CT dataset.

Approach	NCC	SSIM [10]	HD [15]	RMSE [11]
Ideal values	1.000	1.000	0.000	0.000
Selfsupervised FCN [4]	0.824	0.412	980	45.2
DIRnet [6]	0.916	0.538	683	30.239
Voxelmorph [8]	0.921	0.532	632	28.438
Proposed (ϵ^{100})	0.939	0.623	652	**24.483**
Proposed (ϵ^{1000})	**0.947**	**0.632**	**580**	24.489

3.2   Results on Cardiac Dataset

In this subsection, we discuss results of our approach on the cardiac MRI images [14] and compare it with state-of-the-art approaches. The dataset consists of short axis cardiac MR images and the ground truth of their left ventricles endocardial and epicardial segmentations. The cardiac magnetic resonance images are acquired from 33 subjects. Each subject's sequence consist of 20 time points and 8–15 slices along the long axis.

We considered slices of 256×256 voxels of one subject and resized it to 150×150 pixels. In order to generate moving images with different degrees of geometric distortions, we applied deformable or elastic distortion of order 0.8. We have generated 1000 distorted images for a single slice giving a total of 10000 image pairs. For the validation purpose, we split the dataset into training and test sets of size 5000 and 3000 image pairs, respectively. We first show the visual results of few challenging cases as shown in Fig. 4 and then compare the reference image with estimated registered images using our approach as well as

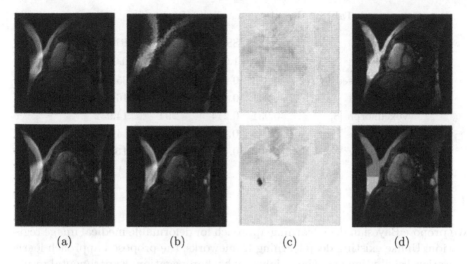

Fig. 4. A few challenging cases in cardiac MRI dataset [14]. (a) reference images, (b) moving images, (c) estimated DVF and (d) corresponding registered images using proposed deep Bayesian architecture.

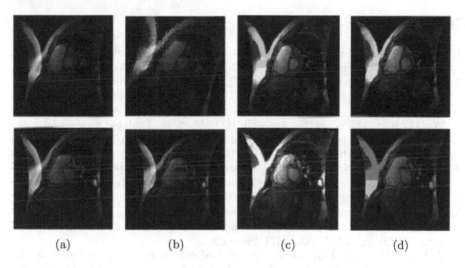

Fig. 5. Comparative results on the challenging cases of cardiac MRI results [14]. (a) and (b) are reference and moving images, respectively, (c) DIRnet [6], and (d) proposed.

recent state-of-the-art approaches for the test data of 3000 image pairs. We show visual comparison results in Fig. 5 while calculated average values of similarity metrics are as listed in the Table 2.

Table 2. Results on cardiac MRI dataset.

Approach	NCC	SSIM [10]	HD [15]	RMSE [11]
Ideal values	1.000	1.000	0.000	0.000
Selfsupervised FCN [4]	0.816	0.419	320	35
DIRnet [6]	0.941	0.436	439	16.23
Voxelmorph [8]	0.931	0.244	**291**	14.932
Proposed (ϵ^{100})	0.943	**0.415**	389	14.27
Proposed (ϵ^{1000})	**0.956**	0.467	365	**13.451**

4   Conclusion

We propose Bayesian deep learning approach for deformable medical image registration. Unlike existing deep learning frameworks, the proposed approach learns posterior distribution over the weights by backpropagation. Experimental results demonstrate the efficacy of proposed architecture when compared to state-of-the-art approaches.

References

1. Sotiras, A., Davatzikos, C., Paragios, N.: Deformable medical image registration: a survey. IEEE Trans. Med. Imaging **32**(7), 1153–1190 (2013)
2. Jahne, B.: Displacement vector fields. In: Jahne, B. (ed.) Digital Image Processing, pp. 297–317. Springer, Heidelberg (1991). https://doi.org/10.1007/978-3-662-11565-7_16
3. Wu, G., Kim, M., Wang, Q., Munsell, B.C., Shen, D.: Scalable high-performance image registration framework by unsupervised deep feature representations learning. IEEE Trans. Biomed. Eng. **63**(7), 1505–1516 (2016)
4. Li, H., Fan, Y.: Non-rigid image registration using self-supervised fully convolutional networks without training data. In: IEEE International Symposium on Biomedical Imaging (ISBI), pp. 1075–1078 (2018)
5. Jaderberg, M., Simonyan, K., Zisserman, A., Kavukcuoglu, K., : Spatial transformer networks. In: Proceedings of the 28th Neural Information Processing Systems (NIPS), pp. 2017–2025. MIT Press (2015)
6. de Vos, B.D., et al.: End-to-end unsupervised deformable image registration with a convolutional neural network. In: Cardoso, M.J., et al. (eds.) DLMIA/ML-CDS -2017. LNCS, vol. 10553, pp. 204–212. Springer, Cham (2017). https://doi.org/10.1007/978-3-319-67558-9_24
7. Mahapatra, D., et al.,: Deformable medical image registration using generative adversarial networks. In: IEEE Conference on International Symposium on Biomedical Imaging (ISBI), pp. 1449–1453 (2018)
8. Balakrishnan, G., et al.,: An unsupervised learning model for deformable medical image registration. In: IEEE Conference Computer Vision and Pattern Recognition (CVPR), Salt Lake City, UT, USA, pp. 8340–8348 (2018)
9. Blundell, C., et al.: Weight uncertainty in neural networks. In: Proceedings of the 32nd International Conference on Machine Learning, vol. 37, pp. 1613–1622. JMLR.org, France (2015)

10. Bovik, A.C., Sheikh, H.R., Simoncelli, E.P.: Image quality assessment. IEEE Trans. Image Process. **13**(4), 600–612 (2004)
11. Wang, Z., Bovik, A.C.: Mean squared error: love it or leave it? A new look at signal fidelity measures. IEEE Signal Process. Mag. **26**(1), 98–117 (2009)
12. Rasmussen, C.E., Christopher, K.I.: Gaussian Processes for Machine Learning (Adaptive Computation and Machine Learning). The MIT Press, Cambridge (2005)
13. Castillo, R., et al.: A framework for evaluation of deformable image registration spatial accuracy using large landmark point sets. Phys. Med. Biol. **54**(7), 1849–1870 (2009)
14. Andreopoulos, A., Tsotsos, J.K.: Efficient and generalizable statistical models of shape and appearance for analysis of cardiac MRI. Med. Image Anal. **12**(3), 335–357 (2008)
15. Huttenlocher, D.P., Klanderman, G.A., Rucklidge, W.J.: Comparing images using the Hausdorff distance. IEEE Trans. Pattern Anal. Mach. Intell. **15**(9), 850–863 (1993)

Design and Analysis of an Isotropic Wavelet Features-Based Classification Algorithm for Adenocarcinoma and Squamous Cell Carcinoma of Lung Histological Images

Manas Jyoti Das$^{(\boxtimes)}$ and Lipi B. Mahanta

Institute of Advanced Study in Science and Technology, Paschim Boragaon,
Guwahati, India
manas.ork@gmail.com

Abstract. One of the most prevailing types of lung cancer is non-small cell lung cancer (NSCLC). Differential diagnosis of NSCLC into adenocarcinoma (ADC) and squamous cell carcinoma (SCC) is important because of prognosis. Histological images are taken from a database consisting of 72 lung tissue samples collected indigenously with a core needle biopsy. In this work, a novel method has been developed where the features of ADC and SCC for a histological image are taken from various statistical and mathematical models implemented on the coefficients of the wavelet transform of an image. The method provides a precision of 95.1% and 96.2% in classifying malignant and non-malignant tissue type respectively. This methodology of classifying ADC and SCC without coding clinical diagnostic features into the system is a necessary step forward towards an autonomous decision system.

Keywords: Adenocarcinoma · Squamous cell carcinoma · Histological · Wavelet · Colour transformation · L*a*b*

1 Introduction

The prevalence of Lung cancer in India can be observed by the fact that for the year 2018 a total of 67,795 new cases were registered for both the sexes and 63,475 deaths related to lung cancer were reported [1]. Non-small cell lung cancer (NSCLC) accounts for more than 80% of lung cancer spectrum. NSCLC can be further subtyped into two major types, viz. adenocarcinoma (ADC) and Squamous Cell Carcinoma (SCC).

It is imperative for differentiating the NSCLC into subtypes as the prognosis of adenocarcinoma (ADC) is found out to be low [2] as compared to the other subtype. Also, for NSCLC, disease management of different subtype is different. SCC outcome is worst [3] among the two NSCLC subtypes. A tissue biopsy can reveal whether the nodule is malignant or benign as well as the subtype if it is malignant. Tissue samples can be extracted by an invasive surgical procedure, core needle biopsy or fine needle aspiration biopsy. Needle biopsy is often favored over surgical biopsy [4]. The classification of lung tissue into ADC or SCC is not straightforward since the architecture that distinguishes them are complex and depends on the grade of the tumour [5].

© Springer Nature Switzerland AG 2019
B. Deka et al. (Eds.): PReMI 2019, LNCS 11942, pp. 50–60, 2019.
https://doi.org/10.1007/978-3-030-34872-4_6

The histopathological images are captured in RGB colour space. It is found in the literature that converting it into other colour space can improve the result. RGB to L*a*b* colour space is done for breast [6], lung [7] and prostate [8]. HSV (Hue, Saturation and Value), for head and neck cancer detection [9] and lung tissue type classification [7]. However, there are other colour spaces like YCbCr, which is used for prostate [10] and CMY for breast cancer [11].

The method described in this work is based on texture characterization of the lung tissue. Texture characterization of tissue can confirm the malignancy of a tumour [12] as filter banks are used for detecting breast cancer [11] and grading of prostate cancer [13]. The fractal dimension, along with the variance of the wavelet coefficient is implemented for prostate cancer grading [10]. Further, the wavelet coefficient as one of the feature to a support vector machine (SVM) for classification of prostate lesions [14] is implemented. Wavelets have been used to analyse texture effectively as they provide multiple scale partition of the image spectrum [15]. We have selected Continuous Wavelet Transform (CWT) as opposed to discrete wavelets as CWT gives a high degree of frequency selectivity [16] of a texture.

To the best of our knowledge, classification of ADC and SCC is done only using Raman scattering microscopy where they use domain specific clinical diagnosis knowledge as feature sets [17]. As the classification of these two subtypes is very complex, coding the domain specific rules into an algorithm like morphology feature may not always produce higher classification rate [17]. This paper is focused on classifying the two subtypes of NSCLC by quantitatively extracting the feature from each subtype automatically. Two dimensional Marr and isotropic Morlet wavelet are used to transform the image into wavelet domain so as to characterize the texture of the image. The wavelet coefficients are modelled with Generalised Gaussian Distribution (GGD). SVM as a classifier, is used, the features for SVM are selected through Recursive Feature Elimination (RFE) [18] method.

2 Methods and Materials

2.1 Data Set

The slides containing the lung tissue are collected from an NABL (National Accreditation Board for Testing and Calibration, India) laboratory. The slide containing lung tissue were prepared from core needle biopsy samples, these samples are stained with Hematoxylin and Eosin (H&E) and are sectioned 5 μm thickness. The images are captured using Leica ICC50 HD digital microscope, digitisation is done at magnification of 20x which forms image pixel resolution of 0.32 μm, the images are of 2048 × 1536 resolution and each pixel is of 24 bit to incorporate 3 channels i.e. Red (R), Green (G) and Blue (B) of 8 bit each. The histopathological images of the lung contain various cellular-based tissue types, like but not limited to: the tumour, red blood cells, fibrosis, necrosis, carbon particles, normal cells.

A surgical pathologist delineates the malignant tumour portion on the digital image, this Region of Interest (ROI) is stored as a binary mask. The subtype of the NSCLC is also mentioned and stored as a label for the binary mask, the rest of the portion of the image containing tissue structures are labelled as Unspecified Lung Region (ULR). As the aim of the study is to classify subclasses of NSCLC from the background, i.e. the portion of the image not containing any lung tissue structure is also labelled as ULR. The ROI and the ULR are segmented into overlapping blocks of size 256 × 256, it is empirically found out that this size gives the highest level of classification for both the classifiers. The number of blocks for ROI and ULR is almost equal to avoid biasness in classification [19]. A visual inspection is done on all the images and only focused images are taken (due to human error unfocused images are sometimes produced). In total slides from 72 core needle biopsy are used of which 34 are of SCC and 38 are of ADC type. Table 1 illustrate the data set.

Table 1. The spread of data for ADC and SCC of the lung.

	# of slides	# of Images (2048 × 1536)	ROI blocks	ULR blocks
ADC	34	330	1572	720
SCC	38	400	1640	810

2.2 Colour Space Transform and Normalization

In H&E stain the nucleus of the cell is stained blue whereas cytoplasm and extracellular materials are stained with varying degree of the colour pink. The colour appearance of the tissue in light microscopy varies due to a wide range of factors, the sensor type of the digital camera, H&E reagents from different manufacturers or from different batches, the concentration of the stain, time for which stains are applied, and many more factors are there. If most of the procedure of staining is standardised still the colour fades with time. Normalization of colour is a necessary pre-processing for histopathological images, in this work we have selected a method that nonlinearly maps a source image (an image that needs to be normalized) to a target image (an image that is used to train the system) [20]. This method shows stable representations as it is not sensitive to the imaging condition and digital sensor used.

The captured digital images of lung histology are in RGB colour space. The 3 colour channels of the RGB colour space are not independent and change in one channel changes other channels. To avoid such pitfall of RGB colour space and to closely resemble the colour perception of human L*a*b* colour space is chosen. The L* channel corresponds to luminance, a* and b* channel represent the variance of red to green and yellow to blue respectively, a* and b* together define the chrominance. Figure 1, shows the representation of different channels. Only two levels of headings should be numbered. Lower level headings remain unnumbered; they are formatted as run-in headings.

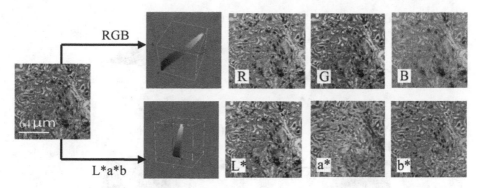

Fig. 1. A lung histopathological image captured at 20x magnification and stained with H&E stain is used to represent different channels in RGB and CIE L*a*b* colour spaces. (Color figure online)

2.3   Wavelet Coefficients Modelling and Similarity Measurement

Marr wavelet is a real, rotation invariant wavelet. In some literature, it is also known as Mexican hat wavelet. A 2D Marr wavelet is defined as (see Fig. 2(a) for representation):

$$\psi(x, y) = \left(2 - x^2 - y^2\right) \exp\left[-\frac{1}{2}\left(x^2 + y^2\right)\right] \tag{1}$$

Marr wavelet was selected for its good localization feature and also its affinity towards representing the nucleus in an effective way in the transform domain as evident from Fig. 2(c) and (d). As nucleus features are essential and they are available only in the L* and b* colour channels since L* is for luminance and nucleus are coloured blue/purple with H&E stain which is darker than the pale pink/red the colour of cytoplasm.

The b* channel record changes from bluish to yellowish colour component thus nucleus are available in this channel (as evident from Fig. 1). Marr wavelet is used for these two colour channels.

Isotropic Morlet wavelet is a complex-valued wavelet. A simple Morlet wavelet can be a plain wave modulated by a Gaussian envelope with a well-localized frequency domain with power only near its fundamental frequency. An isotropic wavelet is given by [21]:

$$\psi(x, y) = \pi^{-1/4} \exp\left[-i\omega_0(x + y)\right] \exp\left[-\left(x^2 + y^2\right)/2\right] \tag{2}$$

where $\omega_0 = (0, \omega_0)$ is a wave vector with $\omega_0 > 5.5$. Phase information is important for texture characterisation since Morlet wavelet are complex valued we can compute the phase as well as magnitude information.

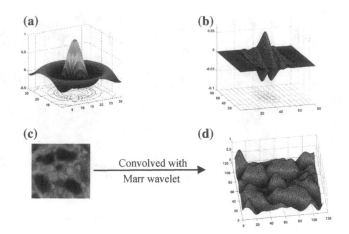

Fig. 2. (a) A 2D Marr wavelet (b) Isotropic Morlet wavelet (c) An cropped image of lung histopathology image (d) The 3D plot of the coefficient results from convolving the image with a Marr wavelet ($a = 6$). (Color figure online)

This wavelet is used on the a* channel, which codes non nucleus information. Since a* is for change of colour information from reddish to greenish and most non nucleus material are pale pink and pale red in appearance (see Fig. 1).

Let I_h be an image, the 2D CWT of the image is given by:

$$C_\psi(a, x', y') = \int\limits_{R^2} I_h(x, y)\psi_{a,x',y'}(x, y)dxdy \tag{3}$$

where $C_\psi(a, x', y')$ is the wavelet coefficient at location x', y' and having scale a ($a > 0$) (for a M × N image MN number of coefficients are extracted). ψ is the complex conjugate of those defined in Eqs. (1) and (2). The scale parameter range plays and important part in our analysis. Since it is computationally intensive to use many scale parameters and also not all scale parameter represents the details of the lung histopathology, small scale and large scale parameters will have over or low detailed information about the image. We empirically found $a = 3$ to 6 suited for our work.

2.4 Wavelet Coefficients Modelling and Similarity Measurement

The marginal distribution of the Marr and isotropic Morlet wavelet coefficients are long tailed, bell-shaped and centered around zero (see Fig. 3). To model such a distribution GGD [22] is used. Two varying parameters can be used to approximate the coefficient of the wavelet transform as shown below:

$$p(x; \alpha, \beta) = \frac{\beta}{2\alpha\Gamma(1/\beta)}e^{-(|x|/\alpha)^\beta} \tag{4}$$

Where Γ is a gamma function and α is a scale parameter which model the width of the Probability density function(PDF) peak, while β is the shape parameter which is inversely proportional to the decreasing rate of the peak. In Fig. 3(b) the distributions are created from the images using isotropic Morlet wavelet, to model shapes described in the figure Laplace distribution can be use. Since Laplace distribution is a special case of GGD (β = 0), so only the latter distribution is considered. The parameters α and β are estimated by maximum likelihood estimator (MLE) [22]. The scale and shape parameters are used as features for the classifier. Various statistical measures [23] of the GGD such as variance, kurtosis and entropy are used also used as an input feature vector to the classifier.

(a)   **(b)**

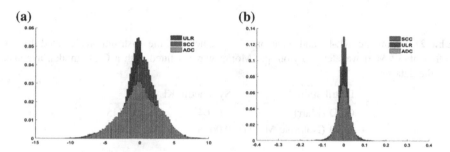

Fig. 3. (a) Marr wavelet coefficient for a particular subband is represented as a histogram for a random 256 × 265 image in L* colour space and selected from ADC and SCC data set. (b) Histogram representing the magnitude of isotropic Morlet wavelet coefficient for a subband, random images of size 256 × 256 each in a* colour space for ADC, SCC and ULR is used.

To quantify the difference between two empirical distributions a distance measure was used. The distance between two distributions was calculated using Kullback-Leibler divergence (KL-D), KL-D cannot be used as a metric since it is not symmetric and a symmetric version [24] of KL-D was implemented for this work. Jensen-Shannon divergence (J-divergence) with multiple probability distributions [25] is used to calculate the similarity of more than two distributions, J-divergence is symmetric. To quantify the goodness of fit of the GGD model to the observed distribution, symmetric KL-D and the χ^2 test is used.

3 Results and Discussion

This section provides evidence that the method that is proposed in this work is implemented correctly and the features that are used for classification are indeed can classify different lung tissue textures. The results reveal the accuracy of our method in classifying ADC, SCC and ULR from a data set. The specifics of the data set are tabled in Table 1, the SVM classifier is used to classify the subtype of the NSCLC with input feature vectors obtained from various methods used in this study.

3.1   Goodness of Fit

The goodness of fit of the model into the empirical distribution of the coefficients is calculated with symmetric KL-divergence and χ^2 test at 5% confidence level. The model which we were trying to fit to the observed distribution was assumed as the null hypothesis. Considering all the data 96.75% have accepted the null hypothesis, i.e. the chi square values in these cases are found to be lower than the upper limit of χ^2 distribution i.e. $\chi^2_{(0.05)} = 3.841$ with degree of freedom equals 1. Table 2 represent the goodness of fit of the data represented by symmetric KL-divergence and Pearson's χ^2 values of the distribution. The GGD model fits the isotropic Morlet distribution more accurately since the distributions produced by isotropic Morlet for the given data set have near symmetric values on the both sides around zero (see Fig. 3(b)).

Table 2. Symmetric KL-D and Pearson's χ^2 values for the distributions created by the coefficients of Marr wavelet and isotropic Morlet wavelet fitted with a GGD model, averaged over the data set.

Distribution	Symmetric KL-D	χ^2
GGD (Marr)	0.0864	0.0445
GGD (Isotropic Morlet)	0.0698	0.0384

3.2   Similarity Measurement of Empirical Distributions of Various Classes

The similarity of distributions within a class (ADC, SCC or ULR) is high since the J-Divergence (J-D) of all the distributions of a class using a particular wavelet function and scale parameter is low as shown in Table 3.

Table 3. Intraclass similarity calculated from J-divergence for multiple distributions for Marr and isotropic Morlet wavelet with different colour channels and scales.

Scales	Marr						Isotropic Morlet		
	L*			b*			a*		
	ADC	SCC	ULR	ADC	SCC	ULR	ADC	SCC	ULR
a = 4	0.0965	0.1089	0.0657	0.0876	0.1268	0.0324	0.0879	0.0868	0.0612
a = 5	0.0834	0.0884	0.0452	0.0721	0.9292	0.0456	0.1064	0.0958	0.0458
a = 6	0.0656	0.0898	0.0341	0.0679	0.8334	0.0478	0.0723	0.0736	0.0326
a = 7	0.0878	0.0956	0.0469	0.8363	0.9187	0.0235	0.0849	0.0747	0.0312

As intra class similarity is high to calculate inter class similarity a distribution need not be compared with the all the distribution of the comparing class, a distribution from one class is taken and KL-D was applied with n' (we take the value of n' such that $n' << N$, where N is the total number of distribution for the comparing class) number of distribution from the other class and the average of these KL-D values is taken as the KL-D value of the distribution with the other class. Table 4 shows the inter class KL-D variations, as variations exists these values were used as feature vector for the classifier.

Table 4. Similarity between various classes represented as KL-D value.

	Marr						Isotropic Morlet		
	L*			b*			a*		
	ADC Vs SCC	ADC Vs ULR	SCC Vs ULR	ADC Vs SCC	ADC Vs ULR	SCC Vs ULR	ADC Vs SCC	ADC Vs ULR	SCC Vs ULR
KL-D at a = 4	0.1854	0.1847	0.2722	0.2055	0.1878	0.5727	0.1436	0.1527	0.5434
KL-D at a = 5	0.1512	0.2341	0.1396	0.1308	0.1751	0.4326	0.1332	0.1607	0.3450
KL-D at a = 6	0.1123	0.2882	0.3198	0.1139	0.1884	0.5712	0.1552	0.1408	0.3002
KL-D at a = 7	0.0885	0.1271	0.6394	0.2153	0.1589	0.3211	0.1488	0.1544	0.2601

3.3 Classification Results

The features for the SVM are selected through Recursive Feature Elimination (RFE), there are eight distinct features: shape and size parameter from the GGD modelling, three symmetric KL-D values for the three class and three statistical measures (variance, kurtosis and entropy). These eight values are extracted for four scale (a = 3, 4, 5, 6) and each colour channel. Data set as defined in Sect. 2.1 is used and to validate the system a ten-fold cross validation method is employed. The effect of different combinations of features for classifying ULR and malignant tissue (ADC and SCC) is shown in Fig. 4(a).

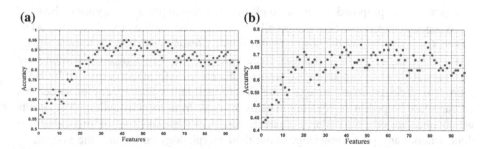

Fig. 4. (a) Accuracy for different features calculated by SVM with RFE, a total of 96 features are used to classify ULR from malignant (ADC and SCC) tissue. (b) The accuracy level of SVM for classifying ADC Vs SCC, using a RFE technique for feature selection.

(a)

	ULR	Malignant Tissue (ADC & SCC)
ULR	96.2	3.8
Malignant Tissue (ADC & SCC)	4.9	95.1

(b)

	ADC	SCC
ADC	77.2	22.8
SCC	24.2	75.8

Fig. 5. (a) Accuracy of classifying ULR and malignant tissue using SVM with 41 features. (b) classification accuracy of SVM in identifying ADC and SCC using 62 features.

To classify the ULR, a subset of 41 features from 96 features is used. Using the 41 features an accuracy of 96.2% for ULR and 95.1% malignant tissue is achieved (Fig. 5 (a)). Few texture structure of ULR might be similar to malignant texture representation since a ULR consists of many different tissue elements some texture represented by mild necrosis while other may be normal tissue but get damaged due to sample preparation or undergoing mitosis.

Classification accuracy of SVM is maximum with 62 features out of total 96 features used for classifying ADC and SCC, Fig. 4(b) shows the variation of classification accuracy between ADC and SCC for various feature sets. An accuracy of 77.2% is achieved in classifying ADC and the method gives an accuracy of 75.8% for classifying SCC, refer Fig. 5(b). These results are satisfactory since conclusive diagnosis of these two subtypes of NSCLC is even contradictory to different pathologists, since the complex organisation of the tissue structures can be seen for different stages of cancer. To have a concrete diagnostic answer often molecular analysis is carried out.

Proposed method is also compared with Gray Level Co-occurrence Matrix (GLCM), with angle ($0°, 45°, 90°$ and $135°$) and four properties viz. energy, contrast, correlation and homogeneity. The accuracy for ULR, Malignant tissue, ADC, SCC are 78.8%, 81.1%, 65.9%, 67.8% respectively for GLCM.

The effect of shape and scale parameter of the GGD on the classification accuracy is very acute, features from L* and b* plays an important role in differentiating ULR from ADC and SCC in our proposed method.

4   Conclusion

In this work, we proposed a method to classify the two important subtypes of NSCLC i.e. Adenocarcinoma and Squamous cell carcinoma. The features for classifying ADC and SCC are not clinical diagnostic features, rather features extracted automatically by a wavelet function from an image. Since colour plays a role in understanding the histological slides by a pathologist, we used the colour information provided by H&E stain. The digitized colour images were transformed into a L*a*b* colour space, this colour space helped in segregating nucleus of a cell from its surrounding. The results we obtained are very promising as characterization of these subtypes of lung cancer are done without any prior knowledge about their morphology coded into the system.

References

1. Bray, F., Jacques, F., et al.: Global cancer statistics 2018: GLOBOCAN estimates of incidence and mortality worldwide for 36 cancers in 185 countries. CA Cancer J. Clin. **68**, 394–424 (2018)
2. Ma, L.H., Li, G., et al.: The effect of nonsmall cell lung cancer histology on survival as measured by the graded prognostic assessment in patients with brain metastases treated by hypofractionated stereotactic radiotherapy. Radiat. Oncol. **11**, 92 (2016)
3. Yano, M., Yoshida, J., et al.: The outcomes of a limited resection for nonsmall cell lung cancer based on differences in pathology. World J. Surg. **40**(11), 2688–2697 (2016)
4. Yao, X., Gomes, M.M., et al.: Fine-needle aspiration biopsy versus core-needle biopsy in diagnosing lung cancer: a systematic review. Curr. Oncol. **19**(1), 16–27 (2012)
5. Webb, W.R., Muller, N.L., Naidich, D.P.: High–Resolution CT of the Lung. Lippincott Williams & Wilkins, Philadelphia (2001)
6. Dundar, M.M., Badve, S.: Computerized classification of intraductal breast lesions using histopathological images. IEEE Trans. Biomed. Eng. **58**(7), 1977–1984 (2011)
7. Sieren, J.C., Weydert, J., et al.: An automated segmentation approach for highlighting the histological complexity of human lung cancer. Ann. Biomed. Eng. **38**(12), 3581–3591 (2010)
8. Nguyen, K., Sabata, B., Jain, A.K.: Prostate cancer grading: gland segmentation and structural features. Pattern Recognit. Lett. **33**(7), 951–961 (2012)
9. Mete, M., Xu, X., et al.: Head and neck cancer detection in histopathological slides. In: 6th IEEE International Conference on Data Mining—Workshops (2006)
10. Tabesh, A., Teverovskiy, M.: Multifeature prostate cancer diagnosis and Gleason grading of histological images. IEEE Trans. Med. Imaging **26**(10), 1366–1378 (2007)
11. Chekkoury, A., Khurd, P., et al.: Automated malignancy detection in breast histopathological images. In: Medical Imaging 2012: Computer-Aided Diagnosis, San Diego, California, vol. 8315 (2012)
12. Jafari-Khouzani, K., Soltanian-Zadeh, H.: Multiwavelet grading of pathological images of prostate. IEEE Trans. Biomed. Eng. **50**(6), 697–704 (2003)
13. Khurd, P., Bahlmann, C., Gibbs-Strauss, S.: Computer-aided Gleason grading of prostate cancer histopathological images using Texton forests. In: 2010 IEEE International Symposium on Biomedical Imaging: From Nano to Macro (2010)
14. Wang, W., John, A., et al.: Detection and classification of thyroid follicular lesions based on nuclear structure from histopathology images. Cytometry Part A **77**(5), 485–494 (2010)
15. Smith, J.R., Chang, S.F.: Transform features for texture classification and discrimination in large image databases. In: Proceedings of the IEEE International Conference on Image Processing (1994)
16. Scheunders, P., Livens S., et al.: Wavelet-based texture analysis. Int. J. Comput. Sci. Inf. Manag. (1997)
17. Gao, L., Li, F., Thrall, M.J.: On-the-spot lung cancer differential diagnosis by label-free, molecular vibrational imaging and knowledge-based classification. J. Biomed. Opt. **16**(9), 096004 (2011). https://doi.org/10.1117/1.3619294
18. Sambl, M.L., Camara1, F.: A novel RFE-SVM-based feature selection approach for classification. Int. J. Adv. Sci. Technol. **43**, 27–36 (2012)
19. Batuwita, R., Palade, V.: Class imbalance learning methods for support vector machines. In: He, H., Ma, Y. (eds.) Imbalanced Learning: Foundations Algorithms and Applications. Wiley, New York (2013)

20. Khan, A.M., Rajpoot, N.: A nonlinear mapping approach to stain normalization in digital histopathology images using image-specific color deconvolution. IEEE Trans. Biomed. Imaging **61**(6), 1729–1738 (2014)
21. Kumar, P.: A wavelet based methodology for scale-space anisotropic analysis. Geophys. Res. Lett. **22**(20), 2777–2780 (1995)
22. Do, M.N., Vetterli M.: Wavelet-based texture retrieval using generalized Gaussian density and Kullback–Leibler distance. IEEE Trans. Image Process. **11**(2), 146–158 (2002)
23. Nadarajah, S.: A generalized normal distribution. J. Appl. Stat. **32**(7), 685–694 (2005)
24. Johnson, D., Sinanovic, S.: Symmetrizing the Kullback-Leibler distance. IEEE Trans. Inf. Theory (2000)
25. Lin, J.: Divergence measures based on the Shannon entropy. IEEE Trans. Inf. Theory **37**(1), 145–151 (1991)

Optimized Rigid Motion Correction from Multiple Non-simultaneous X-Ray Angiographic Projections

Abhirup Banerjee[1,2,3]([✉]) [ID], Robin P. Choudhury[1,2], and Vicente Grau[3] [ID]

[1] Oxford Acute Vascular Imaging Centre, Oxford, UK
[2] Radcliffe Department of Medicine, Division of Cardiovascular Medicine,
University of Oxford, Oxford, UK
{abhirup.banerjee,robin.choudhury}@cardiov.ox.ac.uk
[3] Department of Engineering Science,
Institute of Biomedical Engineering, University of Oxford, Oxford, UK
vicente.grau@eng.ox.ac.uk

Abstract. X-ray angiography is the most commonly used medical imaging modality for the high resolution visualization of lumen structure in coronary arteries. Since the interpretation of 3D vascular geometry using multiple 2D image projections results in high intra- and inter-observer variability, the reconstruction of 3D coronary arterial (CA) tree is necessary. The automated 3D CA tree reconstruction from multiple 2D projections is challenging due to the existence of several imaging artifacts, most importantly the respiratory and cardiac motion. In this regard, the aim of the proposed work is to remove the effects of motion artifacts from non-simultaneous angiographic projections by developing a new iterative method for rigid motion correction. Our proposed approach is based on the optimal estimation of rigid transformation, occurred due to motion in the 3D tree, from each projection. The performance of the technique is qualitatively and quantitatively demonstrated using multiple angiographic projections of the left anterior descending, left circumflex, and right coronary artery from 15 patients.

Keywords: Motion correction · Rigid motion · 3D coronary tree reconstruction · X-ray angiograms

1 Introduction

Invasive coronary angiography (ICA) is the most commonly used imaging modality for the detection of coronary stenoses. Its advantages include simplicity, high spatial and temporal resolution of lumen structure, and most importantly, its utility to guide coronary interventions in real time [11]. However, despite these

This work was supported in part by British Heart Foundation (BHF) Project Grant (no. HSR00410).

clinical advantages, x-ray angiograms pose several challenges, especially in relation to visualizing lesion adequately and judging lesion severity. The 2D projections of 3D vascular structure in different image planes contain significant motion artifacts, including cardiac, respiratory, and patient or device movement, which makes the interpretation of 3D geometry difficult for the cardiologists. This leads to high intra- and inter-observer variability in understanding the anatomical structure and, in turn, affects the lesion severity assessment [6]. In addition, potential adverse effects of higher amount of radiographic contrast agent and exposure to x-rays limit the number of image acquisitions. To overcome these inherent limitations of ICA, 3D reconstruction of CA trees from multiple 2D x-ray projections has been attempted by a number of research groups [1–3,7].

Several of these 3D CA tree reconstruction methods can deal with problems such as vessel overlap, foreshortening, suboptimal projection angles, and tortuosity and eccentricity [3,4]. In order to obtain the acquisition geometry, some reconstruction methods rely on a prior calibration step [13], while others prefer non-calibrated data for reducing table movements during image acquisition and noise in calibrated parameters [2,5]. In cardiac interventions, single-plane systems are usually preferred over biplane systems due to the lower cost and the possibility of generating multiple x-ray angiograms. However, single-plane systems generate non-simultaneous projections, which are prone to several motion artifacts, including mainly patient or imaging device related movements, respiratory motion, and cardiac motion. Although the patient or imaging device related movements can be minimized by following a careful protocol during image acquisition [2], this is difficult to achieve for respiratory motion and not possible for heart beating motion. Even if short acquisitions can be captured during breath-hold, potential misregistration artifacts still remain, affecting geometry conditions [8]. Additionally, the patients are often not able to follow the breath-hold protocol. To overcome the motion artifacts due to cardiac movements, retrospective gating is commonly utilized, where image frames are selected from the same cardiac phase. This is usually achieved using the ECG signal acquired simultaneously with the angiograms. In most of the existing reconstruction methods, the frames are selected from the end of the diastolic phase when heart motion is minimal, such that no cardiac motion can be assumed between projections [10,12]. However, this assumption does not hold in a single-plane x-ray system.

In this regard, the purpose of our work is to generate a retrospective method for rigid motion correction from multiple x-ray projections. The optimal rigid transformation is iteratively estimated for each angiographic acquisition, to adjust for the relative rigid motion, generated from respiration and patient or device movements. The proposed method is formulated using a theorem regarding the geometry of a 3D object. The performance of the proposed motion correction technique is qualitatively and quantitatively evaluated over angiographic projections of both left coronary artery (LCA) and right coronary artery (RCA) from 15 patients, admitted in the hospital for suspected coronary stenosis. The algorithm results in average reprojection error of 0.504 mm after rigid motion correction, from left anterior descending (LAD), left circumflex (LCx) and RCA.

2 Proposed Rigid Motion Correction

2.1 Proposed Theorem

Theorem 1. *After rigid transformation of a 3D object, the 2D projection will remain equal if the source and projection plane are modified using the same rigid transformation.*

Proof. Let us assume the 3D object be A. The location of the x-ray source be S, any point on the projection plane be M, and the normal to the projection plane be N. So, any line passing through S and A is given by

$$Q = S + \alpha(A - S). \tag{1}$$

Since Q lies on the projection plane with a point M and the normal N, $\alpha = \frac{(M-S)\cdot N}{(A-S)\cdot N}$. Now, A is generated after rigid transformation (R, t), rotation R and translation t, on 3D object A^*. So, $A = RA^* + t$. Hence, (1) is rewritten as

$$Q = R(S^* + \alpha(A^* - S^*)) + t, \quad \text{where } S^* = R^{-1}(S - t)$$

Additionally, $\alpha = \frac{(M^*-S^*)\cdot N^*}{(A^*-S^*)\cdot N^*} = \alpha^*$, where $M^* = R^{-1}(M - t)$ and $N^* = R^T N$. So, $Q = R\,Q^* + t$, where $Q^* = S^* + \alpha^*(A^* - S^*)$. Hence, Q^* is a 3D object passing though 3D object A^* and source S^*. Now, we just need to show: Q^* lies on a 2D plane with a point M^* and normal N^*.

$$(Q^* - M^*) \cdot N^* = (R^{-1}(S - M + \alpha(A - S)))^T (R^T N) = (Q - M) \cdot N = 0$$

Hence, if Q is the projection of 3D object A from source S to a plane with point M and normal N, then, for any rigid transformation (R, t), $R^{-1}(Q - t)$ is the projection of $R^{-1}(A - t)$ from source $R^{-1}(S - t)$ on plane with point $R^{-1}(M - t)$ and normal $R^T N$. Since 3D rigid transformation of a 2D plane in 3D does not change the geometry, $R^{-1}(Q - t)$ continues to be the same 2D projection. Additionally, (R, t) transforms any plane with point M and normal N to the plane with point $R^{-1}(M - t)$ and normal $R^T N$. Hence, it suffices to say that, after rigid transformation, 2D projection of a 3D object remain equal if the source and projection plane are modified using the same rigid transformation.

2.2 Rigid Motion Estimation in Object-Domain

One major issue in 3D CA tree reconstruction is the latent motion in acquired angiograms from different views. Even after choosing end-diastole frames for non-simultaneous acquisition, the selected frame from a discrete set retains some temporal misalignment, leading to deformations in coronary vessels. Respirations cause, mostly rigid, transformation in heart and hence, in vessels. Movements in patient or device during acquisitions result in similar rigid motion.

In order to remove the rigid motion artifacts, the proposed approach estimates the optimal transformations from all angiographic acquisitions, minimizing the orthogonal distance between corresponding landmarks in object domain.

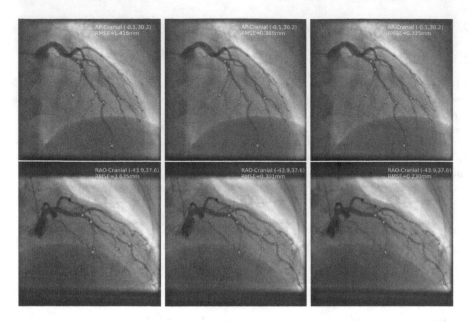

Fig. 1. Removing rigid motion from two projections (red: motion corrected landmarks, green: original landmarks). Left to right: initialization, iteration 5, and at convergence. (Color figure online)

Our proposed rigid motion correction method involves the identification of few (preferably 4–6) corresponding landmarks from all projections. In our method, the vessel bifurcations are manually identified for this purpose. Although, in 3D space, 3 landmarks are theoretically sufficient for estimating a rigid transformation, we prefer to use at least 4, due to high amount of motion artifacts.

Let us assume r rigid motion corrected 3D landmarks as $B_i; i = 1, \cdots, r$ and the projected landmark B_i on jth image plane as b_{ij} for $j = 1, \cdots, s$. Let us also assume the locations of x-ray source, a representative point on projection plane j, and the normal to the plane are defined as F_j, M_j, and N_j, respectively. Since B_i's are unknown, they are initially estimated as the nearest point of intersection of 3D projection lines $\overrightarrow{F_j b_{ij}}, j = 1, \cdots, s$,

$$B_i = \arg\min_p \sum_{j=1}^{s} D(p, \overrightarrow{F_j b_{ij}}), \tag{2}$$

where $D(p, \overrightarrow{F_j b_{ij}})$ denotes the orthogonal distance from any point p to $\overrightarrow{F_j b_{ij}}$. Let us also define $B_{i,j}$ as the nearest point on $\overrightarrow{F_j b_{ij}}$ from B_i. Note that, $B_{i,j}$ is the location of the rigid motion affected 3D landmark B_i during jth image acquisition. Our aim is to measure the optimal rigid transformation (translation t_j and rotation R_j) for each acquisition $j = 1, \cdots, s$, so that the 3D landmarks

$B_{i,j}, i = 1, \cdots, r$ match with the corresponding 3D landmarks B_i, i.e.

$$\underset{t_j, R_j}{\arg\min} \sum_{i=1}^{r} \|B_i - (t_j + R_j B_{i,j})\| \quad \forall j = 1, \cdots, s. \tag{3}$$

Algorithm 1. Proposed Rigid Motion Correction

   Input   : Image frames at end-diastole from each projection plane
   Output: Rigid motion corrected image frames
1 Select n vessel landmarks (preferably $4 - 6$) from all projection planes;
2 **do**
3 | Estimate the rigid motion corrected 3D landmarks as the nearest orthogonal
       | point of 3D projection lines connecting vessel landmarks on projection
       | planes with the source (using (2) and (4));
4 | Estimate the optimal rigid transformation for each acquisition so that the
       | 3D landmarks match with rigid motion corrected landmarks (using (3));
5 **while** *the 3D landmarks for each image acquisition do not coincide;*
6 Return the optimal transformation that minimizes the effect of rigid motion.

The optimal transformation in Eq. (3) is estimated by Horn's quaternion-based method [9]. As the original 3D rigid motion corrected landmarks are not known, an iterative algorithm is developed, where, in every iteration, the 3D landmarks are measured using Eq. (2) and then the optimal rigid transformations are estimated using Eq. (3). In each iteration, the sources and landmarks on projection planes are updated based on the rigid transformation

$$F_{j'} = t_j + R_j F_j \quad \text{and} \quad b_{ij'} = t_j + R_j b_{ij}, \tag{4}$$

as per the theorem developed in previous subsection. An example of the result of the iterative minimization of rigid motion artifact from two projections (AP-cranial and RAO-cranial) of the LCA is depicted in Fig. 1. The pseudo-code of the proposed approach is presented in Algorithm 1.

3 Experimental Results

The performance of the proposed rigid motion correction algorithm is qualitatively as well as quantitatively demonstrated on 15 patients enrolled in clinical studies. For each patient, 2–3 angiographic projections are captured for RCA, while 3–6 angiographic projections are generated for LCA. Since the LAD and LCx arteries are generally not optimally visible at the same angiographic projection, different sets of LCA images are used for LAD and LCx arteries. The LAD, LCx, and RCA arteries are visible from at least two projections in 14, 13, and 12 patients, respectively, out of the total 15 patients. The proposed rigid motion

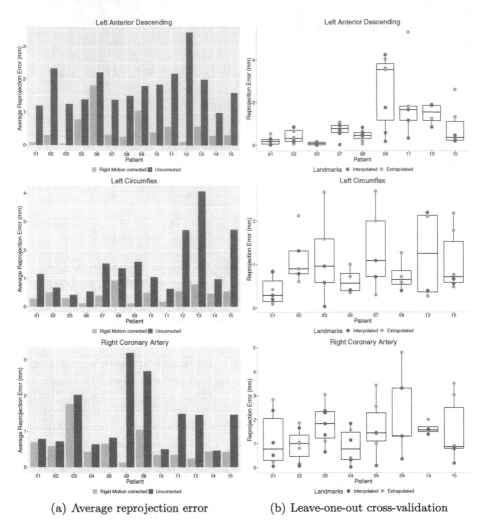

(a) Average reprojection error   (b) Leave-one-out cross-validation

Fig. 2. (a) Average reprojection errors of the landmarks after rigid motion correction; (b) Box plot of the reprojection errors of landmarks for leave-one-out cross-validation.

correction approach requires at least 4–6 bifurcation points, as landmarks, for its optimal performance and typically takes 10–25 iterations to converge.

The performance of the proposed technique is quantitatively measured by comparing the 2D landmarks with the back-projections of motion-corrected 3D landmarks on each of the projection plane. For initial quantitative evaluation, the average reprojection errors of the landmarks after rigid motion correction, along with the same without motion correction, are presented separately for LAD, LCx, and RCA as multiple-bar diagram in Fig. 2(a). The average reprojection errors of complete patient cohort after rigid motion correction are measured as 0.490 mm, 0.438 mm, and 0.591 mm for the LAD, LCx, and RCA, respectively.

The overall reprojection error combining all three arteries is measured as 0.504 mm. From every case presented in Fig. 2(a), it is clearly visible that the proposed approach significantly reduces the motion artifacts from all 2D projections of all coronary arteries.

For the final quantitative evaluation, we used the leave-one-out cross-validation technique. For each artery of every patient, one landmark point is selected and the rigid motion correction is performed based on the rest of the landmarks. Finally, on the rigid motion corrected projection planes, the reprojection error is estimated from the remaining landmark point. The procedure is applied on all landmark points for each of the LAD, LCx, and RCA and the results, in form of box plot, is presented in Fig. 2(b). Since the landmarks are situated either inside the arteries or at the end locations (proximal or distal), their locations are either interpolated or extrapolated after the rigid transformation. Hence, the reprojection errors corresponding to interpolated and extrapolated landmarks are represented using different colors in Fig. 2(b). Since our proposed algorithm requires at least 4 landmark points for its optimal performance, the leave-one-out cross-validation technique is only applicable when we have selected at least 5 landmarks. Hence, the number of cases have been reduced (9, 8, and 8 for LAD, LCx, and RCA, respectively) during our second analysis.

4   Discussion and Conclusions

Several 3D CA tree reconstruction methods have been developed in the past two decades. However, most of the existing methods either did not consider the effects of motion artifacts or followed the breath-hold and no patient movement protocol during image acquisition. In this regard, our objective is to propose a new retrospective rigid motion correction technique, keeping the standard clinical image acquisition protocol. The main advantage of the proposed approach is that it does not actually try to remove the effects of motion artifacts in projection domain, but re-orient the 3D x-ray C-arm system for optimal estimation of rigid transformation between angiogram acquisitions. The proposed algorithm tries identify how the same 2D angiographic projections can be generated if no motion artifact is involved, by optimally orienting the 3D vascular structure, along with the x-ray sources and projection planes.

In our experimental analysis, the average reprojection error for the RCA is higher than that of both LAD and LCx. The underlying reason is that, since the RCA passes along the atrioventricular groove, the effects of residual non-rigid motion artifacts (heart stretching and shrinking due to cardiac motion) in RCA is comparatively higher than in LCA. The height of box plots, presented in Fig. 2(b), is often quite high (high interquartile range) in our data. The reason behind is that the box plots are computed over reprojection errors of both interpolated and extrapolated landmarks and hence, has high degree of variation. Also, since the number of landmarks at each artery are very limited (5–8), the extrapolated landmarks, which tend to have comparatively higher reprojection errors, are considered inside the data distribution, instead of being treated as an outlier. Despite that, the box plots corresponding to LAD have less dispersion (except one), demonstrating the effectiveness of proposed approach.

The objective of the proposed rigid motion correction algorithm is to enable accurate identification of point correspondences between vessels, or vessel centerlines, from multiple x-ray angiograms. This will allow the development of a completely automated 3D CA tree reconstruction method that, when embedded in an x-ray angiogram system, will generate the 3D rendering of coronary vascular structure during cardiac intervention, as well as automatically estimate physiological information, such as severity of the coronary stenoses.

References

1. Banerjee, A., Kharbanda, R.K., Choudhury, R.P., Grau, V.: Automated motion correction and 3D vessel centerlines reconstruction from non-simultaneous angiographic projections. In: Pop, M., et al. (eds.) STACOM 2018. LNCS, vol. 11395, pp. 12–20. Springer, Cham (2019). https://doi.org/10.1007/978-3-030-12029-0_2
2. Cañero, C., Vilariño, F., Mauri, J., Radeva, P.: Predictive (un)distortion model and 3-D reconstruction by biplane snakes. IEEE Trans. Med. Imaging **21**(9), 1188–1201 (2002)
3. Çimen, S., Gooya, A., Grass, M., Frangi, A.F.: Reconstruction of coronary arteries from X-ray angiography: a review. Med. Image Anal. **32**, 46–68 (2016)
4. Chen, S.J., Schäfer, D.: Three-dimensional coronary visualization, part 1: modeling. Cardiol. Clin. **27**(3), 433–452 (2009)
5. Cong, W., Yang, J., Ai, D., Chen, Y., Liu, Y., Wang, Y.: Quantitative analysis of deformable model-based 3-D reconstruction of coronary artery from multiple angiograms. IEEE Trans. Biomed. Eng. **62**(8), 2079–2090 (2015)
6. Eng, M.H., et al.: Impact of three dimensional in-room imaging (3DCA) in the facilitation of percutaneous coronary interventions. J. Cardiol. Vasc. Med. **1**, 1–5 (2013)
7. Galassi, F., Alkhalil, M., Lee, R., Martindale, P., et al.: 3D reconstruction of coronary arteries from 2D angiographic projections using non-uniform rational basis splines (NURBS) for accurate modelling of coronary stenoses. Plos One **13**(1), 1–23 (2018)
8. Holland, A., Goldfarb, J., Edelman, R.: Diaphragmatic and cardiac motion during suspended breathing: preliminary experience and implications for breath-hold MR imaging. Radiology **209**(2), 483–489 (1998)
9. Horn, B.K.P.: Closed-form solution of absolute orientation using unit quaternions. J. Opt. Soc. Am. A **4**(4), 629–642 (1987)
10. Husmann, L., Leschka, S., Desbiolles, L., Schepis, T., et al.: Coronary artery motion and cardiac phases: dependency on heart rate - implications for CT image reconstruction. Radiology **245**(2), 567–576 (2007)
11. Mark, D.B., et al.: ACCF/ACR/AHA/NASCI/SAIP/SCAI/SCCT 2010 expert consensus document on coronary computed tomographic angiography: a report of the american college of cardiology foundation task force on expert consensus documents. J. Am. Coll. Cardiol. **55**(23), 2663–2699 (2010)

12. Shechter, G., Resar, J.R., McVeigh, E.R.: Displacement and velocity of the coronary arteries: cardiac and respiratory motion. IEEE Trans. Med. Imaging **25**(3), 369–375 (2006)
13. Wiesent, K., Barth, K., Navab, N., Durlak, P., et al.: Enhanced 3-D-reconstruction algorithm for C-arm systems suitable for interventional procedures. IEEE Trans. Med. Imaging **19**(5), 391–403 (2000)

Group-Sparsity Based Compressed Sensing Reconstruction for Fast Parallel MRI

Sumit Datta[ID] and Bhabesh Deka[✉][ID]

Department of Electronics and Communication Engineering,
Tezpur University, Tezpur 784028, Assam, India
`bdeka@tezu.ernet.in`

Abstract. Compressed sensing (CS) in parallel magnetic resonance imaging (pMRI) has the potential to reduce the MRI scan time by many folds. Due to the application of CS, conventional linear reconstruction techniques would not work. To reconstruct MR images from undersampled measurements one needs to solve highly nonlinear optimization problems. Practical implementation of CS in pMRI involves reconstruction quality or accuracy and computational time as its major trade-offs. Since clinical multi-dimensional pMRI requires significant amount of raw data, sequential implementation of complex optimization algorithms would not meet the time constraints set for clinically feasible reconstructions. In this paper, we propose a CS based parallel MRI reconstruction method using wavelet forest sparsity and joint total variation as sparsity inducing regularization constraints. The model is applied to multi-slice multi-channel MRI data. Simulations are carried out in the hybrid CPU-GPU environment for reconstruction of complex valued multi-slice multi-coil MR data within a clinically feasible reconstruction time.

Keywords: Compressed sensing · Parallel MRI · Group-sparsity · Forest sparsity · GPU

1 Introduction

Magnetic resonance imaging (MRI) offers several advantages, like, good contrasts for soft-tissue imaging, non-invasiveness, no ionizing radiations, etc. compared to other well known medical imaging modalities, like, X-ray, ultrasound, and computed tomography (CT). However, MRI suffers from two major problems, long scan time and acoustic noise due to gradient variations. Although, multi-channel or parallel MRI (pMRI) significantly reduces the acquisition time, yet it requires 20 to 40 min for imaging depending on the field-of-view (FoV) [12].

Compressed sensing (CS) in MRI demonstrates that it is possible to reconstruct MR images with diagnostic quality from a fewer measurements than that of the conventional approach which follows the Nyquist-Shannon sampling theorem [1,6,10,18]. MR images possesses some favorable characteristics for signal

© Springer Nature Switzerland AG 2019
B. Deka et al. (Eds.): PReMI 2019, LNCS 11942, pp. 70–77, 2019.
https://doi.org/10.1007/978-3-030-34872-4_8

processing such as they are compressible or s-sparse over a transform domain, like, wavelet transform. MRI signals naturally in the frequency domain or k-space which is to some extent incoherent with the wavelets.

Combination of CS with pMRI (CS-pMRI) is able to reduce MRI scan time significantly [23]. In CS-MRI traditional linear reconstruction is not possible. Since acquired data are highly undersampled, nonlinear reconstruction techniques need to be used for reconstruction. Computational time and quality of reconstruction are two major research issues in CS-MRI.

In 3D MRI, multiple 2D thin slices are acquired with a fixed and negligible inter-slice gaps followed by some post-processing techniques to generate a 3D volume for clinical studies. Since slices are thin (2–3 mm) and inter-slice gaps are very less, adjacent slices are highly correlated [7,9]. In pMRI multiple receiver coils work in parallel. Data acquired by different coils are weighted differently due to the distinct spatial sensitivity profile of respective coils. Since target FoV is same for all receiver coils, structures and edges appear at same positions with different intensity values. To exploit data correlations that exist in multi-slice multi-channel MRI, we have used group-sparsity in both wavelet as well as spatial domains [8]. In this paper, we have proposed a fast calibrationless CS-pMRI reconstruction model using group-sparsity as regularization constraint. Since computational time is a major factor for the practical implementation of CS-pMRI reconstruction, we use a hybrid CPU-GPU platform for acceleration in CS reconstruction.

The rest of the paper is organized as follows: Sect. 2 briefly presents the background. Section 3 details the proposed group-sparsity based CS pMRI reconstruction technique. Next, experimental results and performance evaluations are done in Sect. 4, followed by conclusions in Sect. 5.

2   Background

Commonly, CS based MRI reconstruction model involves wavelet domain sparsity and spatial domain gradient sparsity as regularization constraints [18]. A standard CS pMRI reconstruction problem may be described as follows- suppose $\mathbf{y}_q \in \mathcal{C}^m$ is measured k-space data corresponding to c^{th} coil image, $\mathbf{x}_q \in \mathcal{C}^n$ i.e. $\mathbf{y}_q = \mathbf{F}_u \mathbf{x}_q$, where $m \ll n$, $q = 1, 2, ..., Q$; Q is number of coils, and $\mathbf{F}_u \in \mathcal{C}^{m \times n}$ represent a partial Fourier operator [16]. To reconstruct image corresponding to q^{th} receiver coil, $\mathbf{x}_q$ from $\mathbf{y}_c$, need to solve following optimization problem:

$$\hat{\mathbf{x}}_q = \underset{\mathbf{x}_q}{\text{argmin}} \; \tfrac{1}{2} ||\mathbf{F}_u \mathbf{x}_q - \mathbf{y}_q||_2^2 + \lambda_1 ||\Psi \mathbf{x}_q||_1 + \lambda_2 ||\mathbf{x}_q||_{\text{TV}}, \tag{1}$$

where λ_1 and λ_2 are regularization parameters to balance between data fidelity and regularization terms, Ψ wavelet transform operator, and $||\mathbf{x}_q||_{\text{TV}}$ is the TV norm of $\mathbf{x}_q$ i.e. $||\mathbf{x}_q||_{\text{TV}} = \sum_{i,j} \sqrt{\{(\nabla_h \mathbf{x}_q)_{i,j}\}^2 + \{(\nabla_v \mathbf{x}_q)_{i,j}\}^2}$ where ∇_h and ∇_v are the first-order gradients in horizontal and vertical directions, respectively [11].

Reconstruction approaches may be classified into two categories depending on how coil sensitivity profiles of different receiver coils are used for reconstruction. First category explicitly require coil sensitivity profile information for reconstruction, like, in the sensitivity encoding (SENSE) [20] and the simultaneous acquisition of spatial harmonics (SMASH) [21]. SENSE gives optimal reconstruction performance if the coil information is accurate. In the second category, the coil sensitivity profile information is used implicitly during reconstruction, like, in the GRAPPA [13] and the AUTO-SMASH [15]. These methods estimate the coil sensitivity profile from the acquired raw data and known as auto-calibrating methods. Some of the well known auto-calibrating methods are- the SPIRiT [17], the CS-SENSE [16], and the ESPIRiT [22]. The main disadvantage associated with these methods is that in practice, it is very difficult to accurately estimate coil sensitivity profiles. A small error in estimation may lead to non-removable artifacts during reconstruction. Besides the above, there is another kind of CS based pMRI reconstruction approach which is relatively new, do not use coil sensitivity information explicitly or implicitly, known as the calibrationless methods, like, the CalM MRI [19].

According to Chen and Huang [2], wavelet coefficients of MR images are not only compressible but also have quadtree structure. The children coefficients in the finer scale are highly correlated with the parent coefficient in the adjacent coarser scale and may be modeled to follow the tree sparsity. In pMRI, images corresponding to different receiver coils of a particular FoV are similar. So, wavelet coefficients of different coil images of similar positions are expected to be similar. Chen *et al.* [3] exploit inter-channel similarity in wavelet domain during pMR image reconstruction. The concept of tree sparsity in a single image is extended to pMRI, where multiple wavelet trees of identical positions corresponding to different receiver coils are considered. They named it as the forest sparsity and corresponding reconstruction as FCSA Forest.

3   Proposed Method

In this paper, we have proposed a calibrationless pMRI reconstruction technique using group-sparsity in both wavelet as well as spatial domains. To exploit intra-channel, inter-channel/slice similarity in wavelet domain we have used forest sparsity i.e. wavelet tree groups of identical locations from multiple channels and slices. This particular grouping arrangement enforces similarity among the group members of the respective forest group. Similarly, we have used joint total variation (JTV)-norm to exploit data redundancy in spatial domain gradients as reported in JTV MRI [4].

Proposed multi-slice pMRI problem is as follows: suppose the measured undersampled k-space data $\mathbf{Y} = \{(\mathbf{y}_{1,1}, \mathbf{y}_{1,2}, \cdots, \mathbf{y}_{1,Q}); (\mathbf{y}_{2,1}, \mathbf{y}_{2,2}, \cdots, \mathbf{y}_{2,Q}); \cdots ; (\mathbf{y}_{P,1}, \mathbf{y}_{P,2}, \cdots, \mathbf{y}_{P,Q})\}$, corresponding to the underlying spatial domain multi-slice multi-channel data
$$\mathbf{X} = \{(\mathbf{x}_{1,1}, \mathbf{x}_{1,2}, \cdots, \mathbf{x}_{1,Q}); (\mathbf{x}_{2,1}, \mathbf{x}_{2,2}, \cdots, \mathbf{x}_{2,Q}); \cdots ; (\mathbf{x}_{P,1}, \mathbf{x}_{P,2}, \cdots, \mathbf{x}_{P,Q})\}$$
such that $\mathbf{Y} = \mathbf{F}_u \mathbf{X}$ where P and Q is the number of slices and chan-

nels, respectively. We extended the model in Eq. 1 for joint reconstruction of multi-slice multi-channel data $\mathbf{X}$ from $\mathbf{Y}$ as follows:

$$\widehat{\mathbf{X}} = \arg\min_{\mathbf{X}} \frac{1}{2} \sum_{p=1}^{P} \sum_{q=1}^{Q} ||\mathbf{F}_u \mathbf{x}_{p,q} - \mathbf{y}_{p,q}||_2^2 + \lambda_1 \sum_{g_i \in G} ||(\Psi\mathbf{X})_{g_i}||_2 + \lambda_2 ||\mathbf{X}||_{\text{JTV}}, \quad (2)$$

First regularization term is the wavelet forest sparsity which is mathematically denoted by $\ell_{1,2}$-norm, where g_i contains the indices of wavelet coefficient of ith group and G contains indices of all such groups, i.e. $G = [g_1, \ldots, g_i, \ldots, g_p]$. Next JTV-norm is given by- $||\mathbf{X}||_{\text{JTV}} = \sum_{i=1,j=1}^{\sqrt{n},\sqrt{n}} \sqrt{\sum_{p=1,q=1}^{P,Q} \left\{ (\nabla_h \mathbf{x}_{p,q})_{i,j}^2 + (\nabla_v \mathbf{x}_{p,q})_{i,j}^2 \right\}}$. We decompose this problem into two smaller subproblems using the concept of variable splitting, first subproblem is based on wavelet forest sparsity and second subproblems is JTV based, as follows:

$$\mathbf{Z} = \arg\min_{\mathbf{Z}} \lambda_1 \sum_{g_i \in G} ||\mathbf{Z}_{g_i}||_2 + \frac{\beta}{2} ||\mathbf{Z} - G\Psi\mathbf{X}||_2^2 \quad (3)$$

$$\widehat{\mathbf{X}} = \arg\min_{\mathbf{X}} \frac{1}{2} \sum_{p=1}^{P} \sum_{q=1}^{Q} ||\mathbf{F}_u \mathbf{x}_{p,q} - \mathbf{y}_{p,q}||_2^2 + \frac{\beta}{2} ||\mathbf{Z} - G\Psi\mathbf{X}||_2^2 + \lambda_2 ||\mathbf{X}||_{\text{JTV}} \quad (4)$$

Above problems can be solved by the FCSA [14], the WaTMRI [2] and the FCSA Forest [3] algorithms as detailed in Algorithm 1.

Algorithm 1. Proposed Algorithm

Input: $\Psi, \mathbf{F}_u, \mathbf{Y}, \lambda_1, \lambda_2, \beta, L_f$
Initialization: $\mathbf{X}^0 \leftarrow \mathbf{F}_u^T \mathbf{Y}, \mathbf{r}^1 \leftarrow \mathbf{X}^0, t^1 \leftarrow 1, k \leftarrow 0$

1: **while** not converge **do**
2:     $k \leftarrow k + 1$
3:     $\mathbf{Z}^k \leftarrow \text{shrinkgroup}\left(G\Psi\mathbf{X}^{k-1}, \frac{\lambda_2}{\beta}\right)$
4:     $\mathbf{R}^k \leftarrow \mathbf{r}^k - \frac{1}{L_f}\left[\sum_{p=1,p=1}^{P,Q} \mathbf{F}_u^T \left(\mathbf{F}_u \mathbf{r}_{p,q}^k - \mathbf{Y}_{p,q}\right) + \beta\Psi^T G^T \left(G\Psi\mathbf{r} - \mathbf{Z}^k\right)\right]$
5:     $\mathbf{X}^k \leftarrow \arg\min_{\mathbf{X}} \frac{L}{2} ||\mathbf{X}^{k-1} - \mathbf{R}^k||_2^2 + \lambda_1 ||\mathbf{X}^{k-1}||_{\text{JTV}}$
6:     $t^{k+1} \leftarrow \left(\frac{1 + \sqrt{1 + 4(t^k)^2}}{2}\right)$
7:     $\mathbf{r}^{k+1} \leftarrow \mathbf{X}^k + \frac{t^k - 1}{t^{k+1}}\left(\mathbf{X}^k - \mathbf{X}^{k-1}\right)$
8: **end while**
Output: $\widehat{\mathbf{X}} \leftarrow \mathbf{X}^k$

---

4   Experimental Results and Simulations

Simulation works are carried out in the MATLAB running in a workstation having Intel Xeon processor E5-2650, 128 GB of memory, and NVIDIA Quadro P5000 GPU. We have collected several magnitude and raw complex MRI datasets from different sources. Since raw complex datasets are realistic for clinical CS-MRI implementations, we have considered two raw complex datasets for simulations of the proposed work. Dataset I: NYU machine learning data[1] ($768 \times 770 \times 15 \times 31$) and Dataset II: Stanford Fully sampled 3D FSE Knees[2] ($320 \times 320 \times 8 \times 172$). Results are compared with other techniques, we have considered three well known auto-calibrating techniques, namely, the CS-SENSE [16], the SPIRiT [17], and the ESPIRiT [22]; and four calibrationless techniques, namely, the CaLM MRI [19], the JTV MRI [4], the FCSA-Forest [3], and the CaL JS CS SENSE [5]. We have considered two well-known image quality assessment (IQA) metrics, namely, the signal to noise ratio (SNR) and the mean structural similarity index (MSSIM) besides visual results and computational time. In the simulation, we have used 25% undersampling ratio for both datasets. We use "db2" wavelet and 4 decomposition levels for sparse representation.

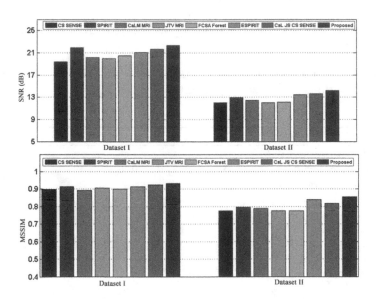

Fig. 1. Comparison of reconstruction performances in terms of SNR and MSSIM for different datasets

To check the feasibility of practical implementations, we have done parallel implementation in a hybrid CPU-GPU simulation environment. 4D problems for

---

[1] http://mridata.org/list?project=NYU%20machine%20learning%20data.

[2] http://mridata.org/list?project=Stanford%20Fullysampled%203D%20FSE %20Knees.

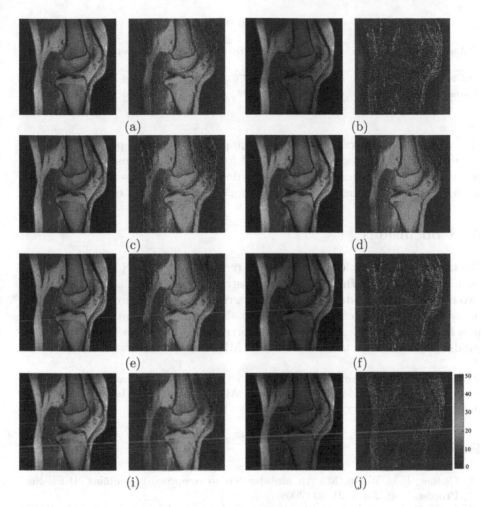

Fig. 2. Comparison of a reconstructed image (slice #10) from dataset I and corresponding error images using different techniques at 25% undersampling ratio. (a)–(j) output and error images using the CS SENSE, the SPIRiT, the CaLM MRI, the JTV-MRI, the FCSA Forest, the ESPIRiT, the CaL JS CS SENSE, and the proposed technique, respectively.

each dataset is split into multiple smaller 4D subproblems. Each such subproblem contains three adjacent slices corresponding to 8 or 15 channels. Reconstruction performances in terms of SNR and MSSIM are shown in Fig. 1. Proposed technique outperforms all other methods for both datasets. Reconstructed images using different techniques corresponding to slice #10 of Dataset I are shown in Fig. 2.

It is clearly visible that the reconstructed image using the proposed method gives the least visible artifacts. Reconstruction time using different methods are

Table 1. Reconstruction time in terms of CPU Time (min.)

Dataset	SPIRiT	CaLM MRI	JTV MRI	FCSA Forest	ESPIRiT	CaL JS CS SENSE	Proposed	Proposed GPU
Dataset I	236	404	145	244	439	832	184	3.35
Dataset II	186	350	110	195	383	540	135	2.93

shown in Table 1. In case of sequential implementation the proposed method takes the least computational time except for JTV MRI because the latter does not use the forest sparsity in the wavelet domain. The proposed parallel implementation method takes only 2 to 3 min for reconstruction of clinical datasets.

5   Conclusions

We have proposed a joint reconstruction technique for 4D CS MRI reconstruction. Performance of the proposed reconstruction technique is evaluated with two raw complex MRI datasets. Simulation results show that proposed technique outperforms the state-of-the-art. We have also implemented proposed technique in a hybrid CPU-GPU platform and observe that proposed method can able to produce reconstructed multi-dimensional MRI data within few minutes.

Acknowledgement. Authors would like to thank "Visvesvaraya PhD Scheme for Electronics and IT" under MeitY, GoI and AICTE, New Delhi, India for providing financial support to carry out this research work.

References

1. Candes, E.J., Wakin, M.: An introduction to compressive sampling. IEEE Sig. Process. Mag. **25**(2), 21–30 (2008)
2. Chen, C., Huang, J.: Exploiting the wavelet structure in compressed sensing MRI. Mag. Reson. Imaging **32**, 1377–1389 (2014)
3. Chen, C., Li, Y., Huang, J.: Forest sparsity for multi-channel compressive sensing. IEEE Trans. Sig. Process. **62**(11), 2803–2813 (2014)
4. Chen, C., Li, Y., Huang, J.: Calibrationless parallel MRI with joint total variation regularization. In: Mori, K., Sakuma, I., Sato, Y., Barillot, C., Navab, N. (eds.) MICCAI 2013. LNCS, vol. 8151, pp. 106–114. Springer, Heidelberg (2013). https://doi.org/10.1007/978-3-642-40760-4_14
5. Chun, I.Y., Adcock, B., Talavage, T.M.: Efficient compressed sensing SENSE pMRI reconstruction with joint sparsity promotion. IEEE Trans. Med. Imaging **35**(1), 354–368 (2016)
6. Datta, S., Deka, B.: Efficient adaptive weighted minimization for compressed sensing magnetic resonance image reconstruction. In: Tenth Indian Conference on Computer Vision, Graphics and Image Processing, ICVGIP-2016, Guwahati, India, pp. 95:1–95:8 (2016)
7. Datta, S., Deka, B.: Magnetic resonance image reconstruction using fast interpolated compressed sensing. J. Opt. **47**(2), 154–165 (2017)

8. Datta, S., Deka, B.: Multi-channel, multi-slice, and multi-contrast compressed sensing MRI using weighted forest sparsity and joint TV regularization priors. In: Bansal, J.C., Das, K.N., Nagar, A., Deep, K., Ojha, A.K. (eds.) Soft Computing for Problem Solving. AISC, vol. 816, pp. 821–832. Springer, Singapore (2019). https://doi.org/10.1007/978-981-13-1592-3_65

9. Datta, S., Deka, B.: Efficient interpolated compressed sensing reconstruction scheme for 3D MRI. IET Image Process. **12**(11), 2119–2127 (2018)

10. Deka, B., Datta, S.: Weighted wavelet tree sparsity regularization for compressed sensing magnetic resonance image reconstruction. In: Kalam, A., Das, S., Sharma, K. (eds.) Advances in Electronics, Communication and Computing. LNEE, vol. 443, pp. 449–457. Springer, Singapore (2018). https://doi.org/10.1007/978-981-10-4765-7_48

11. Deka, B., Datta, S., Handique, S.: Wavelet tree support detection for compressed sensing MRI reconstruction. IEEE Sig. Process. Lett. **25**(5), 730–734 (2018)

12. Deka, B., Datta, S.: Compressed Sensing Magnetic Resonance Image Reconstruction Algorithms. Springer Series on Bio- and Neurosystems. Springer, Singapore (2019). https://doi.org/10.1007/978-981-13-3597-6

13. Griswold, M.A., et al.: Generalized autocalibrating partially parallel acquisitions (GRAPPA). Mag. Reson. Med. **47**(6), 1202–1210 (2002)

14. Huang, J., Zhang, S., Metaxas, D.N.: Efficient MR image reconstruction for compressed MR imaging. Med. Image Anal. **15**(5), 670–679 (2011)

15. Jakob, P.M., Grisowld, M.A., Edelman, R.R., Sodickson, D.K.: AUTO-SMASH: a self-calibrating technique for SMASH imaging. Magn. Reson. Mater. Phys. Biol. Med. **7**(1), 42–54 (1998)

16. Liang, D., Liu, B., Wang, J., Ying, L.: Accelerating SENSE using compressed sensing. Magn. Reson. Med. **62**(6), 1574–1584 (2009)

17. Lustig, M., Pauly, J.: SPIRiT: iterative self-consistent parallel imaging reconstruction from arbitrary k-space. Magn. Reson. Med. **64**(2), 457–471 (2010)

18. Lustig, M., Donoho, D., Pauly, J.M.: Sparse MRI: the application of compressed sensing for rapid MR imaging. Magn. Reson. Med. **58**, 1182–1195 (2007)

19. Majumdar, A., Ward, R.K.: Calibration-less multi-coil MR image reconstruction. Magn. Reson. Med. **30**(7), 1032–1045 (2012)

20. Pruessmann, K.P., Weiger, M., Scheidegger, M.B., Boesiger, P.: SENSE: sensitivity encoding for fast MRI. Magn. Reson. Med. **42**(5), 952–962 (1999)

21. Sodickson, D.K., Manning, W.J.: Simultaneous acquisition of spatial harmonics (SMASH): fast imaging with radiofrequency coil arrays. Magn. Reson. Med. **38**(4), 591–603 (1997)

22. Uecker, M., et al.: ESPIRiT-an eigenvalue approach to autocalibrating parallel MRI: where SENSE meets GRAPPA. Magn. Reson. Med. **71**(3), 990–1001 (2014)

23. Vasanawala, S.S., Lustig, M.: Advances in pediatric body MRI. Pediatr. Radiol. **41**(2), 549–554 (2011)

Sparse Representation Based Super-Resolution of MRI Images with Non-Local Total Variation Regularization

Bhabesh Deka$^{(\boxtimes)}$, Helal Uddin Mullah, Sumit Datta, Vijaya Lakshmi, and Rajarajeswari Ganesan

Department of Electronics and Communication Engineering,
Tezpur University, Tezpur 784028, Assam, India
bdeka@tezu.ernet.in

Abstract. Diffusion-weighted and Spectroscopic MR images are found to be very helpful for diagnostic purposes as they provide complementary information to that provided by conventional MRI. These images are also acquired at a faster rate, but with low signal-to-noise ratio. This limitation can be overcome by applying image super-resolution techniques. In this paper, we propose a single-image super-resolution (SISR) technique via sparse representation for diffusion-weighted (DW) and spectroscopic MR (MRS) images. It is based on non-local total variation approach to regularize an ill-posed inverse problem of SISR. Experiments are conducted for both DW and MRS test images and the results are compared with other recent regularization-based methods using sparse representation. The comparison also validates the potential of the proposed method for clinical applications.

Keywords: Super-resolution · NLTV regularization · DW MRI · MRSI · Sparse representation

1 Introduction

Diffusion-weighted imaging (DWI) and magnetic resonance spectroscopy imaging (MRSI) are important techniques for brain imaging. DWI is a specific MR imaging method based on mapping of diffusion process of water molecules in tissues [1]. It is an effective technique, which provides functional information of the brain tissues. DWI is a faster acquisition technique as compared to conventional MRI and it does not require any contrast agent as well. Most of the obtained images are acquired using high speed protocols with a low spatial resolution, reducing the patient stress but also producing low quality images [1]. On the other hand, MRSI is a non-invasive technique which gives information about the biochemical components within the tissues [7]. MRSI is particularly of great help in early diagnosis of brain lesions on the basis of the spectra obtained from

© Springer Nature Switzerland AG 2019
B. Deka et al. (Eds.): PReMI 2019, LNCS 11942, pp. 78–86, 2019.
https://doi.org/10.1007/978-3-030-34872-4_9

different metabolite concentrations. MRSI also has the same advantage of fast scan time as DWI. In spite of advantages of DWI and MRSI, the rise in the cost of scans and poor signal-to-noise ratio limit their clinical use. These limits can be overcome by image super-resolution (SR) methods [2]. SR methods are categorized as either single-image SR (SISR) or multiple-image SR. SISR is more preferable in medical imaging as multiple images are difficult to be acquired in a particular cross-section, taking considerable scan time. Sparse representation approach has proved to be very effective in case of single-image SR. But the stabilization for the solution of inverse problems is a major issue in sparsity approach, which can be overcome by regularization. In this paper, we focus to develop a SISR method for MR images using sparse representation along with non-local total variation (NLTV) regularization.

The remaining part of the paper is written in following sections: In Sect. 2, a brief literature about of the regularization based SR methods is provided. Section 3 will describe the proposed methodology. Experimental results on different datasets of diffusion-weighted and spectroscopic MR images are explained in Sect. 4. Lastly, Sect. 5 gives a conclusion about the paper.

2   Background

Owing to the resolution limitations, there is a need to create high resolution images in a short acquisition time using post-processing techniques. Even though, basic interpolation methods are available, but these methods introduce blurring artifacts. Taking the disadvantages of the interpolation based techniques into consideration, SR had been developed, which has shown very influential and efficient results in HR image reconstruction. The SR methods tries to perform mapping between the LR and HR images via learning some prior information regarding similarity of the two images with their own image space [9]. Earlier works apply SR algorithm on medical imaging to generate HR images using multiple LR images with different angular views, but because of the insufficient number of low resolution images and unknown blurring operators, it becomes highly ill-posed in nature.

For solving the ill-posed problem of image SR, many recent regularization techniques are developed to get a more stabilized solution [6,8]. Recently, sparse representation based SR reconstruction came into picture as a powerful tool for image restoration [4,5]. Sparse representation is mainly concerned with the solution of inverse problems. In medical image processing, total variation (TV) regularization methods are successfully used as it shows excellent edge preservations. Despite its high effectiveness, problems like over-smoothing of image textures and odd artifacts may limit such methods. Lately, the non local means (NLM) based regularization approaches are also introduced for solving inverse problems like, image restoration, super-resolution and denoising, etc.,

3   Proposed Method

The performances of sparse representation algorithms for image SR are related to several phenomenons, like, quality of the dictionary trained, effectiveness of the constraint term selected for regularization, etc. The proposed method for SR reconstruction from a set of LR MR images is discussed in the following subsections. It consist of two parts: first, learning of LR and HR dictionaries and second, reconstruction of HR output image utilizing the learned dictionaries. The reconstruction algorithm is again can be divided into the following subtasks: first, extraction of high-frequency features of the patches, then solving a sparse prior based regularization and secondly, a non-local total variation regularization to restore the textural details remove the undesirable staircase artifact to recover the fine details and textures. Finally, a global image regularization is done that helps in incorporating the given LR image's point spread function into the reconstructed HR images by utilizing the image acquisition model constraint.

3.1   Sparsity Based Image Super-Resolution

In the beginning, overlapping patches of size $k \times k$ are extracted from the input LR image Y. The sparse coefficients α corresponding to each low-resolution patch y is found with respect to the trained dictionaries D_l and D_h. Next, these sparse coefficients are combined with high-resolution dictionary D_h to find high-resolution patches x.

The solution of the following sparse regularization problem gives the sparse coefficients corresponding to each low-resolution patch y:

$$\alpha^* = \arg\min_{\alpha} \left\| D'\alpha - \tilde{y} \right\|_2^2 + \lambda \left\| \alpha \right\|_1, \tag{1}$$

where $D' = \begin{bmatrix} D_\ell \\ TD_h \end{bmatrix}$, $\tilde{y} = \begin{bmatrix} y \\ w \end{bmatrix}$, and λ is the regularization parameter. T is the overlap region extraction operator which finds the region which is common to both the presently reconstructing patch and the latest HR patch generated; w represents the overlapped pixels contained in the previously reconstructed HR image.

Following the computation of sparse coefficients α^* using Eq. 1, HR patches x are obtained by solving the following relation which supports the fact that the HR and LR dictionaries shares the same sparse representation.

$$x \approx D_h \alpha^* \tag{2}$$

Arranging all the individual HR patches reconstructed by Eq. 2 on a single grid will yield to an intermediate HR image X_0. Before finding the initial HR reconstructed image X_0 by the minimization problem, non local total variation regularization is performed so that the patches to be reconstructed fit properly in the above minimization formulation. Again due to measurement errors, X_0 may not fit the generalized model, $Y = WX$, where Y is the input LR image, X is the desired HR image and W is the image sampling operator. For overcoming these limitations due to noise, a global reconstruction constraint is imposed by solving a minimization problem:

$$X^* = \arg\min_X ||WX - Y||_2^2 + \lambda||X - X_0||_2^2 \qquad (3)$$

The above Eq. 4 is the gradient descent method, which is minimized iteratively to find the final reconstructed image X_0.

3.2   NLTV Regularization

The non-local means (NLM) filtering implies the weighted average of the surrounding pixels within a search window for the computation of the new filtered pixel value. Two blocks of the HR image X_0 having central pixels at x_i and x_j contributes to weight w_{ij} which is the gaussian distance l_2 between the blocks [12]. Consider x_i and x_j denote the pixel at the center of $b_s \times b_s$ blocks and it is assumed that x_j lies in the search window of x_i. Weight w_{ij} is computed by:

$$w_{ij} = \exp(-||x_j - x_i||_2^2/f^2)/c_i \qquad (4)$$

where f and c_i are controlling parameter and normalization factors, respectively. The new filtered pixel value is denoted by NLM (x_i), and this approach of filtering has lead to a new approach of regularization, known as nonlocal regularization. Consider all the pixels in the center organized as a column vector, represented as r and all the weights are also organized as column vector, represented as w. Mathematically, this nonlocal regularization can be represented as:

$$\sum_{x_i \in x} ||x_i - w_i^T r_i||_2^2 \qquad (5)$$

These weights are updated iteratively and before the implementation of gradient descent method for final reconstruction, the nonlocal total variation (NLTV) regularization is implemented, the formulation of this approach is as:

$$\min_x ||D_h\alpha|| + \alpha||x - Wx||_2^2 \qquad (6)$$

The solution of the above formulation is the basis of the NLTV regularization. The HR image X_0 obtained after the regularization is used in Eq. 4, which undergoes minimization iteratively to obtain the final SR image X^*.

3.3   Dictionary Learning

Two training image patch pairs, set of high-resolution patches is represented by $X^h = x_1, x_2, ..., x_k$ and set of low-resolution patches is represented by $Y^l = y_1, y_2, ..., y_k$. These two dictionaries are jointly trained with the condition that both HR and LR image patches have a common sparse representations among them. A joint sparse representation regularization can be formulated involving the LR and HR image patches simultaneously. Mathematically,

$$\min_{\{D_h, D_l, Z\}} \frac{1}{R} ||X^h - D_h Z||_2^2 + \frac{1}{S} ||Y^l - D_l Z||_2^2 + \lambda(\frac{1}{R} + \frac{1}{S})||Z||_1 \qquad (7)$$

where LR and HR patches in vector form have dimensions S and R respectively. $||Z||_1$ is a ℓ_1-norm term that enforces sparsity into both the dictionaries. Equation 7 is solved iteratively for three parameters simultaneously to obtain the HR and LR dictionaries D_h and D_l.

4 Experiments and Evaluations

Simulations of the proposed work is carried out using MATLAB (R2015b) environment on PC having configurations as follows: OS- Windows 7, Processor: Intel core $i5$ (2.2 GHz), and RAM: 8 GB. The diffusion-weighted MRI data has been acquired from a GE HDx 1.5 T with the following parameters: TR/TE: 4225/76.6 ms; Slice thickness: 5 mm, spacing between scans: 5 mm; Field of view (FOV): 100×100; Flip angle: 90^0. The spectroscopic MRI images have also been acquired from a GE HDx 1.5 T with the following parameters: TR/TE: 150/1.372 ms; Slice thickness: 8 mm; spacing between scans: 5 mm; Field of view (FOV): 100×100; Flip angle: 70^0.

4.1 Simulations

First, the LR dictionary D_l and the HR dictionary D_h are trained jointly where both consist of 512 atoms in each. For training, a number of 1,00,000 LR/HR patch pairs are selected from about 30 standard MR images. The regularization parameter for the dictionary has been considered as $\lambda = 0.15$. This dictionary has been trained as per the approach proposed by Yang et al. [11].

Next, for the super-resolution reconstruction, two upscale factors have been considered, i.e., 2 and 3. For both the upscale factors, size of the LR input is 128×128. Size of the output HR image is 256×256 and 384×384 for upscale factor 2 and 3 respectively. The results of the proposed method and some other SR based methods has been given with their magnified view. In Tables 1 and 2, DWI results represent the SR results of diffusion-weighted MRI images and MRSI results represent the SR results of Spectroscopic MRI images.

4.2 Evaluations

The simulation results obtained are evaluated both visually and quantitatively. In Figs. 1, 2, 3, 4, 5, 6, 7 and 8, it is clearly seen that the fine details such as edges have been preserved efficiently in the proposed method. From Tables 1 and 2, it can be seen that evaluation parameters obtained for the proposed method are better in terms of peak signal-to-noise ratio (PSNR) as well as mean structural similarity index (MSSIM) compared to the traditional bicubic interpolation

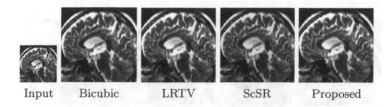

Input   Bicubic   LRTV   ScSR   Proposed

Fig. 1. Results of DW MRI by using different SR techniques for upscale factor 2

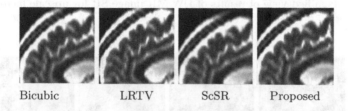

Bicubic   LRTV   ScSR   Proposed

Fig. 2. Magnified view of results of DW image SR for upscale factor 2

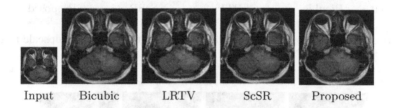

Input   Bicubic   LRTV   ScSR   Proposed

Fig. 3. Results of spectroscopic image SR by different techniques for upscale factor 2

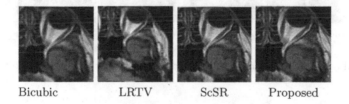

Bicubic   LRTV   ScSR   Proposed

Fig. 4. Magnified view of results of spectroscopic image SR for upscale factor 2

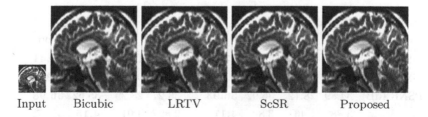

Input   Bicubic   LRTV   ScSR   Proposed

Fig. 5. Results of DW MR image SR by different techniques for upscale factor 3

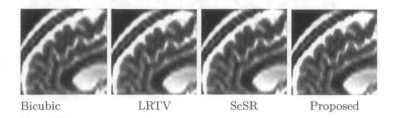

Bicubic   LRTV   ScSR   Proposed

Fig. 6. Magnified view of results of DW MR image SR for upscale factor 3

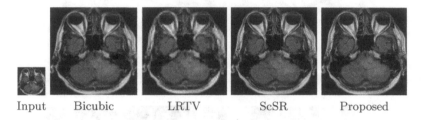

Input   Bicubic   LRTV   ScSR   Proposed

Fig. 7. Results of spectroscopic image SR by different techniques for upscale factor 3

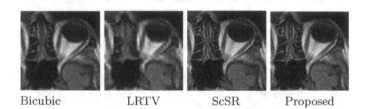

Bicubic   LRTV   ScSR   Proposed

Fig. 8. Magnified view of results of Spectroscopic image SR for upscale factor 3

Table 1. Quantitative evaluations of the DW image SR for different methods using upscale factor 2 and 3

Parameters	Upscale factor 2				Upscale factor 3			
	BCI	LRTV	ScSR	Proposed	BCI	LRTV	ScSR	Proposed
MSE	28.36	78.92	22.49	**17.79**	41.36	130.33	24.06	**16.41**
MSSIM	0.974	0.883	0.977	**0.982**	0.951	0.872	0.962	**0.971**
PSNR (dB)	34.89	28.12	35.92	**36.93**	33.27	28.29	35.62	**36.98**
MI	3.88	1.46	3.89	**4.11**	2.897	4.01	**4.15**	3.517

Table 2. Quantitative evaluations of the spectroscopic image SR for different methods using upscale factor 2 and 3

Parameters	Upscale factor 2				Upscale factor 3			
	BCI	LRTV	ScSR	Proposed	BCI	LRTV	ScSR	Proposed
MSE	24.98	44.09	14.32	**11.66**	41.36	130.33	24.06	**16.41**
MSSIM	0.966	0.878	0.976	**0.980**	0.954	0.798	0.965	**0.972**
PSNR (dB)	35.44	32.99	37.91	**38.92**	35.67	29.29	37.73	**38.67**
MI	3.517	2.568	3.726	**3.827**	3.521	2.327	3.661	**3.789**

techniques. The proposed method has also shown better results as compared to ScSR proposed by Yang *et al.* [11] and LRTV proposed in [10]. Compared to the ground truth, bi-cubic interpolation, and LRTV methods produce blurry results. As far as ScSR is concerned, it does not produce blurring artifacts, but in comparison to the proposed method, it does not preserve equivalent edge details. We have compared the image quality using two more metrics mean-square error (MSE) and mutual information (MI) [3] between the ground truth and the SR result image. For better image quality, MSE should be less and MI should be more. All the images used in the experiments are brain images. The magnified views of the results clearly show the zoomed view of a specified portion of the results. It can be seen that fine details are recovered by the proposed method.

5   Conclusion

In this paper, we have shown that the implementation of non-local TV regularization for solving the regularization issues of the sparsity based approach can be a viable solution to the issues. This combination provides better consistency of patches, thereby giving better results. Quantitative comparisons show that the proposed method outperforms the existing regularization based approaches. Proposed method is computationally expensive due to the iterative process of regularization. As a future work, this can be extended to multi-core processing for computationally efficient results.

Acknowledgements. Authors would like to thank All India Council for Technical Education (AICTE) for providing funds under the project (File No. 8-15/RIFD/RPS/POLICY-1/2016-17) and Ministry of Electronics and Information Technology (MeiTY), GoI for providing financial support under the Visvesvaraya Ph.D. Scheme for Electronics & IT (Ph.D./MLA/ 04(41)/2015-16/01) which helped in smooth conduction of the above research work.

References

1. Altay, C., Balci, P.: The efficiency of diffusion weighted MRI and MR spectroscopy on breast MR imaging. J. Breast Health **10**(4), 197–200 (2014)

2. Baker, S., Kanade, T.: Limits on super-resolution and how to break them. IEEE Trans. Pattern Anal. Mach. Intell. **24**(9), 1167–1183 (2002)
3. Ceccarelli, M., di Bisceglie, M., Galdi, C., Giangregorio, G., Ullo, S.L.: Image registration using non-linear diffusion. In: IEEE International Geoscience and Remote Sensing Symposium, IGARSS 2008 (2008)
4. Datta, S., Deka, B., Mullah, H.U., Kumar, S.: An efficient interpolated compressed sensing method for highly correlated 2D multi-slice MRI. In: 2016 International Conference on Accessibility to Digital World (ICADW), pp. 187–192, December 2016
5. Deka, B., Datta, S.: Compressed Sensing Magnetic Resonance Image Reconstruction Algorithms: A Convex Optimization Approach. Springer Series on Bio- and Neurosystems. Springer, Singapore (2019). https://doi.org/10.1007/978-981-13-3597-6
6. Glasner, D., Bagon, S., Irani, M.: Super-resolution from a single image. In: 12th International Conference on Computer Vision, pp. 349–356. IEEE, Japan (2009)
7. Jain, S., et al.: Patch based super-resolution of MR spectroscopic images. In: 2016 IEEE 13th International Symposium on Biomedical Imaging (ISBI) (2016)
8. Mullah, H.U., Deka, B.: A fast satellite image super-resolution technique using multicore processing. In: Abraham, A., Muhuri, P.K., Muda, A.K., Gandhi, N. (eds.) HIS 2017. AISC, vol. 734, pp. 51–60. Springer, Cham (2018). https://doi.org/10.1007/978-3-319-76351-4_6
9. Park, S., Park, M., Kang, M.: Super-resolution image reconstruction: a technical overview. IEEE Sig. Process. Mag. **20**, 21–36 (2003)
10. Shi, F., Cheng, J., Wang, L., Yap, P., Shen, D.: LRTV: MR image super-resolution with low-rank and total variation regularizations. IEEE Trans. Med. Imaging **34**, 2459–2466 (2015)
11. Yang, J., Wright, J., Huang, T.S., Ma, Y.: Image super-resolution via sparse representation. IEEE Trans. Image Process. **19**(11), 2861–2873 (2010)
12. Zhang, J., Liu, S., Xiong, R., Ma, S., Zhao, D.: Improved total variation based image compressive sensing recovery by nonlocal regularization. In: IEEE International Symposium on Circuits and Systems, ISCAS 2013, Beijing, China (2013)

QIBDS Net: A Quantum-Inspired Bi-Directional Self-supervised Neural Network Architecture for Automatic Brain MR Image Segmentation

Debanjan Konar[1,3]([⊠]), Siddhartha Bhattacharyya[2],
and Bijaya Ketan Panigrahi[3]

[1] Department of Computer Science and Engineering,
Sikkim Manipal Institute of Technology,
Sikkim Manipal University, Majitar, Sikkim, India
konar.debanjan@gmail.com
[2] Department of Information Technology,
RCC Institute of Information Technology, Kolkata, India
dr.siddhartha.bhattacharyya@gmail.com
[3] Department of Electrical Engineering,
Indian Institute of Technology, Delhi, New Delhi, India
bkpanigrahi@ee.iitd.ernet.in

Abstract. A Quantum-Inspired Bidirectional Self-Organizing Neural Network (QIBDS Net) architecture operated by Quantum-Inspired Multi-level Sigmoidal (QIMUSIG) activation function suitable for fully automatic segmentation of T1-weighted contrast enhanced (T1-CE) MR images, is proposed in real time. The QIBDS Net architecture comprises input, intermediate and output layers of neurons represented as *qubits* and inter-connected by second order neighborhood based topology. The inter-connections between the intermediate and output layers are effected by means of counter propagation of quantum states without any training or external supervision. Quantum observation is carried out at the end to obtain the segmented tumor from the superposition of quantum states. The proposed self-supervised network architecture has been tested on T1-CE MR images from publicly available data sets and is found to be very efficient while compared with other state of the art techniques.

Keywords: Quantum computing · QBDSONN · SOFM · CNN

1 Introduction

Brain tumour segmentation, isolating the brain lesions from MR images, is one of the tedious tasks owing to wide variations in structure and gray-levels present in MR images. Efficient knowledge based fuzzy clustering [1,2] and assisted techniques [3] offer to distinct the abnormal cells in MRI images. Owing to inherent

© Springer Nature Switzerland AG 2019
B. Deka et al. (Eds.): PReMI 2019, LNCS 11942, pp. 87–95, 2019.
https://doi.org/10.1007/978-3-030-34872-4_10

parallel and adaptive computing properties, fuzzy logic inspired Artificial Neural Networks (ANN) [4–6] gained popularity among the computer vision researchers for MR image segmentation. Notable examples include a multi-class artificial neural network (ANN) classifier by Kumar *et al.* [5] for $T1 - CE$ MR images segmentation and Self-Organizing Feature Map (SOFM) contributed by Ortiz *et al.* [6]. However, to obtain anatomically correct MR image segmentation, ANN architectures are assisted by complex back-propagation algorithm and explicitly rely on intensity and spatial feature information.

The large scale of redundant intensity and spatial feature information prevalent in MR images have been avoided using Convolutional Neural Networks (CNNs) which revolutionized the field of computer vision. A shallow CNN comprising with two convolution layers is proposed by Zikic *et al.* [7] for brain tumour detection. Of late, a complete tumour detection framework using a binary CNN is also suggested by Lyksborg *et al.* [8]. In addition, to avoid the effect of over fitting, Pereira *et al.* [9] developed a CNN assisted by small kernels (3×3). However, in contrast to automated MR image segmentation, the Convolutional Neural Network based approaches often suffer due to lack of labeled MR images for training and high computational complexities and hence fully automatic MR image segmentations received much attention.

The inherent properties of quantum computing enables the quantum-inspired artificial neural networks to evolve in the field of pattern recognition. A quantum back-propagation algorithm assisted Quantum Multi-layer Self-organizing Neural Network (QMLSONN) architecture is proposed by Bhattacharyya *et al.* [10,11] for fast and efficient binary image segmentation. In contrast to the complex quantum back-propagation algorithm employed in QMLSONN, Konar *et al.* [12,13] resorted to quantum counter propagation of the network states and developed a Quantum Bidirectional Self-Organizing Neural Network (QBDSONN) Architecture characterized by a bi-level sigmoidal function suitable for binary image segmentation. Hence, the QBDOSNN architecture motivates the authors to apply on gray scale MR images with functional modification of the bi-level activation function to a multi-level sigmoidal activation function in quantum environment.

2　Basic Concepts of Quantum Computing

A *qubit* [14] is the normalized superposition of classical bits 0 and 1 and is the constituent unit of quantum processing represented as:

$$|\psi\rangle = \gamma_0|0\rangle + \gamma_1|1\rangle = \begin{bmatrix} \gamma_0 \\ \gamma_1 \end{bmatrix} \tag{1}$$

where, $|\gamma_0|^2 + |\gamma_1|^2 = 1$, γ_0 and γ_1 are complex quantities.

Quantum computation is realized by quantum rotation gate which is the basis of quantum algorithms and is reversible in nature. The operation of a quantum rotation gate applicable on a single *qubit* can be shown as:

$$\begin{bmatrix} \gamma_0' \\ \gamma_1' \end{bmatrix} = \begin{bmatrix} \cos\omega & -\sin\omega \\ \sin\omega & \cos\omega \end{bmatrix} \times \begin{bmatrix} \gamma_0 \\ \gamma_1 \end{bmatrix} \tag{2}$$

The set of superimposed quantum states $|\psi_i\rangle$ with $0-1$ basis forms a Hilbert Hyperspace and the quantum system is defined using the wave function ϕ as

$$|\psi\rangle = \sum_k^n p_k|\phi_k\rangle \tag{3}$$

On quantum observation, the quantum systems interacts with the physical system and the true outcome is obtained.

3 Quantum Inspired Bi-Directional Self Organizing Neural Network (QIBDS Net) Architecture

The proposed QIBDS Net architecture replicates the Quantum Bi-Directional Self Organizing Neural Network (QBDSONN) architecture [12,13] with functional modification in the form of a novel quantum inspired multi-level sigmoidal activation (QIMUSIG) function targeted to address the gray levels pertaining to MR Images. Three layers of neurons viz.,input, hidden and output are composed of *qubits* in the suggested QIBDS Net architecture as illustrated in Fig. 1. The QIBDS Net employs the rotation angle for the interconnection strength and the threshold, represented as Φ and ν respectively. The activation $|\tau\rangle$ can be interpreted as

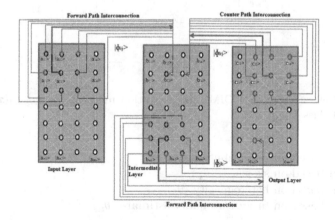

Fig. 1. Quantum Inspired Bi-Directional Self-Organizing Neural Network (QIBDS Net) architecture (Few Inter-layer connections are provided for visibility).

$$|\tau\rangle = \begin{bmatrix} cos\nu \\ sin\nu \end{bmatrix} \tag{4}$$

Without exploring the imaginary section of *qubits*, the input-output dynamics of the network layers is defined as

$$|z\rangle = f_{QIBDSNet}(\sum_{l}^{n} x_l \langle \Phi_l | \tau_l \rangle) = f_{QIBDSNet}(\sum_{l}^{n} x_l cos(\alpha_l - \nu)) \qquad (5)$$

where, x_l is input to the quantum neuron. The activation function, $f_{QIBDSNet}(x)$ is defined as:

$$f_{QIBDSNet}(x) = \frac{1}{\rho_w + e^{-\mu(x-\theta)}} \qquad (6)$$

where, the steepness factor μ and activation θ are represented by *qubits*. The multi-level class response (gray level intensity) is exhibited by ρ_w and is defined as

$$\rho_w = \frac{B_N}{\beta_w - \beta_{w-1}}, 1 \leq w \leq L \qquad (7)$$

where ω is class index; β_w, β_{w-1} are class outputs and the summation of the 8-connected neighborhood gray-scale pixels contribution is denoted as B_N. The QIMUSIG activation function with varying steepness factor μ is shown in Fig. 2. Owing to wide variation of image pixel intensities and pertaining to MR images, four distinct adaptive activation schemes have been introduced in the suggested network architecture over 8-connected neighborhood pixels [15]. These are

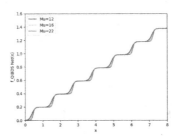

Fig. 2. Multi-class response of QIMUSIG activation function $f_{QIBDSNet}(x)$ for $\mu = 12, 16, 22$ and class $L = 8$

1. Activation based on ϱ-distribution (θ_ϱ)
2. Activation based on skewness (θ_χ)
3. Activation based on heterogeneity (θ_ξ)
4. Activation based on fuzzy cardinality estimate (θ_ν).

Suppose, the response of the intermediate and output layers neurons are denoted as $|In_j\rangle$ and $|Ou_i\rangle$ respectively. In addition, the inter connection weights between input to intermediate layer and intermediate to output layer are expressed as $|\Phi_{kj}\rangle$ and $|\Phi_{ji}\rangle$ respectively with the activations $|\tau_j\rangle$ and $|\tau_i\rangle$ at the intermediate layer and output layer, respectively. The input-output relation of a quantum neuron can be defined as [12]

$$|Ou_i\rangle = f_{QIBDSNet}(\sum_{j}^{N} y_j \langle \Phi_{ji} | \tau_i \rangle) =$$

$$f_{QIBDSNet}(\sum_{j}^{N} f_{QIBDSNet}(\sum_{k}^{L} x_k \langle \Phi_{kj} | \tau_j \rangle) \langle \Phi_{ji} | \tau_i \rangle) \quad (8)$$

$$i.e., |Ou_i\rangle = f_{QIBDSNet}(\sum_{j}^{M} f_{QIBDSnet}(\sum_{k}^{L} x_k cos(\alpha_{kj} - \nu_j)) cos(\alpha_{ji} - \nu_i))$$

Quantum observation allows to evaluate the error incurred by QIBDSNet at the output layer and guided by α and ν as:

$$\xi = \frac{1}{2} \sum \{\vartheta(t+1) - \vartheta(t)\}^2 \quad (9)$$

The inter-connection weights $|\Phi(t)\rangle$ (represented as *qubits*) are transformed into $\vartheta(t)$ at a particular epoch (t) on quantum observation.

4 T1-Weighted CE MR Image Segmentation Using QIBDS Net

The input MR image pixels are received at the input layer of the suggested QIBDS Net as normalized fuzzy intensities and subsequently transformed into quantum states $[0, \frac{\pi}{2}]$ as

$$x_k = \frac{\pi}{2} \times I_k \quad (10)$$

where, normalized MR image pixels are designated by I_k and the transformed quantum states are described by x_k. The proposed QIBDS Net is implemented on $T1$-weighted CE brain MR images from the available data sets [16] in PARAM SHAVAK super computer provided by CDAC, India. In the experimental set up, the steepness hyper-parameter, μ is tailored between 0.24 to 0.25 with step size 0.001 and the three distinct sets $F_{\lambda_{w}L} = \{c_1, c_2, c_3\}$ of class level, $L = \{8\}$ are used. A tiny cluster, considered as tumour is erroneously exposed as outcome after the segmentation and hence to remove these small clusters with threshold $(\sigma = 4)$, a post processing operation has been performed.

5 Results and Discussion

In the current experiments, in comparison to QIBDS Net, a CNN [9] is trained with 1200 MR images from the same data set [16] allowing maximum 100 epochs. 800 T_1-weighted CE MR images are used for validation and testing. Four different types of empirical goodness evaluation metrics (PPV, SS, ACC, DSC) [1] are used in the experiments. A similar post processing technique has been applied on the segmented output images obtained using SOFM [6], fuzzy-clustering method [1] and CNN [9]. Figure 3 shows the input skull-tripped sample MR slices

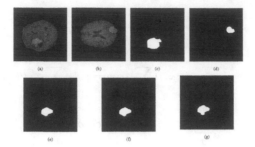

Fig. 3. Skull stripped input MR images (a) slice no: 5 (b) slice no: 68 manually segmented tumor for (c) slice no: 5 and (d) slice no: 68 [16]. Segmented tumours followed by post processing using (e) fuzzy-clustering [1] (f) SOFM [6] and (g) CNN [9] from slice no. 5

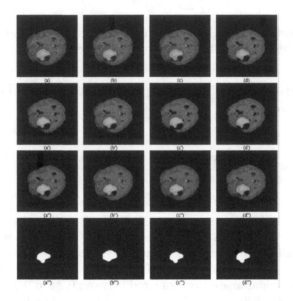

Fig. 4. $(a - d)$ QIBDS Net segmented and followed by post-processing with color map images (Yellow-enhanced tumor region, Green-non-enhanced tumor region and Sky blue-edema region) for slice no. 5 using $L = 8$ transition levels with four different thresholding schemes θ_ϱ $(a - a'')$, θ_χ $(b - b'')$, θ_ξ $(c - c'')$ and $\theta_\nu (d - d'')$ for the three distinct level sets c_1 $(a - d)$, c_2 $(a' - d')$ and c_3 $(a'' - d'')$ (Color figure online)

with the manually segmented ground truth images and the segmented tumour using the state of the art methods. The proposed QIBDS Net segmented output MR images with the post processed tumour regions are demonstrated using the class levels $L = 8; \{c_1, c_2, c_3\}$ and four distinct activations θ_ϱ, θ_χ, θ_ξ and θ_ν, as shown in Fig. 4.

The evaluation metrics: average accuracy (ACC), dice similarity (DSC), positive prediction value (PPV) and sensitivity (SS) using the suggested QIBDS

Net with the class level $L = 8$ and activation (θ_ξ) for three distinct level sets c_1, c_2, c_3, SOFM, CNN and fuzzy-clustering method are reported in Table 1. It is evident from the results that the proposed QIBDS Net outperforms supervised SOFM and unsupervised fuzzy-clustering with respect to evaluation metrics and reports similar accuracy as the supervised CNN. However, the average DSC reported using CNN is superior to the QIBDSN Net which is demonstrated using Box plots as shown in Fig. 5.

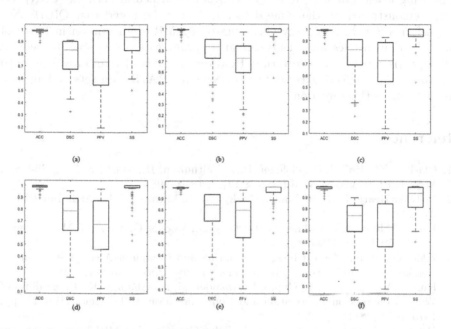

Fig. 5. Box plot of the proposed QIBDS Net with the fixed class level $L = 8$ and class level sets $(a)c_1, (b)c_2$ and $(c)c_3$ and using (d) SOFM (e) CNN and (f) Fuzzy-clustering.

Table 1. A comparative analysis among the proposed QIBDS Net with activation (θ_ξ), Fuzzy-Clustering, SOFM and CNN with one sided two sample KS test (significance level $\alpha = 0.05$ and values marked in bold

Method	Set	ACC	DSC	PPV	SS
QIBDS Net	c_1	0.985	**0.765**	**0.720**	0.906
	c_2	0.984	**0.763**	0.665	**0.970**
	c_3	**0.990**	**0.763**	0.675	0.955
SOFM [6]		0.984	0.738	0.637	**0.962**
CNN [9]		**0.998**	**0.792**	**0.706**	**0.961**
Fuzzy-Clustering [1]		0.982	0.690	0.624	0.898

6 Conclusion

In this paper, a novel quantum-inspired bi-directional self-supervised neural network architecture referred to as QIBDS Net, has been evolved for fully automatic segmentation of MR images. To show the effectiveness of the suggested QIBDS Net, experiments have been performed on $T1$-weighted CE MR images and a comparative analysis with unsupervised fuzzy clustering, supervised self-organizing feature map (SOFM) and convolutional neural network (CNN) has been demonstrated on the same data set. It may be noted that QIBDS Net outperforms SOFM and fuzzy clustering based MR segmentation in terms of all evaluation metrics and yields similar accuracy as CNN in spite of being a self-supervised network architecture. However, it is a subject of investigation to demonstrate the effectiveness of QIBDS Net on BRATS data sets and authors are focusing in this direction.

References

1. Clark, M.C., Hall, L.O., Goldgof, D.B., Velthuizen, R., Murtagh, F.R., Sil-biger, M.S.: IEEE Trans. Medical Imaging **17**(2), 187 (1998)
2. Fletcher-Heath, L.M., Hall, L.O., Goldgof, D.B., Murtagh, F.R.: Artif. Intell. Med. **21**, 43 (2001)
3. Liu, J., Udupa, J.K., Odhner, D., Hackney, D., Moonis, G.: Comput. Med. Imaging Graph. **29**(1), 21 (2005)
4. Zikic, D., et al.: Medical Image Computing and Computer-Assisted Intervention Conference Brain Tumor Segmentation Challenge. Nice, France (2012)
5. Kumar, V., Sachdeva, J., Gupta, I., Khandelwal, N., Ahuja, C.K.: Proceedings of World Congress on Information and Communication Technologies (WICT), pp. 1079–1083 (2011)
6. Ortiz, A., Gorriz, J.M., Ramirez, J., Salas-Gonzalez, D.: MRI brain image segmentation with supervised SOM and probability-based clustering method. In: Ferrández, J.M., Álvarez Sánchez, J.R., de la Paz, F., Toledo, F.J. (eds.) IWINAC 2011. LNCS, vol. 6687, pp. 49–58. Springer, Heidelberg (2011). https://doi.org/10.1007/978-3-642-21326-7_6
7. Zikic, D., et al.: MICCAI Multimodal Brain Tumor Segmentation Challenge (BraTS), pp. 36–39 (2014)
8. Lyksborg, M., Puonti, O., Agn, M., Larsen, R.: An ensemble of 2D convolutional neural networks for tumor segmentation. In: Paulsen, R.R., Pedersen, K.S. (eds.) SCIA 2015. LNCS, vol. 9127, pp. 201–211. Springer, Cham (2015). https://doi.org/10.1007/978-3-319-19665-7_17
9. Pereira, S., Pinto, A., Alves, V., Silva, C.A.: IEEE Trans. Med. Imaging **35**, 5 (2016)
10. Bhattacharyya, S., Pal, P., Bhowmick, S.: Proceedings of Fourth International Conference on Communication Systems and Network Technologies, pp. 512–518 (2014)
11. Bhattacharyya, S., Pal, P., Bhowmik, S.: Appl. Soft Comput. **24**, 717 (2014)
12. Konar, D., Bhattacharya, S., Panigrahi, B.K., Nakamatsu, K.: Appl. Soft Comput. **46**, 731 (2016)

13. Konar, D., Bhattacharyya, S., Das, N., Panigrahi, B.K.: Proceedings of IEEE International Conference on Advances in Computing, Communications and Informatics (ICACCI), pp. 1225–1230 (2015)
14. Mcmohan, D.: Quantum Computing Explained. Wiley, Hoboken (2008)
15. Bhattacharyya, S., Dutta, P., Maulik, U.: Appl. Soft Comput. **11**(1), 946 (2011)
16. Cheng, J.: (2017). https://figshare.com/articles/brain_tumor_dataset/1512427

Colonoscopic Image Polyp Classification Using Texture Features

Pradipta Sasmal[1]([✉]), M. K. Bhuyan[1], Kangkana Bora[1], Yuji Iwahori[2], and Kunio Kasugai[3]

[1] Department of Electronics and Electrical Engineering,
Indian Institute of Technology (IIT) Guwahati,
Guwahati 781039, Assam, India
{s.pradipta,mkb,kangkana.bora}@iitg.ac.in
[2] Department of Computer Science,
Chubu University, Kasugai 487-8501, Japan
iwahori@cs.chubu.ac.jp
[3] Department of Gastroenterology, Aichi Medical University,
Nagakute 480-1195, Japan
kuku3487@aichi-med-u.ac.jp

Abstract. Colonoscopic polyps, which may lead to cancer in the later stage has been claiming lots of lives worldwide. So, an early detection and classification system for such polyps has become crucial for timely diagnosis. Many features learning based classification frameworks have been proposed in the literature. Extreme difficulties have been faced in this field are due to lack of sufficient data set, feature similarity among different classes of polyps and so on. In this paper, a two class classification of polyps namely benign and malignant has been devised. The proposed method uses multi directions and multi frequency texture analysis. Gabor filter banks are used in the initial stage followed by discrete cosine transform (DCT) upon the sub-bands. Local binary pattern (LBP) as texture features is also used. Since malignant polyp generally oozes blood on its surface, color features is also considered. The square of the DCT coefficients calculated on all the sub-bands of the Gabor filter are considered as the energy. The DCT co-efficients derived from all the sub-bands are fused to form the feature vector. Along with these features, LBP features and color features are also added to form the final feature vector. SVM is used as the classifier which performs $k-$fold validation on the data set. The proposed method gives competitive results with recent state-of-the-art methods of colonoscopic polyp classification.

Keywords: Colonoscopic polyps · Gabor filters · DCT · LBP · SVM

1 Introduction

Colonoscopic cancer is becoming a potential threat to the mankind globally. Early detection of such cancer can reduce the mortality rate to a large extent.

© Springer Nature Switzerland AG 2019
B. Deka et al. (Eds.): PReMI 2019, LNCS 11942, pp. 96–101, 2019.
https://doi.org/10.1007/978-3-030-34872-4_11

One such technique is endoscopy, which is used to investigate the conditions of the Gastro-intestinal (GI) tract. For this approach, we are concentrating only on colonic polyp analysis as it is the most common pathological finding of colonoscopy. Polyp generally describes an unnatural growth of mass of tissue protruding into the lumen of the bowel [7]. According to the WHO, a patient with colonic polyp should undergo colonoscopic examination every three years [3]. During this procedure, a device called endoscope is used to scan the entire colon of the patient body from rectum to cecum. When the endoscopists find any abnormal region seeming to be polyp region, they perform visual inspection and/or histological examination to identify the presence of an anomaly. The procedure may be followed by biopsy of the tissue samples to determine the actual nature of the anomaly. Manual categorization of polyp is not an ideal approach due to constraints like observer bias, textural similarity as perceived by human eyes. For this study, automated feature analysis is performed for two classes, i.e. Neoplastic (Malignant) and Non-Neoplastic (Benign) polyp.

During the examination of polyps, the endoscopist looks into the shape, size, presence of blood debris, boundary (smooth or wobble like) and textures of the polyp region. The micro textures on the surface of polyps can be studied to classify such polyps into Benign and Malignant. Since, such texture features can't easily be visualized by naked eye, it is very difficult for the doctor to arrive to any conclusion. Nevertheless, there is minute difference in the textural pattern between the polyps from both the classes. In this work, we are trying to quantify these features using some advanced image processing and machine learning techniques which can represent texture and color features efficiently.

In the proposed work, the Discrete cosine transform (DCT) coefficients of the sub-bands taken from the Gabor filter response are used to form the feature vector. Here, the square of the DCT co-efficients are treated as the energy quantity. Since, malignant polyps have high textural pattern, it is expected to have high energy than the benign polyps. The Local binary pattern (LBP) as the texture feature is also considered here. The S component of the HSV color space is taken as the color feature. All the features are then concatenated to form the final feature representation of the polyps. Finally, classification is performed based on four measures viz. accuracy, recall, precision and f-score using SVM. To avoid over-fitting, four-fold cross-validation technique is used for assessments. Experiments show that the proposed method outperforms some of the existing approaches for Colonic polyp detection. Also, to verify the consistency of the proposed work, the experiments have been performed on the database annotated by an endoscopist.

Rest of the paper is organized as follows - Sect. 2 will give some light on the existing works in this domain, Sect. 3 will portray an overview of the proposed work and explanations on methodologies. Result and discussions are explained in Sect. 4. Finally, conclusions are drawn in Sect. 5.

2   Related Works

From the last many year's researchers have been working on colonoscopy images. Though a lot of work has been done in this domain, still this area is open and needs to be explored. Some have worked on polyp detection and segmentation, and other's work may include only feature extraction, video tracking for best frame selection, and classification using conventional classifiers or deep networks. Since, we are only concentrating on feature extraction, selection and classification, that is why only the concerned literature's in this particular domain are explored.

Hafner *et al.* [5] proposed a method to describe local texture properties within color images with the aim of automated classification of endoscopic images. They suggested a color vector field from an image, then computes the similarity between neighboring pixels. The resulting image descriptor is a compact 1D-histogram which is used for classification using the k-nearest neighbors classifier. Mesejo *et al.* [6] worked on a generated database and try to compute 2D texture features using Invariant gabor texture, rotational invariant LBP; 2D color features using color naming, discriminative color, hue, opponent and color GLCM, and 3D shape features using shape-DNA and kernel-PCA. Color-GLM using SVM classifier is explored by Engelhardt *et al.* [4]. Bag of words descriptors with spatial pyramid matching is done in ref [2]. Muhammad *et al.* [8] proposed to describe the image features using gabor filter and DCT in application to image forgery.

3   Methodology

In the first stage, an RGB endoscopic image is converted into HSV color space. Malignant polyp surface generally contains blood debris *i.e.*, it is highly saturated. So, S component is taken as the color feature.

In the next step, the Gabor filters are applied to the grey images. Gabor filter banks with varied orientation and frequency are used to capture the micro textures in the polyp surface. This transform is a decomposition technique with multi-scale and multi-orientation properties. The mathematical formulation of a 2-dimensional Gabor filter is represented as follows:

$$g(x,y) = \frac{1}{2\pi\sigma_x\sigma_y} \exp\left[-0.5\left(\frac{x^2}{\sigma_x^2} + \frac{y^2}{\sigma_y^2}\right)\right] \exp\left(2\pi jWx\right) \tag{1}$$

Where, σ_x and σ_y defines the width of the gaussian function, W is the central frequency of the sinusoid. This frequency component is called the orientation (rotation angle) of the gaussian envelope. To extract the texture features of an image, Gabor filter bank can be formed with different scale and orientation to encapsulate features in different sub-bands.

In this method, we use Gabor transform with varied scales and orientations. The scales are 1, 2, and 3, and 0, $\pi/4$, $\pi/2$, and $3\pi/4$ are chosen as the rotational angles.

The second step accentuates the textures of the polyps at different sub-bands. To encapsulate such features, DCT is used. Unlike discrete fourier transform (DFT), DCT uses real valued cosine functions as the basis functions for the transformation. An image can be represented as the sum of cosine of different frequencies [1] by the help of DCT. It transforms the image from spatial domain to frequency domain (set of frequency coefficients). In our experiments, we extracted all the DCT coefficients for each subbands. In the fourth step, Local binary pattern (LBP) of each pixel is calculated. LBP is a type of visual descriptor used for classification. The window size of 32 × 32 is taken to form the feature vector. For classification, SVM with RBF kernel is used. With other kernels the results are not satisfactory as with RBF kernel. Performance analysis is done with 4-fold cross validation method. The overall process is given in the Fig. 1.

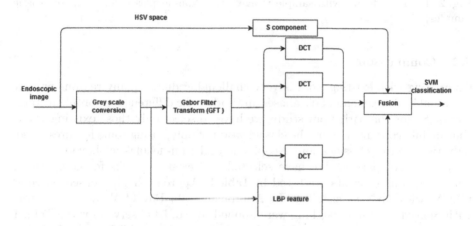

Fig. 1. Block diagram of the proposed method.

4   Results and Discussions

4.1   Data-Set Details

Own database is generated at the Department of Gastroenterology, Aichi Medical University, Nagakute, Japan under the supervision of Konio Kasugai. He has selected only those frames where the polyp is properly visible. Database details are available in Fig. 2. The dataset is composed of colonoscopy videos recorded with Narrow-band Imaging (NBI), White Light (WL) and dye. This database contains 208 benign and malignant images. Ground truths of all the images are verified by doctors. Here in this paper, we are concentrating only on feature analysis and classification. Experiments are performed on 2D images obtained after manual best frame selection by doctors where polyp region is properly visible. Video frame extraction and polyp tracking is not of concern as these are a totally different topic.

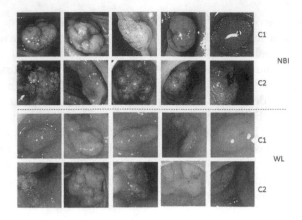

Fig. 2. Data-set details with sample images (C1: Non-neoplastic and C2: Neoplastic samples)

4.2 Comparison

Comparing with existing work is very challenging due to many reasons namely unavailability of generated databases and codes, the differences in problem statements. So for comparing our study, we have to select only those existing works that mainly concentrated on the classification of polyp using some features. Also, only two level of classification (Neoplastic and Non-neoplastic) have been performed for comparison. We have selected two existing works for comparison. The comparative results are listed in Table 1. Approach 1 [4] focused on color GLCM with kNN as classifier. In Approach 2 [2], Bag of Words descriptors with Spatial Pyramid matching were applied. It can be observed for the Table 1 that the proposed work outperform the existing works establishing the fact that the methods used for feature extraction, selection and classification are giving competitive results for colonic polyp classification.

Table 1. Comparison with existing work

	Accuracy	Precision	Recall	F-Score
Proposed	74.035	84.32	85.71	85.00
Approach 1	64.47	75.93	74.55	75.23
Approach 2	73.68	73.97	98.18	84.38

5 Conclusion

In this work, we have tried to quantify texture and color features which help in identifying the dysplasia in a polyp. In doing so, texture features were extracted using DCT upon different sub-bands found using Gabor filter transform (GFT). Another visual descriptor used was LBP features. Final assessment using four different measures was performed to establish the efficiency of the proposed work. Also, the experiments were performed on the dataset provided by Aichi medical and hospital, Aichi, Japan. Also from comparison statistics, it can be concluded that the present work outperforms some of the existing ones establishing the claim made by us. NBI endoscopic images are generally studied as it gives better classification accuracy. In this method images taken from all the three modalities discussed above have been used.

Since this work has done only on single frames selected by doctors so in future we would like to integrate it with video tracking and automated frame selection. Also, future work includes the study of other texture features of the polyp.

References

1. Ahmed, N., Natarajan, T., Rao, K.R.: Discrete cosine transform. IEEE Trans. Comput. **100**(1), 90–93 (1974)
2. Aman, J.M., Summers, R.M., Yao, J.: Characterizing colonic detections in CT colonography using curvature-based feature descriptor and bag-of-words model. In: Yoshida, H., Cai, W. (eds.) ABD-MICCAI 2010. LNCS, vol. 6668, pp. 15–23. Springer, Heidelberg (2011). https://doi.org/10.1007/978-3-642-25719-3_3
3. Atkin, W.S., Saunders, B.P.: Surveillance guidelines after removal of colorectal adenomatous polyps. Gut **51**, v6–v9 (2002)
4. Engelhardt, S., Ameling, S., Wirth, S., Paulus, D.: Features for classification of polyps in colonoscopy. In: Bildverarbeitung für die Medizin (2010)
5. Häfner, M., Liedlgrubera, M., Uhl, A., Vécsei, A., Wrba, F.: Color treatment in endoscopic image classification using multi-scale local color vector patterns. Med. Image Anal. **16**(1), 75–86 (2012)
6. Mesejo, P., et al.: Computer-aided classification of gastrointestinal lesions in regular colonoscopy. IEEE Trans. Med. Imaging **35**, 2051–2063 (2016)
7. Messmann, H.: Atlas of Colonoscopy. Thieme, New York (2006)
8. Muhammad, G., Dewan, M.S., Moniruzzaman, M., Hussain, M., Huda, M.N.: Image forgery detection using Gabor filters and DCT. In: 2014 International Conference on Electrical Engineering and Information & Communication Technology, pp. 1–5. IEEE (2014)

Bioinformatics and Biomedical Signal Processing

GRaphical Footprint Based Alignment-Free Method (GRAFree) for Classifying the Species in Large-Scale Genomics

Aritra Mahapatra[✉] and Jayanta Mukherjee

Department of Computer Science and Engineering,
Indian Institute of Technology, Kharagpur 721302, India
aritra.mhp@iitkgp.ac.in, jay@cse.iitkgp.ac.in

Abstract. In our study, we propose to use novel features from mitochondrial genomic sequences reflecting their evolutionary traits by a novel GRaphical footprint based Alignment-Free method (GRAFree). These features are used to classify a set of species to different classes. A novel distance measure in the feature space is also proposed to measure the proximity of these species in the evolutionary processes. The distance function is found to be a metric. Further we model the evolutionary relationships of these classes by forming a phylogenetic tree. Experimentations were carried out with 157 species covering four different classes such as, Insecta, Actinopterygii, Aves, and Mammalia. We apply our proposed distance function on the selected feature vectors for three different graphical representations of genome. The inferred trees corroborate accepted evolutionary traits. This demonstrates that our proposed distance function and feature representation can be applied to classify different species and to capture the evolutionary relationships among their classes.

Keywords: Classification · Phylogeny · Mitochondrial genome · Graphical footprint · k-nearest neighbor classifier · Hierarchical clustering

1 Introduction

The phylogeny can be considered as the clustering problem. In studying phylogeny of different species using molecular data, mostly the homologous segments of DNA sequences are observed by computational biologists. This approach is sensitive to the selection of segments (e.g. genes, coding segments, etc.) of the sequence. The mitochondrial genomes (mtDNA) are haploid, inherited maternally in most animals, and recombination is very rare event in it [4]. So the changes of mtDNA sequence occur mainly due to mutations. Hence, the evolutionary traits are more conserved in mtDNA. The mtDNA usually consists of a few numbers of non-overlapping fragments called genes.

© Springer Nature Switzerland AG 2019
B. Deka et al. (Eds.): PReMI 2019, LNCS 11942, pp. 105–112, 2019.
https://doi.org/10.1007/978-3-030-34872-4_12

During evolution, the genes of mtDNA very often change their order within the mtDNA and also get fragmented [2]. This violates the collinearity of homologous regions. Apart from this fact, the complexity and versatility of the data make it difficult to develop any simple method in comparative genomics [13]. Conventional methods compute the distance between sequences through computationally intensive process of multiple sequence alignment, which remains a bottleneck in using whole genomic sequences for constructing phylogeny [8]. As a result, there exist a few works to discover evolutionary features in the non-homologous regions of using alignment-free methods [17].

The existing alignment-free methods can be categorized into four types, namely, k-mer frequency based method [1], Substring based method [11], Information theory based method [6], and Graphical representation based method [21].

But they are not suitable to deal with a large number of taxa, and the size of the input sequences is also limited [3]. Due to these difficulties, conventional methods of phylogenetic reconstruction are restricted to working with whole genome sequences as well as large dataset. For the last three decades, several methods have been introduced to represent the DNA sequence with mathematical encoding (both numerically and graphically) [15]. It has been hypothesized that each species carries unique patterns over their DNA sequence which makes a species different from others [10]. There exist various representations of the large genome sequences through line graph by mapping the nucleotides to numeric representations. Considering a genome sequence as a signal (called the **genomic signal**), these methods analyze respective sequences using different signal processing techniques. Several techniques have been proposed to represent DNA sequences graphically from 2D space to higher dimensional space [16]. The graphical representation has a serious limitation of overlapping paths which cause loss of information [16]. GRAFree takes care of this problem by considering the coordinates of each nucleotide.

In this study, we consider three sets of structural groups of nucleotides (purine, pyrimidine), (amino, keto), and (weak H-bond, strong H-bond) separately for representing DNA by a sequence of points in a 2-D integral coordinate space. This point set is called **Graphical Foot Print (GFP)** of a DNA sequence. We propose a technique for extracting features from GFPs and use them to classify the species according to their class.

Experimentations were carried out with a dataset of total 157 species from four different classes, namely, Insecta (insect), Actinopterygii (ray-finned fish), Aves (bird), and Mammalia (Mammal). Our proposed method, GRAFree, classifies the species based on their classes with a high accuracy.

2   Materials and Methods

2.1   Feature Space

Definition 1 (Graphical Foot Print (GFP)). *Let a sequence, $\mathcal{S} \in \Sigma^+$, $\Sigma = \{A, T, G, C\}$. The GFP of $\mathcal{S}$, $\phi(\mathcal{S})$, is the locus of 2-D points in an integral coordinate space, such that (x_i, y_i) is the coordinate of the alphabet s_i, $\forall\ s_i \in \mathcal{S}$, for $i = 1, 2, \ldots, n$, and $x_0 = y_0 = 0$.*

Case-1 or GFP-RY (Φ_{RY}): for Purine (R)/Pyrimidine (Y) [14]

$$
\begin{aligned}
x_i &= x_{i-1} + 1; \; if \; s_i = G & y_i &= y_{i-1} + 1; \; if \; s_i = C \\
&= x_{i-1} - 1; \; if \; s_i = A & &= y_{i-1} - 1; \; if \; s_i = T \\
&= x_{i-1}; \quad otherwise & &= y_{i-1}; \quad otherwise
\end{aligned}
\tag{1}
$$

Case-2 or GFP-SW (Φ_{SW}): for Strong H-bond (S)/Weak H-bond (W) [7]

$$
\begin{aligned}
x_i &= x_{i-1} + 1; \; if \; s_i = C & y_i &= y_{i-1} + 1; \; if \; s_i = T \\
&= x_{i-1} - 1; \; if \; s_i = G & &= y_{i-1} - 1; \; if \; s_i = A \\
&= x_{i-1}; \quad otherwise & &= y_{i-1}; \quad otherwise
\end{aligned}
\tag{2}
$$

Case-3 or GFP-MK (Φ_{MK}): for Amino (M)/Keto (K) [12]

$$
\begin{aligned}
x_i &= x_{i-1} + 1; \; if \; s_i = A & y_i &= y_{i-1} + 1; \; if \; s_i = T \\
&= x_{i-1} - 1; \; if \; s_i = C & &= y_{i-1} - 1; \; if \; s_i = G \\
&= x_{i-1}; \quad otherwise & &= y_{i-1}; \quad otherwise
\end{aligned}
\tag{3}
$$

Definition 2 (Drift of GFP). *For a length L, drift at the ith position is,*
$\delta_i^{(L)} = \phi_{i+L}(\mathcal{S}) - \phi_i(\mathcal{S})$, where $(i+L) \leq |\mathcal{S}|$ and $\phi_i(\mathcal{S})$ the ith coordinate of $\phi(\mathcal{S})$
Considering the drifts for every i^{th}, $i = 1, 2, \ldots, n$, location of the whole sequence, the sequence of drifts is denoted by
$\Delta^{(L)} = [\delta_0^{(L)}, \delta_1^{(L)}, \delta_2^{(L)}, \delta_3^{(L)}, \ldots, \delta_m^{(L)}]$, where $(m + L) = |\mathcal{S}|$

For three different cases, we denote drifts as $\Delta_{RY}^{(L)}$, $\Delta_{SW}^{(L)}$, and $\Delta_{MK}^{(L)}$, respectively.

We also call the elements of $\Delta^{(L)}$ as points, as they can be plotted on a 2-D coordinate system. We call this plot as the scatter plot of the drift sequence. Similarly, we get a scatter plot of a GFP. Compared to $\Phi_i(\mathcal{S})$, $\Delta^{(L)}$ is translation invariant as its set of points does not depend on the starting point of the sequence. It has been observed that in many cases the scatter plots of Δ have similar structure for closely spaced species mentioned in literature. Some typical examples in Fig. 1 show that the species from class insect (Fig. 1(a, b)) have the similar pattern in their drift sequences but the pattern of the drift sequence of fish (Fig. 1(c)) is different from them. This intuitively indicates that the intraclass species are closer than the interclass species.

We represent spatial distribution of these points of Δ by a five dimensional feature descriptor: $(\mu, \Lambda, \lambda, \theta)$, where $\mu = (\mu_x, \mu_y)$ is the center of the coordinates, Λ and λ are major and minor eigen values of the covariance matrix, and θ is the angle between the eigen vector corresponding to Λ and x-axis. We make $\mathcal{F}$ number of non-overlapping equal length fragments from Δ and represent each fragment using the 5D feature descriptor.

2.2 Distance Function and Its Properties

For two sequences $\mathcal{P}$ and $\mathcal{Q}$ with the feature descriptors of $i^{th}, i \leq \mathcal{F}$ fragments $(\mu_{\mathcal{P}i}, \Lambda_{\mathcal{P}i}, \lambda_{\mathcal{P}i}, \theta_{\mathcal{P}i})$ and $(\mu_{\mathcal{Q}i}, \Lambda_{\mathcal{Q}i}, \lambda_{\mathcal{Q}i}, \theta_{\mathcal{Q}i})$, where $\mu_{\mathcal{P}i} = (\mu_{x\mathcal{P}i}, \mu_{y\mathcal{P}i})$

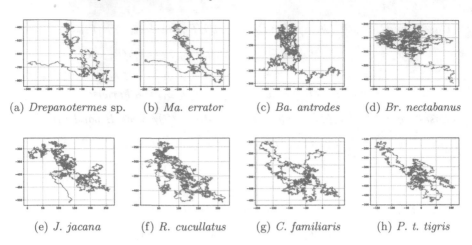

(a) *Drepanotermes* sp.   (b) *Ma. errator*   (c) *Ba. antrodes*   (d) *Br. nectabanus*

(e) *J. jacana*   (f) *R. cucullatus*   (g) *C. familiaris*   (h) *P. t. tigris*

Fig. 1. The Δ_{RY} for $L = 2000$ of (a, b) the insects namely, *Drepanotermes* sp. and *Macrognathotermes errator*, respectively. (c, d) the fishes namely, *Bathygadus antrodes* and *Bregmaceros nectabanus*, respectively. (e, f) the birds namely, *Jacana jacana* and *Raphus cucullatus*, respectively. (g, h) the mammals namely, *Canis familiaris* and *Panthera tigris tigris*, respectively.

and $\mu_{\mathcal{Q}i} = (\mu_{x\mathcal{Q}i}, \mu_{y\mathcal{Q}i})$, we propose the following distance function,

$$D(\mathcal{P}, \mathcal{Q}) = \frac{1}{\mathcal{F}} \sum_{i=1}^{\mathcal{F}} \left(\alpha \sqrt{\mu_{\mathcal{P}i}^T \mu_{\mathcal{P}i} + \mu_{\mathcal{Q}i}^T \mu_{\mathcal{Q}i} - 2\mu_{\mathcal{P}i}^T \mu_{\mathcal{Q}i} cos(\theta_{\mathcal{P}i} - \theta_{\mathcal{Q}i})} \right.$$
$$\left. + (1-\alpha)\sqrt{(\Lambda_{\mathcal{P}i} - \Lambda_{\mathcal{Q}i})^2 + (\lambda_{\mathcal{P}i} - \lambda_{\mathcal{Q}i})^2} \right); \text{where, } \alpha = [0,1] \ (4)$$

The distance D between two sequences is found to be a metric.

2.3 Taxon Sampling and Acquiring Mitochondrial Genome

We have prepared our dataset of 157 species by selecting their mitochondrial genomes sequenced by various researchers. We consider 30, 59, 32, and 36 species from four classes such as mammal [9,22], bird [5], ray-finned fish [18,23], and insect [20], respectively. The selected data have been downloaded from the NCBI database[1]. The average percentage of unrecognized nucleotide of all 157 mtDNA is 0.06% which indicates that the quality of data is good.

2.4 Classification and Phylogenetic Inference

We randomly select ten species from each class and consider the mean of them as the representative of the corresponding class. We consider five such representatives for each class. Finally, we apply k-nearest neighbor classifier based on

---

[1] Website of NCBI database: http://www.ncbi.nlm.nih.gov.

the representatives to classify the species. In k-nearest neighbor classification we apply our proposed distance function (Eq. (4)) for a given value of L, $\mathcal{F}$, and α.

We also apply the same distance function to compute pairwise distances among representatives of four classes.. By applying a hierarchical clustering technique, UPGMA [19], over this distance matrix, we get a phylogenetic tree. We compute phylogenetic trees each separately using representation schemes, such as, GFP-RY, GFP-SW, and GFP-MK (refer to Eqs. (1), (2), (3), respectively).

3   Results and Discussion

3.1   Comparison

For a given number of fragments, $\mathcal{F}$, our proposed features vector is $5\mathcal{F}$ dimensional. We compare our feature vector with an existing k-mer based feature representation [1].

As our input data is genomic sequence which contains only four characters, A, T, G, C, hence, for k length of word, the length of the k-mer based feature is 4^k. So, the k-mer feature vector takes a significantly larger memory space than our proposed feature vector. We have applied Euclidean, Canberra, and Chebyshev distance functions on both k-mer based feature vector and our proposed feature vector for different parameters value. It is observed that, our proposed feature representation using GFP outperforms the Canberra and Chebyshev methods using k-mer based representation and shows a comparable result with k-mer based Euclidean method in classifying the species.

We also compare the performance of our proposed distance function with the same distance functions. It is observed that all of these distance functions perform differently for different parameters value. However, the best performance of the proposed distance function is comparable to the performance of Euclidean and Canberra. The Chebyshev distance function does not perform well in this task of classification.

Apart from that, the phylogenetic trees derived from the proposed distance function are consistent and supports the mostly accepted hypotheses, whereas, the trees derived from Euclidean, Canberra, and Chebyshev distance functions give different relationships among the four selected classes for GFP-RY, GFP-SW, and GFP-MK.

3.2   Observations

Here, we present the clusters generated by GRAFree using the whole mitochondrial genome sequences of the selected species. We enumerate the value of L, $\mathcal{F}$, and α from 50 to 5000, 1 to 200, and 0 to 1, respectively. We consider three different cases, such as GFP-RY, GFP-SW, and GFP-MK (please refer to Eq. (1), (2) and (3), respectively). It is observed that for a given value of the parameters (L, $\mathcal{F}$, and α) the classifier performs differently for GFP-RY, GFP-SW, and GFP-MK. This fact reveals that for GFP-RY, GFP-SW, and GFP-MK, the species contains different signatures in their corresponding drifts.

However, the best accuracy we get 95.5%, 96.2%, and 96.2% for GFP-RY (with $L = 100$, $\mathcal{F} = 150$, and $\alpha = 0.50$), GFP-SW (with $L = 300$, $\mathcal{F} = 55$, and $\alpha = 1$), and GFP-MK (with $L = 50$, $\mathcal{F} = 50$, and $\alpha = 0.50$), respectively. The confusion matrices for the best cases are shown in Table 1. This accuracy is moderately better compared to those obtained using the Euclidean and Chebyshev distance functions and comparable to the Canberra distance function. For the same set of parameters of each case, the accuracy of the Euclidean are 96.82%, 78.34%, and 94.27%, respectively. The accuracy of the Canberra for the same set of parameters are 96.82%, 97.45%, and 98.10%, respectively. The accuracy of the Chebyshev is as low as <70% for all of the three cases.

Table 1. Confusion matrix of four classes Mammal (Mamm), Bird (Bird), Fish (Fish), and Insect (Inse) for GFP-RY, GFP-SW, and GFP-MK

Original	Prediction											
	GFP-RY				GFP-SW				GFP-MK			
	Mamm	Bird	Fish	Inse	Mamm	Bird	Fish	Inse	Mamm	Bird	Fish	Inse
Mamm	28	2	0	0	30	0	0	0	30	0	0	0
Bird	0	58	1	0	4	55	0	0	0	59	0	0
Fish	0	0	32	0	0	0	32	0	1	4	27	0
Inse	2	2	0	32	0	0	2	34	1	0	0	35

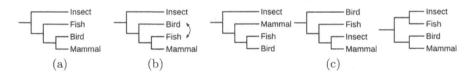

(a)   (b)   (c)

Fig. 2. Derived trees after applying the UPGMA on the representatives of classes by using (a) GRAFree, (b) both Euclidean and Canberra, and (c) Chebyshev. The arrow in Fig. (b) indicates that the position of bird and fish are interchangeable in different topologies derived from both Euclidean and Canberra.

For a particular window of $L = [50, 300]$, $\mathcal{F} = [50, 150]$, and $\alpha = [0.50, 1]$, the accuracy of our method are found to be $(84.36 \pm 4.28)\%$, $(83.42 \pm 3.15)\%$, and $(84.8 \pm 2.91)\%$ for GFP-RY, GFP-SW, and GFP-MK, respectively. So it can be noticed that the proposed technique is very sensitive to the parameter values.

To derive the phylogenetic relationships among the selected classes, we consider all (here five) representatives of each class. We apply our proposed distance measure to derive the pairwise distances followed by the UPGMA. For a given value of L, $\mathcal{F}$, and α the trees are generated for three different cases (GFP-RY, GFP-SW, and GFP-MK). It is observed that for both GFP-RY, GFP-SW, and GFP-MK all the representatives of a particular class placed under a same clade. It is also observed that the interclass relationships of the three trees carry the same topology (Fig. 2) which shows that the GRAFree is robust in deriving the phylogeny among different classes.

Table 2. Time and space complexity to compute distance between two sequences

Methods	Time complexity		Space complexity
	Deriving features	Computing distance	
GRAFree	$\mathcal{O}(M - L + 1)$	$\mathcal{O}((M - L + 1)\mathcal{F})$	$\mathcal{O}(\mathcal{F})$
Euclidean	$\mathcal{O}(k(M - k + 1))$	$\mathcal{O}(4^k)$	$\mathcal{O}(4^k)$
Canberra	$\mathcal{O}(k(M - k + 1))$	$\mathcal{O}(4^k)$	$\mathcal{O}(4^k)$
Chebyshev	$\mathcal{O}(k(M - k + 1))$	$\mathcal{O}(4^k)$	$\mathcal{O}(4^k)$

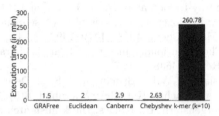

Fig. 3. The execution time for different methods. All the methods are executed in the same system. System configuration: 16 GB RAM, Intel Core i5 processor.

3.3 Complexity Analysis

We have derived both time and space complexities of GRAFree and the other reference methods. It is found that the GRAFree is most time and space economic method among all the reference methods (refer to Table 2). It can also be noticed that the execution time of GRAFree is less than all the other methods (Fig. 3)

4 Conclusion

We have proposed a $5\mathcal{F}$-dimensional feature space and a new metric for classifying the species using large scale genomic features in the method GRAFree. GRAFree uses the graphical representation of the genome. In this study we have selected three graphical representations of a genome considering residues independently. This method can classify the species based on their class (taxonomy rank). The selection of the value of the parameters used in the GRAfree needs further study. We observe presence of evolutionary traits among the selected classes in the proposed feature descriptor extracted from the whole mitochondrial sequences. These exhibit the effectiveness of the proposed feature representation along with the metric for measuring the pairwise distances of species.

References

1. Bernard, G., Ragan, M.A., Chan, C.X.: Recapitulating phylogenies using k-mers: from trees to networks. F1000Research **5**, 2789 (2016)
2. Bernt, M., Braband, A., Schierwater, B., Stadler, P.F.: Genetic aspects of mitochondrial genome evolution. Mol. Phylogenet. Evol. **69**(2), 328–338 (2013)

3. Bourque, G., Pevzner, P.A.: Genome-scale evolution: reconstructing gene orders in the ancestral species. Genome Res. **12**, 26–36 (2002)
4. Eyre-Walker, A., Awadalla, P.: Does human mtDNA recombine? J. Mol. Evol. **53**(4), 430–435 (2001)
5. Fulton, T.L., Wagner, S.M., Fisher, C., Shapiro, B.: Nuclear DNA from the extinct Passenger Pigeon (Ectopistes migratorius) confirms a single origin of New World pigeons. Ann. Anat. - Anatomischer Anzeiger **194**(1), 52–57 (2012). Special Issue: Ancient DNA
6. Gao, Y., Luo, L.: Genome-based phylogeny of dsDNA viruses by a novel alignment-free method. Gene **492**(1), 309–314 (2012)
7. Gates, M.: A simple way to look at DNA. J. Theor. Biol. **119**(3), 319–328 (1986)
8. Huang, Y., Wang, T.: Phylogenetic analysis of DNA sequences with a novel characteristic vector. J. Math. Chem. **49**(8), 1479–1492 (2011)
9. Kumar, V., et al.: The evolutionary history of bears is characterized by gene flow across species. Sci. Rep. **7**, 46487 (2017)
10. Langille, M.G.I., Hsiao, W.W.L., Brinkman, F.S.L.: Detecting genomic islands using bioinformatics approaches. Nat. Rev. Microbiol. **8**(5), 373–382 (2010)
11. Leimeister, C.A., Morgenstern, B.: Kmacs: the k-mismatch average common substring approach to alignment-free sequence comparison. Bioinformatics **30**(14), 2000–2008 (2014)
12. Leong, P., Morgenthaler, S.: Random walk and gap plots of DNA sequences. Bioinformatics **11**(5), 503–507 (1995)
13. Moret, B.M.E.: Phylogenetic analysis of whole genomes. In: Chen, J., Wang, J., Zelikovsky, A. (eds.) ISBRA 2011. LNCS, vol. 6674, pp. 4–7. Springer, Heidelberg (2011). https://doi.org/10.1007/978-3-642-21260-4_3
14. Nandy, A.: A new graphical representation and analysis of DNA sequence structure: I. Methodology and application to globin genes. Curr. Sci. **66**(4), 309–314 (1994)
15. Nandy, A., Harle, M., Basak, S.C.: Mathematical descriptors of DNA sequences: development and applications. ARKIVOC **2006**(9), 211–238 (2006)
16. Randić, M., Novič, M., Plavšić, D.: Milestones in graphical bioinformatics. Int. J. Quantum Chem. **113**(22), 2413–2446 (2013)
17. Ren, J., et al.: Alignment-free sequence analysis and applications. Ann. Rev. Biomed. Data Sci. **1**, 93–114 (2018)
18. Shi, X., Tian, P., Lin, R., Huang, D., Wang, J.: Characterization of the complete mitochondrial genome sequence of the globose head whiptail cetonurus globiceps (gadiformes: macrouridae) and its phylogenetic analysis. PLOS One **11**(4), 688–704 (2016)
19. Sneath, P.H.A., Sokal, R.R.: Numerical Taxonomy. W. H. Freeman and Company, San Francisco (1973)
20. Song, S.N., Tang, P., Wei, S.J., Chen, X.X.: Comparative and phylogenetic analysis of the mitochondrial genomes in basal hymenopterans. Sci. Rep. **6**, 20972 (2016)
21. Xie, G.S., Jin, X.B., Yang, C., Pu, J., Mo, Z.: Graphical representation and similarity analysis of DNA sequences based on trigonometric functions. Acta Biotheoretica **66**(2), 113–133 (2018)
22. Zhang, W., Zhang, M.: Complete mitochondrial genomes reveal phylogeny relationship and evolutionary history of the family Felidae. Genet. Mol. Res. **12**, 3256–3262 (2013)
23. Zhao, L., Gao, T., Lu, W.: Complete mitochondrial DNA sequence of the endangered fish (Bahaba taipingensis): mitogenome characterization and phylogenetic implications. ZooKeys **546**, 181 (2015)

Density-Based Clustering of Functionally Similar Genes Using Biological Knowledge

Namrata Pant and Sushmita Paul[(✉)]

Department of Bioscience and Bioengineering,
Indian Institute of Technology, Jodhpur, India
{pant.1,sushmitapaul}@iitj.ac.in

Abstract. Clustering is used to identify natural groups present in the data. It has been applied widely for analyzing gene expression data to discover gene clusters that might be involved in same biological processes. This information is very important for analyzing data of fatal diseases like cancers and identifying potential diagnostic and prognostic markers. Existing clustering methods used in this regard are computationally efficient, but do not always produce biologically meaningful results. Additionally, they have one or the other shortcomings; either they are not able to deal with arbitrary-shaped clusters, require number of clusters to be specified previously or are not efficient in dealing with noise present in biological data. In this study, a new density-based clustering method specific for gene expression data is introduced that overcomes the above shortcomings and produces biologically enriched clusters of functionally similar genes by incorporating biological information from Gene Ontology (GO). The proposed method integrates the GO semantic similarity information and the correlation information between the genes for obtaining clusters. The clusters are further validated for their biological relevance using Disease Ontology, KEGG Pathway enrichment and protein-protein interaction network analysis.

Keywords: Clustering · Gene expression · Cancer biomarkers

1 Introduction

Unsupervised learning techniques have made it possible to identify ingenuous groups present in a data. One such technique is clustering, that groups data in such a way that the data points within a group are more closely related to each other than the data points in two different groups [3]. Clustering is useful for any discipline that involves analysis of multi-variate data, for example, data mining, image analysis and bioinformatics [9]. Technological advancements in the field of bioinformatics and computational biology has lead to generation of enormous data and growing need for clustering algorithms that can deal with such high-dimensional data [11].

© Springer Nature Switzerland AG 2019
B. Deka et al. (Eds.): PReMI 2019, LNCS 11942, pp. 113–121, 2019.
https://doi.org/10.1007/978-3-030-34872-4_13

Cancer continues to be the leading cause of death worldwide. About 1 in every 6 deaths is due to cancer [1]. The late diagnosis of the disease and low survival rate of the patients with cancer is one of the major reason of this alarming death rate. This can be overcome by developing efficient prognostic methods by identifying certain prognostic molecular markers for different types of cancers. Advancements in bioinformatics and sequencing technologies have made access to genomic data easier. High-throughput RNA sequencing data has been used to identify potential biomarkers for different types of cancers [6]. Identification of set of biomarkers for a particular cancer type from the entire RNA-seq data requires efficient clustering method that can extract the groups of functionally related genes from the data.

Various clustering methods have been used for this purpose. The most commonly used methods are partition based clustering methods like k- means. K-means is fast but requires number of clusters to be specified beforehand and is also not efficient in clustering arbitrary shaped clusters. Density-based clustering methods like DBSCAN [4] is another very popular algorithm for clustering. It groups together dense regions of the data while leaving out noise as region of low density. It has overcome the limitations of k-means, but still has some shortcomings like dependency on the direction in which the data is processed. Also, the determination of dense regions depend entirely upon the distance measure used without involvement of any biological information.

In this study, a new density-based method for clustering specifically developed for gene expression data is proposed. In this method, Gene-Ontology based semantic similarity alongwith correlation between the genes have been used for clustering. A unique unsupervised feature selection step has been used to filterout irrelevant information and noise from the data. This incorporation of biological knowledge in the gene expression data has resulted in generation of highly biologically relevant clusters. The method is tested on four RNA-seq datasets of cervical cancer, stomach cancer, ovarian cancer and breast cancer. It is also compared with commonly used clustering methods like Hard c-means, Robust-Rough Fuzzy C Means (rRFCM) [8], DBSCAN [4] and Self Organizing Maps (SOM) [10], and is shown to perform better in terms of two cluster validity indices, Dunn and beta index as well as biological pathway enrichment analysis.

2   Materials and Method

The various datasets used in the study as well as the proposed method is described in this section.

2.1   Data Sets Used

The study is carried out on four cancer datasets namely cervical cancer, breast cancer, ovarian cancer and stomach cancer. The RNA seq datasets are down-

---

[1] https://www.who.int/news-room/fact-sheets/detail/cancer.

loaded from The Cancer Genome Atlast (TCGA)[2] in the form of matrix containing columns as patient IDs (samples) and rows as genes (features). The genes with more than 60% zero expression are removed from the data. The detailed information about the datasets after filtering genes is as follows:

1. **mRNA TCGA CESC exp HiSeqV2(2015-02-24):** Cervical Cancer data having 309 samples and 17865 features

2. **mRNA Broad GDCA Firehose RNASeq(2015-04-02):** Breast Cancer data having 511 samples and 12233 features

3. **mRNA TCGA Ovarian Serous Cystadenocarcinoma RNASeq(2011-06-29):** Ovarian cancer data having 474 samples and 17578 features

4. **mRNA TCGA STAD HiSeqV2(2018):** Stomach cancer data having 440 samples and 18292 features.

2.2   Method

The proposed method of density-based gene clustering is described in this section. The proposed approach judiciously integrate correlation information among various gene expression and their biological similarity. Next, correlation and GO semantic similarity matrix are described. Later, their integration in the proposed method is described.

Correlation and GO Semantic Similarity Matrix: Both correlation as well as GO semantic similarity have been used as a similarity measure for clustering. A Pearson correlation matrix for all the genes present in the dataset is generated.

Gene Ontology (GO) is used as biological vocabulary to estimate the functional similarity of gene products. GO further comprises of three orthogonal ontologies, which are, biological process (BP), molecular function (MF) and cellular component (CC). In the proposed method, BP was taken into account to obtain semantic similarity matrix using GOSemSim package [7]. It is calculated using Wang method [2] which uses the topology of GO graph structure to calculate semantic similarity.

Integrated Similarity Measure: Having obtained the correlation (Cor) and semantic similarity (SS) matrices, in the next step, these two matrices are integrated using the following relation:

$$Int(x_i, x_j) = \alpha * Cor(x_i, x_j) + (1 - \alpha) * SS(x_i, x_j), \text{ where } 0.1 \leq \alpha \leq 0.95 \quad (1)$$

$Int(x_i, x_j)$ is the integrated similarity, $Cor(x_i, x_j)$ is the correlation and $SS(x_i, j_i)$ is the semantic similarity between two genes x_i and x_j.

---

[2] https://tcga-data.nci.nih.gov.

Dense Region Selection: In this step, the genes are scored on the basis of their integrated similarity with other genes. The high scoring genes are regarded as the region of high density and will be considered for clustering in forthcoming steps.

1. Similarity score S between genes x_i and x_j is computed as follows:

$$S(x_i, x_j) = \begin{cases} 1, \text{if } Int(x_i, j_i) > \delta \\ 0, \text{otherwise} \end{cases} \quad \text{where } 0.6 \leq \delta \leq 0.9 \quad (2)$$

2. The total similarity score N for each gene is computed as follows:

$$N(x_i) = \sum_{j=1}^{n} S(x_i, j_i) \quad (3)$$

3. A threshold score T for the genes is computed as follows:

$$T = \frac{\sum_{i=1}^{n} N_i}{n} \quad (4)$$

The genes having score less that T are regarded as noise or irrelevant features are are not considered further for clustering.

Clustering of Dense Region: The genes of dense region are sorted in descending order of N such that; $N(x_i) > N(x_j).... > N(x_n)$. Genes are picked one by one starting from the highest scoring gene, x_i.

1. If $Int(x_i, x_j) > \delta_2$, where $0.6 \leq \delta_2 \leq 0.9$, then $x_i, x_j \in X_i$, where X_i is the group of genes highly similar to x_i
2. If $|X_i| > 50$, $X_i = C_i$, where C_i is a cluster with gene x_i as the centroid of the cluster. The genes in C_i will not be further assigned to any other cluster
3. The next gene is picked and Steps 1 and 2 are repeated until all the genes are visited once.

Clusters for all the values of parameters α, δ and δ_2 are generated and the optimal value of the parameters is decided on the basis of Dunn index.

2.3   Experimental Setup

In this section, the experimental testbed of the method is described. All the computational operations are carried out in an Intel core i7 processor system with 64 gb RAM. R version 3.5.1 is used in linux (64-bit) platform. The GO semantic similarity matrix is obtained from GOSemSim R package [7].

3   Results and Discussion

In this section, the experimental results are explained.

3.1   Selection of Optimal Values of δ, δ_2 and α

The performance of proposed method is sensitive to α, δ and δ_2 parameters. Therefore, Dunn Index (D) has been used to select optimal value of these parameters. It is defined as:

$$D = \min_i \left\{ \min_{i \neq j} \left\{ \frac{d(C_i, C_j)}{max_k D(C_k)} \right\} \right\} \tag{5}$$

where $1 \leq i, j, k \leq n$. A higher Dunn Index indicates better clustering. The optimal values of the parameters alongwith the data are described in Table 1.

3.2   Performance Analysis of the Proposed Method

Two commonly used cluster validity indices, Dunn Index and Beta Index are considered for comparing the performance of the proposed method with other clustering algorithms. The proposed method is found to perform the best in 3 out of 4 cancer dataset with respect to both the indices, as shown in Fig. 1.

Table 1. Clustering results for different cancer types with the optimal parameters

Cancer type	α	δ	δ_2	No. of genes	No. of clusters
Cervical cancer	0.75	0.8	0.7	124	2
Breast cancer	0.95	0.8	0.7	684	5
Ovarian cancer	0.6	0.7	0.7	6920	2
Stomach cancer	0.4	0.9	0.7	85	2

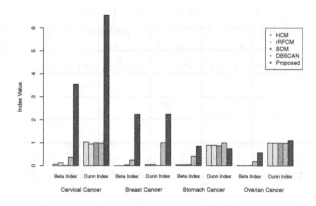

Fig. 1. Comparison of the performance of different clustering methods

3.3   Biological Relevance of Obtained Clusters

The obtained clusters from different datasets are further analyzed to identify their biological role in different cancers. Pathway and disease enrichment analysis is done to find the biological interpretation of the cluster genes. Also, network analysis is performed to find the connectivity between the genes in the clusters.

Pathway Enrichment Analysis: KEGG pathway enrichment analysis is done with the help of clusterProfiler R package [12] to observe the significance of the obtained clusters in various cancers. Top five significant terms from each dataset along-with their p-values are listed in Table 2. It is seen that important terms related to different cancer types are obtained, for example, p53 signalling pathway is known to be disregulated in various cancer types including breast [5] and ovarian cancer [14], which is observed in the result as a significantly associated pathway. Other important pathways specific to different cancer types are seen, for example estrogen signalling pathway in ovarian cancer [1].

Table 2. KEGG pathway enrichment terms obtained from the proposed method

Cancer type	ID	Description	p-value
Cervical cancer	hsa04666	Fc gamma R-mediated phagocytosis	5.48E−05
	hsa05203	Viral carcinogenesis	5.48E−05
	hsa04024	cAMP signaling pathway	3.70E−04
Breast cancer	hsa04514	Cell adhesion molecules (CAMs)	3.10E−20
	hsa04115	p53 signaling pathway	5.05E−05
	hsa04151	PI3K-Akt signaling pathway	1.37E−04
Ovarian cancer	hsa04012	ErbB signaling pathway	2.52E-19
	hsa04115	p53 signaling pathway	3.18E−10
	hsa04915	Estrogen signaling pathway	6.66E−07
Stomach cancer	hsa04024	cAMP signaling pathway	5.32E−06
	hsa04015	Rap1 signaling pathway	4.41E−05
	hsa04022	cGMP-PKG signaling pathway	1.06E−03

Quantification of Biological Relevance: For a quantitative comparison of the pathway terms obtained, annotation Ratio and KEGG Pathway Enrichment Score (KPES) of the gene clusters is calculated. It is defined as:

$$KPES = \frac{c}{n} \sum_{i=1}^{n} -log_2(\text{p-value}) \qquad (6)$$

where c is the total number of enriched clusters, n is the number of enrichment terms obtained. Another quantitative measure annotation ratio (AR) of genes in a cluster is calculated as follows:

$$AR = \sum_{j=1}^{c} \sum_{i=1}^{G} \frac{g}{G} \qquad (7)$$

where c is the total number of clusters, G is the total number of genes in one cluster and g is the gene count associated with an enrichment term. The results are compared with other clustering methods and are reported in Table 3. High annotation ratio suggests that a large fraction of genes in the given set of genes are associated with the obtained pathway terms, whereas high KPES suggests high significance of the pathway terms obtained from the set of genes. It is clear from the table that in terms of both of these quantitative measures, the proposed method has performed better in most of the cases.

Table 3. Comparison of AR and KPES of gene clusters by different methods

Cancer type		HCM	rRFCM	SOM	DBSCAN	Proposed
Cervical cancer	Annotation Ratio	2.06	2.76	1.6	2.11	**3.66**
	KPES	28.31	28.46	23.15	28.46	**48.30**
Breast cancer	Annotation Ratio	2.09	2.67	1.46	1.46	**5.27**
	KPES	50.32	32.87	20.17	22.26	**105.10**
Ovarian cancer	Annotation Ratio	1.67	1.75	1.67	1.82	**2.80**
	KPES	22.58	25.9	23.63	16.5	**79.05**
Stomach cancer	Annotation Ratio	2.02	1.84	1.11	2.13	**3.74**
	KPES	**37.58**	34.43	18.14	28.67	19.67

Disease Ontology Enrichment Analysis: Further, Disease Ontology (DO) enrichment analysis of the cluster genes is done using R package DOSE [13] to identify the disease terms that are significantly associated with the given set of genes. The significant DO terms along with the p-values are reported in Table 4. It is clear from the results that in breast cancer and ovarian cancer datasets highly relevant terms are obtained. The results indicate that the proposed method cannot only identify compact clusters, but also biologically relevant clusters.

Biological Network Analysis: To observe the connectivity of the gene clusters obtained by the proposed method, protein-protein interaction (PPI) network analysis using STRING[3] is done. It is found that networks generated from all

---
[3] https://string-db.org/.

Table 4. Disease Ontology Terms obtained from the proposed method

Cancer type	ID	Description	p-value
Cervical cancer	DOID:0060095	Uterine benign neoplasm	9.87E−04
	DOID:0060086	Female reproductive organ benign neoplasm	1.13E−03
	DOID:18	Urinary system cancer	1.68E−03
Breast cancer	DOID:201	Connective tissue cancer	8.84E−09
	DOID:3355	Fibrosarcoma	2.16E−06
	DOID:3459	Breast carcinoma	1.19E−04
Ovarian cancer	DOID:5683	Hereditary breast ovarian cancer	7.54E−23
	DOID:2394	Ovarian cancer	4.81E−22
	DOID:2151	Malignant ovarian surface epithelial-stromal neoplasm	1.67E−21
Stomach cancer	DOID:104	Bacterial infectious disease	3.94E−07
	DOID:0050338	Primary bacterial infectious disease	8.42E−07
	DOID:4074	Pancreas adenocarcinoma	8.00E−04

the clusters are highly significant having PPI enrichment p-value < 0.05. Some of them are illustrated in Fig. 2. These networks are generated using one of the clusters of cervical cancer and breast cancer studies. It is seen that the genes in the network are densely connected, implying their functional relatedness and involvement in the same biological process.

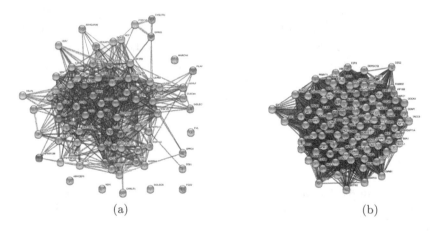

(a)   (b)

Fig. 2. PPINs for most significant clusters obtained by the proposed method for (a) Cervical Cancer (p-value $< 1.0e^{-16}$) and (b) Breast Cancer (p-value $< 1.0e^{-16}$)

3.4   Conclusion

The proposed method is tested on four cancer data sets and has performed better as compared to commonly used clustering methods. It is able to handle high-dimensional as well as noisy data efficiently by its unique unsupervised feature selection step. High biological relevance of the clusters is observed from the pathway enrichment and network analysis results. The gene clusters identified using this method may provide new insights for prognosis and treatment of various cancer types.

Acknowledgement. This work was partially supported by the Department of Science and Technology, Government of India, New Delhi (grant no. ECR/2016/001917).

References

1. Andersen, C.L., et al.: Active estrogen receptor-alpha signaling in ovarian cancer models and clinical specimens. Clin. Cancer Res. **23**, 3802–3812 (2017)
2. Chen, C.F., Wang, J.Z., Yu, P.S., Payattakool, R., Du, Z.: A new method to measure the semantic similarity of GO terms. Bioinformatics **23**(10), 1274–1281 (2007)
3. Daxin, J., Chun, T., Aidong, Z.: Cluster analysis for gene expression data: a survey. IEEE Trans. Knowl. Data Eng. **16**(11), 1370–1386 (2004)
4. Ester, M., Kriegel, H.P., Sander, J., Xu, X.: A density-based algorithm for discovering clusters a density-based algorithm for discovering clusters in large spatial databases with noise. In: Proceedings of the Second International Conference on Knowledge Discovery and Data Mining, KDD 1996, pp. 226–231. AAAI Press (1996)
5. Gasco, M., Shami, S., Crook, T.: The p53 pathway in breast cancer. Breast Cancer Res. **4**, 70 (2002)
6. Hassanein, M., Callison, J.C., Callaway-Lane, C., Aldrich, M.C., Grogan, E.L., Massion, P.P.: The state of molecular biomarkers for the early detection of lung cancer. Cancer Prev. Res. **5**(8), 992–1006 (2012)
7. Li, F., Yu, G., Wang, S., Bo, X., Wu, Y., Qin, Y.: GOSemSim: an R package for measuring semantic similarity among GO terms and gene products. Bioinformatics **26**(7), 976–978 (2010)
8. Maji, P., Paul, S.: Rough-fuzzy clustering for grouping functionally similar genes from microarray data. IEEE/ACM Trans. Comput. Biol. Bioinform. **10**(2), 286–299 (2013)
9. Naegle, K.M., Jimenez, N., Sloutsky, R., Swamidass, S.J.: Accounting for noise when clustering biological data. Brief. Bioinform. **14**, 423–436 (2012)
10. Tamayo, P., et al.: Interpreting patterns of gene expression with self-organizing maps: Methods and application to hematopoietic differentiation. Proc. Natl. Acad. Sci. **96**(6), 2907–2912 (1999)
11. Wang, J., Li, M., Chen, J., Pan, Y.: A fast hierarchical clustering algorithm for functional modules discovery in protein interaction networks. IEEE/ACM Trans. Comput. Biol. Bioinform. **8**(3), 607–620 (2011)
12. Yu, G., Wang, L.G., Han, Y., He, Q.Y.: ClusterProfiler: an R package for comparing biological themes among gene clusters. OMICS: J. Integr. Biol. **16**(5), 284–287 (2012)
13. Yu, G., Yan, G.R., Wang, L.G., He, Q.Y.: DOSE: an R/Bioconductor package for disease ontology semantic and enrichment analysis. Bioinformatics **31**(4), 608–609 (2014)
14. Zhang, Y., Cao, L., Nguyen, D., Lu, H.: Tp53 mutations in epithelial ovarian cancer. Transl. Cancer Res. **5**(6), 650 (2016)

Impact of the Continuous Evolution of Gene Ontology on Similarity Measures

Madhusudan Paul[1,2] , Ashish Anand[1(✉)] , and Saptarshi Pyne[1]

[1] Department of Computer Science and Engineering,
IIT Guwahati, Guwahati, India
anand.ashish@iitg.ac.in
[2] Department of Computer and System Sciences,
Visva-Bharati, Santiniketan, India

Abstract. Gene Ontology (GO) is a taxonomy of biological terms related to the properties of genes and gene products. It can be used to define a similarity measure between two gene products and assign a confidence score to protein-protein interactions (PPIs). GO is being evolved regularly by the addition/deletion/merging of terms. However, there is no study which evaluates the robustness of a particular similarity measure over the evolution of GO. By robustness of a similarity measure, we mean it should either improve or keep its performance similar over the evolution of GO. In this paper, we systematically study the same for the task of scoring confidence of PPIs using GO-based similarity measures. We observe that the performance of similarity measures gets affected due to the regular updates of GO. We find that similarity measures are not robust in all conditions, rather they keep their performance quite similar over the evolution of GO in certain conditions.

Keywords: Gene Ontology (GO) · Protein-protein interaction (PPI) · Similarity measures

1 Introduction

Gene Ontology (GO) [1] is a taxonomy of biological terms to represent the properties of genes and/or gene products (e.g., proteins)[1]. It is organized as a DAG (directed acyclic graph) to describe the relationship among the terms. Gene products are annotated to pertinent GO terms through annotation corpora. There are three GOs: biological process (BP), cellular component (CC), and molecular function (MF). Lord *et al.* [7] did the first pioneering work by utilizing the ontology-based semantic similarity measure (SSM) in the field of genomics. SSM is a quantitative function, $SSM(t_1, t_2)$, that measures the closeness between two terms t_1 and t_2 based on their semantic representations in a given ontology. Subsequently, a variety of GO-based SSMs have been proposed and successfully applied to different genomics applications [4,9].

---

[1] Hereafter we refer to gene products only.

B. Deka et al. (Eds.): PReMI 2019, LNCS 11942, pp. 122–129, 2019.
https://doi.org/10.1007/978-3-030-34872-4_14

The high similarity score between two proteins indicates that either they are annotated with similar cellular components (if CC-based GO is used), or with similar biological processes (if BP-based GO is used). This gives an indirect evidence that the two proteins are likely to be interacting compare to other pairs, which has a low similarity score. Hence several studies have used GO-based SSM between two gene products (involved in a PPI) as a confidence score of the interaction. However, GO is being updated regularly with the addition, deletion, and merging of terms along with their annotations. This may affect similarity score between a protein-pair calculated over different versions of the ontology. However, to the best of our knowledge, there is no study which systematically studies the effect of the evolution of GO over SSMs. In this paper, we systematically study whether changes in GO affect the performance of similarity measures. In particular, we focus on GO-based SSMs. Further, we compare multiple GO-based SSMs under this setting for the task of scoring confidence of PPIs.

Section 2 briefly discusses the necessary backgrounds and terminologies. In Sect. 3, we discuss datasets and different GO versions used along with evaluation metrics. Results are discussed and analyzed in Sect. 4.

2   Background

Semantic Similarity Measure (SSM). SSMs can be categorized mainly into two approaches: *edge- and node-based* [10]. The edge-based approach mainly considers the shared paths between two ontology terms and does not account annotation information of terms. Node-based SSMs compute the similarity between two terms by comparing their properties, common ancestors, and their descendants. This approach is less sensitive to the topological structure of the ontology but more sensitive to change in annotations. SSMs such as [2,14] try to combine both node- and edge-based approaches and are commonly referred to as the *hybrid* approach. Few methods, such as TCSS [4], are developed based on the complex structure of GO DAG.

SSMs are defined for two individual terms, but a protein is annotated with a set of terms. So if two proteins p_1 and p_2 are annotated with a set of terms S and T, respectively, then $SSM(p_1, p_2)$ is calculated as $SSM(S, T)$ which requires combining SSM between individual term-pairs. Generally, three types of strategies are used in the literature: *maximum (MAX)*, *average (Avg)*, and *best-match average (BMA)*. In MAX and Avg strategies, the similarity between S and T is calculated as the maximum and average of the set $S \times T$, respectively. SSMs between two sets of terms can be treated as a matrix. BMA is defined as the average of all maximum similarity scores on each row and column of the matrix.

3   Experimental Design

GOs and SSMs. We consider BP and CC ontologies along with MAX and BMA in the evaluation. These ontologies and strategies are the most relevant for scoring confidence of PPIs [8]. We exclude electronically inferred annotations

(IEA) as they are not verified by human experts. Further, we consider only those PPIs where both the interacting proteins are annotated to at least one GO term other than the root.

We select five different Bioconductor versions of GO and corresponding annotation corpora: 3.0 (2014-09-13), 3.1 (2015-03-13), 3.2 (2015-09-19), 3.3 (2016-03-05), and 3.4 (2016-09-21). We consider six state-of-the-art SSMs proposed by Resnik [12], Lin [6], Schlicker *et al.* [13], Jiang and Conrath [5], Wang *et al.* [14], and Jain and Bader [4], referred to as *Resnik, Lin, Rel, Jiang, Wang*, and *TCSS*, respectively, in the rest of the paper. Resnik and TCSS with MAX strategy have been considered to be the best SSMs for scoring confidence of PPIs by several studies [4,9]. We also consider RDS, RNS, and RES, proposed recently by Paul and Anand [8]. The selected nine SSMs encompass all types of SSMs, as discussed in Sect. 2.

Datasets. We utilize the core subsets of the yeast PPIs from the DIP database (Database of Interacting Proteins) [15] downloaded on 29.10.2015 as positive instances. As done in [4], an equal number of negative PPI instances are generated independently by randomly choosing protein pairs annotated in BP and CC and are not present in the iRefWeb database [11], a combined database of all known PPIs, accessed on 27.11.2015.

Proteins involved in a pathway are more likely to interact among themselves and likely to be annotated to the same or similar GO terms and thus should show high similarity scores. A set of 11 yeast (S. cerevisiae) KEGG pathways is selected as in [8]. During the selection of pathways, the authors of [8] try to maintain a trade-off between functional diversity and computational time required for the experiment.*

Evaluation Metrics. A similarity measure can classify a set of PPIs into two groups: positives and negatives, for a given cutoff similarity score. Hence an SSM can be treated as a binary classifier. We utilize the area under the ROC curve (AUC) as an evaluation metric for binary classifiers.

For each KEGG pathway, an *intra-set average similarity* is computed as the average of all pairwise similarities of proteins within the pathway. An *inter-set average similarity* for every two pathways is computed as the average of all pairwise cross-similarities of proteins between the two pathways. A discriminating power (DP) of a pathway is defined as the ratio between *intra-set average similarity* and the average of all *inter-set average similarities* between that pathway and other pathways as in [3]. Thus the DP quantifies the ability of an SSM to distinguish among various functionally different sets of proteins (e.g., KEGG pathways).

4   Results and Discussion

ROC curve analysis: Table 1 summarizes AUC of the top five SSMs for the different versions of BP ontology. Insignificant change in AUC values for all SSMs indicates that the evolution of GO has no impact on their classification performance. This can be explained easily. An AUC of 1 implies a perfect classifier, while an area of 0.5 indicates a random classifier. So, the practical range of AUC for a reasonably good classifier is very limited (Generally, [0.7, 1]). Unless the majority of the PPIs get affected (due to the changes in GO), it is unexpected to observe high variability in AUCs over the different versions of GO. By affected we mean for a given PPI, an SSM produces different similarity scores for different GO versions. In fact, the majority of PPIs (in the PPI dataset) does not get affected significantly due to the changes in GO.

Table 1. The area under the curves (AUCs) of SSMs for the different GO-BP versions. The best AUC for each strategy is shown in bold.

SSM	Str.	Ver3.0	Ver3.1	Ver3.2	Ver3.3	Ver3.4	Mean
RNS	MAX	0.906	0.908	0.906	0.911	0.911	0.908
	BMA	0.887	0.891	0.887	0.894	0.893	0.890
RES	MAX	0.901	0.903	0.901	0.906	0.904	0.903
	BMA	**0.889**	**0.893**	**0.892**	**0.896**	**0.896**	**0.893**
TCSS	MAX	0.907	0.907	0.908	0.911	0.910	0.909
	BMA	0.859	0.861	0.857	0.863	0.868	0.862
Resnik	MAX	0.908	0.908	0.907	0.912	0.910	0.909
	BMA	0.879	0.879	0.878	0.882	0.882	0.880
Rel	MAX	**0.914**	**0.914**	**0.913**	**0.919**	**0.917**	**0.915**
	BMA	0.882	0.883	0.881	0.885	0.884	0.883

To see the closer picture of the impact, we find those PPIs whose similarity scores change over the versions of GO. For each SSM, we select the common PPIs (more than 99% of PPIs are common) among the five GO versions. For each of the selected PPIs, the standard deviation of the five similarity scores corresponding to the five GO versions is calculated. Then we sort the PPIs according to their standard deviation (in descending order) and select the top 10% PPIs. The selected PPIs are the most affected 10% PPIs due to the changes in GO. An equal number of negative PPIs are selected from the already generated negative PPIs for the corresponding SSM. Finally, AUC is computed for the selected positive and negative PPIs for each GO version. The resultant AUCs of two best performing SSMs for the different versions of GO-BP are demonstrated in Table 2.

Now, the performance variations of SSMs among GO versions are quite visible. For RES, we observe relative changes of approximately 8% and 4% while using MAX and BMA strategies respectively. Similarly, for TCSS, relative changes of approximately 6% and 7% while using MAX and BMA strategies. These changes are observed between versions 3.0 and 3.4. Similar observations are made for the other SSMs and using other ontologies. We also observe that across all measures, the overall variability is higher in CC than BP.

Table 2. The area under the curves (AUCs) of two best performing SSMs for the different GO-BP versions with top 10% most affected PPIs.

SSM	Str.	Ver3.0	Ver3.1	Ver3.2	Ver3.3	Ver3.4	Mean
RES	MAX	0.859	0.887	0.890	0.937	0.928	0.900
	BMA	0.924	0.944	0.943	0.970	0.962	**0.949**
TCSS	MAX	0.887	0.901	0.912	0.964	0.947	**0.922**
	BMA	0.872	0.901	0.883	0.912	0.935	0.901

To find a general pattern of variability among SSMs, we repeat the aforementioned process for different cutoffs (100% to the top 10%) of affected PPIs. Here a cutoff of 100% implies that all PPIs are considered and hence, the majority of them have no change in their similarity score. The mean AUCs (of five GO versions) achieved by SSMs in increasing order of variability of PPIs are shown in Fig. 1.

SSMs with BMA strategy shows robustness compared to MAX strategy. Almost all SSMs with BMA strategy either improve or keep their performance similar from their initial performance as variability increases in both the ontologies. Particularly in BP, the improvement is more smooth and consistent. However, with MAX strategy, the performance is quite fluctuating, and the irregularity is more in CC. Therefore it seems that MAX strategy overestimates in many cases, especially in CC.

All SSMs exhibit higher robustness in BP than CC. If we examine the same for each SSM separately, we get further insights (See Figs. 2 and 3). With all data considered (100%), SSMs with MAX strategy gives better AUC in comparison with BMA. However, as variability increases (by removing PPIs having no changes over GO evolution), SSMs with BMA obtain higher AUCs. In TCSS, although BMA increases its performances continuously, it is unable to cross the performance of MAX, particularly in BP. In fact, the difference of performance between MAX and BMA of TCSS and Resnik is reducing as variability increases, and they show almost similar performances with very high variable PPIs (>50%).

RES-BMA continuously produces the highest AUCs as variability increases. In general, RES, RNS, and TCSS show comparatively high robustness. With the top 10% variable PPIs, the highest mean AUC is 0.949/0.957 (BP/CC) produced by RES-BMA while the second-highest mean AUC is 0.922/ 0.940 (BP/CC) produced by TCSS with MAX or BMA.

Set-discriminating power of KEGG pathways: For each GO versions and SSM, we calculate DP values of each pathway with respect to other 10 pathways. Then we take version-wise (GO) mean DP values. Table 3 shows the mean DP values of all the 11 pathways for each GO-BP version and SSM.

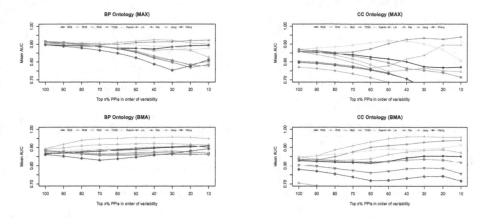

Fig. 1. The mean AUCs of five GO versions achieved by SSMs at different cutoffs of affected PPIs.

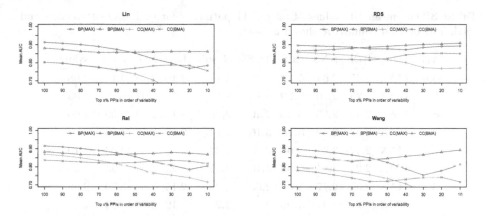

Fig. 2. The mean AUCs achieved by Lin, Rel, RDS, and Wang at different cutoffs of affected PPIs with the plotting of individual SSM.

The majority of SSMs produce quite similar DP values over the evolution of GO since less number of PPIs are affected due to the changes in GO. RES almost continuously produces higher DP values in both the ontologies, particularly, with BMA strategy. TCSS shows competitive performances in both the ontologies while Jiang achieves good DP values in BP only. The significant differences between MAX and BMA strategies, in both BP and CC simultaneously, are observed with RES, TCSS, and some extend with RNS only.

RES-BMA shows continuous and significant improvement over the evolution of GO. We can assume that the newer GO version represents more accurate and complete information than the older, and the robust SSMs should reflect that positively. RES-BMA almost continuously improves its DP value over the evolution of BP ontology (5.59, 5.76, 6.38, 6.58, and 6.50) except for the last

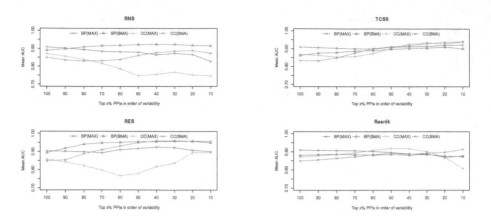

Fig. 3. The mean AUCs achieved by RNS, RES, TCSS, and Resnik at different cutoffs of affected PPIs with the plotting of individual SSM.

Table 3. The mean DP values of all the 11 pathways for each GO-BP version and SSM. The best DP values are shown in bold.

Ver.	Str.	RDS	RNS	RES	TCSS	Resnik	Lin	Rel	Jiang	Wang
3.0	MAX	2.45	2.64	**5.22**	2.46	2.24	2.16	2.38	4.17	2.08
	BMA	2.23	2.70	**5.59**	3.16	2.18	2.16	2.41	5.06	2.03
3.1	MAX	2.44	2.65	**5.39**	2.46	2.20	2.13	2.34	4.09	2.06
	BMA	2.23	2.70	**5.76**	3.16	2.14	2.12	2.36	4.98	2.00
3.2	MAX	2.47	2.78	**5.95**	2.48	2.32	2.23	2.48	4.33	2.10
	BMA	2.26	2.84	**6.38**	3.16	2.25	2.22	2.51	5.43	2.05
3.3	MAX	2.52	2.77	**5.65**	2.66	2.28	2.15	2.36	3.70	2.06
	BMA	2.24	2.94	**6.56**	2.76	2.22	2.19	2.48	4.94	2.01
3.4	MAX	2.49	2.74	**5.56**	2.67	2.28	2.14	2.35	3.73	2.06
	BMA	2.18	2.87	**6.50**	2.78	2.23	2.19	2.50	5.04	2.02

version (Ver. 3.4), whereas other SSMs keep their performances quite similar. In fact, the changes, particularly, in edges, between the two GO-BP versions (Ver. 3.3 to Ver. 3.4) are very less (+0.30%) in comparison with other versions (The avg. successive change is +2.91%). Hence the changes are reflected better way with RES-BMA than the others.

5   Conclusion

In this paper, we systematically study how similarity measures get affected due to the evolution of gene ontology for the task of scoring confidence of PPIs. We observe that the performance of each measure gets affected due to the regular updates of GO. All SSMs exhibit satisfactory robustness with BMA strategy in

BP ontology only. SSMs with MAX strategy have the tendency to overestimate, particularly in CC. Although, RES-BMA, TCSS-BMA and RNS-BMA exhibit comparatively good robustness, the changes in GO is reflected better way with RES-BMA than the others.

References

1. Ashburner, M., et al.: Gene ontology: tool for the unification of biology. Nature Genet. **25**(1), 25–29 (2000)
2. Bandyopadhyay, S., Mallick, K.: A new path based hybrid measure for gene ontology similarity. IEEE/ACM Trans. Comput. Biol. Bioinform. (TCBB) **11**(1), 116–127 (2014)
3. Benabderrahmane, S., Smail-Tabbone, M., Poch, O., Napoli, A., Devignes, M.D.: IntelliGO: a new vector-based semantic similarity measure including annotation origin. BMC Bioinform. **11**(1), 588 (2010)
4. Jain, S., Bader, G.D.: An improved method for scoring protein-protein interactions using semantic similarity within the gene ontology. BMC Bioinform. **11**(1), 562 (2010)
5. Jiang, J.J., Conrath, D.W.: Semantic similarity based on corpus statistics and lexical taxonomy. In: Proceedings of 10th International Conference on Research In Computational Linguistics, ROCLING 1997 (1997)
6. Lin, D.: An information-theoretic definition of similarity. In: Proceedings of the Fifteenth International Conference on Machine Learning, vol. 98, pp. 296–304. Morgan Kaufmann Publishers Inc., San Francisco (1998)
7. Lord, P.W., Stevens, R.D., Brass, A., Goble, C.A.: Investigating semantic similarity measures across the gene ontology: the relationship between sequence and annotation. Bioinformatics **19**(10), 1275–1283 (2003)
8. Paul, M., Anand, A.: A new family of similarity measures for scoring confidence of protein interactions using gene ontology, p. 459107. bioRxiv (2018)
9. Pesquita, C.: Semantic similarity in the gene ontology. In: Dessimoz, C., Škunca, N. (eds.) The Gene Ontology Handbook. MMB, vol. 1446, pp. 161–173. Springer, New York (2017). https://doi.org/10.1007/978-1-4939-3743-1_12
10. Pesquita, C., Faria, D., Falcao, A.O., Lord, P., Couto, F.M.: Semantic similarity in biomedical ontologies. PLoS Comput. Biol. **5**(7), e1000443 (2009)
11. Razick, S., Magklaras, G., Donaldson, I.M.: iRefIndex: a consolidated protein interaction database with provenance. BMC Bioinform. **9**(1), 1 (2008)
12. Resnik, P.: Using information content to evaluate semantic similarity in a taxonomy. In: Proceedings of the 14th International Joint Conference on Artificial Intelligence, pp. 448–453. Morgan Kaufmann Publishers Inc., San Francisco (1995)
13. Schlicker, A., Domingues, F.S., Rahnenführer, J., Lengauer, T.: A new measure for functional similarity of gene products based on gene ontology. BMC Bioinform. **7**(1), 302 (2006)
14. Wang, J.Z., Du, Z., Payattakool, R., Yu, P.S., Chen, C.F.: A new method to measure the semantic similarity of go terms. Bioinformatics **23**(10), 1274–1281 (2007)
15. Xenarios, I., Rice, D.W., Salwinski, L., Baron, M.K., Marcotte, E.M., Eisenberg, D.: DIP: the database of interacting proteins. Nucleic Acids Res. **28**(1), 289–291 (2000)

DEGnet: Identifying Differentially Expressed Genes Using Deep Neural Network from RNA-Seq Datasets

Tulika Kakati[1], Dhruba K. Bhattacharyya[1(✉)], and Jugal K. Kalita[2]

[1] Department of Computer Science and Engineering, Tezpur University, Tezpur 784028, Assam, India
tulika.kakati@gmail.com, dkb@tezu.ernet.in
[2] Department of Computer Science, University of Colorado, Colorado Springs, CO 80918, USA
jkalita@uccs.edu

Abstract. Differential expression (DE) analysis and identification of differentially expressed genes (DEGs) provide insights for discovery of therapeutic drugs and underlying mechanisms of disease. Statistical methods, such as DESeq2, edgeR, and limma-voom produce a number of false positives and false negatives and fail to differentiate between the DEGs as up-regulating (UR) and down-regulating (DR) genes linking them to disease progression. Machine learning (ML) including deep learning (DL) methods to identify DEGs from RNA-seq data face challenges due to smaller sample sizes (n) compared to number of genes (g). In this work, we propose a deep neural network (DNN) called DEGnet to predict the UR and DR genes from Parkinson's disease (PD) and breast cancer (BRCA) RNA-seq datasets. The accuracies we obtained from PD and BRCA were 100% and 87.5% respectively, higher than ML-based methods on the same datasets. However, to the best of our knowledge, we are the first to apply DNN on for classification of DEGs into UR and DR, and identify significant UR and DR genes that play role in progression of a disease. Experimental results show that DEGnet is a good performer and can be applied in other RNA-seq data, despite the $n \ll g$ issue.

Keywords: Deep neural network · RNA-seq · Parkinson's disease · Breast cancer

1 Introduction

Differential Expression (DE) analysis studies the variance of gene expressions across two cell conditions, such as control (or normal) and disease (or tumor). Genes with varied expressions across cell conditions have been implicated in a number of severe diseases. Therefore, DE analysis and identification of differentially expressed genes (DEGs) may provide insights into underlying mechanisms of disease and even into discovery of therapeutic drugs. Recent advances

© Springer Nature Switzerland AG 2019
B. Deka et al. (Eds.): PReMI 2019, LNCS 11942, pp. 130–138, 2019.
https://doi.org/10.1007/978-3-030-34872-4_15

in technologies such as next-generation sequencing have led to development of large-scale repositories of biological data, including gene expression datasets.

Recently developed statistical methods for DE analysis can be divided into two groups, parametric and non-parametric, depending upon whether the data distribution is considered a parameter. Log2 fold change (log2FC) measures the logarithmic scale in base 2 of the ratio of gene expression change in disease condition to the control condition [1]. A few methods such as, DESeq [2], DESeq2 [3], edgeR [4], and voom [5] compute variance (dispersion) in gene expression values. However, these statistical methods produce a high number of false positives and false negatives due to small biases incorporated in the estimation of dispersion for predicting DEGs from RNA-seq data. Here, we take three common methods, namely DESeq2, edgeR, and limma-voom to compare the effectiveness of our proposed model.

Later, with advances in Big Data and machine-learning (ML), ML-based DE analysis was introduced to identify DEGs [6,7], to learn from existing data and predict variations of gene expression patterns. However, application of deep learning models is a challenge for analysis of gene expression data due to smaller sample sizes (n) compared to number of genes (g), unlike image and other datasets found in usual deep learning application areas [8].

In this work, we propose a model based on deep learning which we call DEGnet to identify DEGs. The deep neural network learns from gene expressions in PD and BRCA datasets measured under control and disease conditions with log2FC change labels - 1 for up-regulation and 0 for down-regulation. The main motivation for this work is that the probability of predicting UR and DR genes using the baseline models, DESeq2, edgeR, and limma-voom from biologically validated test data based on the log2FC estimates is low. This is due to the fact the baseline methods produce high false positive rate and false negative rates due to the small biases incorporated in computing the dispersion across samples of RNA-seq data. We argue that the proposed model is generalized because it is trained on the consensus labels based on log2FC estimates of DESeq2, edgeR, and limma-voom. We also demonstrate that it predicts UR and DR genes from biologically validated test data with higher accuracy than the three baseline models. Further, we apply LR [9], KNC [10], SVM [11], GNB [12], DTC [13], and RFC [14] on PD and BRCA data and evaluate their performance in terms of accuracy, sensitivity, specificity, and precision. We found that DEGnet outperforms these traditional ML-based methods. The UR and DR genes are assessed for biological enrichment (GO enrichment and pathway analysis) using web-based tools in the ToppGene Suite [15].

Section 2 of the paper describes the datasets and the description of the strategy used by DEGnet. Section 3 gives experimental results in terms of statistical and biological validation and comparison of the performance of the proposed method with the DL based models. Finally, we conclude by presenting how the method can be developed further in the last section.

2   Method and Materials

The proposed method, DEGnet, runs on two phases. In the first phase, preprocessing, labelling and splitting of the data are done. The second phase consists of
training, fine-tuning, and testing. We use log2FC estimates of statistical models
(baseline models) DESeq2, edgeR, and limma-voom and prior knowledge from
the literature to label the datasets. The use of log2FC estimates and knowledge of prior gene regulation with a DNN enable the capture of the non-linear
patterns from biologically validated gene samples and improve the prediction
performance of our model in determining UR and DR genes. Figure 1a gives the
workflow of our proposed method. We use two datasets: PD (GSE68719) and
BRCA (TCGA) [16] (described in the Dataset subsection).

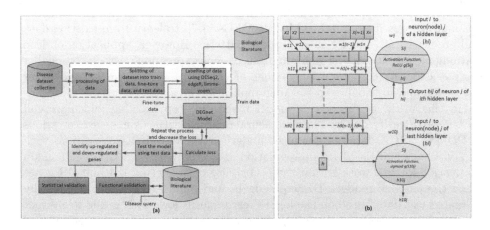

Fig. 1. (a) A representation DEGnet framework. The method has two phases. The first
phase involves preprocessing, splitting, labelling of the RNA-seq data set. The second
phase includes training (first level of training), fine-tuning (second level of training),
testing of DEGnet model, identification of UR and DR genes and validation of identified
DEGs. (b) Architecture of DNN used in DEGnet

DEGnet is a sequential deep neural network (Fig. 1b) with 1 input layer, 1
output layer, and 10 hidden layers, consisting of nodes (neurons). For batch size
equal to 1, the input to the model is a vector N of size $1 \times n$, where n is the total
number of control (or normal) and disease (or tumor) samples. The inputs for PD
and BRCA datasets are two vectors of sizes 1×73 and 1×1215, respectively.
To set the optimal number of hidden layers, we initialized the network with 2
hidden layers, and then add layers until it starts to overfit the training data and
the test loss does not improve. Based on our experiment, we set the optimal
number of hidden layers as 10. The most common rule to set optimal hidden
layer size (number of neurons) is that the hidden layer size should be between
the number of input and output size. The optimal hidden layer sizes for PD
and BRCA datasets are found to be 60 and 1000, respectively based on the

different input sizes (73 and 1215) of the two datasets. In order to regularize the network, we then use dropouts rate (25%). We use rectified linear unit (ReLU) and sigmoid as activation functions, to determine whether a node should activate or not depending upon the sum of inputs-weights products of each layer. ReLU is applied at the sum h and its output is $max(0, h)$, where h is the output of each hidden layer. The sigmoid is another non-linear activation function with smooth gradient and its output range is (0,1). In our model, we predict probability of a gene to be up-regulating or down-regulating across samples. Since the probability is in the range of (0,1), sigmoid is the best choice for activation function and is used in the output layer. Also, ReLU is less computationally expensive and is most widely used activation functions. The model computes the loss that measures the error in predicting the optimal output for a given input x and updates the parameters based on the gradients. For this, we used $optim.Adam()$ as an optimizer and $BSELoss()$ as a loss function, which measures the binary cross entropy between the truth (y) and the predicted output (y^{pred}).

$$loss_{BCE}(y, y^{pred}) = \{l_1, l_2, \ldots, l_g\}^T,$$

$$Here \ l_i = -w_i[y_i^{pred} \cdot log y_i + (1 - y_i^{pred}) \cdot log(1 - y_i)] \ \text{and } g \text{ is the batch size.}$$

If the $\bar{y}$ is the optimal output for test data t and if $\bar{y} < 0.5$ then $\bar{y} = 0$ otherwise 1. Once the model is trained with the consensus labelled train data; we fine tune the model using biologically validated fine-tune data. This trains the model with both log2FC estimates (sample variance) and incorporates prior knowledge of the data. Thereafter, the model is tested with the biologically validated test data. We used confusion matrix to calculate the accuracy, sensitivity (recall), specificity, and precision for evaluating the performance of our model.

2.1 Datasets and Preprocessing

We use the gene-expression datasets for PD RNA-seq (GSE68719) and BRCA (TCGA) [16] with control (or normal) and disease (tumor) samples. The first dataset contains mRNA-seq gene expression and MS3 proteomics with 29 PD and 44 control samples, profiled from human post-mortem BA9 brain tissue for PD and neurologically normal individuals. The second dataset contains the RNA-seq gene expression with 113 normal and 1102 tumor samples, profiled from breast invasive carcinoma (BRCA) expression data using an Illumina HiSeq2000 system. For preprocessing of the datasets, we remove the redundant genes and the rows with NAN values. Batch effects are removed using $removeBatchEf$-$fect()$ of the edgeR package. Since the number of samples is small, we used a different approach, where we split the dataset as train data (98%), validation data or fine-tune data (1%), and test data (1%). We use three baseline methods DESeq2, edgeR, and limma-voom to calculate logFC estimates of each gene of the train data across control and disease samples. The positive logFC estimates are labelled UR and negative logFC estimates as DR. From the three baseline methods, we get three labels (of class 0 or 1) for each gene, and therefore to remove the bias, we use the consensus labels for the genes and label them as UR (1) or DR (0) genes. The fine-tune and test data are labelled using prior knowledge acquired from the literature regarding up-regulation and down-regulation.

3 Results

Here, we show the assessment of our proposed method, DEGnet in terms GO terms enrichment, pathway enrichment and statistical metrics such as accuracy, sensitivity, specificity, and precision. We also compare the performance of DEGnet with six other ML-based methods for both PD and BRCA datasets.

3.1 Functional Validation of UR and DR Genes

In Tables 1 and 2, we show the GO enrichment and pathway enrichment in terms of p and q values for UR and DR genes extracted from the PD dataset. From Table 1, it is seen that UR and DR genes extracted using DEGnet from PD are enriched with GO terms such as activation of MAPK activity (GO:0000187), regulation of apoptotic process (GO:0042981), chemokine receptor binding (GO:0042379), CXCR chemokine receptor binding (GO:0045236), etc. which are associated with differentiation, degradation, and death of cell during pathogenesis of PD. Moreover, the PD associated pathways such as Apoptosis, Programmed Cell Death, IL-17 signaling pathway, Neurodegenerative Diseases, etc. mapped from these UR and DR genes are found to be significant with low p and q values. Similarly for BRCA, the UR and DR genes are biologically enriched. Recent studies say that there is a precise relation between Mitogen-activated protein kinase (MAPK) activation and proliferation, death, invasion of tumor during progression of cancer [17]. In Fig. 2, we show the MAPK pathway mapped from DR genes such as JUND and GADD45B of BRCA dataset identified using DEGnet.

Table 1. Analysis of GO enrichment of UR and DR genes of PD extracted using DEGnet

Disease	UR/DR genes	GO ID	p value	q value
PD	UR	Pyridine N-methyltransferase activity (GO:0030760)	6.434E−4	3.079E−2
		Activation of protein kinase (GO:0032147)	3.616E−5	2.769E−2
		Nuclease activity regulation (GO:0032069)	9.558E−5	2.769E−2
		Activation of MAPK activity (GO:0000187)	1.223E−4	2.769E−2
		Apoptotic process regulation (GO:0042981)	1.751E−4	2.769E−2
		Programmed cell death regulation (GO:0043067)	1.863E−4	2.769E−2
	DR	Cytokine activity (GO:0005125)	6.364E−7	5.761E−5
		Activity of chemokine (GO:0008009)	6.473E−7	5.761E−5
		Binding activity of chemokine receptor (GO:0042379)	1.603E−6	8.064E−5
		Binding of CXCR chemokine receptor (GO:0045236)	1.812E−6	8.064E−5
		Regulation of cell migration (GO:0030334)	2.714E−9	2.838E−6

Table 2. Analysis of pathways mapped from UR and DR genes of PD extracted using DEGnet

Disease	UR/DR genes	Pathways	p value	q value
PD	UR	Apoptosis	4.077E–4	3.319E–2
		Programmed Cell Death	4.286E–4	3.319E–2
		IGF1 pathway	2.649E–4	2.872E–2
		Cytokine Signaling in Immune system	3.263E–3	4.686E–2
		Genes regulating PIP3 signaling in cardiac myocytes	1.521E–3	4.686E–2
		IGF1 pathway	2.649E–4	2.872E–2
	DR	IL-17 signaling pathway	1.782E–8	7.825E–6
		TNF signaling pathway	4.978E–5	2.732E–3
		Neurodegenerative Diseases	1.087E–5	1.193E–3
		Chemokine signaling pathway	3.734E–4	1.091E–2
		Interleukin-4 and 13 signaling	1.315E–3	2.750E–2

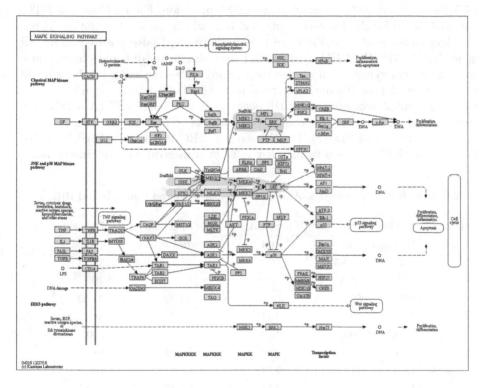

Fig. 2. MAPK pathway mapped from DR genes JUND and GADD45B of BRCA dataset. During pathogenesis of BRCA, there are perturbations in the DR genes JUND and GADD45B which cause significant disturbances in biological activities such as apoptosis, and synthesis of cells.

3.2   Statistical Analysis of UR and DR Genes

Here, we assess the performance of our proposed model with six ML based methods, namely, LR, KNC, SVM, GNB, DTC, and RFC in terms of accuracy, sensitivity, specificity, and precision on the PD and BRCA test data. For PD, except DTC and RFC, other methods find more false positives and false negatives than that of DEGnet method. Similarly, for the BRCA test data, though the DEGnet model does not obtain the best results in terms of statistical parameters, but the false positive rate and false negative rate are lower than other discussed methods. This shows that the proposed method is efficient in identifying potential disease biomarkers from a disease dataset. In Table 3, we compare the performance of DEGNet with the same six other ML-based methods in terms of accuracy, sensitivity, specificity, and precision. We see that for the PD dataset, DEGNet and DTC scores maximum accuracy, sensitivity, specificity, and precision. But for the BRCA dataset, DEGNet outperforms all six methods with 87.5% accuracy, 87.5% sensitivity, 100% specificity, and 100% precision. For DEGnet and DTC, the AUC score for the PD disease dataset is 1, which means that the model has an ideal measure of separability of true positives and false positives. The UR genes RPL3, APOD, PGK1, and PSMC1 show significant difference in functions such as protein synthesis, lipid metabolism, glycolysis pathway, catebolism and modification of proteins during pathogenesis of PD. Similarly, significant gene expression differences are seen in DR genes such as CSE1L, EEF1A, CD74, and SPP1, which are involved in transport, synthesis of protein, immune response during progression of PD [18]. The UR genes in BRCA such as SRCAP, HMGB1, PPIA, and ZNF9 are seen to play key roles in transcription, protein synthesis,

Table 3. Statistical analysis of UR and DR genes filtered from PD and BRCA

Dataset	Method	Accuracy (%)	Sensitivity (%)	Specificity (%)	Precision (%)
PD	DEGnet	100	100	100	100
	LR	53.85	62.5	50	35.71
	KNC	61.54	50	66.67	40
	SVM	50	62.5	44.44	33.33
	GNB	50	100	27.78	38.1
	DTC	100	100	100	100
	RFC	96.15	100	94.44	88.9
BRCA	DEGnet	87.5	87.5	83.74	100
	LR	40	13.04	76.47	42.86
	KNC	40	4.34	88.23	33.33
	SVM	47.5	17.39	81.03	66.67
	GNB	72.5	30.43	70.58	58.33
	DTC	40.15	21.74	82.35	62.5
	RFC	50.5	13.04	94.11	75

transport, and degradation. Similarly, significant differences are seen in the DR genes of BRCA such as TM4SF1, HMGN1, LMNA and SOD2, which participate in significant functions such as adhesion, synthesis of membrane proteins, transcription factors, and metabolism [19].

4   Discussion

The biological data has fewer samples than the number of genes and therefore the use of neural networks is challenging. In our paper, we proposed a model DEGnet, with a deep neural network of one input, multiple hidden layers, and one output layer. The trained model was used to identify UR and DR genes from PD and BRCA datasets with higher statistical and functional significances. The hallmark of the proposed method is that it can identify UR and DR with zero or minimal false positive or false negative rate. Based on the dataset size, the model may be extended later to test on other RNA-seq datasets to find potential biomarkers related to diseases by tuning the hidden layer size.

References

1. Dembélé, D., Kastner, P.: Fold change rank ordering statistics: a new method for detecting differentially expressed genes. BMC Bioinform. **15**(1), 14 (2014)
2. Anders, S., Huber, W.: Differential expression analysis for sequence count data. Genome Biol. **11**(10), R106 (2010)
3. Love, M.I., Huber, W., Anders, S.: Moderated estimation of fold change and dispersion for RNA-seq data with DESeq2. Genome Biol. **15**(12), 550 (2014)
4. Robinson, M.D., McCarthy, D.J., Smyth, G.K.: edgeR: a bioconductor package for differential expression analysis of digital gene expression data. Bioinformatics **26**(1), 139–140 (2010)
5. Law, C.W., Chen, Y., Shi, W., Smyth, G.K.: VOOM: precision weights unlock linear model analysis tools for RNA-seq read counts. Genome Biol. **15**(2), R29 (2014)
6. Wang, L., Xi, Y., Sung, S., Qiao, H.: RNA-seq assistant: machine learning based methods to identify more transcriptional regulated genes. BMC Genom. **19**(1), 546 (2018)
7. Sekhon, A., Singh, R., Qi, Y.: DeepDiff: deep-learning for predicting differential gene expression from histone modifications. Bioinformatics **34**(17), i891–i900 (2018)
8. Kong, Y., Yu, T.: A deep neural network model using random forest to extract feature representation for gene expression data classification. Sci. Rep. **8**(1), 16477 (2018)
9. Kleinbaum, D.G., Klein, M.: Logistic Regression. Springer, New York (2002). https://doi.org/10.1007/b97379
10. Sarkar, M., Leong, T.-Y.: Application of k-nearest neighbors algorithm on breast cancer diagnosis problem. In: Proceedings of the AMIA Symposium, p. 759. American Medical Informatics Association (2000)
11. Polat, K., Güneş, S.: Breast cancer diagnosis using least square support vector machine. Digit. Signal Proc. **17**(4), 694–701 (2007)

12. Soria, D., Garibaldi, J.M., Ambrogi, F., Biganzoli, E.M., Ellis, I.O.: A 'non-parametric' version of the naive Bayes classifier. Knowl.-Based Syst. **24**(6), 775–784 (2011)
13. Singireddy, S., Alkhateeb, A., Rezaeian, I., Rueda, L., Cavallo-Medved, D., Porter, L.: Identifying differentially expressed transcripts associated with prostate cancer progression using RNA-seq and machine learning techniques. In: 2015 IEEE Conference on Computational Intelligence in Bioinformatics and Computational Biology (CIBCB), pp. 1–5. IEEE (2015)
14. Liaw, A., Wiener, M., et al.: Classification and regression by randomForest. R News **2**(3), 18–22 (2002)
15. Chen, J., Bardes, E.E., Aronow, B.J., Jegga, A.G.: ToppGene suite for gene list enrichment analysis and candidate gene prioritization. Nucleic Acids Res. **37**(suppl_2), W305–W311 (2009)
16. Tomczak, K., Czerwińska, P., Wiznerowicz, M.: The Cancer Genome Atlas (TCGA): an immeasurable source of knowledge. Contemp. Oncol. **19**(1A), A68 (2015)
17. Santen, R.J., et al.: The role of mitogen-activated protein (MAP) kinase in breast cancer. J. Steroid Biochem. Mol. Biol. **80**(2), 239–256 (2002)
18. Kim, J.-M., et al.: Identification of genes related to Parkinson's disease using expressed sequence tags. DNA Res. **13**(6), 275–286 (2006)
19. Zucchi, I., et al.: Gene expression profiles of epithelial cells microscopically isolated from a breast-invasive ductal carcinoma and a nodal metastasis. Proc. Natl. Acad. Sci. **101**(52), 18147–18152 (2004)

Survival Analysis with the Integration of RNA-Seq and Clinical Data to Identify Breast Cancer Subtype Specific Genes

Indrajit Saha[1]([✉]), Somnath Rakshit[2,3], Michal Denkiewicz[2,4],
Jnanendra Prasad Sarkar[5,6], Debasree Maity[7], Ujjwal Maulik[6],
and Dariusz Plewczynski[2,4]

[1] Department of Computer Science and Engineering,
National Institute of Technical Teachers' Training and Research, Kolkata, India
indrajit@nitttrkol.ac.in
[2] Centre of New Technologies, University of Warsaw, Warsaw, Poland
[3] School of Information, The University of Texas at Austin, Austin, USA
[4] Faculty of Mathematics and Information Science,
Warsaw University of Technology, Warsaw, Poland
[5] Larsen & Toubro Infotech Ltd., Pune, India
[6] Department of Computer Science and Engineering, Jadavpur University,
Kolkata, India
[7] MCKV Institute of Engineering, Liluah, Howrah, India

Abstract. Breast cancer is one of the most widespread forms of cancer that affects a significant portion of the female population today. Its early detection and subsequent treatment can be life saving. However, it is difficult clinically and computationally to detect breast cancer and its subtypes in their early stages. On the other hand, Next Generation Sequencing (NGS) techniques have significantly accelerated the process of mapping the human genomes by providing high-throughput expression data of RNA. In this work, we study such NGS based expression data of mRNAs with the clinical data in order to (a) rank the genes based on their importance in survival of breast cancer subtypes and (b) find the relation between the up/down regulation of genes and survival probability of a population. In this regard, first volcano plot is used to find the differentially expressed genes for each subtype, and second, such genes are used to perform the Kaplan-Meier survival analysis with the integration of mRNA expression and clinical data to rank the genes by their importance in survival of breast cancer subtypes. These genes are ranked based on the p-value and significant genes are filtered out by considering the cut-off as p-value < 0.05 for each breast cancer subtype. In our analysis, we have found a relation between gene regulation and survival probability, e.g. up and down regulated genes of a population show low rate of survival of that population. Moreover, for the biological significance, PPI network and KEGG Pathway analysis are conducted on a common set of genes that are present in all subtypes. The datasets, code and supplementary materials of this work are provided online (http://www.nitttrkol.ac.in/indrajit/projects/mrna-survival-breastcancer-subtypes/).

© Springer Nature Switzerland AG 2019
B. Deka et al. (Eds.): PReMI 2019, LNCS 11942, pp. 139–146, 2019.
https://doi.org/10.1007/978-3-030-34872-4_16

Keywords: Breast cancer · Gene selection · Kaplan-Meier plot · Next Generation Sequencing · Transcriptomes · Volcano plot

1   Introduction

Breast cancer is one of the most prolific types of cancer in females worldwide, and the most common cause of cancer related deaths regardless of race and ethnicity, responsible in 2018 for more than 626 thousand deaths worldwide [1]. One of the difficulties in seeking treatment is that breast cancer is a heterogeneous form of cancer with multiple subtypes, each requiring different treatment [3]. Thus, it is crucial to identify its subtype-specific biomarkers. To this end, computational approaches are being used, and tools and databases are being developed, yet it is still a challenging task, both computationally and clinically. The RNA-Seq method utilizes the Next Generation Sequencing (NGS) techniques to rapidly sequence and analyze whole transcriptome for potential breast cancer biomarkers. One of the possible biomarker types is mRNA: a type of RNA molecule which carries genetic information from the DNA nucleus to the ribosome, where the amino acid sequence of the protein products of gene expression is specified by mRNA's sequence. The identification of subtype-wise biomarkers using NGS data is crucial for decision making in case of disease management, yet it is difficult to analyze this data because of its high dimensionality and complexity. However, some tools like Oncomine [8], BioXpress [9], etc. have been created to facilitate the search for biomarkers. Another class of tools can perform survival analysis. These include, for example, Kaplan–Meier Plotter [4], BreastMark [6], etc. Integrated solutions like KM-express [2] start to emerge that make use of NGS data and combine many approaches, are being developed as well.

This fact motivated us to study the NGS based high-throughput expression data of mRNAs. In this regard, we first analyze the datasets of breast cancer subtypes viz. Luminal A (LA), Luminal B (LB), HER2-Enriched (HER2-E), Basal-Like (BL) with Control samples separately in order to find the differentially expressed up and down regulated genes using volcano plot. Thereafter, Kaplan-Meier (KM) survival analysis [5] is performed on these genes to rank them based on p-value. Moreover, the significant genes from the ranked lists are identified based on the criterion, p-value < 0.05, for each subtype. From the analysis, it is observed that the survival rate of a population is directly proportional with the up and down regulation of genes of same population. Furthermore, the common set of significant up and down regulated genes for all subtypes are considered to see their biological significance in terms of Protein-Protein Interaction (PPI) network and KEGG Pathway Analysis.

2   Material and Method

This section describes briefly the preparation of the datasets and the proposed framework.

2.1 Dataset Preparation

The RNA-Seq data in form of RSEM (RNA-Seq by Expectation Maximization) contains the expression values of mRNAs. It is obtained from The Cancer Genome Atlas (TCGA) [10]. The data is normalized by transforming to log_2 scale. The dataset contains 17,987 mRNAs of 326 patients. Furthermore, the breast cancer subtype information is collected from [7]. Only the mRNAs that contains less than 1% zero expression values are kept, while the others are discarded in order to minimize the missing data. After completing this step, we obtain 14,831, 14,330, 14,609 and 14,267 mRNAs for LA, LB, HER2-E and BL subtype respectively. The number of samples, average age of patients and average follow-up days in each subtype are mentioned in Table 1.

Table 1. Statistics of the dataset

Subtype	ID	Average age	No. of days to last follow-up	No. of patients
Luminal A	LA	57.71	2520.83	86
Luminal B	LB	52.80	1481.79	39
HER2-Enriched	HER2-E	52.87	1307.13	24
Basal-Like	BL	56.41	1747.63	41
Control	Control	49.42	1928.42	41

2.2 Method

The proposed framework is discussed below and shown in Fig. 1.

Identification of Differentially Expressed Genes Using Volcano Plot: For the purpose of identifying up and down regulated genes, volcano plot technique is used. Volcano plot identifies differential genes using the t-test and fold-change (FC) methods. It plots log2 of fold-change value on the X-axis against -log10 of p-value from the t-test on the Y-axis. Genes having positive and negative fold change are called up and down regulated genes respectively. In the present experiment, the up and down regulated genes are obtained using volcano plot for each subtype.

Survival Analysis Using Kaplan-Meier Method: In order to assess the impact of mRNA expression on patient survival, we use the Kaplan-Meier method, which uses the patient status and days to last follow up information, to estimate the survival function $S(t)$, which represents the probability of an individual from a given population to be alive at time t. The log-rank test can then

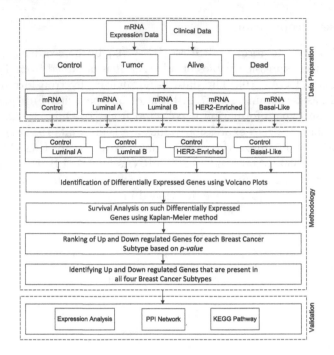

Fig. 1. Steps of the proposed method to rank the genes

be used to compare survival curves between groups. In our experiment, for each subtype, we divide the samples into high-expression and low-expression groups, by median split. Then we compute the subtype-wise p-value of the genes. This gives a p-value to each gene for each subtype. Moreover, KM plots are generated for such genes in each subtype. Thereafter, the genes are ranked subtype-wise by their p-value in ascending order.

Identification of Genes Common to All Subtypes: After obtaining subtype-wise rankings of genes, we seek to find those genes, that are important in all subtypes. This is performed by selecting those genes with p-value < 0.05, hence they are statistically significant for the survival of patients. These four lists of genes are then used to plot a Venn diagram to find the significant common genes that are present in all four subtypes.

3   Experimental Results

3.1   Experimental Testbed

The experiment has been conducted with the use of MATLAB 2018a, Pandas 0.23 and Numpy 1.14 in Python 3.6.5. A computer with Intel i7 processor with

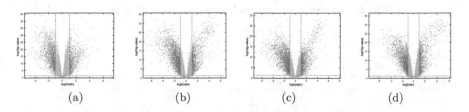

Fig. 2. Volcano plots to determine the differentially expressed genes for (a) LA, (b) LB, (c) HER2-E and (d) BL subtypes of breast cancer

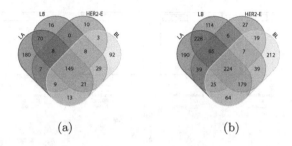

Fig. 3. Venn diagrams showing the belongingness of (a) Up and (b) Down regulated genes across all subtypes

Table 2. Number of subtype-wise up and down regulated genes

Subtype	No. of genes	Up regulated genes	Down regulated genes
LA	14831	457	1014
LB	14330	301	862
HER2-E	14609	194	412
BL	14267	324	769
Common	–	149	224

8 cores and 8 GB RAM are used for all computational purposes. An online tool[1] is used to plot the Venn diagram.

3.2   Results

The proposed framework has been used on the breast cancer subtype-specific datasets. As a result, volcano plot provides the subtype-wise up and down regulated genes as shown in Fig. 2. The number of obtained differentially expressed genes are reported in Table 2. These genes are used further for Kaplan-Meier estimator in order to perform the survival analysis. Table 3 reports top ten up and down regulated genes for each subtype while rest are mentioned in the supplementary. A number of genes well known for their association with breast

---

[1] http://bioinformatics.psb.ugent.be/webtools/Venn/.

Table 3. Top ten genes for each subtype, ranked by p-value

LA Up Gene	P-value	LA Down Gene	P-value	LB Up Gene	P-value	LB Down Gene	P-value	H2 Up Gene	P-value	H2 Down Gene	P-value	BL Up Gene	P-value	BL Down Gene	P-value
CREB3L4	8.81E-06	RBPMS2	1.47E-05	CELSR3	2.28E-04	CCRL1	1.12E-05	CD24	1.28E-02	RNF157	4.18E-04	MAST1	7.42E-05	SLC4A4	8.31E-06
FXYD3	1.19E-05	PRKAR2B	1.92E-05	MB	1.66E-03	C5orf23	3.21E-05	PYCR1	1.96E-02	AKR1C1	4.90E-04	SEPT3	1.17E-04	PDE3B	1.91E-05
TSPAN1	2.99E-05	ADM	7.77E-05	EPN3	2.04E-03	NPR3	3.32E-05	C9orf140	2.12E-02	AKR1C2	4.90E-04	CCDC64	3.82E-04	GPX3	3.31E-05
B4GALNT4	5.49E-05	PKDCC	2.46E-04	C20orf103	3.97E-03	LILRB5	7.74E-05	PGAP3	2.99E-02	ADRB1	9.72E-04	KCNG1	4.32E-04	DDR2	5.15E-05
DBNDD1	8.49E-05	CPM	2.53E-04	STARD10	4.90E-03	SLC4A4	1.25E-04	F12	3.26E-02	C1QTNF7	1.23E-03	FOXP3	4.46E-04	DNAH5	6.90E-05
TMEM125	1.19E-04	HRASLS5	2.54E-04	CLGN	6.43E-03	PDE3B	3.21E-04	FOXP3	3.40E-02	AKR1C3	1.34E-03	BCL2A1	5.87E-04	STC2	1.17E-04
CCDC64	1.26E-04	MAPK10	2.88E-04	LRRC46	7.12E-03	MRC1	3.49E-04	CCDC64	4.13E-02	PLCXD3	1.37E-03	HPDL	6.17E-04	BTG2	1.65E-04
KIAA1211	1.52E-04	FGF10	3.01E-04	SPDEF	9.13E-03	C20orf103	4.51E-04	C20orf103	4.64E-02	GPC3	1.38E-03	APLP1	6.81E-04	MAPK10	1.78E-04
TNFRSF18	1.95E-04	ZNF667	3.40E-04	SORD	9.17E-03	PCSK5	4.97E-04	ORC1L	5.36E-02	ACACB	1.64E-03	SH2D2A	1.25E-03	DLC1	2.83E-04
FBXL16	2.42E-04	MBNL3	4.04E-04	CTXN1	1.38E-02	VWF	5.11E-04	PSMD3	6.33E-02	PCDH9	2.52E-03	PRR7	1.25E-03	PLXNA4	2.84E-04

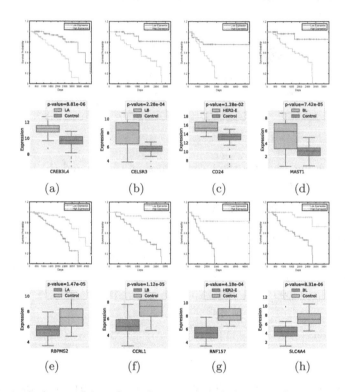

Fig. 4. Survival plots and boxplots of expression of the top subtype specific genes in (a)–(d): Up and (e)–(h): Down regulated genes for LA, LB, HER2-E and BL subtypes respectively. Here, blue color represents low expression and red color represents high expression. (Color figure online)

cancer are seen in the table, such as CREB3L4, CD24, PYCR1, MAST1, etc. The listed genes in each subtype are used to find the common genes in all four subtypes and are shown in the Venn diagrams in Figs. 3(a) and (b). As a result, 149 and 224 common genes have been obtained.

In order to find the relation between the up/down regulation of genes and survival probability of a population, KM plot of top genes and boxplot of expression for each subtype are shown in Fig. 4. It is found from the figure that the

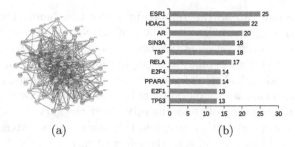

(a)   (b)

Fig. 5. (a) PPI Network of the TFs that are associated with the common up and down regulated genes for all subtypes. (b) The degree of the top ten nodes in the PPI network is shown in the bar plot

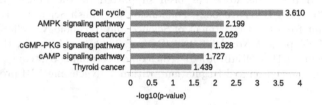

Fig. 6. Bar plot of the significant KEGG pathways

survival probability of a population is very low for the up regulated genes viz. CREB3L4, CELSR3, CD24 and MAST1. Similarly, for the down regulated genes viz. RBPMS2, CCRL1, RNF157 and SLC4A4, the same low survival probability of a population is observed. This is true for the up and down regulated genes.

Furthermore, the common 149 up and 224 down regulated genes are used to find their targeted Transcription Factors (TFs) from the TRRUST database. Using these TFs, a Protein-Protein Interaction network is plotted as shown in Fig. 5(a) with the help of STRING database. Here, a PPI enrichment p-value $< 10^{-16}$ is obtained showing the significance of the network. The degree of top ten nodes of Fig. 5(a) is shown in Fig. 5(b). It is seen that TFs that are well known for their association with breast cancer such as ESR1, HDAC1, AR etc. are found on top. Apart from this, KEGG pathway analysis has also been performed using Enrichr tool for such common up and down regulated genes. The obtained pathways, as shown in Fig. 6 are Cell cycle, AMPK signaling pathway, Breast cancer, cGMP-PKG signaling pathway, cAMP signaling pathway, Thyroid cancer, etc. These pathways are significant for breast cancer. Therefore, the obtained genes are significant as they belong to the same pathways of breast cancer.

4   Conclusion

Detection of breast cancer subtypes in early stages is still a challenging task. It is thus important to identify biomarkers that can accurately help in early detection. NGS techniques can provide opportunities for new biomarker detection by providing high-throughput expression data of mRNA. We presented a systematic approach to identify genes as potential biomarkers for specific subtype of breast cancer using statistical analysis of population of patients' survival data. In order to show our identified sets of genes to be enriched in breast cancer related pathways and functions, we validated them biologically. Moreover, it is seen that the up and down regulation of genes directly affect the survival probability of a population. As a scope of further research, this work can be extended to understand the biological mechanism, e.g., the effect of miRNAs, DNA Methylations and Enhancers to change the expression of such genes which are responsible for the change of survival probability of a population.

Acknowledgements. This work has been supported by the Polish National Science Centre (2014/15/ B/ST6/05082), Foundation for Polish Science (TEAM to DP) and by the grant from the Department of Science and Technology, India under Indo-Polish/Polish-Indo project No.: DST/INT/POL/P-36/2016. Moreover, the work was cosupported by grant 1U54DK107967-01 "Nucleome Positioning System for Spatiotemporal Genome Organization and Regulation" within 4DNucleome NIH program.

References

1. Bray, F., Ferlay, J., Soerjomataram, I., Siegel, R.L., Torre, L.A., Jemal, A.: Global cancer statistics 2018: GLOBOCAN estimates of incidence and mortality worldwide for 36 cancers in 185 countries. CA Cancer J. Clin. **68**(6), 394–424 (2018)
2. Chen, X., Miao, Z., Divate, M., Zhao, Z., Cheung, E.: KM-express: an integrated online patient survival and gene expression analysis tool for the identification and functional characterization of prognostic markers in breast and prostate cancers. Database (2018)
3. Dai, X., et al.: Breast cancer intrinsic subtype classification, clinical use and future trends. Am. J. Cancer Res. **5**(10), 2929–2943 (2015)
4. Györffy, B., et al.: An online survival analysis tool to rapidly assess the effect of 22,277 genes on breast cancer prognosis using microarray data of 1,809 patients. Breast Cancer Res. Treat. **123**(3), 725–731 (2010)
5. Jager, K.J., van Dijk, P.C., Zoccali, C., Dekker, F.W.: The analysis of survival data: the Kaplan-Meier method. Kidney Int. **74**(5), 560–565 (2008)
6. Madden, S.F., et al.: BreastMark: an integrated approach to mining publicly available transcriptomic datasets relating to breast cancer outcome. Breast Cancer Res. **15**(4), R52 (2013)
7. Cancer Genome Atlas Network, et al.: Comprehensive molecular portraits of human breast tumours. Nature **490**(7418), 61–70 (2012)
8. Rhodes, D.R., et al.: Oncomine 3.0: genes, pathways, and networks in a collection of 18,000 cancer gene expression profiles. Neoplasia **9**(2), 166–180 (2007)
9. Wan, Q., et al.: BioXpress: an integrated RNA-seq-derived gene expression database for pan-cancer analysis. Database (2015)
10. Weinstein, J.N., et al.: The cancer genome atlas pan-cancer analysis project. Nat. Genet. **45**(10), 1113–1120 (2013)

Deep Learning Based Fully Automated Decision Making for Intravitreal Anti-VEGF Therapy

Simran Barnwal[1(✉)], Vineeta Das[2], and Prabin Kumar Bora[2]

[1] Department of Physics, Indian Institute of Technology Guwahati, Guwahati, India
simranbarnwal.iitg@gmail.com
[2] Department of Electronics and Electrical Engineering,
Indian Institute of Technology Guwahati, Guwahati, India

Abstract. Intraocular anti-vascular endothelial growth factor (VEGF) therapy is the most significant treatment for vascular and exudative diseases of the retina. The highly detailed views of the retina provided by optical coherence tomography (OCT) scans play a significant role in the proper administration of anti-VEGF therapy and treatment monitoring. With increasing cases of visual impairment worldwide, computer-aided diagnosis of retinal pathologies is the need of the hour. Recent research on OCT-based automatic retinal disease detection has focused on using the state-of-the-art deep convolutional neural network (CNN) architectures due to their impressive performance in image classification tasks. However, these architectures are large in size and take significant time during testing, thus limiting their deployment to machines with ample memory and computation power. This paper proposes a novel deep learning based OCT image classifier, utilizing a small CNN architecture named as SimpleNet. It provides better classification accuracy with 800x fewer parameters, 350x less memory requirement, and is 50x faster during testing compared to state-of-the-art deep CNNs. Unlike other papers focusing on the prediction of specific diseases, we focus on broadly classifying OCT images into needing anti-VEGF therapy, needing simple routine care or normal healthy retinas.

Keywords: Deep learning · Optical coherence tomography · Anti-VEGF therapy

1 Introduction

The macula, located in the central part of the retina, is the primary sensory region associated with vision. Retinal pathologies like choroidal neovascularization (CNV), diabetic retinopathy (DR), age-related macular degeneration (AMD) and diabetic macular edema (DME) affect macular health. These diseases deteriorate normal vision and are also the primary causes of irreversible

© Springer Nature Switzerland AG 2019
B. Deka et al. (Eds.): PReMI 2019, LNCS 11942, pp. 147–155, 2019.
https://doi.org/10.1007/978-3-030-34872-4_17

vision impairment across the world [14]. Intraocular administration of anti-vascular endothelial growth factor (anti-VEGF) has become the standard treatment modality for these vision-threatening diseases [2]. Retinal optical coherence tomography (OCT) imaging provides highly detailed views of the morphology of the retina and is being widely used by retinal specialists in the decision making and treatment monitoring process of anti-VEGF indications [1]. A 3D-OCT volume may contain as many as 1600 B-scans [4] and individually interpreting each scan is a time-consuming task for ophthalmologists. Moreover, the interpretation of retinal OCT scans varies heavily among human experts [3]. Automatic computer-aided diagnosis of OCT scans, employing deep convolutional neural networks (CNNs) have the potential to revolutionize retinal disease diagnosis by performing accurate and rapid classifications on large amounts of OCT data [9].

Recent researches have used deep learning on OCT images for retinal pathology identification based on segmentation or classification methods. Nugroho *et al.* [11] showed that deep neural networks like Resnet50 and DenseNet-169 outperformed handcrafted features based HOG and LBP methods for classification of OCT images. Kermany *et al.* [6] utilized transfer learning, based on InceptionV3 architecture, on OCT scans to classify AMD and DME, to guide the administration of anti-VEGF therapy. Rasti *et al.* [13] proposed a multi-scale deep convolutional neural network consisting of multiple feature-learning branches, each having different input image sizes, to identify dry-AMD and DME from normal retinas. Prahs *et al.* [12] focused on predicting anti-VEGF treatment indication from OCT B-scans, by developing a deep CNN based on GoogLeNet Inception models. The authors classified OCT images into two classes, "injection" and "no injection" groups. Li *et al.* [10] integrated handcrafted features with features extracted from VGG, DenseNet and Xception networks for OCT-based retinal disease classification.

However, these studies [6,10–12] made use of deep CNNs with millions of parameters to extract features from OCT scans, which not only limited their deployment to machines with large memories but also entailed long testing times for each B-scan. Moreover, most studies [6,10,11,13] focused on specific retinal pathology classification without generalizations to other retinal diseases. From an ophthalmologist point of view, it is essential to classify whether anti-VEGF treatment is required or routine care would suffice so that proper strategies and treatment plans can be designed for the patients. Therefore, in this study, we propose a fully automated classification algorithm, utilizing a CNN with fewer parameters for macular OCT B-scans concerning the requirement of anti-VEGF therapy. In this research, OCT scans are classified into "urgent referral", "routine referral" or "normal" categories. Urgent referral corresponds to OCT scans of patients requiring immediate anti-VEGF therapy to prevent irreversible blindness, whereas routine referral corresponds to mild forms of pathologies which do not require anti-VEGF medication but do require routine care. Normal corresponds to patients with healthy eyes. Not only do we predict whether anti-VEGF injections are needed or not, but we also predict whether a patient has other mild forms of eye disease requiring routine care. The main contribution is that a novel CNN with 800x less number of parameters, requiring 10x less memory and 50x less testing time per scan is proposed, which provides better classification performance compared to the well known deep CNNs.

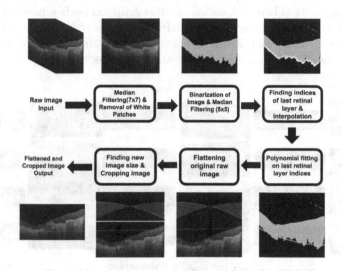

Fig. 1. Data preprocessing: retinal curvature flattening and image cropping

2   The Proposed Method

2.1   Data Preprocessing

Due to various biological structures and acquisition disturbances, retinal layers in raw OCT scans may be shifted or oriented randomly, consequently causing high variability in their locations and axes [13]. Retinal layers in OCT scans also appear naturally curved due to standard OCT image acquisition practices and display [8] which varies between patients and imaging techniques [16]. Retinal curvature flattening and image cropping gives uniformity to images before automated classification and is widely used in literature. To this end, the retinal flattening algorithm proposed by Srinivasan *et al.* [16] is adopted with some modifications.

Figure 1 shows the steps for preprocessing the retinal OCT images. Median filtering with a mask of size 7×7 is initially applied to remove the speckle noises present in the grayscale OCT images. The high-intensity white patches resulting from random orientation in raw OCT scans were removed by replacing pixels with values above 250 with zero. The image is then binarized to obtain image pixels with intensities above the mean intensity of the image, thus, highlighting the retinal layers. The hyper-reflective retinal pigment epithelium (RPE) layer is estimated by locating the last occurrences of high intensities in the binarized image. Interpolation and a 5 × 5 median filter are applied to remove outliers from the estimated RPE layer. A second-order polynomial fit is obtained on the estimated RPE, and each column in the original raw image is shifted up or down to obtain a flat RPE. The diagnostic information in the OCT images is contained in the retinal layers. Therefore, each B-scan is cropped horizontally, considering 220 pixels above and 20 pixels below the RPE to remove the vitreous and choroid-sclera regions. Finally, each image is resized to 60 × 200 pixels. Figure 2 shows raw OCT scans in various classes before and after preprocessing.

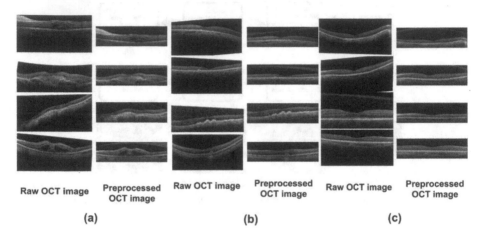

| Raw OCT image | Preprocessed OCT image | Raw OCT image | Preprocessed OCT image | Raw OCT image | Preprocessed OCT image |

(a)   (b)   (c)

Fig. 2. Raw OCT images showing random orientation, retinal curvature, and variability in location. Corresponding preprocessed images after retinal curvature flattening, cropping, and resizing (a) Urgent referral category (b) Routine referral category (c) Normal category

2.2 Proposed Architecture for Classification: SimpleNet

In this paper, a small CNN architecture with 30,793 parameters is used as the classifier. The proposed SimpleNet consists of four convolutional layers, two batch normalization layers, two max-pooling layers, and a final fully connected output layer consisting of three nodes with softmax activation function. The convolutional layers extract features from the preprocessed OCT images using convolution operation performed with convolution filters (kernels). The mathematical formulation of the convolution operation across different layers is given as follows:

$$X_j^l = \sum_i \sigma(X_i^{l-1} * w_{i,j}^l + b_j^l) \tag{1}$$

where $*$ represents the convolution operation, X_j^{l-1} and X_j^l represents the feature maps at the convolutional layer $l-1$ and l respectively, $w_{i,j}^l$ and b_j^l are the trainable weights and biases. σ represents the rectified linear unit (ReLU) activation function given as

$$f(x) = \max(0, x) \tag{2}$$

The dropout layer after the convolution layer prevents the CNN from overfitting on the data while the batch-normalization layer accelerates and stabilizes the training of the CNN [18]. The max-pooling operation takes the highest value from each kernel and reduces the size of feature maps. The fully-connected layer employs a soft-max activation function for the output layer to predict input images into three categories, i.e., urgent referral, routine referral, or normal categories. The details of the SimpleNet model is summarized in Table 1.

Table 1. Details of the proposed SimpleNet architecture

Layers (type)	Number of feature maps	Number of neurons in layer	Size of kernel for each feature map	Stride	Padding	Number of parameters
Input	1	$60 \times 200 \times 1$				
Convolution + ReLU	20	$56 \times 196 \times 20$	$5 \times 5 \times 1$	1	0	520
Convolution + ReLU	20	$52 \times 192 \times 20$	$5 \times 5 \times 20$	1	0	10020
Dropout (10%)		$52 \times 192 \times 20$				
Batch Normalization		$52 \times 192 \times 20$				80
Max-pooling		$13 \times 48 \times 20$	4×4	1		
Convolution + ReLU	40	$11 \times 46 \times 40$	3×3	1	0	7240
Convolution + ReLU	30	$9 \times 44 \times 30$	3×3	1	0	10830
Dropout (20%)		$9 \times 44 \times 30$				
Batch Normalization		$9 \times 44 \times 30$				120
Max-pooling		$2 \times 11 \times 30$	4×4	1		
Flatten		660				
Fully-connected		3				1983
Total =						30793

The training process uses Adam as the optimizer with a learning rate of 0.001, while the "categorical cross-entropy" cost function is used to train the output layer. The training was performed in batches of 128 images per step. Given the imbalance between classes in the training set (Fig. 3(b)), different weights are assigned to loss functions of each class based on the number of images it contains.

3   Experimental Studies and Results

3.1   Database

The publicly available OCT database proposed by Kermany et al. [6] was used in our study. The large database containing 109,309 validated OCT images from 5319 adult subjects, were split into train and test sets of independent patients.

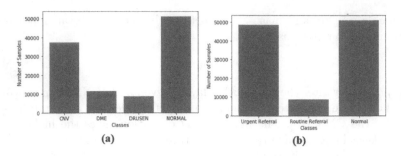

Fig. 3. (a) Original training data distribution per class (b) New distribution of training instances per class

The OCT images in the dataset have four categories; CNV, DME, DRUSEN, and NORMAL. The data distribution of the four classes in the training set is shown in Fig. 3(a). The test set contained 250 images from each category. CNV and DME require urgent care in the form of anti-VEGF medication, whereas drusen, which are lipid deposits under the retina, requires routine medications. Thus, we combined CNV and DME into a new class called "urgent referral," and DRUSEN was labeled as a "routine referral." The new training set distribution is shown in Fig. 3(b).

3.2   Performance Comparison

We compared the proposed method with VGG16 [15], ResNet50 [5] and InceptionV3 [17] based classification. We used the transfer learning technique where we removed the fully connected layers, froze the convolutional layers, and fed the features extracted from it to a fully connected layer with 3 nodes and softmax as the activation function. Batch-normalization layers were, however, trained again for this specific classification problem. 5-fold cross validation was used to present an accurate estimate of the efficacy of each model on unseen data. Each of these models was trained with Adam as the optimizer and categorical cross-entropy as the cost function for the output layer. Class weights of 0.71, 4.19, and 0.74 were given to loss functions of normal, routine referral and urgent referral classes respectively to overcome the imbalance in training set [7]. Table 2 presents the performance metrics for SimpleNet, VGG16, Resnet50, and InceptionV3 for 5-fold cross-validation on the preprocessed dataset. It can be seen that SimpleNet heavily outperforms other architectures in terms of space and testing times while achieving better classification performance (Fig. 4).

Table 3 shows the class wise performance of the proposed SimpleNet architecture. The best SimpleNet model achieves a prediction accuracy of 97.7% with a sensitivity of 100%, 95% and 97.8% for the normal, routine referral and urgent referral classes respectively.

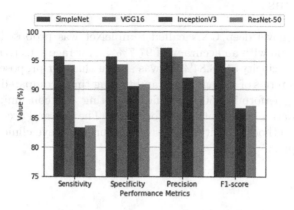

Fig. 4. Graph showing performance metrics of the proposed SimpleNet model against transfer learning based classification methodology according to 5-fold cross-validation on preprocessed data

Table 2. Details and performance metrics of the proposed SimpleNet model against transfer learning based classification methodology based on 5-fold cross-validation. Testing was done on GPU with 5 GB GPU memory and card model Nvidia Tesla K20c. The card was fitted on a Dell Workstation of Model R7610 with 32 GB memory and 600 GB SAS HDD.

Classification method	Number of parameters	Number of trainable parameters	Average testing time per image (s)	Size on disk (MB)	[Sensitivity, Specificity, Precision, F1-score] (%)
SimpleNet	**30,793**	**30693**	**0.0464**	**0.15**	**[95.67, 95.7, 97.12, 95.63]**
VGG16	14,789,955	75,267	2.1444	56.4	[94.15, 94.25, 95.71, 93.89]
InceptionV3	21,956,387	153,603	0.9027	84.4	[83.37, 90.46, 92.00, 86.74]
ResNet-50	23,888,771	354,179	1.4566	73.2	[83.68, 90.86, 92.28, 87.12]

Table 3. Performance metrics of the best SimpleNet model for each class on the test set of 1000 preprocessed images

Categories	Sensitivity (%)	Specificity (%)	Precision (%)	F1-score (%)
Normal	100.0	99.07	97.28	98.62
Routine referral	95.20	98.67	95.97	95.58
Urgent referral	**97.80**	**98.80**	**98.79**	**98.29**

4   Conclusions

In this paper, a novel deep CNN called SimpleNet was proposed, which performed classification with an accuracy of 97.7%, an average sensitivity of 97.67% and an average specificity of 98.84%. It was shown that our proposed model was highly efficient in terms of storage space and testing times compared to the state-of-the-art CNNs, performing 50 times faster testing and consuming 350 times less space, with better classification performance. The fast and accurate features of the proposed method make it highly suitable for use in eye clinics and remote health care centers.

References

1. Cheung, N., Wong, I.Y., Wong, T.Y.: Ocular anti-VEGF therapy for diabetic retinopathy: overview of clinical efficacy and evolving applications. Diabetes Care **37**(4), 900–905 (2014)
2. Cheung, N., Wong, T.Y.: Diabetic retinopathy and systemic vascular complications. Prog. Retinal Eye Res. **27**(2), 161–176 (2008)
3. Ferrara, N.: Vascular endothelial growth factor and age-related macular degeneration: from basic science to therapy. Nat. Med. **16**(10), 1107 (2010)
4. Gabriele, M.L., et al.: Three dimensional optical coherence tomography imaging: advantages and advances. Prog. Retinal Eye Res. **29**(6), 556–579 (2010)
5. He, K., Zhang, X., Ren, S., Sun, J.: Deep residual learning for image recognition. In: Proceedings of the IEEE Conference on Computer Vision and Pattern Recognition, pp. 770–778 (2016)
6. Kermany, D.S., et al.: Identifying medical diagnoses and treatable diseases by image-based deep learning. Cell **172**(5), 1122–1131 (2018)
7. King, G., Zeng, L.: Logistic regression in rare events data. Polit. Anal. **9**(2), 137–163 (2001)
8. Kuo, A.N., et al.: Correction of ocular shape in retinal optical coherence tomography and effect on current clinical measures. Am. J. Ophthalmol. **156**(2), 304–311 (2013)
9. Lee, C.S., Baughman, D.M., Lee, A.Y.: Deep learning is effective for classifying normal versus age-related macular degeneration OCT images. Ophthalmol. Retina **1**(4), 322–327 (2017)
10. Li, X., Shen, L., Shen, M., Qiu, C.S.: Integrating handcrafted and deep features for optical coherence tomography based retinal disease classification. IEEE Access **7**, 33771–33777 (2019)
11. Nugroho, K.A.: A comparison of handcrafted and deep neural network feature extraction for classifying optical coherence tomography (OCT) images. In: 2018 2nd International Conference on Informatics and Computational Sciences (ICICoS), pp. 1–6. IEEE (2018)
12. Prahs, P., et al.: OCT-based deep learning algorithm for the evaluation of treatment indication with anti-vascular endothelial growth factor medications. Graefe's Arch. Clin. Exp. Ophthalmol. **256**(1), 91–98 (2018)
13. Rasti, R., Rabbani, H., Mehridehnavi, A., Hajizadeh, F.: Macular OCT classification using a multi-scale convolutional neural network ensemble. IEEE Trans. Med. Imaging **37**(4), 1024–1034 (2017)

14. Resnikoff, S., et al.: Global data on visual impairment in the year 2002. Bull. World Health Organ. **82**, 844–851 (2004)
15. Simonyan, K., Zisserman, A.: Very deep convolutional networks for large-scale image recognition. arXiv preprint arXiv:1409.1556 (2014)
16. Srinivasan, P.P., et al.: Fully automated detection of diabetic macular edema and dry age-related macular degeneration from optical coherence tomography images. Biomed. Opt. Express **5**(10), 3568–3577 (2014)
17. Szegedy, C., et al.: Going deeper with convolutions. In: Proceedings of the IEEE Conference on Computer Vision and Pattern Recognition, pp. 1–9 (2015)
18. Yamashita, R., Nishio, M., Do, R.K.G., Togashi, K.: Convolutional neural networks: an overview and application in radiology. Insights Imaging **9**(4), 611–629 (2018)

Biomarker Identification for ESCC Using Integrative DEA

Pallabi Patowary[1]([✉]), Dhruba K. Bhattacharyya[1], and Pankaj Barah[2]

[1] Department of Computer Science and Engineering, Tezpur University,
Tezpur, Assam, India
{ppallabi,dkb}@tezu.ernet.in
[2] Department of Molecular Biology and Biotechnology, Tezpur University,
Tezpur, Assam, India
barah@tezu.ernet.in

Abstract. This paper investigates a fatal disease called Esophageal Squamous Cell Carcinoma (ESCC) and is ranked as sixth leading cancer all over the world. The method of integrative analysis is used to identify a set of most responsible genes that may cause the progression of this disease. We consider both microarray and RNA-seq gene expression data. Initially, independent analysis for each type of data is conducted followed by a consensus function to identify a set of significant Differentially Expressed Genes (DEGs) obtained from multiple tools. Further, a set of low preserved modules across the states are identified and the genes belonging to the modules have been carefully examined and validated from various topological and biological approaches. BAG6 and COL27A1 are the two significant genes identified to be associated with ESCC.

Keywords: Microarray · RNA-seq · Differential expression analysis · Co-expression network · Module preservation

1 Introduction

Esophageal Squamous Cell Carcinoma (ESCC) is one of the most common cancers found in India, especially in the North-East, India. The very cause of the disease can be related to the habits of excessive alcohol drinking, cigarette smoking and betel nut chewing [11] and the mortality rate of this disease is also very high. ESCC makes it difficult or impossible for liquid and food to pass through our food pipe. Differential Expression Analysis (DEA) is one of the prominent approaches for investigating such diseases in progression. Over the decade, a good number of useful tools have been introduced to support DEA. However, none of these tools can be considered effective for all cases. Hence, an ensemble approach could help to improve the performance of significant gene identification consistently. Additionally, due to the increasing development of sequencing technology, the generation of a large number of datasets related to such deadly

© Springer Nature Switzerland AG 2019
B. Deka et al. (Eds.): PReMI 2019, LNCS 11942, pp. 156–164, 2019.
https://doi.org/10.1007/978-3-030-34872-4_18

diseases have been possible to obtain. However, each type of dataset has its own expressiveness and limitations. So, it may not be justified to experiment with a single type of dataset towards conclusive identification of a set of responsible biomarker genes for a given disease. It is essential to consider an integrative analysis using datasets given by different technologies supported by appropriate consensus function to identify an unbiased set of biomarkers for a given disease. First, we conduct an independent DEA on both microarray and RNA-seq gene expression data to identify a set of significant DEGs for each type of data. To identify DEGs, we use multiple DEA tools for each dataset and use an appropriate consensus function or towards the generation of unbiased results. From the selected DEGs, some low preserved modules across the states are identified. The genes belonging to the low preserved modules have been carefully investigated and validated from both topological and biological approaches.

2 Related Work

A gene is considered as differentially expressed if a statistically significant difference is observed in read counts or expression levels between two experimental conditions [15]. All the well established DEA methods for microarray data may not be compatible with RNA-seq data since data distribution of RNA-seq differs from microarray data. In this analysis, we use several popular and freely available DEA packages for both RNA-seq and microarray data available in R platform such as DESeq2 [9], edgeR [12], limma-voom [7], limma [14], SAM [16] and EBAM [13]. In a downstream analysis of Gene Expression Data (GED), identifying a list of DEGs is not the outcome of the analysis; further biological insight of the DEGs is necessary. The drastic changes of expression values for the corresponding gene(s) in normal and disease states signify the hidden truth behind the changes of expression patterns. These genes might be helpful for identification or diagnosis of a particular disease.

Gene Co-expression Network (GCN) analysis discovers functions of unknown genes as well as their confidence levels of predictions of associations with diseases. For any GED, we are interested in identifying groups of genes which show similar expression patterns across a group of samples, known as modules. Co-expression analysis helps in finding such modules. It is reported that weighted networks are more efficient and can give reliable output then unweighted one. GCN based on the identified DEGs is constructed using WGCNA [5], freely available in R platform to study their topological behaviours. Before the construction of GCN, the gene similarity matrix is calculated using adjacency function and used as input to find the modules of genes. To retain the strongly correlated genes as well as to ensure scale-free topology of the networks, we consider soft threshold power. It results in an adjacency matrix from which a signed weighted GCN is constructed. More biologically meaningful modules can be found from a signed weighted co-expression network than using an unsigned network [10]. Adjacency matrix is converted to the topological overlapping matrices (TOM) and after this step, average linkage hierarchical clustering algorithm is applied according to the TOM-based dissimilarity measure to find modules of co-expressed genes. An eigengene represents a module. For further analysis, module eigengenes are extracted and some modules are merged by choosing a cutline for module dendrogram.

The module preservation statistics help quantify the preservation of within-module *Zsummary* statistics score is calculated to determine the module preservation. Highly preserved modules have their *Zsummary* score greater than 10 as well as their *Medianrank score* is relatively low [3]. If *Zsummary* score is between 2 and 10 then they are moderately preserved and if *Zsummary* score is below 2, modules are lowly preserved. A module with lower *Medianrank* tends to exhibit stronger observed preservation statistics than a module with a higher median rank [6].

The 'degree' of a node is the most elementary characteristic for a network. The most biological networks are scale-free which means that their degree distribution follows a power law, as most of the nodes have few links and only a few nodes are densely connected [2]. The 'Hub' is the node with the highest degree, plays important role in determining the outcome or phenotype of a disease of interest.

Pathway information helps to represent a group of genes that work together in a biological process. Gene Ontology enrichment and Pathway analysis are performed to interpret the biological significance for a given list of hub genes.

3   Our Approach

In this paper, an ensemble approach is used supported by an effective consensus function to find an unbiased set of DEGs by multiple DEG finding tools. These DEGs are considered as input for GCN construction and for subsequent downstream analysis. GCN analysis is carried out to extract modules from the co-expression network to perform module preservation analysis. The low preserved modules which are identified by module preservation analysis are further investigated using topological, pathway, GO enrichment analysis and in light of relevant literature evidence to identify an interesting set of biomarkers for the microarray data and RNA-seq data. Finally, interesting biomarkers are identified. Our method is comprised of eight major steps which are stated below.

1. Input: D_1: Microarray data, D_2: RNA-seq data; α, β, γ, k: user defined Thresholds;
2. Pre-process D_1 and D_2 to obtain D'_1, D'_2.
3. Execute the following steps to obtain lists of DEGs for both D'_1, D'_2.
   (a) For D'_1, use tools limma, SAM, and EBAM to generate $DGLM_i^\alpha$ i.e. a set of DEGs considered w.r.t. a user defined threshold α, where i = {1, 2, 3}.
   (b) For D'_2, use tools DESeq2, edgeR and limma-voom to generate $DGLR_i^\beta$ i.e. a set of DEGs considered w.r.t. a user defined threshold β, where i = {1, 2, 3}.

4. Obtain a common list of DEGs i.e. S^1_{common} through consensus building. Mathematically,
   (a) Generate the common set of DEGs based on $DGLM_i^\alpha$.

   $$S^1_{common} \leftarrow DGLM_1{}^\alpha \cap DGLM_2{}^\alpha \cap DGLM_3{}^\alpha$$

   (b) Identify top-k other significant (with lower P-value) genes from $DGLM_i$ w.r.t. a user defined threshold (γ) for filtering. Mathematically,

   $$DGLM_i' \leftarrow DGLM_i{}^\gamma - S^1_{common}; i = 1, 2, 3$$

   (c) Obtain the final set of DEGs for all the tools by taking common of S^1_{common} and $DGLM_i'$. Mathematically,

   $$DEGM \leftarrow S^1_{common} \cup DGLM_1' \cup DGLM_2' \cup DGLM_3'$$

   (d) Repeat the steps (a) through (c) for RNA-seq data to obtain DEGR.
5. Perform the following steps for topological and presentation analysis on DEGM and DEGR.
   (a) Construct Co-Expression Networks (CENs) for both DEGM and DEGR.
   (b) Extract sets of modules i.e. M_{DEGM} and M_{DEGR} for the CENs.
6. Perform preservation analysis to find set of low preserved modules say $LPM_j{}^{DEGM}$ and $LPM_j{}^{DEGR}$ (for j = 1, 2..) based on *Zsummary* score.
7. Calculate intra-modular connectivity for each low preserved modules of $LPM_j{}^{DEGM}$ and $LPM_j{}^{DEGR}$ to identify a sets of hub genes G^{DEGM} and G^{DEGR}.
8. Validate each member gene of G^{DEGM} and G^{DEGR} to obtain the useful biomarker genes for both D_1 and D_2.

4   Results

This section presents the descriptions of both the microarray and RNA-seq data, their pre-processing and analysis, and validation using both topological and biological approaches.

4.1   Dataset Description and Pre-processing

We use ESCC RNA-seq data under accession number SRP064894 downloaded from Recount2[1]. This count dataset contains 58,000 transcripts and 29 esophageal samples including 15 non-tumor tissues and 14 esophageal squamous cell carcinoma (ESCC) tumor tissue samples. Also, we use one normalized microarray dataset GSE20347 containing 17 human ESCC samples and 17 matched adjacent normal tissue samples with 22278 genes downloaded from GEO[2]. The RStudio Version 1.1.463 and Bioconductor packages are used to

---

[1] https://trace.ncbi.nlm.nih.gov/Traces/sra/?study=SRP064894.
[2] https://www.ncbi.nlm.nih.gov/geo/query/acc.cgi?acc=GSE20347.

process the ESCC dataset. We remove genes with very low read counts in all samples before using DEA tools. We use TMM normalization method available in edgeR and CPM function to obtain the normalized expression values of our RNA-seq dataset. Finally, the dataset is log2(x+1) transformed to make ready before downstream analysis. ENSEMBLE gene IDs are mapped to SYMBOL IDs. Incase of GSE20347 microarray dataset, we use goodSamplesGenes() function to investigate missing entries and zero-variance genes. Mapping of PROBE IDs is done with SYMBOL IDs.

4.2   Differentially Expressed Gene (DEG) Identification

By considering the results obtained from the various DEA tools for both the datasets, a consensus is built and applied for each dataset to obtain two lists of DEGs and this consensus function are applied separately for both datasets.

Consensus Findings for Microarray Data: For microarray data, we consider *limma, SAM* and *EBAM* for DEG identification, and total 2916, 3594 and 7386 DEGs are identified respectively for a threshold limit $adjPvalue < 0.01$ for *limma* and the false discovery rate $FDR < 0.01$ for *EBAM* and *SAM*. From these genes 1471 DEGs are detected as common among three tools, denoted as S^1_{common}. For the second sets of DEGs, we consider highly significant DEGs. By considering $adjPvalue < 0.001$ for limma and found 2125 DEGs. EBAM and SAM give 2288 and 4186 DEGs which are most significant by considering $FDR < 0.001$. After excluding S^1_{common} from each second set of DEGs, total 977, 3024 and 1129 DEGs are filtered out for limma, SAM and EBAM respectively. In this case, total of 3108 DEGs are reported as common. Finally, we prepare the final list of DEGs by considering all these 3108 and S^1_{common}, total *4579* DEGs are identified for downstream analysis. limma results total 4282 up and 4099 down regulated genes.

Consensus Findings for RNA-seq Data: Similarly, for RNA-seq data, *DESeq2, edgeR* and *limma-voom* tools are applied and total *7722, 5337* and *2308* number of DEGs are obtained respectively for a threshold limit $Pvalue < 0.01$. Total up and down regulated genes (up, down) detected by DESeq2, edgeR and limma-voom are (6087, 4113), (3592, 2383), and (3710, 3945). For the first set of common genes i.e. S^1_{common}, total of 2272 DEGs are found. The DEGs present in DESeq2 but not in S^1_{common} and by choosing $Pvalue < 0.001$ is 2807. Similarly, for edgeR and limma-voom, we get 2198 and 223 DEGs, respectively. Among 2807, 2198 and 223 DEGs, total of 3137 genes are identified as common. After that, we have computed the union between 3137 DEGs and S^1_{common}, finally, 5409 genes are detected as DEGs. After removing of missing entries and filtering of low read counts for co-expression network construction, *5165* genes are considered as the significant DEGs. Total 1807 DEGs are found identical for both the datasets.

4.3   Construction of the Weighted Co-expression Network

For microarray data, downstream analysis is done with 4579 DEGs. Normal and disease samples are separated for 4579 DEGs. Scale-free GCN is constructed with the help of WGCNA package in R for the two datasets normal and disease separately. An appropriate soft threshold power (here, it is 10 for normal and 5 for disease dataset) is selected with which adjacency matrix between all selected DEGs are computed. The adjacency matrix is transformed into a TOM to calculate corresponding dissimilarity (1-TOM). Performed hierarchical clustering method to find modules using dissimilarity measure. For both datasets, 27 different modules are identified using dynamic cut tree algorithm and each module is assigned a unique colour and modules are further analysed by finding eigengenes for each modules. Again eigenmodules are clustered using dissimilarity of module eigengenes and most similar modules are merged (MEDissThres = 0.25). Finally, 22 and 25 modules are extracted from normal and disease datasets respectively. Similarly, for RNA-seq data, WGCNA is applied on two datasets separately for normal and disease samples with 5165 genes. We consider co-expressed modules with a high topological significance which are constructed using soft threshold power as 16 and 18 for control and disease networks and at last, 14 modules from control and 21 modules from disease network are extracted.

4.4   Module Preservation Analysis and Hub Gene Finding

For the microarray data, the white module is detected as the low preserved module with *Zsummary score* 1.1 and *MedianRank* 18 shown in Fig. 1(a). Total 41 genes are present in this module. It is observed that for the RNA-seq data, *Zsummary* value for steelblue module is 1.7, which is the lowest one and *MedianRank* statistics is 13 shown in Fig. 1(b). This module consists of total of 55 genes and are considered for further analysis. Since the hub gene plays an important role in the biological network, we find the hub genes - BAG6 and COL27A1 from the low preserved modules white and steelblue respectively. Since the network is signed, we got the degrees in terms of correlation weight. Weights of the hub genes are: 8.346468982 and 13.52551985 respectively. These hub genes are found from each module with the highest intra-modular connectivity, looking at all genes in the expression data in R. From cBioPortal[3] website, hub genes are found matching with the mutated gene of Esophageal cancer.

4.5   Finding Biological Significance of BAG6 and COL27A1 Genes

The hub genes BAG6 and COL27A1 are found as down-regulated and up-regulated genes for our ESCC datasets. To understand the biological insights of these genes, GO analysis and Pathway analysis are performed in DAVID[4] and

---

[3] https://www.cbioportal.org/.

[4] https://david.ncifcrf.gov/.

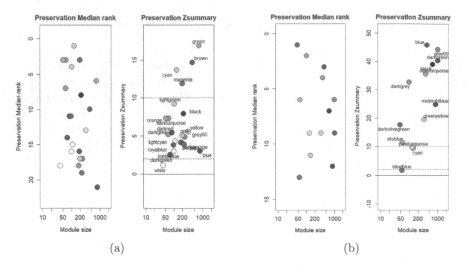

(a)   (b)

Fig. 1. (a) Preservation analyses for microarray data. Here, the white module is the least preserved module. (b) Preservation analyses for RNA-seq data. Here, the steelblue module is the least preserved module. (Color figure online)

GeneAnalytics[5]. Percentage of enrichment in Biological Process, Cellular Component and Molecular Function for the white module are 89.7, 97.4, and 97.4 and for steelblue are 65.2, 63, and 73.9. In DAVID, gene-term enrichment (overrepresented) analysis is done for both modules to identify the most relevant biological terms associated with the given gene list. The enriched terms associated with *steelblue* module are phosphoprotein, extracellular matrix part, collagen, nuclear lumen, intracellular organelle lumen, membrane-enclosed lumen, alternative splicing, adhesion biological adhesion, polymorphism, nucleoplasm part, proteinaceous extracellular matrix, transcription factor binding, Fibronectin type-III 7, epidermis development, regulation of cell motion, ectoderm development, TSPN, Forkhead-associated, metal ion binding, BRCT, cation binding, basement membrane, odontogenesis of dentine-containing tooth, cell cycle, transcription from RNA polymerase II promoter etc. The enriched terms associated with the *white* module are Isopeptide bond, Acetylation, nuclear chromatin, nucleoplasm, collagen catabolic process, Ubl conjugation, cross-link: Glycyl lysine isopeptide, ATP-binding, DNA replication, Hydroxylation, cellular protein modification process, spindle pole, Calcium, Dwarfism, proteasome binding, Cell division, cytoplasm, Phosphoprotein, region of interest: Triple-helical region, synaptonemal complex assembly, Nucleotide-binding, topological domain: Lumenal, endoplasmic reticulum lumen, cellular response to DNA damage stimulus, protein binding, Extracellular matrix, Hydrolase, Mitosis, LamG, Cytoplasm, mitotic nuclear division, Transferase, Catalytic, and Coiled coil.

---

[5] http://geneanalytics.genecards.org/.

For squamous cell carcinomas (SCC), *COL27A1* is reported as the significant mutated gene [8]. *COL27A1* shares Phospholipase-C, Degradation of the extracellular matrix, Collagen chain trimerization, and Integrin Pathways which are the important factors in cancer invasion and metastasis. Wnt/Hedgehog/Notch pathway is shared by *BAG6* which plays key roles in embryogenesis and carcinogenesis. For epidermal growth factor receptor, this pathways acts as activators in ESCC [1]. *BAG6* gene is found to be associated with lung cancer [4].

5   Conclusion

Our analysis reveals several important genes, including *BAG6* and *COL27A1* which have adequate evidence to be involved in ESCC. These two genes identified based on the differential expression analysis and module preservation analysis, may provide scopes for further investigations in the pathogenesis of ESCC. We are working on the development of a robust soft computing enabled outlier based DEG finding method to substitute the present ensemble approach. Further, a multi-objective gene ranking approach is also underway.

References

1. Abbaszadegan, M.R., Riahi, A., Forghanifard, M.M., Moghbeli, M.: WNT and NOTCH signaling pathways as activators for epidermal growth factor receptor in esophageal squamous cell carcinoma. Cell. Mol. Biol. Lett. **23**(1), 42 (2018)
2. Barabási, A.L., Albert, R.: Emergence of scaling in random networks. Science **286**(5439), 509–512 (1999)
3. Chen, J., Wang, X., Hu, B., He, Y., Qian, X., Wang, W.: Candidate genes in gastric cancer identified by constructing a weighted gene co-expression network. PeerJ **6**, e4692 (2018)
4. Etokebe, G.E., et al.: Association of the FAM46A gene VNTRs and BAG6 rs3117582 SNP with non small cell lung cancer (NSCLC) in Croatian and Norwegian populations. PLoS ONE **10**(4), e0122651 (2015)
5. Langfelder, P., Horvath, S.: WGCNA: an R package for weighted correlation network analysis. BMC Bioinform. **9**(1), 559 (2008)
6. Langfelder, P., Luo, R., Oldham, M.C., Horvath, S.: Is my network module preserved and reproducible? PLoS Comput. Biol. **7**(1), e1001057 (2011)
7. Law, C.W., Chen, Y., Shi, W., Smyth, G.K.: VOOM: precision weights unlock linear model analysis tools for RNA-seq read counts. Genome Biol. **15**(2), R29 (2014)
8. Liu, P., et al.: Identification of somatic mutations in non-small cell lung carcinomas using whole-exome sequencing. Carcinogenesis **33**(7), 1270–1276 (2012)
9. Love, M.I., Huber, W., Anders, S.: Moderated estimation of fold change and dispersion for RNA-seq data with DESeq2. Genome Biol. **15**(12), 550 (2014)
10. Mason, M.J., Fan, G., Plath, K., Zhou, Q., Horvath, S.: Signed weighted gene co-expression network analysis of transcriptional regulation in murine embryonic stem cells. BMC Genom. **10**(1), 327 (2009)
11. Morita, M., et al.: Alcohol drinking, cigarette smoking, and the development of squamous cell carcinoma of the esophagus: epidemiology, clinical findings, and prevention. Int. J. Clin. Oncol. **15**(2), 126–134 (2010)

12. Robinson, M.D., McCarthy, D.J., Smyth, G.K.: edgeR: a bioconductor package for differential expression analysis of digital gene expression data. Bioinformatics **26**(1), 139–140 (2010)
13. Schwender, H., Krause, A., Ickstadt, K.: Identifying interesting genes with siggenes. Newslett. R Proj. **6**, 45 (2006). Volume 6/5, December 2006
14. Smyth, G.K.: Linear models and empirical Bayes methods for assessing differential expression in microarray experiments. Stat. Appl. Genet. Mol. Biol. **3**(1), 1–25 (2004)
15. Trapnell, C., et al.: Differential gene and transcript expression analysis of RNA-seq experiments with TopHat and Cufflinks. Nat. Protoc. **7**(3), 562 (2012)
16. Tusher, V.G., Tibshirani, R., Chu, G.: Significance analysis of microarrays applied to the ionizing radiation response. Proc. Natl. Acad. Sci. **98**(9), 5116–5121 (2001)

Critical Gene Selection by a Modified Particle Swarm Optimization Approach

Biswajit Jana and Sriyankar Acharyaa[✉]

Maulana Abul Kalam Azad University of Technology,
Kolkata, West Bengal, India
biswajit.cseng2012@gmail.com, srikalpa8@gmail.com

Abstract. Gene selection is an eminent area of research in computational biology. Identifying disease critical genes analysing large gene expression dataset is a challenging problem in computational biology. Application of meta-heuristics is considered to be efficient to attempt such problem. Particle Swarm Optimization (PSO) is a meta-heuristic technique based on swarm intelligence. An improved version of PSO, namely, New Variant Repository and Mutation based PSO (NVRMPSO) combined with kNN (NVRMPSO-kNN) has been applied here to find the subset of genes (disease critical) from gene expression dataset. The modification has been done on a recent PSO version, referred as, Repository and Mutation based PSO (RMPSO). The kNN algorithm has been used for sample classification. The performance of proposed NVRMPSO-kNN has been compared with RMPSO incorporated with kNN (RMPSO-kNN) in gene selection problem. The RMPSO-kNN and proposed NVRMPSO-kNN have been applied to three different sizes of datasets, numbered as, E-MEXP-1050, GSE60438 and GSE10588 of preeclampsia disease. The experimental results reveal that the efficacy of NVRMPSO-kNN is better than that of RMPSO kNN when Classification Accuracy is considered. Furthermore, the performance of proposed NVRMPSO-kNN has been validated with the state-of-the-art observations.

Keywords: Gene selection · Gene expression · PSO · RMPSO · kNN

1 Introduction

Finding disease critical genes from a large gene expression dataset is a challenging work for the researcher in computational biology. In a gene expression dataset, there is large number of genes (of the order of tens of thousands) with different sample sets. Only few genes, responsible for a particular disease are called disease critical genes. Gene expression profiles are available for research through DNA microarray technology [1].

Meta-heuristic techniques [2, 4–7] have gained great importance in the domain of gene expression analysis like Gene Regulatory reconstruction (GRN) reconstruction, Critical Gene Selection, Classification, and Clustering. The main objective of this research work is to find the subset of genes that together differentiate between normal and diseased samples. This subset may be considered responsible for causing the disease.

Gene identification and selection is normally done employing sample classification. Among the classifiers, k-Nearest Neighbor (kNN) [3] and SVM (Support Vector

© Springer Nature Switzerland AG 2019
B. Deka et al. (Eds.): PReMI 2019, LNCS 11942, pp. 165–175, 2019.
https://doi.org/10.1007/978-3-030-34872-4_19

Machine) [4] have been widely exploited in classification of sample/disease. Meta-heuristic techniques, like, PSO [5], Genetic Algorithms (GA) [4], Harmony search (HS), Honey Bee Mating Optimization (HBMO), Differential Evolution (DE) [2], Bat Swarm Optimization (BAT) [6], Artificial Bee Colony (ABC) [7] have been employed among others to perform gene selection.

Saha et al. [3] applied two meta-heuristics, namely, Simulated Annealing (SA) and PSO with k-NN as classifier. Kar et al. [8] apply PSO algorithm for gene selection and KNN has been for classification purpose. Dashtban et al. [9] proposed a Bio-inspired Multi-objective BAT algorithm, namely MOBBA-LS, for gene identification using microarray datasets. The prediction accuracy of each subset of gene has been evaluated applying four widely-used classification techniques where SVM, KNN, Naïve Bayes (NBY), and Decision Tree (DT) are included. Alomari et al. [6] applied two BAT variant with k-NN for gene selection. Pyingkodi and Thangarajan [10] combined Genetic Algorithm with Levy Flight (GA-LV) and it has been applied for classifying genes causing cancer using microarray gene expression data. Biswas et al. [2] has applied four meta-heuristics, such as, HBMO, HS, DE, GA incorporated with k-NN algorithm to find specific genes responsible of preeclampsia that attack women during gestation.

Here, in this paper, a recent variant of RMPSO [11], namely, New Variant Repository and Mutation Based PSO (NVRMPSO) has been proposed combining k-NN algorithm with it (NVRMPSO-kNN) and it has been applied to gene selection problem. The proposed method has been applied to three different sizes of datasets E-MEXP-1050, GSE60438 and GSE10588 of preeclampsia disease [2]. The experimental results state that the proposed method performs well in E-MEXP-1050 and GSE10588 and competitive to other methods in GSE60438 data set.

The remaining part of the paper is organized as follows. Gene selection by sample classification has been described in Sect. 2. In Sect. 3, the basic PSO has been discussed. NVRMPSO has been discussed in Sect. 4. Section 5 contains the experimental result and discussion and finally Sect. 6 concludes the paper.

2   Gene Selection

In the input microarray dataset (M), rows represent genes and columns represent samples (both disease and normal). The dataset M contains more than 40000 genes with different samples/conditions. The number of disease critical gene is much less compared to the size of the gene set. The main objective is to find a few genes (say $g = 30$) from M that differentiate between normal and disease samples. There are n_{C_g} possibilities to choose a small number (g) of genes from M, where, n = number of genes. The search space is too large to find the appropriate subset of genes if the number of genes is large. So the Meta-heuristic approach is required to deal with this situation.

2.1   Description of Input Dataset M and Expression Sub Matrix M_X

In samples set, N_{sampj} represent the jth normal sample, where $j \in (1, 2, 3, \ldots., maxsampN)$ and D_{sampk} represent the kth disease samples, where $k \in (1, 2, 3, \ldots., maxsampD)$. The maximum number of normal and disease samples is represented by *maxsampN* and *maxsampD* respectively. The total number of samples present in the dataset is $N = maxsampN + maxsampD$. Each entry of gene expression data set M is the normalized expression level $g^C_{i,l}$ of G_i for sample l (class C of sample l may be either normal or diseased, $l \in \{1, 2, \ldots, N\}$.

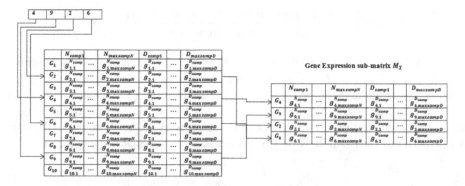

Fig. 1. An example of candidate solution (X) and its expression sub-matrix (M_X)

A candidate solution X contains m number of unique gene indices chosen randomly from the M_{index} $(M_{index} = (1, 2, \ldots, n)$. An example of a candidate solution is presented in Fig. 1. Here 4 genes $(D = 4\{4, 9, 2, 6\})$ are selected randomly from the data set. Each gene has it's gene expression values with specified class level $(g^C_{i,l})$. The gene expression matrix for the subset of genes (here, 4 genes) in Fig. 1 is called gene expression sub-matrix M_X.

2.2   Sample Classification by k-NN

Each sample of M_X is classified employing kNN classifier. Cross-validation [2] is a validation technique, which is used to determine to what extent the samples existing in a data set are possible to be classified. Here in this paper, a type of cross validation, namely, Leave One Out (LOO), has been applied to classify the sample in M_X. In LOO, N validation tests are performed for N sample present in the dataset M to test the class of each sample. In each step of validation test, the sample taken for classification is put in test-set and the rest $(N - 1)$ samples construct the train-set.

Here, kNN algorithm has been used to determine the class possessed by the sample in the test set. For a sample in test set, the distance to every sample present in the train set [2] is calculated. After calculating the distance, the samples are according to their distance value in ascending order and top k (here $k = 3$) samples are chosen and denoted as the nearest neighbors of the sample under test. Among the nearest neighbors, if the majority and the test sample are from the same class, the test sample is called properly classified and the *score* is increased by 1, otherwise it remain same. The objective function in Eq. 1 is the total score in the N validation of a candidate solution (M_X). The main objective is to maximize the number of samples that are properly classified in M_X. The *score* represents the fitness value of a candidate solution.

$$Fit\_kNN(M_X) = \sum_{i=1}^{N} score_i \qquad (1)$$

The quality of a candidate solution is measured with respect to Classification Accuracy (CA) and it is calculated by Eq. 2.

$$CA = \frac{Number\ of\ properly\ classified\ sample\ in\ M_X}{Total\ number\ of\ sample\ in\ M} \times 100\% \qquad (2)$$

The kNN method has been discussed in Algorithm 1. The input of kNN algorithm is a candidate solution M_X. At start, all the samples of M_X are put in a set T_{SAMP}. A cross validation has been performed for each sample in T_{SAMP}. Samples in train set and test set are defined by Eqs. 3 and 4 respectively. A sample J is chosen from T_{SAMP} and it kept in the test set (*Test*) and rest of the samples excluding J are put in the train set (*Train*). For each sample K in Train, the distance is calculated using Eq. 5.

$$J = \left[g_{1,J_{index}}^{C_j}, g_{2,J_{index}}^{C_j}, g_{3,J_{index}}^{C_j}, \cdots, g_{D,J_{index}}^{C_j} \right] \qquad (3)$$

$$K = \left[g_{1,K_{index}}^{C_k}, g_{2,K_{index}}^{C_k}, g_{3,K_{index}}^{k}, \cdots, g_{D,K_{index}}^{k} \right] \qquad (4)$$

Here, C_j and C_k are the classes of sample J and K, respectively. The J_{index} and K_{index} are the indices for J and K in M_X.

$$D(K,J) = \sqrt{\left(g_{1,K_{index}}^{C_k} - g_{1,J_{index}}^{C_j} \right)^2 + \left(g_{2,K_{index}}^{C_k} - g_{2,J_{index}}^{C_j} \right)^2 + \cdots + \left(g_{D,K_{index}}^{C_k} - g_{D,J_{index}}^{C_j} \right)^2} \qquad (5)$$

Algorithm 1: kNN

```
Fit = 0;
T_SAMP = N_samp1, N_samp2, ...., N_maxsampN, D_samp1, D_samp2, ...., D_maxsampD
  for J=1 to N
     Test=J;
     Train=T_SAMP-Test;
     for K= 1to N-1
        Calculate D(K,J)by using eq.5;
     end for
     Sort samples of Train in ascending order based on D(K,J);
     Choose top k samples (k=3)from sorted Train to Neighbor;
     Get class C_j of J;
     for L=1 to k
        Get class C_K ;
     end for
     if class of majority of C_j matches with C_K
        score= 1;
     else
        score=0;
        Fit = Fit + score;
     end if
  end for
  return Fit
```

3 Particle Swarm Optimization (PSO)

PSO [11] is a meta-heuristic approach stimulated by swarm intelligence. A swarm of random particle travels in space of dimension D. The ith particle of D dimensional space represents a member solution $P_i = P_{i1}, P_{i2}, \ldots\ldots, P_{iD}$ and velocity of every particle is represented by $V_i = V_{i1}, V_{i2}, \ldots\ldots, V_{iD}$. Each particle remember two best solutions; one is the best solution achieved so far by the individual particle named as local best ($Pbest_i$) and another is the best obtained by a particle with respect to whole swarm named as global best ($Gbest_{gD}$). Each particle changes its position by the Eqs. 6 and 7.

$$V_{iD}^{t+1} = w * V_{iD}^t + C_1 * rand_1 * (Pbest_{iD} - X_{iD}) + C_2 * rand_2 * (Gbest_{gD} - X_{iD}) \quad (6)$$

$$P_{iD}^{t+1} = P_{iD} + V_{iD}^{t+1} \quad (7)$$

Where, V_{iD}^t is the velocity of particle i at iteration t, w is the weight of inertia, C_1 is coefficient of cognitive acceleration, C_2 is coefficient of social acceleration and $rand_1$, $rand_2$ are random numbers uniform in the range $[0, 1]$.

4 New Variant Repository and Mutation Based PSO (NVRMPSO)

Here, a modified version of RMPSO [11], called New Variant Repository and Mutation Based PSO (NVRMPSO) has been described. In RMPSO, for each solution, fitness is computed after changing the particle's position employing Eqs. 6 and 7. When the fitness obtained is compared with the local best and global best, there is a possibility that the fitness becomes equal to local best or global best or both. The main idea behind the RMPSO is that it stores all the solutions having the same objective (fitness) value. There are two repositories to store the solutions corresponding to local (P_{best_rep}) and global bests (G_{best_rep}) having same objective function value. The objective of this new variant is to improve the exploration strength of the search technique.

The repository stores only those solutions that are distinct and have same fitness. Identical solutions are not stored and there will be no solution in the repository in multiple copies. Local and global best solutions from appropriate repositories will be selected randomly at each iteration. Instead of applying 5 successive mutation strategies on global best solution of RMPSO, here, in NVRMPSO, a two point mutation operation has been activated to both local best ($mutatePbest_i$) and global best solution ($mutateGbest$) to perturb the solution. There are two changes here with respect to the earlier version: one is to apply mutation on local best also which was not done in earlier version and other is to reduce the number of mutations. Application of mutation to local best enhances the exploration strength of the technique and reduction in the number of mutations reduces the overhead of function evaluations.

The main drawback of PSO lies in its suffering from early convergence when the state space is not explored enough. The mutation strategy helps explore the search space and rescue the search process from local optima.

Algorithm 2. NVRMPSO

```
Set RMPSO parameters t_max , w , Pop , C₁ , C₂ ,D;
Initialize    Gbest, Vᵢ, P_best_rep, G_best_rep;
for t = 1 to MaxIt
  for i = 1:Pop
    Update the particle velocity (Vᵗᵢ)ᵢ using eq. 6;
    Update the particle position (Pᵗᵢ)using eq.7 ;
    Calculate the fitness value of updated Mₓ, based on Pᵗᵢ using Fit_kNN(Mₓ);
    if Fit_kNN(Mₓ) < Pbest
      Clear repository;
      P_best_repᵢ = Pᵗᵢ;
      P_best_rep = Fit_kNN(Mₓ);
    elseif Fit_kNN(Mₓ) == Pbest
      Increase the size of repository;
      Add positions Pᵗᵢ and Pbestᵢ into P_best_repᵢ if is not present in P_best_repᵢ;
    end if
    Choose Pbestᵢ and it's corresponding Pbest in randomly from P_best_repᵢ;
    Apply two point mutation to mutate the Pbestᵢ (mutatePbestᵢ),mutatePbestᵢⱼ ∈ M_index;
    Create gene expression sub-matrix Mₓ for mutatePbestᵢ;
    mutateMₓ=Mₓ;
    if Fit_kNN(mutateMₓ) < Pbest
      Pbestᵢ= mutatePbestᵢ;
      Pbest= Fit_kNN(mutateMₓ);
    end if
    if Fit_kNN(Mₓ) < Gbest
      Clear repository;
      Gbestₘ = Pᵗᵢ;
      Gbest = Fit_kNN(Mₓ);
    elseif Fit_kNN(Mₓ) == Gbest
      Increase the size of repository ;
      Add both positions Pᵗᵢand Gbestₘ into G_best_rep if is not present in G_best_rep;
    end if
  end for
  Choose Gbestₘ and its corresponding Gbest randomly from G_best_rep;
  Apply two point mutation to mutate the global best solution (mutateGbest) ;
  Create gene expression sub-matrix Mₓ for mutateGbest;
  mutateMₓ=Mₓ;
  if Fit_kNN(mutateMₓ) < Gbest
      Gbestₘ= mutateGbest;
      Gbest= Fit_kNN(mutateMₓ);
  end if
end for
```

Based on RMPSO, a variant, namely, NVRMPSO has been proposed here. Each particle P_i in population Pop represents a candidate solution X, containing D number of genes which is randomly selected from M_{index}. A gene expression sub-matrix M_X has been created from candidate solution P_i. The fitness of each candidate solution has been calculated with the objective (fitness) function $Fit\_kNN(M_X)$. Each particle P_i remembers its own local best position ($Pbest_i$). The position of a particle P_i is changed using Eqs. 6 and 7. If the current fitness of a particle P_i is greater than its previous fitness of local best, then store the current position in $Pbest_i$ nd store its corresponding fitness in $Pbest$. If the current fitness of a particle P_i is equal its previous fitness, then store both solutions in local best repository ($P_{best\_rep_i}$). After that a two point mutation strategy has been applied to $Pbest_i$ ($mutatePbest_i$) to mutate the $Pbest_i$. Create gene expression sub-matrix M_X for *mutatePbest$_i$* and store it to a variable *mutateM$_X$*. Calculate the fitness of *mutateM$_X$* using fitness function *Fit_kNN(mutateM$_X$)*. If the mutated solution's fitness is greater than that of *Pbest$_i$*, then *Pbest$_i$* and *Pbest* have been updated by *mutatePbest$_i$* and *Fit_kNN(mutateM$_X$)*. The same procedure has been applied on Global best solution *Gbest$_g$* also.

5   Experimental Result

The experimental results of gene selection by sample classification using RMPSO-kNN and NVRMPSO-kNN are discussed in the section. The specifications of hardware and software are: processor = Intel Core i7, speed = 3.4 GHz, RAM = 8 GB, OS = 64-bit Windows 7, MATLAB 2015a. For RMPSO-kNN and NVRMPSO-kNN, population size (*Pop*) is 30, dimension (number of gene selected randomly from M) *D* is 30, upper limit of iterations (*MaxIt*) = 200, initial value of weight (*w*) = 0.9 and it has been decreased linearly by Eq. 8. The acceleration coefficient C_1 and social coefficient C_2 is set to 2.

$$w = (w_{max} - w_{min}) * \left(\frac{t_{max} - t}{t_{max}} \right) + w_{min} \tag{8}$$

Where w_{max} = maximum inertia weight = 0.9 and w_{min} = minimum inertia weight = 0.4.

The RMPSO-kNN and proposed NVRMPSO-kNN has been applied in three different datasets of preeclampsia disease i.e., E-MEXP-1050, GSE10588 and GSE60438. These three datasets have been presented in Table 1. All the datasets undergo a preprocessing phase before final use, called normalization and which is performed according to Eq. 9.

$$Norm(g_{i,l}) = \frac{g_{i,l} - G_i^{min}}{G_i^{max} - G_i^{min}} \tag{9}$$

Where, G_i^{min} and G_i^{max} are minimum and maximum gene expression value of each gene (G_i) over all samples respectively.

Table 1. Description of datasets

Dataset accession no.	Collected from (site)	Normal samples	Disease samples	Total samples	No. of genes
E-MEXP-1050	ArrayExpress	17	18	35	8,793
GSE10588	GEO, NCBI	26	17	43	32,878
GSE60438	GEO, NCBI	42	35	77	47,323

The RMPSO-kNN and proposed NVRMPSO-kNN has been executed 25 times in each dataset. The experimental results have been discussed in Table 2. It shows that with respect to the average number of properly classified samples (average fitness) NVRMPSO-kNN outperforms RMPSO-kNN.

Table 2. Experimental results of RMPSO-kNN vs NVRMPSO-kNN on datasets

Dataset name	Algorithms	Min fitness	Max fitness	Fitness std	Average fitness	Median fitness	CA over mean (%)	Max CA (%)
E-MEXP-1050	**NVRMPSO-kNN**	**32**	**35**	**0.8205**	**33.44**	**33**	**95.54**	**100**
	RMPSO-kNN	30	34	0.9183	31.480	31	89.943	97.143
GSE10588	**NVRMPSO-kNN**	**43**	**43**	**0**	**43**	**43**	**100**	**100**
	RMPSO-kNN	42	43	0.3741	42.840	43	99.628	100
GSE60438	**NVRMPSO-kNN**	**64**	**69**	**1.5133**	**66.04**	**66**	**85.76**	**89.61**
	RMPSO-kNN	60	65	1.4224	62.240	62	80.831	84.416

The overall accuracy has been calculated based on average fitness value of 25 runs and it has been computed using Eq. 2. In all datasets, it is observed that the average performance of NVRMPSO-kNN is better than RMPSO-kNN (Table 2).

The observed results also have been validated against the state-of-the-art [2] and it has been discussed in Table 3. Here, Classification Accuracy over mean fitness value is the criteria for comparison.

Table 3. Performance comparison of NVRMPSO-kNN with state-of-the-art

Dataset Name	Algorithms	Average fitness	CA over mean (%)	Best fitness	Std dev. of fitness
E-MEXP-1050	**NVRMPSO-kNN**	**33.44**	**95.54**	**35**	**0.82057**
	RMPSO-kNN	31.48	89.943	34	0.918331
	HBMO-kNN	**35**	**100.0000**	**35**	**0**
	HS-kNN	33.28	95.0857	34	0.5416
	DE-kNN	**34.80**	**99.4286**	**35**	**0.4082**
	SGA-kNN	32.36	92.4571	34	0.9074

(*continued*)

Table 3. (*continued*)

Dataset Name	Algorithms	Average fitness	CA over mean (%)	Best fitness	Std dev. of fitness
GSE10588	**NVRMPSO-kNN**	**43.00**	**100**	**43**	**0**
	RMPSO-kNN	**42.84**	**99.628**	**43**	**0.374165**
	HBMO-kNN	**43.00**	**100**	**43**	**0**
	HS-kNN	**42.20**	**98.1395**	**43**	**0.5000**
	DE-kNN	**43.00**	**100**	**43**	**0**
	SGA-kNN	**41.68**	**96.9302**	**43**	**0.9452**
GSE60438	NVRMPSO-kNN	66.040	85.76	69	1.5133
	MLPSO-kNN	62.240	80.831	65	1.42243
	HBMO-kNN	**76.72**	**99.6364**	**77**	**0.4583**
	HS-kNN	**75.36**	**97.8701**	**77**	**0.7572**
	DE-kNN	**76.88**	**99.8442**	**77**	**0.3317**
	SGA-kNN	73.92	96.0000	76	1.2220

It shows that NVRMPSO-kNN (modified RMPSO), HBMO-kNN (modified HBMO) and DE-kNN (modified DE) outperform other methods in GSE10588 dataset. For dataset, E-MEXP-1050, NVRMPSO-kNN has achieved the better accuracy compared to HS-kNN and SGA-kNN. For dataset GSE60438, NVRMPSO-kNN achieves moderate accuracy compared to other algorithms.

The fitness convergence graph of RMPSO-kNN vs NVRMPSO-kNN in three datasets has been presented in Figs. 2, 3 and 4.

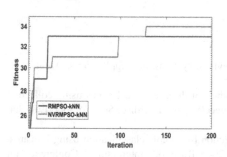

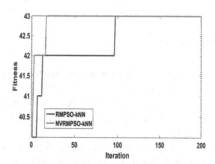

Fig. 2. Convergence graph of RMPSO-kNN vs NVRMPSO kNN on E-MEXP-1050

Fig. 3. Convergence graph of RMPSO-kNN vs NVRMPSO- kNN on GSE1058.

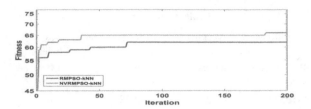

Fig. 4. Convergence graph of RMPSO-kNN vs NVRMPSO-kNN on GSE60438

For the disease Preeclampsia, following genes have been found in our predicted set of genes which have already been clinically tested: MAP2, HDAC, FLT1, STS, MTHFR [12, 13]. The other genes predicted by us need to be clinically verified whether they are genuine or not.

6   Conclusion

Here, a modified version of RMPSO, namely, NVRMPSO-kNN is proposed to identify the disease critical genes. To improve the efficiency of search by improving the exploration in the state space, two repositories (one for local best and another for global best) have been used to store the solutions having same objective function value. Furthermore a two point mutation operator has been applied to the two best solutions: local and global, to mutate the current solution. kNN algorithm has been used here for sample classification. The RMPSO-kNN and proposed NVRMPSO-kNN have been applied to three different preeclampsia datasets i.e. E-MEXP-1050, GSE60438 and GSE10588. It is observed that the proposed method has outperformed the earlier method (RMPSO-kNN) in all datasets. Furthermore, the performance of NVRMPSO-kNN has been compared to some literature methods like, HS-kNN, DE-kNN, HBMO-kNN, SGA-kNN. There is a future scope for comparing the proposed method with other meta-heuristics. It can also be applied to other diseases datasets.

References

1. Heller, M.J.: DNA microarray technology: devices, systems, and applications. Annu. Rev. Biomed. Eng. **4**(1), 129–153 (2002)
2. Biswas, S., Dutta, S., Acharyya, S.: Identification of disease critical genes using collective meta-heuristic approaches: an application to preeclampsia. Interdisc. Sci. Comput. Life Sci. 1–16 (2017)
3. Saha, S., Biswas, S., Acharyya, S.: Gene selection by sample classification using k nearest neighbor and meta-heuristic algorithms. In: 2016 IEEE 6th International Conference on Advanced Computing (IACC), pp. 250–255. IEEE, February 2016
4. Chen, X.W.: Gene selection for cancer classification using bootstrapped genetic algorithms and support vector machines. In: Proceedings of the 2003 IEEE Bioinformatics Conference. CSB 2003 on Computational Systems Bioinformatics, CSB 2003, pp. 504–505. IEEE, August 2003

5. Alba, E., Garcia-Nieto, J., Jourdan, L., Talbi, E.G.: Gene selection in cancer classification using PSO/SVM and GA/SVM hybrid algorithms. In: 2007 IEEE Congress on Evolutionary Computation, pp. 284–290. IEEE, September 2007

6. Alomari, O.A., Khader, A.T., Al-Betar, M.A., Abualigah, L.M.: Gene selection for cancer classification by combining minimum redundancy maximum relevancy and bat-inspired algorithm. Int. J. Data Min. Bioinform. **19**(1), 32–51 (2017)

7. Moosa, J.M., Shakur, R., Kaykobad, M., Rahman, M.S.: Gene selection for cancer classification with the help of bees. BMC Med. Genomics **9**(2), 47 (2016)

8. Kar, S., Sharma, K.D., Maitra, M.: Gene selection from microarray gene expression data for classification of cancer subgroups employing PSO and adaptive K-nearest neighborhood technique. Expert Syst. Appl. **42**(1), 612–627 (2015)

9. Dashtban, M., Balafar, M., Suravajhala, P.: Gene selection for tumor classification using a novel bio-inspired multi-objective approach. Genomics **110**(1), 10–17 (2018)

10. Pyingkodi, M., Thangarajan, R.: Informative gene selection for cancer classification with microarray data using a metaheuristic framework. Asian Pac. J. Cancer Prev. APJCP **19**(2), 561 (2018)

11. Jana, B., Mitra, S., Acharyya, S.: Repository and Mutation based Particle Swarm Optimization (RMPSO): a new PSO variant applied to reconstruction of Gene Regulatory Network. Appl. Soft Comput. **74**, 330–355 (2019)

12. Ching, T., et al.: Genome-wide hypermethylation coupled with promoter hypomethylation in the chorioamniotic membranes of early onset pre-eclampsia. Mol. Hum. Reprod. **20**(9), 885–904 (2014)

13. Vaiman, D., Miralles, F.: An integrative analysis of preeclampsia based on the construction of an extended composite network featuring protein-protein physical interactions and transcriptional relationships. PLoS ONE **11**(11), e0165849 (2016)

Neuronal Dendritic Fiber Interference Due to Signal Propagation

Satyabrat Malla Bujar Baruah$^{(\boxtimes)}$, Plabita Gogoi, and Soumik Roy

Department of Electronics and Communication Engineering, Tezpur University,
Napam, Tezpur, Assam, India
baruah.satyabrat@gmail.com, plabita.gogoi6@gmail.com, xoumik@tezu.ernet.in

Abstract. In the proposed model, the mathematical modeling of a pair of dendritic fibers using the famous cable equation has been carried out to simulate the effects of signal interference. The response of one fiber due to propagating signal in another fiber has been simulated using the modeled equations which show some interesting results. The simulation shows that the baseline wandering in propagating signal as one of the probable reason for interfiber signal interference. The results also show low-frequency noise while modeling for interfiber signal interference in a pair of dendritic fibers which seems to be an effect of phase-shifted signal, propagation in nearby fiber. In similar simulation, with on phase signal propagation in the two nearby fiber, the results shows shift in the signal baseline potential accompanied by baseline wandering in the signal, which shows the baseline shift, baseline wandering and noise as probable function of interference due to propagating signal in a nearby fiber.

Keywords: Neuron signal interference · Neuronal signal artifacts · Hodgkin and Huxley model · Membrane properties · Inter-fibre interference · Cable equation

1 Introduction

Cable equation and its implementation in modeling and simulation of neuronal signal processing have been playing significant role. Its implication is found in modeling of either signal propagation in a neuronal fiber or extracellular field potential models thus giving theoretical insight into complex neuronal processes. The technical understanding of varying action potential shape and change in velocity in a neuronal fiber of varying diameter has been put forward in [6, 15, 17] using cable equation approach, which results in Rall's equivalent cylinder and Rall's 3/2 rule, implementable in branching dendritic fibers for a class of neuronal morphology. Cable model in [19] combined with Rall's 3/2 rule and Agmon-Snir [1] extension model of cable equation shows its implementation in axonal regions and passive dendrites. On similar notes, in [4, 10, 13, 16, 18, 19] cable models found its uses in understanding of different dendritic morphology and structures. Even

B. Deka et al. (Eds.): PReMI 2019, LNCS 11942, pp. 176–183, 2019.
https://doi.org/10.1007/978-3-030-34872-4_20

extended models of the infamous cable equation have found its place in modeling effects of local field potentials (LFP) and their respective effects in neuronal activities and cognition [3].

Numerous research has been put forward in order to understand the role of complex neuronal arbors, either in understanding their role in signal processing or in the understanding of LPF dynamics due to complex structures. In [9], incorporating ephaptic interactions between neuronal fibers, Holts et al. modeled ephaptic depolarization and discussed their probable functional relevance. In literature [11], the neuronal signal noises are interpreted due to channel properties, thermal noise of membrane and random background activity, considering the fluctuation of membrane conductance to be significantly lower than the resting conductance. In Hasselmo et al. [8], it has been found from extracellular recordings and computational modeling that the sub-threshold membrane potential interaction with different branches plays important role in theta frequency oscillations whereas Anastassiou [2] discusses the capability of ephaptic coupling of extracellular field and its role in low-frequency oscillations in cortical neuron. Similarly, literature [7,20] also discusses the capability of extracellular field in mediating blocked action potential, correlates the extracellular potential as one of the important aspects of neuronal signal processing [12]. Such models of sustained cortical oscillations and ephaptic coupling of neuronal fibers have influenced to further investigate the cause and effects of the extracellular potential on neuronal signal processing. An attempt is made to model and understands the effect of extracellular potential on nerve membrane using a detailed cable model. In the methods section, representation of the cable model corresponding to a pair of passive dendritic fiber has been discussed and equivalent membrane potential equations are designed whereas, in the results section, the simulated results of the membrane equations are discussed.

2 Methods: Interference Model and Its Cable Representation

Figure 1 is the representation of two parallel dendritic fibres where Ra_1 and Ra_2 are the axial endoplasmic resistance, Re is the exoplasmic resistance, Ri_1 and Ri_2 are the membrane resistance and C_m is the membrane capacitance. The corresponding membrane constants are considered for a length of Δx as seen in Fig. 1.

Considering Fig. 1, the node potential for the external and internal section of the fiber is represented as Vi_1, Vi_2, Ve and V_{m1}, V_{m2} being the membrane potential of the two fibers, applying Kirchhoff's voltage and current law at the system circuit gives

$$\frac{dV_{i1}}{dx} = -Ri_1 Ia_1, \quad \frac{dV_{i2}}{dx} = -Ri_2 Ia_2, \quad \frac{dV_e}{dx} = Re Ie \qquad (1)$$

$$I_{T1} = \frac{dIa_1}{dx}, \quad I_{T2} = \frac{dIa_2}{dx}, \quad I_e = I_{T1} + I_{T2} \qquad (2)$$

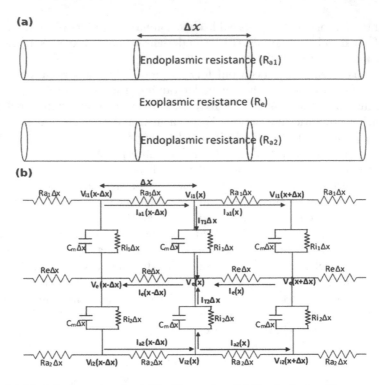

Fig. 1. (a) Model of passive dendritic fiber pairs. (b) Equivalent cable representation circuit considering endoplasmic resistance, exoplasmic resistance, membrane resistance, membrane capacitance, and voltage and current effects in the circuit due to interference between the two fibers during signal transmission/propagation.

Using Eqs. 1 and 2, the membrane potential for individual fibers are given as

$$\frac{d^2V_{m1}}{dx^2} = Ra_1Cm_1\frac{dV_{m1}}{dt} + (Ra_1I_{ionic1} + ReI_{ionic2}) \tag{3}$$

$$\frac{d^2V_{m2}}{dx^2} = Ra_2Cm_2\frac{dV_{m2}}{dt} + (Ra_2I_{ionic2} + ReI_{ionic1}) \tag{4}$$

Simplifying Eqs. 3 and 4, considering the fiber to be equipotential over small length and equating for time response of the system of equation for a definite length of the dendritic fibers as a function of membrane constants, the system of equation can be described as

$$\frac{dV_{m1}}{dt} = \frac{Re}{(Ra_1Ra_2 - Re^2)Cm_1}(Ra_2I_{ionic2} + ReI_{ionic1}) \tag{5}$$

$$\frac{dV_{m2}}{dt} = \frac{Re}{(Ra_1Ra_2 - Re^2)Cm_2}(Ra_1I_{ionic1} + ReI_{ionic2}) \tag{6}$$

where I_{ionic1} and I_{ionic2} are ionic currents due to membrane leakage or membrane transport.

Considering the two dendritic fibers being triggered due to active membrane on injecting two currents I_{in1}, I_{in2} on either of the fiber, the passive propagating response of the system is given as

$$\frac{dV_{m1}}{dt} = I_{inj1} + \frac{Re}{(Ra_1 Ra_2 - Re^2) Cm_1} (Ra_2 I_{ionic2} + Re I_{ionic1}) \qquad (7)$$

$$\frac{dV_{m2}}{dt} = I_{inj2} + \frac{Re}{(Ra_1 Ra_2 - Re^2) Cm_2} (Ra_1 I_{ionic1} + Re I_{ionic2}) \qquad (8)$$

3   Results and Discussion

All results and data are generated in the XPP solver for solving differential equations. XPP solver is basically a differential equation solver, initially designed and written by Bard Ermentrout and John Rinzel to simulate differential equations associated with nerve membrane dynamics and phaseplane analysis. Current version incorporates number of solvers that can solve differential equations, delay

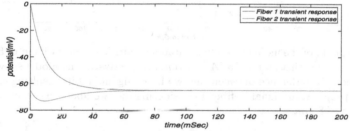

(a) Transient response of a pair of dendritic fibers when potential decay affects the resting state of the other dendritic fiber.

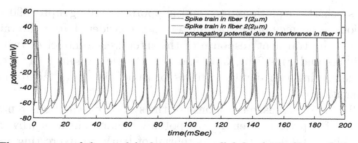

(b) The response of the model when two parallel dendritic fibers of diameter $2\mu m$ are triggered with stimulus $15nA$ and $70nA$ respectively. The model showing the effect of interference between the two fibers signal and its resulting effect on the propagating signal in fiber 1 considering the two fibers to overlay over small length.

Fig. 2. Interference responses from the model.

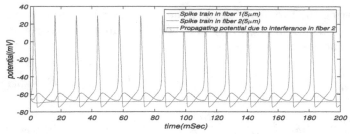

(a) Response of the model when two parallel dendritic fibers of diameter $5\mu m$ are triggered with stimulus $15nA$ and $0nA$ respectively. The model showing the effect of interference between the two fibers signal and its resulting effect on the propagating signal in fiber 2 considering the two fibers to overlay over small length.

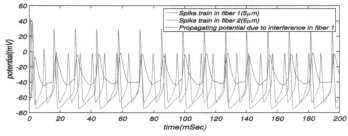

(b) Response of the model when two parallel dendritic fibers of diameter $5\mu m$ are triggered with stimulus $15nA$ and $70nA$ respectively. The model showing the effect of interference between the two fibers signal and its resulting effect on the propagating signal in fiber 1 considering the two fibers to overlay over a length of $80\mu m$.

Fig. 3. Interference responses from the model on baseline shift and baseline wondering.

equations stochastic equations etc. and supports handling upto 590 differential equations. The membrane parameters are considered from literature [5]. Figure 2(a) is the transient response of the interference model of two parallel dendritic fibres and Figs. 2(b), 3(a) and (b) are the responses of interference model implementing two dendritic fibres of same diameters facilitating spike train propagation.

Figure 2(a) is the transient response of the system of equations representing the response of a passive fiber due to potential change in a nearby fiber. As a dendritic fiber stabilizes from a potential of 0 mV to its resting potential, the membrane potential of a nearby fiber, initially at rest, is perturbed due to interference from the first fiber. Figure 2(b) is the model response considering each fiber ($2\,\mu$m *diameter*) facilitating spike train propagation that results in inter-fiber interference. Figure 2(b) shows the interference response of fiber 1 due to signal propagation in fiber 2, which results in small noise like structures convolving with the original signal, which suggests that the high-frequency noise

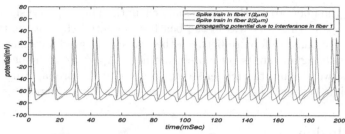

(a) Response of the model when two parallel dendritic fibers of diameter $2\mu m$ are triggered with stimulus $15nA$ and $20nA$ respectively. The model showing the interference results in amplification of propagating signal as the phase shift between two action potential increases in fiber 1 considering the two fibers to overlay over a small length.

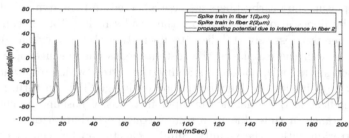

(b) Response of the model when two parallel dendritic fibers of diameter $2\mu m$ are triggered with stimulus $15nA$ and $20nA$ respectively. The model showing the interference results in amplification of propagating signal as the phase shift between two action potential increases in fiber 2 considering the two fibers to overlay over a small length.

Fig. 4. Interference responses from the model showing amplification of the propagating signal.

in the propagating signal might be the result of interference from another nearby fiber facilitating signal transmission at same instance of time.

Also, Fig. 3(a) shows the model response of fiber 2 due to spike train propagation in fiber 1, where single fiber facilitates spike train propagation. Due to spike train propagation in fiber 1, sustained oscillations are produced in the fiber 2 which was initially devoid of any spiking activity. Again, Fig. 3(b) is the response of interference model, considering two fiber of diameter 5 μm each. This response model shows significant shift in the baseline of the propagating signal along with minor baseline wondering as discussed in [14], which suggests that the baseline potential shift and baseline wondering might also be the consensus result due to propagating signal in nearby fiber when interference surface exposed is over a longer distance.

Figure 4(a) and (b) are the responses of two fibers (2 μm diameter each) showing the effect of on phase and off phase signals on the interference model, considering the fibers to overlay over smaller length. The response shows the on phase signals affecting very less in the baseline potential shift of the two fibers whereas, as the phase difference between the two signals gradually increases, the baseline wandering and shifts get more prominent. Apart from the baseline shift and baseline wondering, the model seems to amplify the propagating signal as the phase shift between the two signals in the nearby fibers increases.

4   Conclusion

As the endoplasm of a neuron is more negatively charged than the exoplasm, the membrane potential of a neuron is negative. During the application of stimulation, the ion density of the inside with respect to the outside of the membrane changes significantly that gives rise to the action potential. Due to such dynamics of the cytoplasm, the nearby fibers also get affected and the dynamic behavior of the nearby fibers also changes. From the results in Fig. 2(a), it can be seen that as a fiber progressively stabilizes to its resting state by exchanging Na^+ and K^+ ions with the exoplasm, the instantaneous change in exoplasmic ion concentration starts ionic motion to stabilize the system which results in attraction of new negative ions to balance the ion exchange. This, in turn, results in interference effect in the other nearby fiber in terms of baseline potential shift for small instance of time due to instantaneous perturbation of extracellular ion concentration within the vicinity. Similar effects can be seen in Figs. 2(b), 3(a), (b) etc. which has some additional effects such as interference noise, signal amplification and baseline wondering. Results in Fig. 4(a) and (b) also suggests that the amplification and baseline shift is minimal when two fibers spikes are at same phase whereas the amplification and baseline shift increases with increase in phase shift between the two respective signals.

Acknowledgement. This publication is an outcome of the R&D work undertaken project under the Visvesvaraya Ph.D. Scheme of Ministry of Electronics & Information Technology, Government of India, being implemented by Digital India Corporation.

References

1. Agmon-Snir, H.: A novel theoretical approach to the analysis of dendritic transients. Biophys. J. **69**(5), 1633–1656 (1995)
2. Anastassiou, C.A., Perin, R., Markram, H., Koch, C.: Ephaptic coupling of cortical neurons. Nat. Neurosci. **14**(2), 217 (2011)
3. Buzsáki, G., Anastassiou, C.A., Koch, C.: The origin of extracellular fields and currents—EEG, ECoG, LFP and spikes. Nat. Rev. Neurosci. **13**(6), 407 (2012)
4. Cao, B.J., Abbott, L.: A new computational method for cable theory problems. Biophys. J. **64**(2), 303–313 (1993)
5. Gerstner, W., Kistler, W.M.: Spiking Neuron Models: Single Neurons, Populations, Plasticity. Cambridge University Press, Cambridge (2002)
6. Goldstein, S.S., Rall, W.: Changes of action potential shape and velocity for changing core conductor geometry. Biophys. J. **14**(10), 731–757 (1974)

7. Guo, S., Wang, C., Ma, J., Jin, W.: Transmission of blocked electric pulses in a cable neuron model by using an electric field. Neurocomputing **216**, 627–637 (2016)
8. Hasselmo, M.E., Giocomo, L.M., Zilli, E.A.: Grid cell firing may arise from interference of theta frequency membrane potential oscillations in single neurons. Hippocampus **17**(12), 1252–1271 (2007)
9. Holt, G.R., Koch, C.: Electrical interactions via the extracellular potential near cell bodies. J. Comput. Neurosci. **6**(2), 169–184 (1999)
10. Major, G., Evans, J.D., Jack, J.: Solutions for transients in arbitrarily branching cables: I. Voltage recording with a somatic shunt. Biophys. J. **65**(1), 423 (1993)
11. Manwani, A., Koch, C.: Detecting and estimating signals in noisy cable structures, I: neuronal noise sources. Neural Comput. **11**(8), 1797–1829 (1999)
12. Martinez-Banaclocha, M.: Ephaptic coupling of cortical neurons: possible contribution of astroglial magnetic fields? Neuroscience **370**, 37–45 (2018)
13. Monai, H., Omori, T., Okada, M., Inoue, M., Miyakawa, H., Aonishi, T.: An analytic solution of the cable equation predicts frequency preference of a passive shunt-end cylindrical cable in response to extracellular oscillating electric fields. Biophys. J. **98**(4), 524–533 (2010)
14. Onslow, A.C.E., Hasselmo, M.E., Newman, E.L.: DC-shifts in amplitude in-field generated by an oscillatory interference model of grid cell firing. Front. Syst. Neurosci. **8**, 1 (2014)
15. Rall, W.: Theory of physiological properties of dendrites. Ann. N. Y. Acad. Sci. **96**(4), 1071–1092 (1962)
16. Rall, W.: Time constants and electrotonic length of membrane cylinders and neurons. Biophys. J. **9**(12), 1483–1508 (1969)
17. Rall, W.: Core conductor theory and cable properties of neurons. Compr. Physiol. 39–97 (2011)
18. Ramón, F., Joyner, R., Moore, J.: Propagation of action potentials in inhomogeneous axon regions. In: Moore, J.W. (ed.) Membranes, Ions, and Impulses FASEBM, vol. 5, pp 85–100. Springer, Boston (1975). https://doi.org/10.1007/978-1-4684-2637-3_8
19. Rinzel, J.: Voltage transients in neuronal dendritic trees. In: Moore, J.W. (ed.) Membranes, Ions, and Impulses. FASEBM, vol. 5, pp. 71–83. Springer, Boston (1975). https://doi.org/10.1007/978-1-4684-2637-3_7
20. Yi, G.S., Wang, J., Wei, X.L., Tsang, K.M., Chan, W.L., Deng, B.: Neuronal spike initiation modulated by extracellular electric fields. PLoS ONE **9**(5), e97481 (2014)

Two-Class *in Silico* Categorization of Intermediate Epileptic EEG Data

Abhijit Dasgupta[1], Ritankar Das[1], Losiana Nayak[1], Ashis Datta[2], and Rajat K. De[1(✉)]

[1] Machine Intelligence Unit, Indian Statistical Institute,
203 Barrackpore Trunk Road, Kolkata 700108, India
rajat@isical.ac.in
[2] Epilepsy and Electrophysiology Division, Institute of Neurosciences,
185/1 A.J.C. Bose Road, Kolkata 700017, India

Abstract. Epilepsy treatment depends on multiple instances of EEG recordings. Often, clinicians encounter intermediate/borderline EEG signals in the recordings, and as a result, inconclusiveness arises regarding the epilepsy status of a patient. In this paper, we have addressed this issue with a computational solution. We have classified and created class-specific clusters of the EEG data belonging to epileptic patients and normal individuals using a scaled conjugate feed forward neural network (FNN) and average silhouette based k-means clustering algorithm respectively. Thereafter, we have categorized the intermediate data into the clusters of these two classes for a better clinical decision making using minimum squared Euclidean distance. The methodology proposed here can help the clinicians in dealing with intermediate EEG signals found in individuals suspected as suffering from epilepsy. It will also help in categorizing intermediate EEG data, and in turn facilitate clear diagnosis and better patient care in the case of undecided patients.

Keywords: Epilepsy · Epileptic network · Feed forward neural network · k-means clustering · Borderline EEG data

1 Introduction

Seizure is a common neurological disorder caused by abnormal electrical activities in the brain. Most often a seizure begins with micro-seizures starting asynchronously from small neuronal clusters. These micro-seizures then construct a macroscale hypersynchronization, *i.e.*, a seizure in combined form [11]. Epilepsy is a chronic disorder of the brain that can be caused by some underlying disease mechanism and is usually characterized by occurrence of seizure. However, the exact cause remains unknown in about 50% of patients [15].

It is estimated that about 70% of people living with epilepsy could live seizure-free if proper diagnosis and treatment are followed. However, in developing countries, epilepsy is still not considered as a disease of importance. The

B. Deka et al. (Eds.): PReMI 2019, LNCS 11942, pp. 184–192, 2019.
https://doi.org/10.1007/978-3-030-34872-4_21

negligence in treatment due to lack of awareness, resources, and manpower as caregivers often leads to premature mortality of epileptic patients [3].

Epilepsy is characterized by recurrent epileptic seizures. Electroencephalogram (EEG) is helpful in evaluating a patient with epilepsy [13]. It is usually interpreted by an experienced neurologist, epileptologist or an electrophysiologist. Often, the intermediate or borderline EEG recordings of individuals create a lot of confusion in clinical decision making. Whether these recordings behave as normal or epilepsy cannot be specifically determined, and thus it affects treatment. Clinicians are usually inclined towards further observation and treatment as and when required. Automated analysis and classification of such EEG data with the help of computational techniques can be helpful in interpreting such recordings as normal or epilepsy. Although a considerable amount of work has been done to classify EEG signals as either normal or epilepsy [1,2,14], none of them involves categorization of intermediate EEG data.

In this paper, we have analyzed EEG data of 91 healthy volunteers and 119 epilepsy patients to categorize 14 intermediate individuals. We have extracted 120 significant features from adjacency matrices obtained from the EEG data of all the categories. Thereafter, scaled conjugate gradient (SCG) feed forward neural network (FNN) is employed to classify the data in a frame of two-class problem. Silhouette based k-means clustering has been used to extract clusters of normal and epileptic EEG data followed by superimposition of the intermediate data to either epileptic or normal classes using minimum squared Euclidean distance.

2   Data

We have collected Electroencephalographic (EEG) data [8] of epileptic, intermediate and normal individuals, recorded using 16 channels/electrodes placed on the scalp. The location and naming of the electrodes follow the internationally accepted Modified Combinatorial Nomenclature (MCN) system [10].

Here, we have collected data from 91 healthy volunteers comprising 36 males and 55 females with a mean age of 19.1 years. Besides 119 patients suffering from epilepsy have been considered. Among them, 53 are males and 66 are females with a mean age of 18.75 years. Intermediate data (data which cannot be rightly classified as normal or epilepsy) has been collected from 14 individuals. Here, a total of 73 EEG recordings have been selected from 7 females and 7 males with a mean age of 44.21 years.

EEG Data have been recorded using Recorders & Medicare Systems Pvt. Ltd. (RMS) computerized EEG machine and Nihon Kohden Neurofax (NKN) EEG-1200, for an interval of 20–30 min each. During recording, all the participants have been asked to relax in a no-thinking state, and to lay motionless as far as possible keeping their eyes open.

3   Method

The methodology involves data filtering, data pre-processing, computation of adjacency matrix, creation of the final dataset, classifying normal and epileptic data and finally categorization of intermediate data as depicted in Fig. 1.

Data Filtering and Pre-processing

Data filtering steps include resampling, re-referencing, baseline correction and Independent Component Analysis (ICA) as per our previous investigations [5,6]. Here we have used EEGLAB toolbox [7] version 13 in MATLAB R2018a platform to implement the data filtering steps.

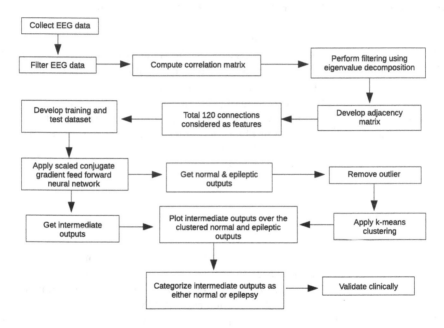

Fig. 1. Flowchart of the overall methodology.

In addition, we have employed a sliding window based eigenvalue decomposition technique to remove unwanted noise from the filtered time series data so that an adjacency matrix of the network, connecting positional nodes of the electrodes, can be generated. A vector of length l has been defined to measure the potential of each electrode on a continuous time series. Here, we have designed a window of size s to slide over the entire length l to divide the data into N_{trial} number of smaller segments/trials for each class, such as normal, epileptic and intermediate. Thus, we have obtained 1085, 834 and 499 trial data from 91 healthy volunteers, 119 epileptic and 14 intermediate patients respectively. It helps in increasing the number of inputs applied to the learning algorithm. Here, the sliding window has an overlap of 20%.

Computing Correlation Matrix. We have computed Pearson's correlation coefficient to measure the linear dependency of one channel to another. Here, for each window of size s, and two channels $\boldsymbol{x}^p = [x_1^p, x_2^p, \ldots, x_s^p]^T$ and $\boldsymbol{x}^q = [x_1^q, x_2^q, \ldots, x_s^q]^T$, the Pearson's correlation coefficient (c_{pq}) can be calculated as

$$c_{pq} = \frac{1}{s-1} \sum_{i=1}^{s} \left(\frac{x_i^p - \mu_{x^p}}{\sigma_{x^p}} \right) \left(\frac{x_i^q - \mu_{x^q}}{\sigma_{x^q}} \right) \tag{1}$$

Here, the terms μ_{x^p} and μ_{x^q} represent the mean values, while σ_{x^p} and σ_{x^q} stand for standard deviations of $\boldsymbol{x}^p$ and $\boldsymbol{x}^q$ respectively.

Thus, we obtain a correlation matrix $\mathbf{C} = [c_{pq}]_{n \times n}$ for each trial of each class, where $n = 16$ in our study. Here $c_{pq} = 1$ if $p = q$, otherwise c_{pq} lies between $+1$ and -1. Positive values represent positive correlation and negative values represent negative correlation, while a value of zero represents no linear correlation between channels. Higher the value, higher is the correlation between channels/electrodes.

Filtering Correlation Matrix. In order to get a more appropriate noise free correlation among channels, we have carried out eigenvalue decomposition [9] on the correlation matrix. The eigenvalue spectrum of the correlation matrix contains a global part $\mathbf{C}^{global}$ with the largest eigenvalue, regional part $\mathbf{C}^{regional}$ of our interest with intermediate eigenvalues, and finally a random part $\mathbf{C}^{random}$ which contains a bulk of small eigenvalues. Thus, the correlation matrix (C) can be expressed as

$$C = \boldsymbol{\nu}_0 \lambda_0 \boldsymbol{\nu}_0^T + \sum_{i=1}^{n_h} \boldsymbol{\nu}_i \lambda_i \boldsymbol{\nu}_i^T + \sum_{j=n_h+1}^{n-1} \boldsymbol{\nu}_j \lambda_j \boldsymbol{\nu}_j^T \tag{2}$$

Here, the term n_h determines the separating boundary between $\mathbf{C}^{regional}$ and $\mathbf{C}^{random}$. The terms $\boldsymbol{\nu}_0$ and λ_0 stand for the eigenvector and eigenvalue of the global part respectively, while $\boldsymbol{\nu}_i$ and λ_i, for $i = 1, \ldots, n_h$, represent the eigenvectors and eigenvalues of the regional part respectively. Similarly, the terms $\boldsymbol{\nu}_j$ and λ_j, for $i = n_{h+1}, \ldots, n-1$, represent the eigenvectors and eigenvalues of the random part respectively.

Filtering the correlation matrix involves removal of both the global and random part. The global part contributes to the highest eigenvalues and affects all the channels during the recording of EEG. On the other hand, random noise that lies at the bottom of the eigenvalue spectrum, may interfere with one or more channels but does not affect all the channels as in the case of global noise. Thus, the filtered correlation matrix can be written as

$$C_{filtered} = \sum_{i=1}^{n_h} \boldsymbol{\nu}_i \lambda_i \boldsymbol{\nu}_i^T \tag{3}$$

Here, n_h has been selected in such a way that it can be possible to nullify the effect of the random part by filtering out the eigenvectors with minimum eigenvalues. There are 16 eigenvalues (λ_i, for $i = 1, \cdots, 16$) for each correlation matrix of each trials for each class. The highest eigenvalue in the eigenvalue spectrum represents the global part of each correlation matrix. Besides, the random part contains the smallest eigenvalues. For epilepsy, we have chosen n_h to be 6, *i.e.*, after removing the global part, the next six highest eigenvalues contribute to the regional part. Similarly, we have set n_h as 7 and 8 for normal and intermediate trials respectively.

The determination of n_h is difficult but obtaining an exact value is not crucial, as a small change of n_h does not affect the result. In the eigenvalue spectrum, the eigenvalues corresponding to the regional part, are confined to a small portion. Therefore, if we neglect the eigenvalues that lie close to the boundary of the random region, it does not affect the final result much.

Computation of Adjacency Matrix

We set a cutoff value α from the filtered correlation matrix to obtain an adjacency matrix **A**. The value α can be calculated as

$$\alpha = \frac{1}{N_{ind}} \sum_{i=1}^{N_{ind}} \left(\frac{1}{N_{trial}} \sum_{j=1}^{N_{trial}} \left(\frac{1}{n \times n} \sum_{k,l} |c_{kl}| \right) \right) \tag{4}$$

where N_{ind} represents number of individuals belonging to each class, and N_{trial} is the number of trials carried out on each individual. Subsequently, c_{kl} is an element of the $n \times n$ matrix $C_{filtered}$. Thus, for $(a, b)^{th}$ element of $C_{filtered}$ for each trial, $|c_{ab}| \geq |\alpha|$ and $a \neq b$, we consider an edge between a^{th} and b^{th} electrode with the corresponding correlation value ($|c_{ab}|$) as weights, otherwise we have considered no edge between the two electrodes.

Creation of the Final Dataset

The resultant adjacency matrix provides a weighted undirected graph $\mathbf{G}(V, E)$, where V is a finite non-empty set of electrodes and E is the set of edges. There exists a function from $V \times V \rightarrow E$ such that, $E = \{\{u, v\} | u, v \in V, \text{ and } u \neq v\}$. Thus, for a set of 16 electrodes, there are maximum of $(16 \times 15)/2 = 120$ significant connections possible. These connections are considered as features. Besides, the weight value assigned to each connection represents a feature value. Thus, the training input matrix for the learning algorithm (a feed forward neural network) is represented by $\mathcal{I}_{N_{trial} \times 121}$ such that $e_{ij} \in \mathcal{I}$ ($j \neq 121$) represents j^{th} feature of ith trial and $e_{ij} \in \mathcal{I}$ ($j = 121$) represents the class level of ith trial. Similarly, the test input matrix for the learning algorithm is denoted by $\mathcal{I}'_{N_{trial} \times 120}$ such that $e_{ij} \in \mathcal{I}'$ represents j^{th} feature of ith trial.

We have used 1919 samples/trials to be classified into two classes, *viz.*, epilepsy and normal. Out of the total number of samples 70% (1343 samples) are randomly selected for training while 10% (192 samples) are considered for validation. The remaining 20% (384 samples) are used to test the prediction accuracy.

Classifying Normal and Epileptic Data

Here, we have classified epileptic and normal samples using a scaled conjugate gradient (SCG) [4,12] feed-forward neural network for its quick and better convergence rate. Besides, SCG has shown better classification accuracy compared to self-organizing feature map, multilayer perceptron and support vector machine.

We have randomly chosen an initial weight vector and set an initial conjugate vector to the steepest descent vector. Thereafter, we have calculated the Hessian matrix approximation and the scaling factor. The scaling factor has been readjusted based on the value of error approximation, whenever the Hessian matrix is not positive definite. Subsequently, the step size has been calculated. The process has been repeated iteratively until it meets the predefined performance goal or maximum validation failure.

During implementation of the aforesaid algorithm, we have set the maximum validation failures as 10, whereas the minimum performance gradient has been considered as $1/e$. Subsequently, we have considered the learning rate of 0.1. Here, the hidden layer consists of 100 nodes.

Categorization of Intermediate Data
The normal and epileptic data have again been fed into the trained feedforward neural network (FNN) to generate all the corresponding points on the output space. In order to eliminate outliers from the output space under consideration, we have used curve fitting toolbox implemented in MATLAB R2018a. Here, we have considered the data points on the output space as outliers those lie above or below 1.5 times the standard deviation from the mean.

The outlier removed data are then clustered into k clusters using average silhouette based k-means clustering algorithm. Further, the trained FNN generated output points for intermediate trials have been projected on resulting k clusters for normal and epileptic dataset. Here, the squared Euclidean distance from each intermediate output point d to each cluster center ψ_i, *for $i = 1, \ldots, k$*, can be calculated as

$$\gamma^2 = (\psi_i - d) * (\psi_i - d)^T \tag{5}$$

Now, the point d has been clubbed into the cluster corresponding to minimum γ^2. Similarly, all other output points for intermediate epileptic trails have been categorized as either normal or epilepsy. For each individual patients, if more than 50% of trails have been categorized as normal/epilepsy, then the individual recording has been treated as that of the same category. Whereas, if exactly 50% of trials have been categorized as normal/epilepsy, we have left the individual recording as undetermined.

4   Results and Discussion

Here, applying scaled conjugate gradient feed forward neural network with 100 nodes in the hidden layer, we have correctly classified 781 epilepsy and 1049 normal trials out of total 1919 trials obtained from 91 normal and 119 epileptic individuals. Subsequently, 53 epileptic trials have wrongly been classified as normal, while 36 normal trials have wrongly been classified as epileptic. Thus, it has shown an overall accuracy of 95.36%.

190   A. Dasgupta et al.

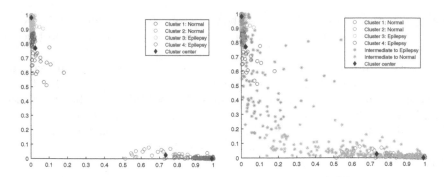

Fig. 2. The outlier removed output points from feed forward neural network for the two classes, *i.e.*, normal and epilepsy, have been plotted on the output space and clustered into four different clusters (left). Depending on minimum squared Euclidean distance from the cluster centres, the output points for intermediate data have been categorized as either normal or epilepsy (right).

Thereafter, intermediate data have been applied as inputs to the previously trained aforementioned FNN model. The output points for intermediate data have been used to categorize intermediate epileptic patients as either normal individuals or epileptic patients. Here, the output obtained from the trained FNN for all normal and epileptic trials have been clustered to serve as a template for categorizing the intermediate data as depicted in Fig. 2. In this context, it should be noted that out of 73 intermediate EEG recordings involving 14 individuals, clinicians are able to identify 38 recordings (obtained from nine individuals) as belonging to either epilepsy or normal category. Here, 31 predictions obtained by the proposed methodology have matched with the clinical recommendation. However, one recording have not be identified, while six recordings have wrongly been predicted, thus giving us an accuracy of 81.58% for intermediate recordings. On the other hand, while categorizing intermediate individuals into either normal or epilepsy class and comparing the results with the diagnosis by clinicians, we have found that the proposed methodology has correctly identified 7 individuals while 2 individuals have wrongly been categorized.

Table 1. Clinical view and computational prediction for intermediate recordings

Categorized as	Clinical view	Prediction through proposed methodology on recordings			
		Correct	Wrong	Undetermined	Accuracy (in %)
Normal	12	8	2	1	**81.58**
Epilepsy	26	23	4	0	
Undetermined	35	0	0	0	
Total	**73**	**31**	**6**	**1**	

Table 2. Clinically and computationally predicted decisions for intermediate individuals

Categorized as	Clinical view	Prediction through proposed methodology on individuals			
		Correct	Wrong	Undetermined	Accuracy (in %)
Normal	2	1	1	0	**77.78**
Epilepsy	7	6	1	0	
Undetermined	5	0	0	0	
Total	**14**	**7**	**2**	**0**	

Henceforth, we obtain an accuracy of 77.78% for intermediate individuals. Although clinicians have failed to identify the remaining 35 recordings, the proposed method is able to categorize 12 of them as epileptic and leave another one as undetermined, while rest have been categorized as normal. Thus, three clinically unidentified individuals have computationally been categorized as normal while two have been categorized as epileptic. Here, an individual has been assigned to a particular class containing his/her maximum recordings. The summary of our findings are found in Tables 1 and 2.

5   Conclusion

The primary objective of this article is to categorize intermediate EEG data into two classes; namely, normal and epilepsy. It would assist clinicians in their decision making regarding patients who cannot be rightly categorized. Clinically these individuals are usually treated according to the symptoms as and when observed. Therefore, it delays the process of disease identification and treatment.

Here we have tried to employ different machine learning tools to categorize intermediate recordings as either normal or epilepsy. It is quite successful in categorizing each intermediate recordings and has shown 77.78% accurate result for individuals. The prediction made by the methodology can thus help clinicians in accurately diagnosing individuals as normal, epilepsy or intermediate.

This study has been conducted on individuals above the age of 10 years. Among them, most of the people are right-handed. All recordings have been taken at no sleep and eye open state. Similar studies can be conducted on children and/or during sleep. Further, obtaining regional brain signatures from right brain dominant individuals and comparing them with our study may provide a more clear picture of brain regions that are usually affected during epilepsy.

Acknowledgement. AD acknowledges Digital India Corporation (formerly Media Lab Asia), Ministry of Electronics and Information Technology, Government of India, for providing him a Senior Research Fellowship under the Visvesvaraya Ph.D. scheme for Electronics and IT. RD acknowledges Council for Scientific and Industrial Research (CSIR), India for providing him a Senior Research Fellowship (09/093(0182)/2018 EMR-I). LN acknowledges University Grants Commission, India for a UGC Post-Doctoral Fellowship (No. F.15-1/2013-14/PDFWM-2013-14-GE-ORI-19068(SA-II)).

References

1. Bajaj, V., Pachori, R.B.: EEG signal classification using empirical mode decomposition and support vector machine. In: Deep, K., Nagar, A., Pant, M., Bansal, J.C. (eds.) Proceedings of the International Conference on Soft Computing for Problem Solving (SocProS 2011) December 20-22, 2011. AISC, vol. 131, pp. 623–635. Springer, New Delhi (2012). https://doi.org/10.1007/978-81-322-0491-6_57
2. Bajaj, V., Pachori, R.B.: Separation of rhythms of EEG signals based on Hilbert-Huang transformation with application to seizure detection. In: Lee, G., Howard, D., Kang, J.J., Ślęzak, D. (eds.) ICHIT 2012. LNCS, vol. 7425, pp. 493–500. Springer, Heidelberg (2012). https://doi.org/10.1007/978-3-642-32645-5_62
3. Bell, G.S., Neligan, A., Sander, J.W.: An unknown quantity-the worldwide prevalence of epilepsy. Epilepsia 55(7), 958–962 (2014)
4. Chel, H., Majumder, A., Nandi, D.: Scaled conjugate gradient algorithm in neural network based approach for handwritten text recognition. In: Nagamalai, D., Renault, E., Dhanuskodi, M. (eds.) CCSEIT 2011. CCIS, vol. 204, pp. 196–210. Springer, Heidelberg (2011). https://doi.org/10.1007/978-3-642-24043-0_21
5. Dasgupta, A., Das, R., Nayak, L., De, R.K.: Analyzing epileptogenic brain connectivity networks using clinical EEG data. In: 2015 IEEE International Conference on Bioinformatics and Biomedicine (BIBM), pp. 815–821. IEEE (2015)
6. Dasgupta, A., Nayak, L., Das, R., Basu, D., Chandra, P., De, R.K.: Feature selection and fuzzy rule mining for epileptic patients from clinical EEG data. In: Shankar, B.U., Ghosh, K., Mandal, D.P., Ray, S.S., Zhang, D., Pal, S.K. (eds.) PReMI 2017. LNCS, vol. 10597, pp. 87–95. Springer, Cham (2017). https://doi.org/10.1007/978-3-319-69900-4_11
7. Delorme, A., Makeig, S.: EEGLAB: an open source toolbox for analysis of single-trial EEG dynamics including independent component analysis. J. Neurosci. Methods 134(1), 9–21 (2004)
8. Ebersole, J., Milton, J.: The electroencephalogram (EEG): a measure of neural synchrony. In: Milton, J., Jung, P. (eds.) Epilepsy as a Dynamic Disease. BIOMEDICAL, pp. 51–68. Springer, Heidelberg (2003). https://doi.org/10.1007/978-3-662-05048-4_5
9. Kim, D.H., Jeong, H.: Systematic analysis of group identification in stock markets. Phys. Rev. E 72(4), 046133 (2005)
10. Klem, G.H., Lüders, H.O., Jasper, H., Elger, C., et al.: The ten-twenty electrode system of the International Federation. Electroencephalogr. Clin. Neurophysiol. 52(3), 3–6 (1999)
11. Lowenstein, D.H.: Decade in review-epilepsy: edging toward breakthroughs in epilepsy diagnostics and care. Nat. Rev. Neurol. 11(11), 616 (2015)
12. Møller, M.F.: A scaled conjugate gradient algorithm for fast supervised learning. Neural Netw. 6(4), 525–533 (1993)
13. Noachtar, S., Rémi, J.: The role of EEG in epilepsy: a critical review. Epilepsy Behav. 15(1), 22–33 (2009)
14. Subasi, A.: EEG signal classification using wavelet feature extraction and a mixture of expert model. Expert Syst. Appl. 32(4), 1084–1093 (2007)
15. Zunt, J.R., et al.: Global, regional, and national burden of meningitis, 1990–2016: a systematic analysis for the Global Burden of Disease Study 2016. Lancet Neurol. 17(12), 1061–1082 (2018)

Employing Temporal Properties of Brain Activity for Classifying Autism Using Machine Learning

Preetam Srikar Dammu[(✉)] and Raju Surampudi Bapi

School of Computer and Information Sciences, University of Hyderabad,
Hyderabad, India
preetam.srikar@gmail.com, raju.bapi@iiit.ac.in

Abstract. Exploration of brain imaging data with machine learning methods has been beneficial in identifying and probing the impacts of neurological disorders. Psychopathological ailments that disrupt brain activity can be discerned with the help of resting-state functional magnetic resonance imaging (rs-fMRI). Research has revealed that brain connectivity is dynamic in nature and that its dynamic properties are affected by brain disorders. In the literature, numerous approaches have been proposed for identifying the presence of Autism Spectral Disorder (ASD), yet most of them do not consider brain dynamics in their diagnostic process. Significant amount of knowledge can be procured by taking the evolution of brain connectivity over time into account. In this work, we propose a new approach that leverages brain dynamics in the classification of autistic and neurotypical subjects using rs-fMRI data. We examined the proposed method on a large multi-site dataset known as ABIDE (Autism Brain Imaging Data Exchange) and have achieved state-of-the-art classification results with an accuracy of 73.6%. Our work has shown that taking the temporal properties of brain connectivity into account improves the classification performance.

Keywords: rs-fMRI · Autism detection · Dynamic FC

1 Introduction

Machine learning is being used extensively for identifying diseases of all kinds using biological data in the recent times. Neuroimaging of the brain produces high dimensional data which possesses abundant information regarding a person's mental health, emotions and cognitive activities. Disturbances in brain activity patterns caused by ailments which are imperceptible to humans owing to high dimensionality of the data can be recognized by machine learning methods.

Functional magnetic resonance imaging (fMRI) uses blood-oxygen-level dependent (BOLD) signals [10] to measure the level of activity of brain regions. It has been established by various studies that anomalies in cognitive functions

© Springer Nature Switzerland AG 2019
B. Deka et al. (Eds.): PReMI 2019, LNCS 11942, pp. 193–200, 2019.
https://doi.org/10.1007/978-3-030-34872-4_22

are induced by neurological diseases, and these aberrations are reflected in brain activity patterns [3,12,13,15,18,24]. Hence, the information obtained from fMRI scans can be used to identify the presence of psychopathological disorders.

Functional Connectivity (FC) of the brain evolves over time and has multiple distinct states characterized by unique connectivity patterns [1,4,11].

Autism Spectrum Disorder (ASD) is a developmental disorder known to alter brain connectivity [3,12,13,15,24]. Taking brain dynamics into account has increased the accuracy in identifying subjects with ASD [22] which demonstrates that brain dynamics bear notable information about the disorder.

In this work, we leverage the temporal properties of brain connectivity to classify autistic and typically developing (TD) subjects from the ABIDE dataset [6]. We observed significant improvement in performance when compared to methods that do not consider brain dynamics.

2   Data Description

The Autism Brain Imaging Data Exchange (ABIDE) [6] is a collaborative project which consists of data from 17 different imaging sites. It is a diverse dataset comprising of data generated by different machines and scan parameters. The demographics of the participants varies highly from one site to another, as can be seen in Table 1. The heterogeneous nature of ABIDE dataset poses a challenge but also provides for better generalization [20].

2.1   Data Preprocessing

The Configurable Pipeline for the Analysis of Connectomes (CPAC) is a pipeline provided by Preprocessed Connectomes Project (PCP) which performs several preprocessing on the fMRI data. The Automated Anatomical Labeling (AAL) atlas has 116 region of interests (ROIs). Time series fMRI data used in this study uses the AAL parcellation and is preprocessed with the CPAC pipeline. Information regarding the steps involved in the CPAC pipeline is available on the ABIDE website (http://preprocessed-connectomes-project.org/abide/).

2.2   Participants

To sufficiently account for the temporal properties of brain connectivity, data from sites which have scan duration less than five minutes have not been considered. This step has resulted in the exclusion of four sites (Caltech, OHSU, Pitt & Trinity) leaving us with 14 different imaging sites and 869 participants (423 ASD, 446 TD). All of the subjects from the remaining sites included in a previous study which reported state-of-the-art accuracy have been included [8]. The demographics of the participants are presented in Table 1.

Table 1. Phenotypic details of the participants. [8] (†: Information unavailable, SD: Standard Deviation, FD: Mean Framewise Displacement, ADOS: Autism Diagnostic Observation Schedule [17])

Site	ASD			TD		FD
	Age mean (SD)	ADOS mean (SD)	Count	Age mean (SD)	Count	
CMU	26.4 (5.8)	13.1 (3.1)	M 11, F 3	26.8 (5.7)	M 10, F 3	0.29
KKI	10.0 (1.4)	12.5 (3.6)	M 16, F 4	10.0 (1.2)	M 20, F 8	0.17
LEUVEN	17.8 (5.0)	† (†)	M 26, F 3	18.2 (5.1)	M 29, F 5	0.09
MAX MUN	26.1 (14.9)	9.5 (3.6)	M 21, F 3	24.6 (8.8)	M 27, F 1	0.13
NYU	14.7 (7.1)	11.4 (4.1)	M 65, F 10	15.7 (6.2)	M 74, F 26	0.07
OLIN	16.5 (3.4)	14.1 (4.1)	M 16, F 3	16.7 (3.6)	M 13, F 2	0.18
SBL	35.0 (10.4)	9.2 (1.7)	M 15, F 0	33.7 (6.6)	M 15, F 0	0.16
SDSU	14.7 (1.8)	11.2 (4.3)	M 13, F 1	14.2 (1.9)	M 16, F 6	0.09
STANFORD	10.0 (1.6)	11.7 (3.3)	M 15, F 4	10.0 (1.6)	M 16, F 4	0.11
UCLA	13.0 (2.5)	10.9 (3.6)	M 48, F 6	13.0 (1.9)	M 38, F 6	0.19
UM	13.2 (2.4)	† (†)	M 57, F 9	14.8 (3.6)	M 56, F 18	0.14
USM	23.5 (8.3)	13.0 (3.1)	M 46, F 0	21.3 (8.4)	M 25, F 0	0.14
YALE	12.7 (3.0)	11.0 (†)	M 20, F 8	12.7 (2.8)	M 20, F 8	0.11

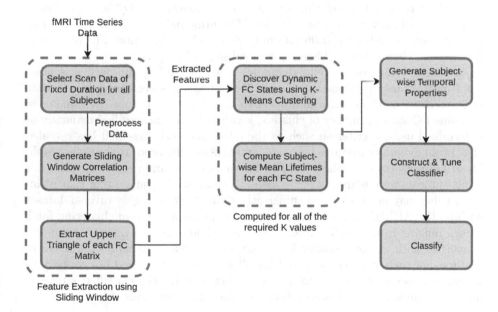

Fig. 1. Flowchart of the proposed Dynamic FC classification pipeline.

3 Methods

3.1 Proposed Dynamic FC Approach

In this work, we propose a new strategy to extract the temporal properties of brain connectivity and consequently use them as features to test for presence of autism.

When working with dynamic FC using sliding window correlation, it is essential that the duration of scan of all subjects are equal. It is also preferable if the scan parameters such as repetition time between each frame acquisition (TR) is identical for all of the subjects. Since ABIDE consists of data from various imaging sites, we do not have a single uniform TR for the entire dataset, rather we work with different scan parameter settings. The most common TR among the sites present in ABIDE is 2 s while a few of them have a different TR value. To handle this situation, we use linear interpolation on data from sites that have TR values other than 2 s and bring all of the data to a uniform TR setting (TR = 2 s).

For every subject, frames corresponding to the first 308 s (approximately 5 min) were selected and adjusted to 2 s TR, if required. This resulted in 154 frames (154 frames * 2 s TR = 308 s) for each subject. A sliding window of size 22 frames was slid in steps of 1 frame, resulting in 132 windowed correlations. Since these resultant correlation matrices are symmetric, only the upper triangular values of each matrix are extracted. The principal diagonal is also dropped since it contains self-correlation information of each brain region. The remaining features are then vectorized and used in subsequent steps.

Learning brain connectivity dynamics using clustering approach was practised in previously published studies [1,25] and was observed to yield adequate characterization. However, a limitation to this approach is that the number of dynamic FC states (number of clusters, k value) has to be specified manually or determined using a criterion such as the elbow method. Research has revealed that brain dynamics matures with age, and is also influenced by the demographics of subjects being studied [19,23]. The number of brain states revealed for adult subjects were higher in comparison to children [23], suggesting that a single k value may not be able to sufficiently characterize a highly diverse dataset such as the ABIDE dataset. Therefore, we implemented group clustering for k values ranging from 6 to 31 on all of the participant data to generate temporal properties and let the classifier learn the optimal features. Subject-wise mean lifetimes for each state is computed for all k values and stacked horizontally to generate a vector of size 481 for every subject. Finally, the classifier is trained on the dynamic temporal properties generated in the previous step (Fig. 1).

3.2 Classifier Methods

Machine learning classification methods, namely Support Vector Machines (SVM) [26], Random Forests (RF) [9], Extreme Gradient Boosting (XGB) [5,7] and Deep Neural Networks (DNN) [16] were trained on both dynamic and static features. Hyper-parameters were tuned over a sufficiently large search space. The classifiers were first tuned using random search over a large search space and then followed by a grid search on a finer and smaller search space consisting of high performing parameters.

4   Results and Discussion

Significant improvement in classification performance has been observed when dynamic features were used. The 10-fold cross-validation scores for all classifiers using both dynamic and static features are presented in Table 2. When dynamic features were used, all of the classifiers achieved accuracies greater than 71%, XGB being the highest performing classifier with 73.6% accuracy. In the case of static features, the classifiers demonstrated accuracies in the range of 65% to 70% and XGB outperformed the rest with an accuracy of 70.1%. It has to be noted that further optimization might increase performance of the classifiers, especially in the case of DNNs where the number of hyper-parameters are high and comprehensive optimization of the neural networks architecture is computationally expensive.

Table 2. Comparing classification performances of machine learning methods using Static FC and Dynamic FC features. (Classifier with the highest accuracy is boldfaced.)

Method	Accuracy	Sensitivity	Specificity	F1 score
SVM, Static	0.674	0.710	0.635	0.689
RF, Static	0.651	0.699	0.599	0.670
XGB, Static	0.701	0.719	0.682	0.710
DNN, Static	0.663	0.676	0.650	0.676
SVM, Dynamic	0.735	0.764	0.704	0.748
RF, Dynamic	0.717	0.759	0.673	0.733
XGB, Dynamic	**0.736**	**0.750**	**0.720**	**0.744**
DNN, Dynamic	0.710	0.713	0.706	0.714

To ascertain that there is a significant increase in classification accuracy, we performed a paired sample t-test on the 10-fold cross-validation accuracies generated for both static and dynamic features. We have 10 scores per each group, i.e. 10 scores generated using static features and 10 scores generated using dynamic features, and this gives a degree of freedom of 9 ($df = n - 1$). The results of the paired t-test are tabulated in Table 3 and we observe that in every case, the p-value obtained is less than the significance level (0.05). This demonstrates that using dynamic features has yielded a considerable increase in the classification performance.

Table 3. Results of paired sample t-test on the 10-fold cross-validation accuracies of methods using Static FC and Dynamic FC features. (2nd and 3rd columns show the mean 10-fold cross-validation accuracies of the corresponding method.)

Method	Static features accuracy	Dynamic features accuracy	P-value	T-value
SVM	0.674	0.735	0.001	4.549
RF	0.651	0.717	0.035	2.465
XGB	0.701	0.736	0.036	2.448
DNN	0.663	0.710	0.007	3.391

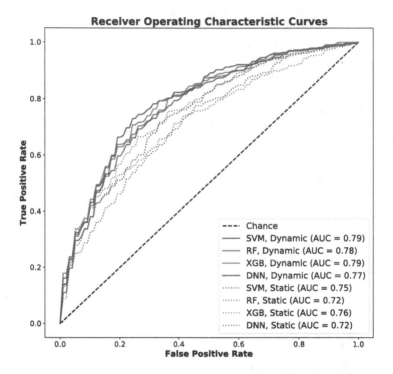

Fig. 2. ROC curves of methods using Static FC and Dynamic FC features.

Receiver operating characteristic (ROC) curves for all of the methods using both static and dynamic features are plotted in Fig. 2, along with their respective area under the curves (AUC). From the plot, we clearly notice that each method when using dynamic features outperforms its counterpart using static features. The AUCs ranged from 0.77 to 0.79 when dynamic features were used as opposed to 0.72 to 0.76 when static features were used.

The ability to harness the temporal properties of FC and use them in classification is what sets our approach apart from the previously reported ones. Most of the approaches relied on stationary FC for classification [8,14,21]. A classification method which took dynamic FC into consideration was published earlier in the literature but the experiments presented in the paper were conducted on a relatively small dataset with 60 subjects (30 ASD, 30 TD) [22]. Our method has shown to be capable of handling large datasets while being relatively less expensive computationally.

The previously reported state-of-the-art accuracy was 70% [8] and the study was conducted on data from 17 different imaging sites with 1035 subjects (505 ASD, 530 TD). It has to be noted that fewer site-wise variations may lead to an increase in classification performance, and the accuracy drops as the population size increases [2,8,20]. Another recent study reported state-of-the-art accuracy of 71.4% using the AAL parcellation on a sample size of 774 subjects (379 ASD, 395 TD) [14].

Evidently, our approach has improved the state-of-the-art results by achieving an accuracy of 73.6% on a sample of 869 participants (423 ASD, 446 TD). However, as it has been discussed earlier that the site-wise variations and sample demographics have an impact on the performance of the model, caution should be taken while comparing studies conducted on different samples.

5   Conclusion

In this paper, we proposed a new framework that leverages brain dynamics for classifying autistic individuals and neurotypicals. Our work has shown that considering the dynamic nature of brain connectivity can yield improved results in the identification of autism. Usage of more advanced techniques for modeling the brain dynamics could further improve the diagnostic process.

References

1. Allen, E.A., Damaraju, E., Plis, S.M., Erhardt, E.B., Eichele, T., Calhoun, V.D.: Tracking whole-brain connectivity dynamics in the resting state. Cereb. Cortex **24**(3), 663–676 (2014)
2. Arbabshirani, M.R., Plis, S., Sui, J., Calhoun, V.D.: Single subject prediction of brain disorders in neuroimaging: promises and pitfalls. Neuroimage **145**, 137–165 (2017)
3. Aylward, E.H., et al.: MRI volumes of amygdala and hippocampus in non-mentally retarded autistic adolescents and adults. Neurology **53**(9), 2145–2145 (1999)
4. Chang, C., Glover, G.H.: Time-frequency dynamics of resting-state brain connectivity measured with fMRI. Neuroimage **50**(1), 81–98 (2010)
5. Chen, T., Guestrin, C.: XGBoost: a scalable tree boosting system. In: Proceedings of the 22nd ACM SIGKDD International Conference on Knowledge Discovery and Data Mining, pp. 785–794. ACM (2016)
6. Craddock, C., et al.: The neuro bureau preprocessing initiative: open sharing of preprocessed neuroimaging data and derivatives. Neuroinformatics (41) (2013)
7. Friedman, J.H.: Greedy function approximation: a gradient boosting machine. Ann. Stat. **29**(5), 1189–1232 (2001)
8. Heinsfeld, A.S., Franco, A.R., Craddock, R.C., Buchweitz, A., Meneguzzi, F.: Identification of autism spectrum disorder using deep learning and the ABIDE dataset. NeuroImage Clin. **17**, 16–23 (2018)

9. Ho, T.K.: Random decision forests. In: Proceedings of 3rd International Conference on Document Analysis and Recognition, vol. 1, pp. 278–282. IEEE (1995)
10. Huettel, S.A., Song, A.W., McCarthy, G., et al.: Functional magnetic resonance imaging, vol. 1. Sinauer Associates, Sunderland (2004)
11. Hutchison, R.M., et al.: Dynamic functional connectivity: promise, issues, and interpretations. Neuroimage **80**, 360–378 (2013)
12. Just, M.A., Keller, T.A., Kana, R.K.: A theory of autism based on frontal-posterior underconnectivity. In: Development and Brain Systems in Autism, pp. 35–63 (2013)
13. Kana, R.K., Keller, T.A., Cherkassky, V.L., Minshew, N.J., Just, M.A.: Atypical frontal-posterior synchronization of Theory of Mind regions in autism during mental state attribution. Soc. Neurosci. **4**(2), 135–152 (2009)
14. Khosla, M., Jamison, K., Kuceyeski, A., Sabuncu, M.R.: 3D convolutional neural networks for classification of functional connectomes. In: Stoyanov, D., et al. (eds.) DLMIA/ML-CDS -2018. LNCS, vol. 11045, pp. 137–145. Springer, Cham (2018). https://doi.org/10.1007/978-3-030-00889-5_16
15. Koshino, H., Carpenter, P.A., Minshew, N.J., Cherkassky, V.L., Keller, T.A., Just, M.A.: Functional connectivity in an fmri working memory task in high-functioning autism. Neuroimage **24**(3), 810–821 (2005)
16. LeCun, Y., Bengio, Y., Hinton, G.: Deep learning. Nature **521**(7553), 436 (2015)
17. Lord, C., Rutter, M., DiLavore, P.C., Risi, S., Gotham, K., Bishop, S., et al.: Autism diagnostic observation schedule: ADOS. Western Psychological Services, Los Angeles, CA (2012)
18. Ma, S., Calhoun, V.D., Phlypo, R., Adalı, T.: Dynamic changes of spatial functional network connectivity in healthy individuals and schizophrenia patients using independent vector analysis. NeuroImage **90**, 196–206 (2014)
19. Naik, S., Subbareddy, O., Banerjee, A., Roy, D., Bapi, R.S.: Metastability of cortical bold signals in maturation and senescence. In: 2017 International Joint Conference on Neural Networks (IJCNN), pp. 4564–4570. IEEE (2017)
20. Nielsen, J.A., et al.: Multisite functional connectivity mri classification of autism: ABIDE results. Front. Hum. Neurosci. **7**, 599 (2013)
21. Plitt, M., Barnes, K.A., Martin, A.: Functional connectivity classification of autism identifies highly predictive brain features but falls short of biomarker standards. NeuroImage Clin. **7**, 359–366 (2015)
22. Price, T., Wee, C.-Y., Gao, W., Shen, D.: Multiple-network classification of childhood autism using functional connectivity dynamics. In: Golland, P., Hata, N., Barillot, C., Hornegger, J., Howe, R. (eds.) MICCAI 2014. LNCS, vol. 8675, pp. 177–184. Springer, Cham (2014). https://doi.org/10.1007/978-3-319-10443-0_23
23. Ryali, S., et al.: Temporal dynamics and developmental maturation of salience, default and central-executive network interactions revealed by variational bayes hidden Markov modeling. PLoS Comput. Biol. **12**(12), e1005138 (2016)
24. Schipul, S.E., Williams, D.L., Keller, T.A., Minshew, N.J., Just, M.A.: Distinctive neural processes during learning in autism. Cereb. Cortex **22**(4), 937–950 (2011)
25. Surampudi, S.G., Misra, J., Deco, G., Bapi, R.S., Sharma, A., Roy, D.: Resting state dynamics meets anatomical structure: temporal multiple kernel learning (tMKL) model. NeuroImage **184**, 609–620 (2019)
26. Vapnik, V.: The support vector method of function estimation. In: Suykens, J.A.K., Vandewalle, J. (eds.) Nonlinear Modeling, pp. 55–85. Springer, Boston (1998). https://doi.org/10.1007/978-1-4615-5703-6_3

Time-Frequency Analysis Based Detection of Dysrhythmia in ECG Using Stockwell Transform

Yengkhom Omesh Singh[1] ⓘ, Sushree Satvatee Swain[1](✉) ⓘ,
and Dipti Patra[2](✉) ⓘ

[1] IPCV Lab, Department of Electrical Engineering, NIT Rourkela, Rourkela, India
omeshyeng@gmail.com, ssatvatee16@gmail.com
[2] Department of Electrical Engineering, NIT Rourkela, Rourkela, India
dpatra@nitrkl.ac.in

Abstract. Dysrhythmia is the abnormality in rhythm of our cardiac activity. Dysrhythmia is mainly caused by the re-entry of the electric impulse resulting abnormal depolarization of the myocardium cells. Sometimes, such activity causes life threatening ailments. Several myocardial diseases have been studied and detected by help of time frequency representation based techniques effectively. So, a powerful tool called Stockwell transform (ST) has been evolved to provide better time-frequency localization. Wavelet transform based methods arose as an challenging tool for the analysis of ECG signals with both the temporal and the frequency resolution levels. In this study, S-Transform based time-frequency analysis is adopted to detect the Dysrhythmia in ECG signal at high frequencies, which is difficult to study by using Continuous Wavelet transform (CWT). The time frequency analysis is performed over 3 frequency ranges namely low frequency (1–15 Hz) zone, mid-frequency (15–80 Hz) zone and high-frequency (>80 Hz) zone and their respective Integrated time-frequency Power (ITFP) are calculated. The patients with Dysrhythmia has higher ITFP in the high frequency zone than the healthy individuals. The accuracy of the detection of Dysrhythmia is found out to be 88.09% using ST whereas CWT method provides only 58.62% detection accuracy.

Keywords: Dysrhythmia · Stockwell transform (ST) · Continuous Wavelet transform (CWT) · Integrated time-frequency Power (ITFP)

1 Introduction

ECG signal is the outcome of electrical activity occuring in the atria and ventricles of the heart. The presence of the myocardial disease involves the changes in the pathological and morphological behaviour of the ECG signal. The abnormality in the rhythm or electrical activity of ECG signal is the cause of Dysrhythmia [16]. The P-waves are likely to be affected in Atrial, Mobitz 1 and Mobitz 2 Dysrhythmias. In 3rd degree Heart block, ventricle depolarizes independently and it

© Springer Nature Switzerland AG 2019
B. Deka et al. (Eds.): PReMI 2019, LNCS 11942, pp. 201–209, 2019.
https://doi.org/10.1007/978-3-030-34872-4_23

originates from AV node of heart, where QRS complex can be affected [12]. In some scenarios, Dysrhythmia is followed by cardiac arrest. For physicians, it is quite difficult to see the long ECG records manually and identify the presence of anomaly in it.

The time-frequency analysis of ECG signal using transformation based methods helps to detect the myocardial disease. Short time Fourier transform (STFT) analyses small portion of the signal at a time in both the time and the frequency resolution [4]. But due to its fixed window length, it is limited to give good resolution in both time and frequency [15]. WT has capability of only probing the local amplitude or power spectra of the ECG signal [13]. Wavelet analysis of signal is non-adaptive in nature and the same basis function has to be used for analyzing all the ECG data. ST has got advantages over WT due to its progressive resolution characteristics, frequency invariant amplitude response and absolutely referenced phase information. WT analyses a signal in time-scale plot which is not suitable for intuitive visual analysis, but ST analyses a signal in time-frequency distribution of the signal so that the components of the signal can be isolated and processed independently [15]. These advantages of ST motivate us to adopt it for detection of Dysrhythmia. Generally ST is STFT with variable window and upgraded version of wavelet transform [9,17]. It can provide better frequency resolution in low frequency and better time resolution in high frequency [10,15]. Apart from time-frequency resolution, ST can also give good power spectrum [10].

The presence of myocardial disease changes the time-frequency power distribution over the frequencies present in the ECG signal [2,11,14]. The presence of high frequency and power in QRS complex of Ischemic cardiomyopathy ECG signal is detected [6]. The change in frequency components between 20 Hz and 40 Hz before and after angioplasty of patients with right and left coronary stenosis are studied in [2]. The QRS complex shows greater ITFP values which is mentioned in [3,5]. The change in ITFP depends on the age of the subjects and ventricular fibrosis was detected by means of time-frequency power using CWT is described in [11].

The current paper emphasizes on time-frequency analysis using S-transform and the ITFP changes in the respective frequencies have been analyzed thoroughly. Taking ITFP into consideration Dysrhythmia in ECG signal has been detected. Simulation results show better detection accuracy of dysrhythmia in case of S-transform.

2   Materials and Methods

In this study, PTB diagnostic ECG database of Physionet is used for detection of Dysrhythmia. The ECG signals are extracted from 12-lead ECG records. The ECG signal are collected from volunteered Healthy control and the myocardial diseased patients at the Department of Cardiology, University Clinic Benjamin Franklin, Berlin, Germany. The ECG signals are sampled with 1000 Hz. ECG data of 11 dysrhythmic patients and 17 healthy control patients are analysed in this study. Myocardial diseases affect each beat of the entire ECG record in similar manner. When Dysrhythmia happens all the ECG beats get affected uniformly. So, analysis of changes in features in a single beat are enough for the detection of Dysrhythmia [8].

The respiration, body movement and the electrode impedance changes due to perspiration are responsible for the baseline wander in ECG signal. Baseline wander mainly affects the R peaks and the ST segments. So, removal of baseline wander is necessary for the accurate detection of dysrhythmia. In this study, the process of removing baseline wander is same as in [1]. For baseline wander removal the raw ECG signal is subjected to WT. From WT the detail coefficient (D) and approximate coefficient (A) are extracted. Then the approximate coefficient is reconstructed and it is subtracted from raw ECG to obtain baseline wander free signal. The methods for the segmentation of ECG signal given in [7] are as followed.

Algorithm 1. Beat segmentation

1: Find all the maximum points on the ECG signal which are greater than the threshold value = mean of maximum and minimum value of the ECG signal. This points are the R peaks of the ECG signal.
2: Find the location of the R peaks in the time axis and say locations are $R_0, R_1, R_2, R_3, ...R_{n-1}, R_n, R_{n+1}, ...$
3: Evaluate Start point : $R_n - 1/3[R_n - R_{n-1}]$.
4: Evaluate Ending point : $R_n - 2/3[R_{(}n + 1) - R_n]$.
5: Plot the ECG signal from Start point to End point.

2.1 Stockwell Transform (ST)

ST is the combination of Gabor transform with the progressive scale resolution feature of WT [17]. ST is useful to analyse localized signal and provide unique time-frequency localization of the signal. The ST of continuous signal is expressed as [15, 17]:

$$ST(\tau, f) = \int_{-\infty}^{+\infty} s(t) w_s(|f|(t - \tau)) e^{-2\pi j f t} dt = F[s] \otimes F[w_s] \tag{1}$$

where $s(t)$ is the signal, $F[.]$ represents Fourier Transform and $(w_s(|f|(t - \tau)) = \frac{|f|}{\sqrt{2\pi}} e^{-\frac{(\tau - t)^2 f^2}{2}}$ is the window.

The progressive resolution characteristic of ST gives a precise assessment of the signal properties [10]. Because of the direct relation of ST and FT, computation of ST is comparatively easier.

Algorithm 2. Calculation of ST of a signal

1: FFT of the signal,$s(t)$: $s(t) \rightarrow S(u)$;
2: For every frequency f (where $f \neq 0$), start:
    a) Compute frequency-domain localizing Gaussian window at the current frequency f: $e^{-\frac{2\pi^2 u^2}{f^2}} \rightarrow W(u)$.
    b) Change the Fourier spectrum: $S(u) \rightarrow S(u+f)$.
    c) Calculate the point-wise multiplication of the shifted Fourier spectrum $S(u+f)$ and the frequency-domain window $W(u)$, and the result is denoted as $M(u)$.
    d) $IFFT$ of $M(u)$ to give the temporal locations of the events that corresponds to frequency f: $IFFT[M(u)] \rightarrow \widetilde{S}(\tau)$.
3: For Zero frequency ($f = 0$), assign mean of $s(t)$ to $\widetilde{S}(\tau)$.

Relation Between CWT and ST

CWT of a function, $s(t)$, is given as

$$CWT(\tau, a) = \int\limits_{-\infty}^{+\infty} s(t)\omega(t - \tau, a)dt \tag{2}$$

where τ and a are the translation and scaling parameter, $\omega(t, a)$ is mother wavelet. Let us multiply (2) by $e^{-j2\pi ft}$. Then, (2) becomes

$$S_0(\tau, f) = \int\limits_{-\infty}^{+\infty} CWT(\tau, a)e^{-j2\pi ft} = \int\limits_{-\infty}^{+\infty} s(t)\omega(t - \tau, a)e^{-j2\pi ft}dt \tag{3}$$

If $\omega(t, a) = \frac{|f|}{\sqrt{2\pi}} e^{-\frac{t^2 f^2}{2}}$, then (3) becomes

$$S_0(\tau, f) = \int\limits_{-\infty}^{+\infty} s(t)\frac{|f|}{\sqrt{2\pi}} e^{-\frac{(t^2 - \tau)f^2}{2}} e^{-j2\pi ft}dt \tag{4}$$

We can write (4)

$$S_0(\tau, f) = \int\limits_{-\infty}^{+\infty} s(t)w_s(|f|(t - \tau))dt = ST(\tau, f) \tag{5}$$

3 Results and Discussions

3.1 Time-Frequency Analysis

In this study, from the whole ECG record, only one beat is extracted for the analysis. The Fig. 1(a) gives time-frequency representation of the lead I ECG data of a healthy individual using ST. The Fig. 1(b) is the 3-D time-frequency representation of the Fig. 1(a) showing the changes in magnitude of ECG signal. The Fig. 1(c) depicts the highest frequency present in the same ECG signal. The ITFP is calculated for all ECG signals (healthy individuals = 17 and Dysrhythmic patients = 11) and then compared. In this investigation, the leads I, II, AVL, AVF, V1, V2 and V3 are considered. In most of the healthy ECG signals, the dominant frequency power lies between 5 Hz to 100 Hz. The ITFP of the frequencies higher than 100 Hz are lower in Healthy individual than that of Dysrhythmia.

The Fig. 2 shows the time-frequency representation of ST-ECG signals of two different Dysrhythmic patients. Figure 2(a) is Atrial Fibrillation and Fig. 2(b) is Congenital complete AV-block. From the representation, the presence of frequency higher than 250 Hz are very significant in Dysrhythmic ECG but the frequencies higher than 200 Hz are not significant in Healthy individual ECG signal. Again,it is observed clearly that the high frequencies (>250 Hz) in Dysrhythmic ECG signal is longer than that of healthy individual ECG signal in the QRS complex. The Fig. 2(c) shows 3D ITFP distribution of ST ECG signal over time and frequency.

In healthy cases, ITFP decreases smoothly with increases in frequency in MF zone. In some ECG signals of healthy individuals, the frequency more than 200 Hz are observed but the ITFP distribution in this frequency range is very low (insignificant). In Dysrhythmic ECG, the frequencies more than 200 Hz are observed in almost all the leads under discussion. In Dysrhythmic case, the abrupt increase in ITFP is observed in HF zone in all the leads under discussion. In all the Dysrhythmic ECG signal, the time-frequency representation is longer in the QRS complex than that of healthy case. The abnormality in the P-wave shows the vibration of atrium resulting in high frequency. The high frequencies more than 200 Hz are observed in the QRS complex in ST-ECG signal, which is depicted in Fig. 3(a). This indicates the presence of Atrial fibrillation. The abnormality which results ventricular vibration can also be captured using ST-ECG. Independent depolarization of ventricles results in 3rd degree heart block which leads to increase in ITFP after the QRS complex is depicted in Fig. 2(b).

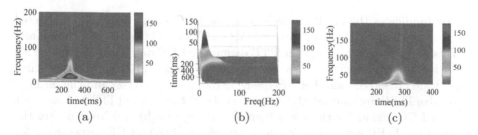

Fig. 1. (a) Time-frequency representation of Healthy individual ECG signal of lead I using ST, (b) 3-D representation of Fig. 1(a) and (c) Highest frequency present in the ECG signal

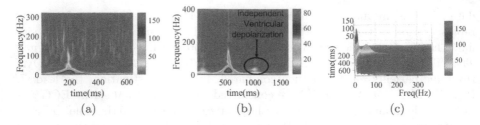

Fig. 2. (a) ST-ECG signal of Atrial fibrillation, (b) ST-ECG signal of congenital complete AV-Block, and (c) 3D representation of Dysrhythmic ST-ECG signal

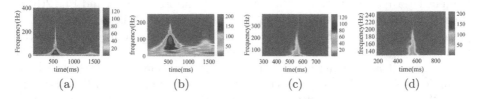

Fig. 3. (a) ST-ECG signal of Dysrhythmic patient, (b) CWT-ECG signal of same Dysrhythmic patient, (c) Highest frequency present in Fig. 3(a) and (d) Highest frequency present in Fig. 3(b)

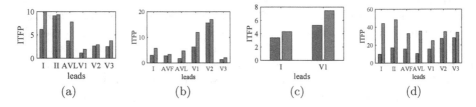

Fig. 4. (a) Total mean ITFP of Healthy individual ECG signals and Dysrhythmic ECG signals, (b) Total LF ITFP of healthy individual and Dysrhythmic patient, (c) Total MF ITFP of healthy individuals and Dysrhythmic patients, and (d) Total HF ITFP of healthy individuals and Dysrhythmic patients

3.2   Power Analysis

Figure 4 indicates the ITFP of one beat of the ECG signal. We have considered the powers of leads I, II, AVF, AVL, V1, V2 and V3 for 17 healthy individuals and 11 Dysrhythmic patients from the PTB diagnostic ECG database. The frequency range is classified into 3 zone: Low-frequency (LF) zone (<15 Hz), Mid-frequency (MF) zone (15–80 Hz) and High-frequency (HF) zone (>80 Hz) and the mean of respective ITFP are calculated. Figure 4(a) shows the mean of ITFP of one beat of each ECG signal for the leads mentioned. Figure 4(b), (c) and (d) are the mean of the ITFP segmented frequency zones LF, MF and HF respectively for all the ECG signal. The blue and orange color bars denote the healthy and Dysrhythmic cases respectively.

In the leads I, II, AVL, V1, V2 and V3, the total mean ITFP of all the frequencies in one beat (PQRST) is found to be higher in Dysrhythmic patients than that of healthy individuals which is shown in Fig. 4(a). For the low frequency power in Fig. 4(b), the leads I, AVF, AVL, V1, V2 and V3 are considered for the investigation. When total mean of LF ITFP are compared, it is observed that the Dysrhythmic patients have higher values than that of healthy individuals. Among the leads mention above I, AVL and V1 have significant changes. Most of the powers are distributed equally in mid-frequency range in both healthy individuals and Dysrhythmic patients. However, the difference in total mean ITFP in MF zone can be observed in lead I and V1 only in Healthy and Dysrhythmic case as depicted in Fig. 4(c). High-frequency range is the most important in this study because of presence of high frequency in the PQRST complex in the Dysrhythmic ECG. The abrupt increase in ITFP around 80 Hz or >80 Hz gives higher ITFP in HF zone in Dysrhythmic case. So, the higher ITFP in HF zone of Dysrhythmic ECG is the main criteria for the detection of Dysrhythmia. In the leads I, II, AVF, AVL, V1, V2 and V3, the presence of higher ITFP in Dysrhythmic ECG signal is observed in HF zone than that of the healthy individual ECG signal which is shown in Fig. 4(d).

In our investigation, the same analysis is done using CWT. It is found that the presence of higher frequency can be detected better in the ST than that of CWT which is depicted in Fig. 3(a) and (b). The highest frequency which can be detected by CWT is 245 Hz but for the same signal, the highest frequency localized by ST is 350 Hz. The presence of high frequencies are the main key for the detection of Dysrhythmia. So, ITFP of the high frequency using ST is significant than that of CWT. Because of limitation of high frequency localization in CWT, detection of the Dysrhythmia is better using ST. The calculated ITFP are compared with a threshold value for each individual signals. The values which are greater than the threshold value are kept separated from the values which are lesser than the threshold value. It is found that the ITFP of high frequency of the Dysrhythmic ECG signal is higher than the ITFP values of the high frequency of the Healthy individual ECG signal. The assessment analysis is given in Table 1, where TPR = True Positive Rate, TNR = True Negative Rate, PPV = Positive Predictive Value, FNR = False Negative Rate, FDR = False discovery Rate, ACC = Accuracy.

Table 1. Feature analysis of different leads

Method	TPR	TNR	PPV	NPV	FNR	FDR	F1	ACC
CWT	66.67%	53.70%	46.64%	72.86%	32.66%	53.36%	54.80%	58.62%
ST	90.91%	86.27%	81.23%	93.99%	9.09%	18.76%	85.59%	88.09%

4   Conclusion

From the investigation, it is clear that ST is better than CWT in higher frequency analysis. The ITFP of one beat of ECG was calculated using ST for healthy individuals (n = 17) and Dysrhythmic patients (n = 11). The presence of high frequency is detected in the Dysrhythmic ECG signals by time frequency analysis. The ITFP of the Dysrhythmic ECG signal using ST has higher value than the healthy individual ECG signal in all the segmented frequency zones. Accuracy of the detection of Dysrhythmia is found to be better using ST than that of CWT.

References

1. German-Sallo, Z.: ECG signal baseline wander removal using wavelet analysis. In: Vlad, S., Ciupa, R.V. (eds.) International Conference on Advancements of Medicine and Health Care through Technology. IFMBE, vol. 36, pp. 190–193. Springer, Heidelberg (2011). https://doi.org/10.1007/978-3-642-22586-4_41
2. Gramatikov, B., Brinker, J., Yi-Chun, S., Thakor, N.V.: Wavelet analysis and time-frequency distributions of the body surface ECG before and after angioplasty. Comput. Methods Programs Biomed. **62**(2), 87–98 (2000)
3. Gramatikov, B., Georgiev, I.: Wavelets as alternative to short-time Fourier transform in signal-averaged electrocardiography. Med. Biol. Eng. Comput. **33**(3), 482–487 (1995)
4. Lu, W.K., Zhang, Q.: Deconvolutive short-time Fourier transform spectrogram. IEEE Sig. Process. Lett. **16**(7), 576–579 (2009)
5. Morlet, D., Peyrin, F., Desseigne, P., Touboul, P., Rubel, P.: Wavelet analysis of high-resolution signal-averaged ECGs in postinfarction patients. J. Electrocardiol. **26**(4), 311–320 (1993)
6. Reynolds Jr., E.W., Muller, B.F., Anderson, G.J., Muller, B.T.: High-frequency components in the electrocardiogram: a comparative study of normals and patients with myocardial disease. Circulation **35**(1), 195–206 (1967)
7. Sadhukhan, D., Mitra, M.: Detection of ECG characteristic features using slope thresholding and relative magnitude comparison. In: 2012 Third International Conference on Emerging Applications of Information Technology, pp. 122–126. IEEE (2012)
8. Sadhukhan, D., Pal, S., Mitra, M.: Automated identification of myocardial infarction using harmonic phase distribution pattern of ECG data. IEEE Trans. Instrum. Meas. **67**(10), 2303–2313 (2018)
9. Stockwell, R., Mansinha, L., Lowe, R.: Localisation of the complex spectrum: the S transform. J. Assoc. Explor. Geophys. **17**(3), 99–114 (1996)
10. Stockwell, R.G.: A basis for efficient representation of the S-transform. Digit. Sig. Process. **17**(1), 371–393 (2007)
11. Takano, N.K., Tsutsumi, T., Suzuki, H., Okamoto, Y., Nakajima, T.: Time frequency power profile of QRS complex obtained with wavelet transform in spontaneously hypertensive rats. Comput. Biol. Med. **42**(2), 205–212 (2012)
12. Thomas, L.J., Clark, K.W., Mead, C.N., Ripley, K., Spenner, B., Oliver, G.: Automated cardiac dysrhythmia analysis. Proc. IEEE **67**(9), 1322–1337 (1979)

13. Tripathy, R.K., Paternina, M.R., Arrieta, J.G., Zamora-Méndez, A., Naik, G.R.: Automated detection of congestive heart failure from electrocardiogram signal using Stockwell transform and hybrid classification scheme. Comput. Methods Programs Biomed. **173**, 53–65 (2019)
14. Tsutsumi, T., et al.: Time-frequency analysis of the QRS complex in patients with ischemic cardiomyopathy and myocardial infarction. IJC Hear. Vessel. **4**, 177–187 (2014)
15. Wang, Y.H., et al.: The tutorial: S transform. Graduate Institute of Communication Engineering, National Taiwan University, Taipei (2010)
16. Xia, Y., et al.: An automatic cardiac arrhythmia classification system with wearable electrocardiogram. IEEE Access **6**, 16529–16538 (2018)
17. Zhu, H., et al.: A new local multiscale Fourier analysis for medical imaging. Med. Phys. **30**(6), 1134–1141 (2003)

Block-Sparsity Based Compressed Sensing for Multichannel ECG Reconstruction

Sushant Kumar, Bhabesh Deka$^{(\boxtimes)}$, and Sumit Datta

Department of Electronics and Communication Engineering, Tezpur University,
Tezpur 784028, Assam, India
bdeka@tezu.ernet.in

Abstract. Multichannel electrocardiogram (MECG) provides significant information for the detection of cardiovascular diseases. Compressed sensing (CS) is a simultaneous sensing and reconstruction technique from a few compressed measurements with low level of distortion. CS promises to lower energy consumption of sensing nodes for wireless body area network (WBAN) in continuous ECG monitoring. In this paper, we propose an energy efficient novel block-sparsity based compressed sensing for MECG reconstruction which exploits both spatial and temporal correlations in the wavelet domain effectively. Experimental results show that the proposed method achieve MECG data compression and reconstruction better than others.

Keywords: Multichannel ECG · Compressed Sensing ·
Block-sparsity · Wireless Body Area Network · Energy efficient sensing

1 Introduction

According to the World Health Organizations (WHO), cardiovascular diseases (CVD) is the number one cause of early deaths in human globally[1]. Recently, new technologies such as low cost wireless body area networks (WBAN) [2] enable the real-time ambulatory electrocardiogram (ECG) monitoring for healthcare service providers and users. However, a major challenge in such a system is due to the sensor nodes which lack energy efficiency for continuous WBAN-based ECG monitoring systems as huge amount of data are collected by them.

Compressed Sensing (CS) is a simultaneous signal acquisition and reconstruction technique from a relatively smaller set of linear projections in the transform domain [1]. CS has been applied in real-time ECG data compression for WBAN applications [6,9,15,16] which offers significant advantages by low complexity data encoding and thereby making the WBAN system highly energy efficient.

Multichannel ECG (MECG) data provide more feature information, which can be very useful for diagnostic purpose compared to single-channel ECG [13]. MECG signals in different channels have both spatial and temporal correlations

---

[1] https://www.who.int/news-room/fact-sheets/detail/cardiovascular-diseases-(cvds).

© Springer Nature Switzerland AG 2019
B. Deka et al. (Eds.): PReMI 2019, LNCS 11942, pp. 210–217, 2019.
https://doi.org/10.1007/978-3-030-34872-4_24

either in the wavelet domain or in the time domain as reported in [12]. Thus, exploiting spatial and temporal correlations of MECG signals simultaneously gives better reconstruction of signals.

CS-based ECG compression works reported in the literature have exploited either temporal [6, 8, 9, 15, 16] also called as single measurement vector (SMV) approach or spatial [5, 11] correlations also called as multiple measurement vector (MMV) approach but very few works have been reported that exploit both types of correlations [12].

In this paper, we have employed a CS model which takes advantages of spatial as well as temporal correlations simultaneously while modeling a block-sparsity property of MECG signals. A novel block sparsity-based l_2/l_{1-2} minimization model is proposed for MECG reconstruction. Main contributions of this work are as follows:

Contributions

- The proposed algorithm exploits both temporal as well as spatial correlations of MECG signals in the wavelet domain.
- A rakeness-based sensing matrix suitable for MECG acquisition and a mix norm called l_2/l_{1-2} for CS reconstruction are applied for effective MECG reconstruction.
- Block-sparsity of MECG signals is exploited for CS reconstruction.
- Simulation results are evaluated based on Physikalisch-Technische Bundesanstalt (PTB) diagnostic ECG database. Performances of the proposed method are compared with traditional CS techniques.

The rest of the paper in order as follows: Sect. 2 gives a short background on CS based MECG compression. The proposed work is explained in Sect. 3 followed by experimental results in Sect. 4. Finally, conclusions are drawn in Sect. 5.

2 Background

MECG signals are not strictly sparse, but can be sparsified in the wavelet domain because most of the wavelet coefficients have very low magnitudes, near to zero for the finest scale. The compressed measurements of sparse MECG signals are obtained as

$$Y = \Phi X = \Theta \alpha \tag{1}$$

where $\Theta = \Phi \Psi$; $\Psi \epsilon \mathbb{R}^{N \times N}$ is an orthonormal matrix, $\Phi \epsilon \mathbb{R}^{M \times N}$ the sensing matrix, $X \epsilon \mathbb{R}^{N \times L}$ the MECG signals from L channels, $Y \epsilon \mathbb{R}^{M \times L}$ the compressed measurement vector, and $\alpha \epsilon \mathbb{R}^{N \times L}$ the unknown sparse coefficient vector, respectively.

In [7], authors shows that rakeness-based CS (Rake-CS) is suitable for ECG signals and similar to binary sensing matrices. The prime advantage of the Rake-CS compared to the general CS techniques is that the amount of projections needed for a certain quality of service without changing the adopted family of architectures for the encoder stage, and compatibility with the hardware-friendly constraint of having $\boldsymbol{\Phi}$ made only of antipodal symbols. Employing a low cost compression scheme will help in saving node power consumption in a WBAN system and in turn will extend the network lifetime supported by battery power.

3   Proposed Methodology

3.1   CS-based MECG Data Compression

We used rakeness-based sensing matrix for efficient encoding of MECG data. In this approach, the sensing matrix $\boldsymbol{\Phi}$ is generated in such a way that randomness is imposed on each row of $\boldsymbol{\Phi}$ keeping in view that total energy of the signal is concentrated within a few samples of the signal and at the same time ensures that mutual coherence between the representation basis $\boldsymbol{\Psi}$ and $\boldsymbol{\Phi}$ improves.

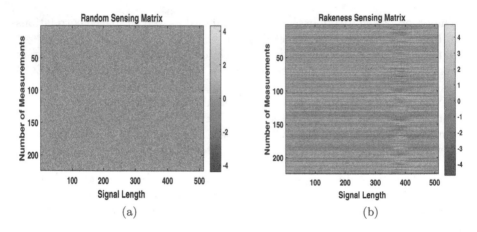

(a)   (b)

Fig. 1. Antipodal sensing matrix based on: (a) random and (b) rakeness projections from MECG signals

Figure 1(a) shows a completely random sensing matrix and Fig. 1(b) shows a rakeness-based sensing matrix simulated for MECG signals considered for our simulations. Each matrix is of size 224×512. It is clearly observed that the latter follows some periodic structures as opposed to the former, which further reduces the number of projections of the input for CS reconstruction. This correlation structure is also same as that of the MECG signals.

3.2   MECG Reconstruction Using Block-Sparsity Based l_2/l_{1-2} Minimization

A WBAN-based ECG healthcare monitoring system having multiple sensors continuously acquires and transmits signals in real time. Our aim is efficient compression and reconstruction of MECG signals with less level of distortions. Since there exists spatio-temporal correlations of MECG signals in the wavelet domain, we need to impose block sparsity property in $\boldsymbol{\alpha}$. Mathematically, we solve the following block l_2/l_{1-2} minimization problem:

$$min\frac{1}{2} \parallel \mathbf{Y} - \boldsymbol{\Theta\alpha} \parallel_2^2 + \lambda(\parallel \boldsymbol{\alpha}_{g_i} \parallel_{2,1} - \parallel \boldsymbol{\alpha}_{g_i} \parallel_2) \tag{2}$$

where the second term indicates the l_2/l_{1-2} -norm, which heavily relies on the sensing matrix with different block coherence to function well, and therefore reconstruction of signals will result in less distortions [14], λ is a positive regularization parameter to achieve a tradeoff between sparsity and data consistency and $g_i \in \{g_1, g_2, ...,g_p\}$ is the index corresponding to the ith group that $\boldsymbol{\alpha}_{g_i} \in \mathbb{R}^{(N \times L)_i}$ denotes the subvector of $\boldsymbol{\alpha}$ indexed by g_i. Let us assume there are p blocks with block size $b = N/p$ in $\boldsymbol{\alpha}$. If $\boldsymbol{\alpha}$ has at most s nonzero blocks, we refer to it as the block s-sparse signal. An accurate sparse representation improves the signal reconstruction further at a given number of compressed measurements. In order to solve the model given in Eq. 2, we use the alternating direction method of multipliers (ADMM) algorithm for block sparsity detailed in [14]. The estimated coefficient matrix $\boldsymbol{\alpha}$ consists of wavelet domain MECG signals which can be transformed back into time domain using inverse wavelet transform.

4   Experimental Setup and Results

All experiments are performed in the MATLAB (R2015a) computing environment on a computer with 3.40 GHz-i7 core CPU, 10 GB of RAM. Dataset used for experiments are obtained from PTB diagnostic MECG database [4]. It consists of 15-channel MECG datasets from 290 patients and each signal is sampled at fs = 1 kHz with 16-bit resolution. To evaluate the performance of the proposed method, a dataset that comprises of 10 ECG records: s0014lrem, s0015lrem, s0016lrem, s0017lrem, s0020arem, s0021arem, s0027lrem, s0028lrem, s0029lrem, and s0031lrem is formed. ECG matrix $\mathbf{X} \in \mathbb{R}^{512 \times 10}$ has been built from ten channels (leads I - V4) present in each record for our experiments. Performance of the proposed reconstruction algorithm are evaluated based on percentage root-mean-square difference (PRD) and signal-to-noise ratio (SNR) [5]. For ECG reconstruction quality, in [17] authors had classified quality of ECG signals based on the values of PRD and SNR.

To generate the sensing matrix by rakeness method, we use the gaussian random vector thresholding technique [7, Sect. IIA] and apply the source codes available at[2]. In each experiment, all the CS algorithms use the same sensing matrix to compress MECG recordings. For SOMP implementation, we use the solver "somp" from the simultaneous sparse approximation toolbox [10] and for BSBL we use codes available at[3]. Daubechies-6 ("db6") wavelet [3] is used as orthonormal bases $\boldsymbol{\Psi}$ for representation of ECG signals which has six vanishing moments.

---

[2] http://cs.signalprocessing.it/.
[3] https://sites.google.com/site/researchbyzhang/software.

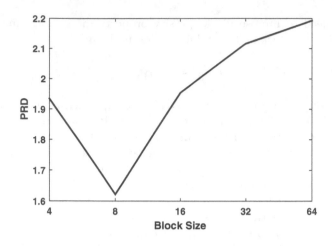

Fig. 2. Average PRD of proposed method versus block size for MECG signals

Table 1. Channelwise average PRD of different methods at different measurements

M	Methodology	Channels									
		I	II	III	aVR	aVL	aVF	V1	V2	V3	V4
96	**Proposed**	**4.540**	**16.142**	**5.130**	**16.304**	**3.426**	**8.599**	**1.680**	**2.070**	**1.769**	**1.284**
	BSBL	6.086	18.494	6.9974	19.930	4.2064	10.778	2.1363	2.136	2.4009	1.973
	SOMP	87.26	85.48	91.27	67.21	90.47	90.88	89.69	89.85	89.73	90.16
128	**Proposed**	**3.723**	**9.610**	**3.884**	**11.925**	**2.651**	**5.757**	**0.924**	**1.255**	**1.133**	**1.075**
	BSBL	4.692	18.164	6.775	20.009	3.7929	10.682	1.7074	2.201	1.612	1.182
	SOMP	135.66	116.13	139.98	79.11	140.41	136.10	137.29	136.77	137.15	138.94
160	**Proposed**	**2.782**	**7.225**	**2.846**	**8.889**	**2.215**	**4.132**	**0.742**	**0.988**	**0.780**	**0.787**
	BSBL	4.352	16.246	6.181	17.817	3.490	9.602	1.544	1.949	1.492	1.034
	SOMP	4.0048	10.693	4.5752	12.378	3.4003	6.504	1.01	1.455	1.074	1.140
192	**Proposed**	**2.516**	**6.172**	**2.499**	**7.075**	**1.953**	**3.557**	**0.517**	**0.690**	**0.528**	**0.550**
	BSBL	3.857	15.705	5.919	17.218	3.140	9.269	1.578	1.936	1.454	0.915
	SOMP	3.737	9.363	3.507	11.688	2.603	5.305	0.880	1.135	0.894	0.922
224	**Proposed**	**2.036**	**4.554**	**1.810**	**5.499**	**1.454**	**2.605**	**0.412**	**0.584**	**0.479**	**0.515**
	BSBL	3.769	15.569	5.949	16.855	3.315	9.201	1.452	1.839	1.356	0.848
	SOMP	2.986	7.131	3.125	8.505	2.467	4.348	0.617	0.803	0.628	0.747
256	**Proposed**	**2.039**	**3.566**	**1.504**	**5.066**	**1.456**	**2.003**	**0.335**	**0.453**	**0.348**	**0.394**
	BSBL	3.446	14.213	5.409	15.379	2.918	8.494	1.400	1.779	1.339	0.809
	SOMP	2.733	5.709	2.463	7.206	2.083	3.390	0.553	0.732	0.600	0.724

Figure 2 describes the average PRD against varying block sizes, i.e. 4, 8, 16, 32, 64 (MECG signals segment length is 512) and we can see that a block size of 8 gives the best results over different block sizes in terms of PRD value. Averaged PRD values per channel using the proposed algorithm, the BSBL and the SOMP at different number of measurements for 10 channels of PTB database are shown in Table 1. The proposed algorithm performs better than the BSBL and the SOMP with the improvement of average PRD at almost all levels of measurements for different channels. This is because the proposed algorithm improves the reconstruction quality by exploiting inter- and intra-channel correlations simultaneously.

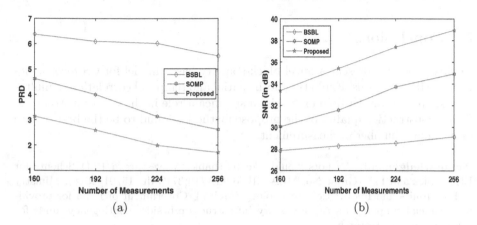

Fig. 3. Average PRD and SNR at different number of measurements for different algorithms

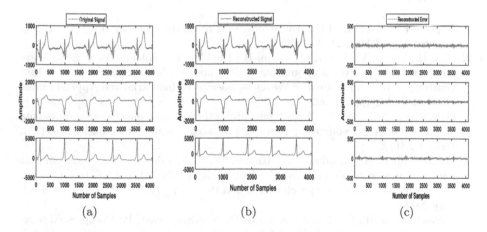

Fig. 4. Reconstructed signals using the proposed method for different channels at M = 1536. Original MECG signals (4096 samples) are from PTB dataset - s0014lre

Averaged PRD and SNR over 10 channels are also plotted in Fig. 3 for different number of measurements. PRD values of the proposed method is less than the BSBL and the SOMP algorithms. Thus, the proposed method can significantly improve the compression ratio without sacrificing the quality of recovered signals.

Figure 4 shows that the reconstruction quality of the proposed method is better. We also show corresponding error signal plots in the figure. It is evident from different waveforms that clinical information of myocardial infarction present in the QT wave segment are well preserved in case of the proposed method. Above experimental results for PTB databases demonstrate that exploiting both spatial and temporal correlations is able to produce accurate MECG reconstruction for different levels of measurements.

5   Conclusions

In this paper, we present a novel block-sparsity based model for CS reconstruction of MECG signals. Exploiting both spatial and temporal correlations simultaneously in the wavelet domain carries a significant role in the CS reconstruction. The reconstruction quality of the propose method is found to be the best in case of minimum number of measurement.

Acknowledgement. Authors would like to thank "Visvesvaraya PhD Scheme for Electronics and IT" (Grant No. MLA/MUM/GA/10(37)B dt. 15/01/2018), Ministry of Electronics and Information Technology (MeitY), Government of India for providing financial support to setup necessary infrastructure besides contingency funds for carrying out this research.

References

1. Candes, E.J., Wakin, M.B.: An introduction to compressive sampling. IEEE Sig. Process. Mag. **25**(2), 21–30 (2008)
2. Cao, H., Leung, V., Chow, C., Chan, H.: Enabling technologies for wireless body area networks: a survey and outlook. IEEE Commun. Mag. **47**(12), 84–93 (2009)
3. Daubechies, I.: Ten Lectures on Wavelets. Society for Industrial and Applied Mathematics, Philadelphia (1992)
4. Goldberger, A.L., et al.: Physiobank, physiotoolkit, and physionet: components of a new research resource for complex physiologic signals. Circulation **101**(23), E215–E220 (2000)
5. Mamaghanian, H., Ansaloni, G., Atienza, D., Vandergheynst, P.: Power-efficient joint compressed sensing of multi-lead ECG signals. In: 2014 IEEE International Conference on Acoustics, Speech and Signal Processing (ICASSP), pp. 4409–4412, May 2014
6. Mamaghanian, H., Khaled, N., Atienza, D., Vandergheynst, P.: Compressed sensing for real-time energy-efficient ECG compression on wireless body sensor nodes. IEEE Trans. Biomed. Eng. **58**(9), 2456–2466 (2011)
7. Mangia, M., Pareschi, F., Cambareri, V., Rovatti, R., Setti, G.: Rakeness-based design of low-complexity compressed sensing. IEEE Trans. Circ. Syst. I Regul. Pap. **64**(5), 1201–1213 (2017)
8. Polania, L.F., Carrillo, R.E., Blanco-Velasco, M., Barner, K.E.: Compressed sensing based method for ECG compression. In: 2011 IEEE International Conference on Acoustics, Speech and Signal Processing (ICASSP), pp. 761–764, May 2011

9. Polania, L.F., Carrillo, R.E., Blanco-Velasco, M., Barner, K.E.: Exploiting prior knowledge in compressed sensing wireless ECG systems. IEEE J. Biomed. Health Inform. **19**(2), 508–519 (2015)
10. Rakotomamonjy, A.: Surveying and comparing simultaneous sparse approximation (or group-lasso) algorithms. Sig. Process. **91**(7), 1505–1526 (2011)
11. Singh, A., Dandapat, S.: Exploiting multi-scale signal information in joint compressed sensing recovery of multi-channel ECG signals. Biomed. Sig. Process. Control **29**, 53–66 (2016)
12. Singh, A., Dandapat, S.: Block sparsity-based joint compressed sensing recovery of multi-channel ECG signals. Healthc. Technol. Lett. **4**(2), 50–56 (2017)
13. Surawicz, B., Knilans, T.: Chous Electrocardiography in Clinical Practice, 6th edn. Elsevier, Philadelphia (2008)
14. Wang, W., Wang, J., Zhang, Z.: Block-sparse signal recovery via ℓ_2/ℓ_{1-2} minimisation method. IET Sig. Process. **12**(4), 422–430 (2018)
15. Zhang, J., Gu, Z., Yu, Z.L., Li, Y.: Energy-efficient ECG compression on wireless biosensors via minimal coherence sensing and weighted l_1 minimization reconstruction. IEEE J. Biomed. Health Inform. **19**(2), 520–528 (2015)
16. Zhang, Z., Jung, T., Makeig, S., Rao, B.D.: Compressed sensing for energy-efficient wireless telemonitoring of noninvasive fetal ECG via block sparse Bayesian learning. IEEE Trans. Biomed. Eng. **60**(2), 300–309 (2013)
17. Zigel, Y., Cohen, A., Katz, A.: The weighted diagnostic distortion (WDD) measure for ECG signal compression. IEEE Trans. Biomed. Eng. **47**(11), 1422–1430 (2000)

A Dynamical Phase Synchronization Based Approach to Study the Effects of Long-Term Alcoholism on Functional Connectivity Dynamics

Abir Chowdhury[✉], Biswadeep Chakraborty, Lidia Ghosh, Dipayan Dewan, and Amit Konar

Artificial Intelligence Laboratory, Department of Electronics and Telecommunication Engineering, Jadavpur University, Kolkata, India
abir.chwdhry@gmail.com, biswadeep.ju.etce@gmail.com,
lidiaghosh.bits@gmail.com, diiipayan93@gmail.com,
amit.konar@jadavpuruniversity.in

Abstract. The paper attempts to investigate the changes in the brain network dynamics between alcoholic and non-alcoholic groups using electroencephalographic signals. A novel entropy-based technique is proposed in this study to understand the dynamics of the neural network for the two groups. To do this, we have examined for the two groups their time-varying instantaneous phase synchronization events when the subjects are engaged in an object recognition task. Next, we have characterized the complexity of the phase synchronization using Fuzzy Sample Entropy over different time scales, referred to as Multiscale Fuzzy Sample Entropy (MFSampEn). The temporal dynamics of phase synchronization arise due to the presence of deterministic characteristics in the time series and the results of surrogate analyses confirm that. Lastly, we check the applicability of Multiscale Fuzzy Sample Entropy (MFSampEn) as entropy-based features in the context of alcoholic and non-alcoholic object recognition classification.

Keywords: Alcoholism · EEG · Fuzzy sample entropy · Multiscale Fuzzy Sample Entropy · SVM · Complexity · Dynamical Phase Synchronization

1 Introduction

The human brain comprises four basic modules: which often interact among themselves to execute a cognitive task, such as reasoning, learning, planning, and sensory-motor control/coordination. The present paper aims at studying the biological underpinnings of the brain dynamics with respect to functional connectivity between activated brain regions in alcoholics and non-alcoholics in an object recognition problem. The fact that the functional brain connectivity is

© Springer Nature Switzerland AG 2019
B. Deka et al. (Eds.): PReMI 2019, LNCS 11942, pp. 218–226, 2019.
https://doi.org/10.1007/978-3-030-34872-4_25

not static and that it varies over time and for different pathological disorders has attracted the broad attention of scientists [15]. It is known that even a moderate amount of alcohol consumption may affect the brain causing slurred speech, decrease in anxiety and motor skills, etc. Therefore, it is important to compare the brain network dynamics of the two groups to investigate whether long term alcohol abuse causes any permanent damage of one/more brain regions resulting in significant deterioration of cognitive skills.

In recent years, considerable amount of researches [2,3,17] have focused on investigating how long term alcohol consumption affects the brain functionality. The existing literature also reveals that heavy drinking has negative implications on the short term memory and may sometimes even lead to blackouts [2]. It has also come to our notice that the drinking onset age may have some influence on alcohol-related neuro-toxicity [3]. However, all these findings have either physiological or philosophical basis of analysis. Understanding the biological insight of the brain functionality changes due to alcohol consumption is the new research era.

Although there exist recent works on functional Magnetic Resonance Imaging (fMRI) based study for discriminating the functional brain connectivity of the alcoholics as compared to the healthy controls [11], there is a dearth of literature which investigates the essence of integrating brain dynamics in functional connectivity analysis. The present work fills the void by an electroencephalographic (EEG) means of brain response acquisition. There are a considerable amount of work on EEG Event-Related Potentials (ERP) for alcoholic and non-alcoholic discrimination. For instance, beta power synchronization is observed among alcoholics as compared to non-alcoholics [17] in most of the existing studies. Numerous studies on gamma band analysis indicate that gamma oscillation is responsible for binding features of an object [17].

In this study, we set out to examine Electroencephalography (EEG) based time-varying instantaneous phase synchronization (referred to as dynamical phase synchronization [15]) of 10 alcoholic and 10 non-alcoholic subjects engaged in an object recognition task. We then investigate their dynamical pattern using multiscale fuzzy sample entropy analysis [6,12,23]. Furthermore, our experience of working with electroencephalography (EEG) signals [8,9] reveals that for a given brain lobe the features extracted from the acquired EEG response vary widely across experimental instances of a given subject for similar stimulation. Due to the presence of fuzzy properties in the experimental instances, the logic of fuzzy sets can, therefore, be used to capture the dynamics of the brain. In the case of sample entropy (SampEn) calculation, a Heaviside function is used whose boundary is rigid. That is, the data points inside the boundary are considered equally, whereas the points marginally outside it are neglected. The rigid boundary condition causes discontinuity, and this may lead to abrupt fluctuations in the entropy values with a small deflection in the tolerance level, and may even fail to provide SampEn value if no template match can be found for a small tolerance value [23]. To circumvent this problem, Chen et al. developed a new complexity metric, Fuzzy Sample Entropy (FSampEn), and used a fuzzy membership function instead of the Heaviside function. In contrast to the Heaviside function, a fuzzy membership function has no rigid boundary. Furthermore, Fuzzy Sample Entropy is continuous and will not fluctuate dramatically with a slight change in the location parameter, as given by the rule of fuzzy membership function. Real biomedical results from [23] show that FSampEn handles noisy data better than its state-of-the-art competitors and is more capable of determining hidden complexity. The paper presents a novel strategy to enjoy the composite benefit of brain dynamics and FSampEn at different time scales (MFSampEn) [6,7,15] for changes in functional connectivity dynamics due to long-term alcoholism.

The paper is divided into 4 sections. In Sect. 2, we provide in detail the principles and methodology of the proposed technique. Section 3 deals with experiments on EEG data acquisition and results obtained thereof, and Sect. 4 discusses the conclusion.

2   Principles and Methodologies

2.1   Dynamical Phase Synchronization (DPS)

Phase Difference Eliminating Phase Slips. After preprocessing of the EEG signals acquired from the subjects' scalp, the phase information of the EEG signals are extracted using the Hilbert transform. The instantaneous frequency (in Hz) is then derived from the obtained phase signal $\phi(t)$ using first-order differences. The activity in the gamma band (30–50 Hz) shows prominent oscillations with a narrow bandwidth and is thus ideal for estimating instantaneous frequency taking into account the background noise [4]. Also, primarily, gamma-band activity is considered for effects of addiction in the behavior of people [17]. Thus, in this study we have focused on the gamma band oscillations.

However, there may lie a directional trend in the phase differences which affects the probability of two sets of simultaneous data points of length m having distance $<r$. To circumvent this, we derived an unwrapped phase signal $\Phi(t)(-\infty < \Phi(t) < \infty)$. The phase differences across the electrodes indexed in time is then determined as:

$$\Delta\Phi_{i,j}(t) = \Phi_i(t) - \Phi_j(t) \tag{1}$$

where i, j denote the electrodes.

In order to eliminate the trend while calculating the multiscale fuzzy sample entropy (MFSampEn) as in [15], the obtained phase differences are wrapped into units of radians, as $\mathrm{mod}(\Delta\phi_{i,j}, 2\pi)$. If we expand the plot of $\mathrm{mod}(\Delta\phi_{ij}, 2\pi)$ versus time (in a sec) for an electrode pair, we obtain a pattern of dynamical synchronous and asynchronous phase differences as DPS [15].

2.2   Multiscale Fuzzy Sample Entropy

Let the input be a data matrix X of size $(N - m) \times m$, length N, pattern length m and location parameter R.

$$X = \begin{bmatrix} x(1) & x(2) & \cdots & x(m) \\ x(2) & x(3) & \cdots x(m+1) \\ \vdots & \vdots & \vdots \\ x(N-m) \ x(N-m+1) & & x(N-1) \end{bmatrix}$$

In sample entropy, self-matches are eliminated and X has $N - m$ rows in order to match dimension in the m and $m + 1$ pattern lengths. These measures remove the bias in the estimations.

In order to determine the vectors' similarity, we use a Gaussian fuzzy membership function [23], given by $\mu(d(i,j), R, c)$, where $d(i,j)$ denotes the distance between $(N-m)$ vectors as the maximum difference of their corresponding scalar components, R is the location parameter determining the location of the function, and c is the shape parameter determining the steepness of the function, with larger c implying a higher relaxation.

The $B(i)$ term derived, represents the probability that there exists a pattern in the pattern space X that is similar in nature to the ith pattern.

Hence, $\phi_m =$ Probability that a given pair of patterns of length m are similar with each other. The output of the system, $\ln\phi_m - \ln\phi_{m+1}$, represents the rate at which new information is created by the underlying system.

Algorithm 1 Fuzzy Sample Entropy

1: **procedure** FSAMPEN(X,N,m,R)
2:     **for** $i=1$ to $N-m$ **do**
3:         **for** $j=1$ to $N-m$ **do**
4:             **if** $i!=j$ **then**
5:                 $d(i,j) = \max(|X(i,1) - X(j,1)|, |X(i,2) - X(j,2)|, \cdots, |X(i,i+m-1) - X(j,i+m-1)|)$
6:                 $D(i,j)=\mu(d,R,c)$ $\triangleright \mu(d(i,j),R,c)=\exp\left(-\frac{d^{\ln(\ln2^c)/\ln R}}{c}\right)$
7:
8:                 $B(i)= \frac{1}{N-m-1}\sum_{j=1}^{N-m+1}D(i,j)$
9:
10:                $\phi_m = \frac{1}{N-m}\sum_{i=1}^{N-m}B(i)$
11:                **return** $FSampEn(x,m,R)=\ln\phi_m-\ln\phi_{m+1}$ $\triangleright$
     The output of fuzzy sample entropy

Finally, in order to get the multiscale framework of fuzzy sample entropy [23], we take an approach as in [6,7] to put the original time series through a "coarse-graining" process. The process is done in the sense that multiple coarse-grained time series are obtained by estimating the mean of the data points within non-overlapping windows of increasing length, τ (where τ represents the scale factor and $1 < j \leq \frac{N}{\tau}$.) The length of each coarse-grained time series is $\frac{N}{\tau}$. The rescaled coarse-grained time series can be calculated using the equation,

$$y_j^{(\tau)} = \frac{1}{\tau} \sum_{i=(j-1)\tau+1}^{j\tau} x_i \tag{2}$$

Next, the fuzzy sample entropy (FSampEn) [1,23] is calculated from each coarse-grained time series and is plotted against each scale factor. The calculation of FSampEn for different coarse-grained time series have been done choosing the values of m and R as 2 and 0.15 respectively [6], where m denotes embedding dimension and R denotes location parameter.

2.3 Surrogate Analysis

For each subject, a surrogate data time series is estimated using the iterated amplitude-adjusted Fourier transform (IAAFT) [19] against the original phase difference $\Delta\phi_{i,j}(t)$ to search for the presence of nonlinear deterministic processes in DPS. IAAFT is an iterative variation of amplitude-adjusted Fourier transform (AAFT) [22] and making use of the iterative scheme to achieve the arbitrarily close approximation to the autocorrelation and the amplitude distribution [19]. The surrogate method follows a process of randomization of phase components of the signals.

2.4 Power Spectrum Analysis

A fast Fourier transform based average power calculation has been undertaken here. For the calculation, a Hanning window approach [10] has been employed for each epoch of 20 s. Power spectral density (PSD) (dB/Hz) has been calculated.

2.5 Classification of Alcoholic and Non-alcoholic Brain Responses Using Multiscale Fuzzy Sample Entropies as EEG Features

To classify the brain responses of alcoholic and non-alcoholic group, the multiscale fuzzy sample entropy (MFSampEn) [23] information are used as features to train and test a SVM classifier.

3 Experiments and Results

3.1 Experimental Dataset

The experimental data used in this paper has been obtained from the public EEG database [13,14,24] of the State University of New York Health Center and owned by Henri Begleiter. The recordings were collected from a pool of ten alcoholic and ten non-alcoholic (control) subjects while being engaged in a visual object recognition task. The data-set contains EEG recordings acquired with 64 electrodes, which includes 3 reference electrodes and the sampling rate of the EEG device used is 256 Hz. Total 260 objects were used to produce the memory set, where the object images were depicted from 1980 Snodgrass and Vandwart Picture set [20]. The subjects for the experiment were exposed to two consecutive pictures of objects from the picture set in such a way that the first picture differs from the second picture with regards to its semantic category, i.e., four footed animal, fruit, weapon, etc. The subjects' mental task involved identifying whether the pictures shown are matching or non-matching. The detailed framework is explained in [24].

3.2   Data Pre-processing

In the pre-processing stage, the EEG signals are first band-pass filtered using elliptical band pass filter of order 6 with a pass band of 30–50 Hz. In the next step, eye-blinking artifacts are removed manually by checking for $100\,\mu V$ potential lasting for a period of 250 ms. Other environmental and physiological artefacts are removed following Independent Component Analysis (ICA) [5].

3.3   Feature Level Discrimination Between Alcoholic and Non-alcoholic Groups

The experiment is carried out to find a clear discrimination between alcoholic and non-alcoholic brain responses from the phase difference information obtained using Dynamical Phase Synchronization method. The experiment is carried out into two steps. In the first step, we extract the instantaneous phase information by Hilbert transform and instantaneous frequency by dividing the phase signal by $2n\pi$. The instantaneous phase and frequency obtained for the electrode Fp_1 and Fp_2 for both the alcoholic and non-alcoholic groups are presented in Fig. 1.

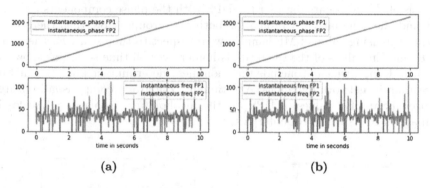

Fig. 1. Instantaneous phase and frequency of (a) Alcoholic and, (b) Non-alcoholic control group

In the second step, we compute the phase differences before and after phase-wrapping around $\mathrm{mod}(\Delta\theta, 2\pi)$, and frequency differences between each electrode pair. For instance, the phase difference before and after phase wrapping, and frequency difference between Fp_1 and Fp_2 for alcoholic and non-alcoholic control group are depicted in Fig. 2. It is evident from Fig. 2 that phase wrapping reduces the noise as a result of the removal of the trend, present in the EEG data, thus providing better discrimination between two classes: alcoholic and non-alcoholic.

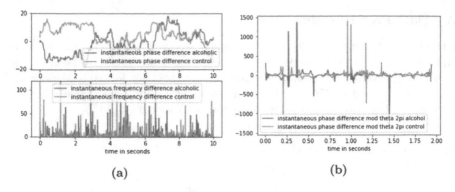

(a)   (b)

Fig. 2. (a) Instantaneous phase difference before phase wrapping and instantaneous frequency difference and, (b) Instantaneous phase difference after phase wrapping of alcoholic and non-alcoholic control group

3.4   Experimental Results

Multiscale Fuzzy Sample Entropy Analysis of DPS. The MFSampEn analysis of the surrogate data for the DPS with the phase components randomized and the original DPS profile for both the groups show that there is an overall increasing value of MFSampEn with respect to time scale. Furthermore, the FSampEn values of the surrogate is higher over all time scales, as shown in Fig. 3(a). Hence, we can conclude that nonlinear deterministic processes dictate DPS in brain network dynamics. Additionally, the group comparison indicated elevated FSampEn values of DPS for the alcoholic group across all scales in comparison to its control counterpart.

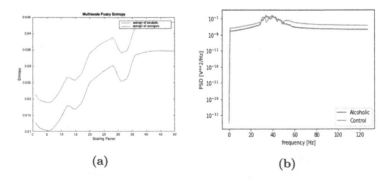

(a)   (b)

Fig. 3. (a) MFSampEn plot for the DPS of alcoholic group and its surrogate, and (b) Variation of PSD for the alcoholic and non-alcoholic group for Fp_1 electrode

Power Spectral Density Analysis. After observing for each electrode, it is observed that the power spectral density of the control group is greater than the power spectral density of the alcoholic group. Figure 3(b) illustrates the variation of PSD for the alcoholic and non-alcoholic groups for Fp_1.

Classifier Performance. In this section, we attempt to judge the feature extraction ability of Multiscale Fuzzy Sample Entropy (MFSampEn) [23] as features from alcoholic and non-alcoholic groups. Here, we have employed Support Vector Machine (SVM) [21] to compare the classification performances of MFSampEn with other entropy-based features, eg., FSampEn, MSampEn, SampEn [18], and ApEn [16]. The classification performances obtained are listed in Table 1.

Table 1. Relative study of SVM classifier accuracies (in %) for four entropy based features

Entropy based features	Classification accuracy
Multiscale fuzzy sample entropy	97.75
Multiscale sample entropy	91.87
Sample entropy	79.65
Approximate entropy	77.13

4   Conclusion

This paper examines the complexity and power analysis of alcoholic and non-alcoholic subjects in the context of visual object recognition task. The complexity metric employed in this study is multiscale FSampEn (MFSampEn) whose merits lie in the fact that it gives more accurate entropy definition than MSampEn and ApEn. MFSampEn has more consistency and has less dependence on data. Experimental results indicate that the complexity of alcoholic subjects is more than those of non-alcoholic subjects. When both the groups are compared, it has been observed that the power spectral density of the alcoholic group is less than its non-alcoholic counterpart.

References

1. Azami, H., Fernández, A., Escudero, J.: Refined multiscale fuzzy entropy based on standard deviation for biomedical signal analysis. Med. Biol. Eng. Comput. **55**(11), 2037–2052 (2017)
2. Balli, T., Palaniappan, R.: On the complexity and energy analyses in EEG between alcoholic and control subjects during delayed matching to sample paradigm. Int. J. Comput. Intell. Appl. **7**(03), 301–315 (2008)
3. Brust, J.: Ethanol and cognition: Indirect effects, neurotoxicity and neuroprotection: a review. Int. J. Environ. Res. Public Health **7**(4), 1540–1557 (2010)

4. Cohen, M.X.: Fluctuations in oscillation frequency control spike timing and coordinate neural networks. J. Neurosci. **34**(27), 8988–8998 (2014)
5. Comon, P.: Independent component analysis, a new concept? Sig. Process. **36**(3), 287–314 (1994)
6. Costa, M., Goldberger, A.L., Peng, C.K.: Multiscale entropy analysis of complex physiologic time series. Phys. Rev. Lett. **89**(6), 068102 (2002)
7. Costa, M., Goldberger, A.L., Peng, C.K.: Multiscale entropy analysis of biological signals. Phys. Rev. E **71**(2), 021906 (2005)
8. Ghosh, L., Konar, A., Rakshit, P., Nagar, A.K.: Hemodynamic analysis for cognitive load assessment and classification in motor learning tasks using type-2 fuzzy sets. IEEE Trans. Emerg. Top. Comput. Intell. **3**(3), 245–260 (2018)
9. Ghosh, L., Konar, A., Rakshit, P., Ralescu, A.L., Nagar, A.K.: EEG induced working memory performance analysis using inverse fuzzy relational approach. In: 2017 IEEE International Conference on Fuzzy Systems (FUZZ-IEEE), pp. 1–6. IEEE (2017)
10. Harris, F.J.: On the use of windows for harmonic analysis with the discrete Fourier transform. Proc. IEEE **66**(1), 51–83 (1978)
11. Herting, M.M., Fair, D., Nagel, B.J.: Altered fronto-cerebellar connectivity in alcohol-naive youth with a family history of alcoholism. Neuroimage **54**(4), 2582–2589 (2011)
12. Humeau-Heurtier, A.: The multiscale entropy algorithm and its variants: a review. Entropy **17**(5), 3110–3123 (2015)
13. Ingber, L.: Statistical mechanics of neocortical interactions: canonical momenta indicators of electroencephalography. Phys. Rev. E **55**(4), 4578 (1997)
14. Ingber, L.: Statistical mechanics of neocortical interactions: training and testing canonical momenta indicators of EEG. Math. Comput. Model. **27**(3), 33–64 (1998)
15. Nobukawa, S., Kikuchi, M., Takahashi, T.: Changes in functional connectivity dynamics with aging: a dynamical phase synchronization approach. NeuroImage **188**, 357–368 (2019)
16. Pincus, S.M., Gladstone, I.M., Ehrenkranz, R.A.: A regularity statistic for medical data analysis. J. Clin. Monit. **7**(4), 335–345 (1991)
17. Porjesz, B., Begleiter, H.: Alcoholism and human electrophysiology. Alcohol Res. Health **27**(2), 153–160 (2003)
18. Richman, J.S., Moorman, J.R.: Physiological time-series analysis using approximate entropy and sample entropy. Am. J. Physiol. Heart Circ. Physiol. **278**(6), H2039–H2049 (2000)
19. Schreiber, T., Schmitz, A.: Improved surrogate data for nonlinearity tests. Phys. Rev. Lett. **77**(4), 635 (1996)
20. Snodgrass, J.G., Vanderwart, M.: A standardized set of 260 pictures: norms for name agreement, image agreement, familiarity, and visual complexity. J. Exp. Psychol. Hum. Learn. Mem. **6**(2), 174 (1980)
21. Subasi, A., Gursoy, M.I.: EEG signal classification using PCA, ICA, LDA and support vector machines. Expert Syst. Appl. **37**(12), 8659–8666 (2010)
22. Theiler, J., Eubank, S., Longtin, A., Galdrikian, B., Farmer, J.D.: Testing for nonlinearity in time series: the method of surrogate data. Phys. D **58**(1–4), 77–94 (1992)
23. Xie, H.B., Chen, W.T., He, W.X., Liu, H.: Complexity analysis of the biomedical signal using fuzzy entropy measurement. Appl. Soft Comput. **11**(2), 2871–2879 (2011)
24. Zhang, X.L., Begleiter, H., Porjesz, B., Wang, W., Litke, A.: Event related potentials during object recognition tasks. Brain Res. Bull. **38**(6), 531–538 (1995)

Information Retrieval

RBM Based Joke Recommendation System and Joke Reader Segmentation

Navoneel Chakrabarty$^{(\boxtimes)}$, Srinibas Rana, Siddhartha Chowdhury,
and Ronit Maitra

Jalpaiguri Government Engineering College, Jalpaiguri, West Bengal, India
{nc2012,srinibas.rana,sc2024,rm2004}@cse.jgec.ac.in

Abstract. In the recent scenario, consumers are bared to a variety of information as well as commodities which leads to a variance in their choices. Recommender systems are a way to endure this challenge. An appropriate approach in recommending jokes on the basis of their preferences will be of substantial help in the future reference of the probable jokes. It is followed by segmentation of readers that has the potential to allow analysts to address each Joke-Reader in the most efficient way. This study aims at the development of a Joke Recommendation System based on Collaborative Filtering and Joke-Reader Segmentation based on the similarities in their preference patterns. A Bernoulli Restricted Boltzmann Machine (RBM) Model is implemented for constructing the Joke Recommender and k-means Clustering Model is deployed for achieving the Joke-Reader Segments. Considering the recommendation operation altogether, it is observed that the Joke-Reader Segmentation is firmly associated with the recommended ratings.

Keywords: Recommender systems · Segmentation · Collaborative Filtering · Joke Reader Segmentation · Bernoulli Restricted Boltzmann Machine · k-means Clustering

1 Introduction

In today's contemporary world, people tend to rely upon technology for foraging of the next relatable content that they would probably like it online rather than manually finding recommendations from people they know because of the tedious nature of that very work. For example, an online book recommendation system will probably recommend books that people have already bought/read adjoining the same book that a user would certainly buy/read but that will not be enough. This paper extensively proposes a strong and rational framework of a recommender system that recommends jokes to its users who haven't read them yet, based on the previous rating pattern of jokes by other users. So, Collaborative filtering approach is chosen to recommend the next joke the users will like based on the preferences and tastes of many other users. Hence, users are to be divided into specific groups (clusters) depending upon their preference patterns in order to:

© Springer Nature Switzerland AG 2019
B. Deka et al. (Eds.): PReMI 2019, LNCS 11942, pp. 229–239, 2019.
https://doi.org/10.1007/978-3-030-34872-4_26

- relate the Joke-Readers in each cluster for magnifying the value of each Joke-Reader.
- help Joke-Writers to create jokes that will draw maximum attention and are suitable for specific groups of readers.

In this paper, the Bernoulli Restricted Boltzmann Machine (RBM) based Recommendation System takes the user's likes and dislikes as input, analyzes and correlates the rating patterns of other users and gives the possible like or dislike rating as output. After that, k-means Clustering Model is trained only on the recommended set of ratings of all the jokes for all users and applied on the true set of ratings for the jokes that are read by the users and recommended ratings only for those jokes that are not read by them till now.

This paper has been organized as follows: Sect. 2 discusses the related works forming the Literature Review, Sect. 3 illuminates the Proposed Methodology followed by Sect. 4 elucidating the Implementation Details, Sect. 5 showcases the Results, Visualization and Performance Analysis, Sect. 6 highlights the future scope followed by Sect. 7 concluding the article.

2   Literature Review

There are many related works involving recommendation and segmentation attempted in the recent past.

Some of the existing contributions on Recommendation Systems associated to Jokes are as follows:

- Goldberg et al. introduced Eigentaste Collaborative Filtering Algorithm which is of constant time and implemented it to build a Joke Recommendation System [5].
- Babu et al. applied Bayesian Networks, Pearson & Cosine Correlations, Clustering and Horting for constructing Joke Recommender System [2].
- Polatidis et al. proposed a Multi-Label Collaborative Filtering Algorithm and tested its efficiency by demonstrating experiments using 5 real datasets: MovieLens 100 thousand, MovieLens 1 million, Jester (joke dataset), Epinions and MovieTweetings [6].
- Corso et al. formulated a Non-negative Matrix Factorization (NMF) of the rating matrix for dealing with Recommendation System problems and applied on MovieLens Dataset, Jester Dataset (jokes) and Amazon Fine Foods Dataset [3].

There also have been attempts in which Recommendation System is developed banking on Segmentation:

- Rezaeinia et al. extracted and calculated the weights associated with the RFM (Recency, Frequency and Monetary) attributes of the clients. The weighted RFM and clustering algorithms based on closest k-neighbours are used for obtaining independent recommendations for each cluster [7].

- Vandenbulcke et al. proposed a way of segmenting consumers in the retail zone based on shopping behaviours. Firstly products to be recommended to a unique consumer is resolved and secondly, consumers are segmented according to the recommendations [10]
- Eskandanian et al. developed a user segmentation technique for developing personalized recommendation system and also highlighted the impact of segmentation on the originality of recommendations [4].
- Shi et al. proposed a smart recommendation system based on customer segmentation. Usefulness is substantiated and the fidelity of customers (or customer net worth) is also graded by FCM clustering approach [9].
- Rodrigues et al. proposed a hybrid recommender model combining content-based, collaborative filtering and data mining techniques in order to take care of the varying customer behaviour [8].

But here, Joke-Reader Segmentation is done on the basis of Recommendation Results given by the Recommender System and true values of likes and dislikes.

3   Proposed Methodology

3.1   Dataset

The Jester Dataset containing 4.1 million anonymous ratings from 73,421 users of 100 specific jokes [1] is used for developing the Joke Recommender and Joke-Reader Segmentation Models. Each and every rating in the dataset is a value between -10.00 to $+10.00$. The dataset was prepared by Goldberg et al. [5] at UC Berkeley. In this dataset, every row represents ratings of certain number of jokes (among the 100 specific jokes) by a unique user with some jokes left unrated by the users. The unrated jokes are actually those jokes which are not read by the user and hence, has to be recommended and are denoted by the number 99 in the dataset.

3.2   Data Preprocessing

The Data Pre-processing steps employed in the Jester Dataset are as follows:

1. All the jokes with user rating between $+7.00$ to $+10.00$ (both inclusive) are considered as liked by the user and marked as '1'.
2. All the jokes with user ratings between -10.00 to $+6.00$ (both inclusive) are considered as disliked by the user and marked as '0'.
3. All the jokes with no user ratings and with default marker '99' are replaced by new marker, '-1'.

3.3   Development of Joke-Recommender System Using Bernoulli Restricted Boltzmann Machine (RBM)

Bernoulli Restricted Boltzmann Machine is an Energy-based Probabilistic Graphical Deep Learning Model. In RBM, there is a Visible Layer representing the Input Nodes and a Hidden Layer representing the Hidden Nodes. The hidden nodes in the RBM intuitively helps in Internal Feature Identification of the jokes which is necessary in any Collaborative Filtering Approach. There is an energy associated with the model at different states of training and is governed by weights connecting the Visible and Hidden Layer synapses (nodes) and biases. At each step of RBM training, these weights and biases are updated in such a way that the model attains its minimum energy possible. A Graphical Interpretation and formulation of the same is given in Fig. 1.

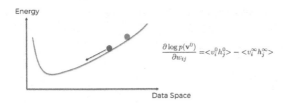

$$\frac{\partial \log p(\mathbf{v}^0)}{\partial w_{ij}} = <v_i^0 h_j^0> - <v_i^\infty h_j^\infty>$$

Fig. 1. RBM: an energy-based Probabilistic Graphical Model in which v^0 represents initial energy, v^0 & v^∞ denotes the vector of visible nodes initially & finally respectively, h^0 & h^∞ denotes the vector of hidden nodes initially & finally respectively and w is the weight matrix

Gibbs Sampling is done for Training the RBM following 20-Step Contrastive Divergence as the Learning Algorithm.

The RBM Architecture consists of

1. 100 visible nodes corresponding to the 100 specific jokes and 20 hidden nodes.
2. The weight matrix of dimension 20×100 is randomly initialized according to Standard Normal Distribution with Mean of 0 and Std Deviation of 1.
3. The bias for the probability, $p(h = 1|v)$ is randomly initialized according to Standard Normal Distribution (with Mean of 0 and Standard Deviation of 1) as a vector of size 20.
4. The bias for the probability, $p(v = 1|h)$ is randomly initialized according to Standard Normal Distribution (with Mean of 0 and Standard Deviation of 1) as a vector of size 100.

A sample RBM Architecture with 6 visible nodes and 5 hidden nodes is shown in Fig. 2.

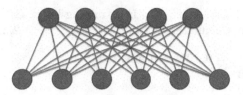

Fig. 2. RBM Architecture with 6 visible nodes denoted by blue colour and 5 hidden nodes denoted by red colour (Color figure online)

Gibbs Sampling: The steps employed in Gibbs Sampling are as follows:

1. Given the visible layer nodes forming the 1st sample of visible nodes, the probability $p(h = 1|v)$ and values of the hidden nodes are calculated.
   So, this gives a vector of size 20 which represents the hidden layer values forming the 1st sample of hidden nodes.
2. Now, given the sample of hidden nodes created in the previous step, the probability $p(v = 1|h)$ and values of the visible nodes are calculated.
   So, this gives a vector of size 100 which represents the visible layer values forming the 2nd sample of visible nodes. Considering this sample, steps 1 and 2 are repeated 19 times more.

Here, Step 1 creates Gibbs Samples of Hidden Layer nodes and Step 2 creates Gibbs Samples of Visible Layer Nodes.

This whole process of Gibbs Sampling, with steps 1 and 2 repeated 20 times (in total) is known as **20-step Contrastive Divergence**. An illustrative diagram of Gibbs Sampling is shown in Fig. 3.

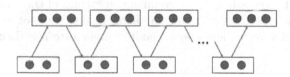

Fig. 3. Illustrative explanation of Contrastive Divergence

Training the Bernoulli RBM Model: The dataset is split into Training and Validation Sets such that 80% of the 73,421 instances i.e., 58,736 instances are used for Training the RBM and remaining 14,685 instances are used for Validation and Performance Analysis. Here, 20-Step Contrastive Divergence is used as the Learning Algorithm for Training the RBM with 10 as the number of epochs. The k-Step Contrastive Divergence is given in Algorithm 1.

– Input: RBM($v_1, ..., v_m, h_1, ... , h_n$) in a training batch B
– Output: Gradient Calculation- $\Delta w_{ij}, \Delta b_j, \Delta a_i$; $i = 1,2,..,n$ and $j = 1,2..,m$

1. **initialize** $\Delta w_{ij} = 0, \Delta b_j = 0$ **and** $\Delta a_i = 0$
2. **for all** v **in B**
3. $v^0 = v$
4.   **for** $t = 0$ **to** $k - 1$:
5.   **for** $i = 1$ **to** n:
6.   **Gibbs Sample** $h_i^t \sim p(h_i|v^t)$
7.   **for** $j = 1$ **to** m:
8.   **Gibbs Sample** $v_j^{t+1} \sim p(v_j|h^t)$
9.   **for** $i = 1$ **to** n:
10.   **for** $j = 1$ **to** m:
11. $\Delta w_{ij} = \Delta w_{ij} + p(H_i = 1|v^0) * v_j^0 - p(H_i = 1|v^k) * v_j^k$
12.   **for** $j = 1$ **to** m:
13. $\Delta b_j = \Delta b_j + v_j^0 - v_j^k$
14.   **for** $i = 1$ **to** n:
15. $\Delta a_i = \Delta a_i + p(H_i = 1|v^0) - p(H_i = 1|v^k)$

Algorithm 1. k-Step Contrastive Divergence

For Validation of remaining 14,685 instances, blind walk by 1-step Contrastive Divergence is done.

3.4   Joke-Reader Segmentation Using k-Means Clustering

Now, the recommended ratings obtained in both Training and Validation Set are combined to form a Dataset D1. And the training set and test set with original values of joke-ratings and recommended ratings of the unrated jokes are combined to form a Dataset D2. Now, for Joke-Reader Segmentation, a k-Means Clustering Model is to be developed. The k-means clustering algorithm is given in Algorithm 2.

– Input:
    1. Number of clusters - k
    2. Training Set: $x^1, x^2, x^3, ..., x^m$
– Output: Each instance is assigned a cluster among the k clusters

1. **Randomly initialize** k **cluster centroids,** $u_1, u_2, ..., u_k$
2. **Repeat**{
3.   **for** $i = 1$ **to** m:
4. $c_i :=$ **index(from 1 to** k**) of cluster centroid closest to** x_i
5.   **for** $j = 1$ **to** k:
6. $u_j :=$ **average of the points assigned to cluster j**
7. **}**

Algorithm 2. k-Means Clustering

Before training, optimal number of clusters needs to be obtained. Elbow Method is employed for obtaining the optimal number of clusters. In the Elbow Method, the k-Means Clustering Model is trained multiple times on a dataset for different values of k and the corresponding Clustering Cost i.e., sum of the distances of the samples to the closest cluster centroid are plotted. This plot of Clustering Cost vs k value is known as Elbow Curve. From the curve, the value of k is taken as the optimal number of clusters that represents a 'perfect elbow'. The Elbow Method is applied taking Datasets D1 and D2 separately and Elbow Curves for the same are shown in Fig. 4.

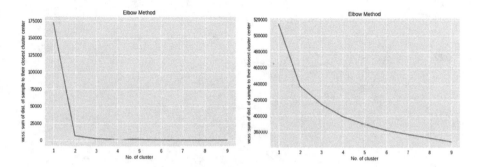

Fig. 4. Elbow curves for datasets D1 and D2

From Fig. 4, the elbow curve for Dataset D1 gives a 'perfect elbow' at $k = 3$. On the contrary, from Fig. 4, no such perfect elbow is obtained and moreover the Clustering Cost using Dataset D2 as Training Set, is much higher than that of D1. Hence, Dataset D1 is chosen as the Training Set for the k-Means Clustering Model.

The trained k-Means Clustering Model is applied on Dataset D2 for Joke-Reader Segment Visualization in Sect. 5.2.

4   Implementation Details

The Deep Learning Model (RBM) is deployed in PyTorch with each and every vector and matrix used as Tensor and the Machine Learning Model (k-Means) is implemented using Python's Scikit-Learn Machine Learning Toolbox on a TPU configured Google Colab Notebook. The Elbow Curves and Training-Loss Curve (shown in Figs. 4 and 5 respectively) are generated by Python's Data Visualization Library, Matplotlib. The Joke-Reader Segment (Cluster) visualizations in Fig. 6 is done using Microsoft Excel's Line Charts.

5   Results, Visualization and Performance Analysis

5.1   Bernoulli RBM for Joke Recommender System

The evaluation of the performance of Bernoulli RBM Model is done on the following metrics:

1. Training Loss: It is the Mean Absolute Error on the Training Set.
2. Validation/Test Loss: It is the Mean Absolute Error on the Test Set.

Also, the Mean Absolute Error is calculated only with respect to the jokes that are read and rated (marked as liked or disliked) initially excluding the ones that were unrated (marked as '−1') The Performance Analysis of the Bernoulli-RBM for Joke Recommender is tabulated in Table 1.

The training-loss history curve is given in Fig. 5.

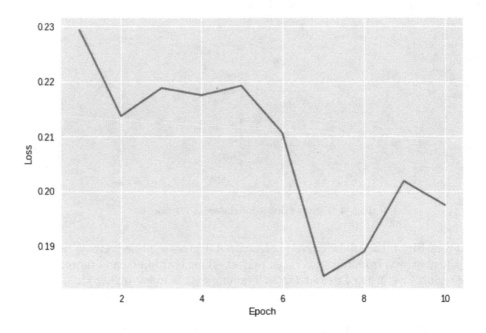

Fig. 5. Training-loss history curve

Table 1. RBM model performance analysis

Performance metrics	Result
Training loss (MAE)	**0.1974**
Validation loss (MAE)	**0.1786**

5.2   Visualization of the Joke-Reader Segments

The Performance Analysis of the k-Means Clustering Model is done by visualizing the clusters of Joke-Readers with Line Charts representing Preference

Patterns of Jokes of each segment (cluster) of Joke-Readers. Each of the 3 clusters, so obtained from D2 (after validation) is characterized by a 100-sized vector (for 100 jokes) of Preference Values in which,

$$Preference(c_i, j_k) = \frac{Number\ of\ Readers\ belonging\ to\ cluster\ c_i\ and\ liking\ the\ joke\ j_k}{Total\ Number\ of\ Readers\ belonging\ to\ cluster\ c_i}$$

Now, the 3 vectors of preference values, each representing a cluster are graphically plotted into Line Charts in order to generate the Preference Patterns or the shape of the Line Charts characterizing each cluster shown in Fig. 6.

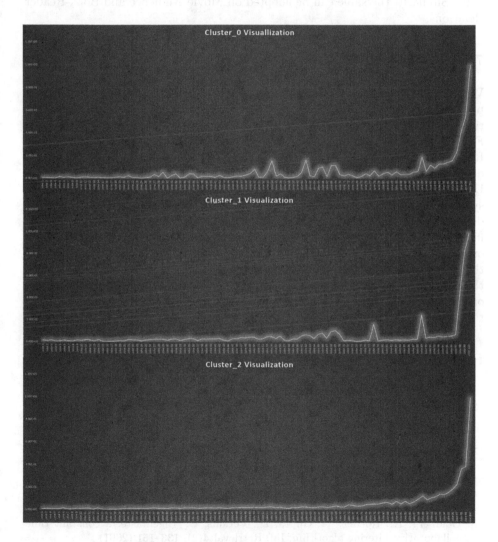

Fig. 6. Joke-reader segment visualization

In Fig. 6, the preference pattern curves representing the 3 clusters are of entirely different shapes but the preference values for certain jokes or the preference patterns of certain consecutive jokes are quite similar. This reflects the degree of overlapping among the 3 clusters.

6 Future Scope

This methodology can be further utilized on Product-Customer Relation where products can be recommended and customer segmentation can be done accordingly. Similarly, the same can be adopted on Movie-Audience and Book-Reader Relations.

7 Conclusion

We have proposed a methodology for recommending jokes to readers as per their preferences as well as the Joke-writers for writing new jokes to draw the attention of readers. The key highlights of the composed methodology are as follows: Firstly, our model uses Collaborating Filtering approach for developing the Joke Recommendation System banking on which Segmentation of Joke-Readers is done. Secondly, Bernoulli Restricted Boltzmann Machine (RBM), which is an Unsupervised Deep Learning Approach is used for building the recommendation system. Thirdly, the Performance Analysis of the k-Means Clustering Model is done by visualizing the clusters of Joke-Readers with Line Charts each representing Preference Pattern of Jokes of each segment (cluster) of Joke-Readers. Moreover, the methodology might prove useful for the Joke-writers as well because they can get a rough idea about the jokes they should compose depending upon the mass preferences of all the Jokes-readers for Generalized Entertainment and of joke-readers within the specific segments (clusters) for Personalized Entertainment.

References

1. Jester dataset. https://goldberg.berkeley.edu/jester-data/
2. Babu, M.S.P., Kumar, B.R.S.: An implementation of the user-based collaborative filtering algorithm. Int. J. Comput. Sci. Inf. Technol. (IJCSIT) **2**(3), 1283–1286 (2011)
3. Del Corso, G.M., Romani, F.: Adaptive nonnegative matrix factorization and measure comparisons for recommender systems. Appl. Math. Comput. **354**, 164–179 (2019)
4. Eskandanian, F., Mobasher, B., Burke, R.D.: User segmentation for controlling recommendation diversity. In: RecSys Posters (2016)
5. Goldberg, K., Roeder, T., Gupta, D., Perkins, C.: Eigentaste: a constant time collaborative filtering algorithm. Inf. Retrieval **4**(2), 133–151 (2001)
6. Polatidis, N., Georgiadis, C.K.: A multi-level collaborative filtering method that improves recommendations. Expert Syst. Appl. **48**, 100–110 (2016)

7. Rezaeinia, S.M., Rahmani, R.: Recommender system based on customer segmentation (RSCS). Kybernetes **45**(6), 946–961 (2016)
8. Rodrigues, F., Ferreira, B.: Product recommendation based on shared customer's behaviour. Proc. Comput. Sci. **100**, 136–146 (2016)
9. Shi, Z., Wen, Z., Xia, J.: An intelligent recommendation system based on customer segmentation. Int. J. Res. **2**(11), 78–90 (2015)
10. Vandenbulcke, V., Lecron, F., Ducarroz, C., Fouss, F.: Customer segmentation based on a collaborative recommendation system: application to a mass retail company. In: Proceedings of the 42nd Annual Conference of the European Marketing Academy (2013)

TagEmbedSVD: Leveraging Tag Embeddings for Cross-Domain Collaborative Filtering

M. Vijaikumar[✉], Shirish Shevade, and M. N. Murty

Department of Computer Science and Automation,
Indian Institute of Science, Bangalore, India
{vijaikumar,shirish,mnm}@iisc.ac.in

Abstract. Cross-Domain Collaborative Filtering (CDCF) mitigates data sparsity and cold-start issues present in conventional recommendation systems by exploiting and transferring knowledge from related domains. Leveraging user-generated tags (e.g. ancient-literature, military-history) for bridging the related domains is becoming a popular way for enhancing personalized recommendations. However, existing tag based models bridge the domains based on common tags between domains and their co-occurrence frequencies. This results in capturing the syntax similarities between the tags and ignoring the semantic similarities between them. In this work, to address these, we propose TagEmbedSVD, a tag-based CDCF model to cross-domain setting. TagEmbedSVD makes use of the pre-trained word embeddings (word2vec) for tags to enhance personalized recommendations in the cross-domain setting. Empirical evaluation on two real-world datasets demonstrates that our proposed model performs better than the existing tag based CDCF models.

Keywords: Cross-Domain Collaborative Filtering · User-generated tags

1 Introduction

Recommendation systems play an important role in filtering out irrelevant information and suggest contents that interest users. Examples include book recommendation in LibraryThing, movie recommendation in Netflix. Collaborative Filtering (CF) [9] has been successful in predicting user needs to be based on the like-minded users' interests using their historical records. However, in practice, users rate a very few as compared to available items. Besides, new users and/or items are added at regular intervals. These two phenomena lead to sparsity and cold-start problems.

To deal with the data sparsity and cold-start problems, researchers have exploited Cross-Domain Collaborative Filtering (CDCF) [3,7,8,15], which leverages the knowledge extracted from related domains. For example, users who

© Springer Nature Switzerland AG 2019
B. Deka et al. (Eds.): PReMI 2019, LNCS 11942, pp. 240–248, 2019.
https://doi.org/10.1007/978-3-030-34872-4_27

watch movies from adventure genre will most likely be interested in reading books written on adventure travel. However, in most cases, users and/or items across domains do not overlap. In such a scenario, exploiting user-generated tags [2,4,5,11,13–16] (e.g. tags like ancient-literature or military-history) for bridging the related domains is becoming a popular way for enhancing personalized recommendations. That is, though we do not know the exact mapping of the users and/or items across domains, we establish the connections with the following assumption: *the users who use similar tags are similar, and the items which are assigned with similar tags are similar.* Nevertheless, existing tag based CDCF models work based on common tags and its co-occurrence count alone to bridge the related domains [5,13,14]. Hence, these models capture the syntax similarities between the tags and ignore the semantic relationships between them. We illustrate these by the following toy example.

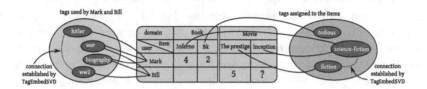

Fig. 1. Illustration of tag based CDCF

Let us assume that a cross-domain system has users (*Bill* and *Mark*), items (*The prestige* and *Inception* from Movie domain, and *Inferno* and *Bk* from book domain, respectively), ratings (in the scale of 1–5, where 1 being low and 5 being high) and their tag assignments as shown in Fig. 1.

Treating tags as tokens/lexicons, most of the existing models connect the users *Bill and Mark* by only the tags *biography and war* based on their direct usage. These models completely ignore the relationship between the tags *science-fiction* and *fiction* (as well as *ww2* and *hitler*). However, these tags provide a lot of knowledge about the similarities between the movies *The Prestige* and *Inception*, and also the movie *The Prestige* and the book *Inferno* since most users who are interested in movies or books related to *fiction* might also be interested in movies or books related to *science-fiction*. This can be well captured by word-vector embeddings such as word2vec [12], because the cosine similarity between the embedding vectors *fiction* and *science-fiction* is 0.61[1] which is considerably high. Besides the above facts on the similarity between items that can be established through tags, the similarity between the users *Mark* and *Bill* can also be strongly established with the help of semantic word embeddings, because the users are interested in similar topics *fiction* and *science-fiction* based on the items they rated. Accounting all of the above makes us predict the missing rating of the user *Bill* on the movie *Inception* more accurately.

---

[1] Here Google's pre-trained word2vec model trained on google news dataset is used for similarity calculation.

Contributions. We propose a novel extension of SVD++, TagEmbedSVD, to the cross-domain setting where there is no assumption that users and items overlap across domains. TagEmbedSVD leverages user-generated tags to bridge the related domains. That is, TagEmbedSVD employs word2vec – a word vector representation for finding the semantic relationship between the tags to bridge the domains and enhance the recommendation performance. We perform comprehensive experiments on two real-world datasets – LibraryThing and MovieLens, and show that our proposed model outperforms existing tag based CDCF models, particularly in sparse and cold-start settings. Our implementation is available at https://github.com/mvijaikumar/TagEmbedSVD.

2   Proposed Model

Problem Formulation. Suppose we have a set of ratings $[r_{uj}^S]_{m_S \times n_S}$ and $[r_{uj}^T]_{m_T \times n_T}$ from source and target domains. Here $r_{uj}^S, r_{uj}^T \in [a, b] \cup \{0\}$, and $a, b > 0$ denote the minimum and maximum ratings of user u on item j and 0 denotes unavailable ratings, respectively. Further, m_S, n_S, m_T and n_T represent the number of users and items from source and target domains respectively. In addition, we are given tags associated with every user u and item j denoted by sets $\mathcal{T}_u$ and $\mathcal{T}_j$ respectively. Here, the tag can be a word or a set of words or a phrase (for example, philosophy, mind-blowing, one time watchable).

Let U^S, U^T, I^S and I^T denote the sets of users and items from source and target domains respectively. Note that, $U = U^S \cup U^T$ and $I = I^U \cup I^T$, and $U^S \cap U^T = \varnothing$ and $I^S \cap I^T = \varnothing$. Let $\mathcal{I}_u$ and $\mathcal{U}_j$ be the set of items rated by user u and the set of users who rated item j respectively. Let $\Omega^S = \{(u, j) : r_{uj}^S > 0\}$ and $\Omega^T = \{(u, j) : r_{uj}^T > 0\}$ be the sets indicating the available ratings and $\Omega = \Omega^S \cup \Omega^T$. Our goal here is to predict the unavailable ratings for the users on items in the target domain with the help of available ratings and tag information from both domains. Formally, we want to predict ratings $r_{uj}^T, \forall(u, j) \notin \Omega^T$ in the target domain using $r_{uj}^S, \forall(u, j) \in \Omega^S, r_{uj}^T, \forall(u, j) \in \Omega^T$ and $\mathcal{T}_u, \mathcal{T}_j, \forall u \in U, \forall j \in I$. To avoid notational clutter, wherever the context is clear from Ω^S and Ω^T, we drop the superscripts S and T and combine the ratings from source and target domains together.

2.1   TagEmbedSVD

In this section, we explain our proposed model – TagEmbedSVD, in detail. TagEmbedSVD is an extension of SVD++ [9]. The main objective here is to incorporate knowledge learned from tags into the SVD++ model for cross-domain settings. In this way, tags are able to not only provide additional knowledge to understand the system but also able to bridge the users and items across domains. Let $p_u \in \mathbb{R}^d$ be user (u) embedding and $q_j \in \mathbb{R}^d$ and $y_j \in \mathbb{R}^d$ be item (j) embeddings, $\mu \in \mathbb{R}, b_u \in \mathbb{R}$ and $b_j \in \mathbb{R}$ be the mean value of all the available ratings, user bias and item bias, respectively. Let $t_k \in \mathbb{R}^c$ be embedding vector associated with tag k, and c denote the embedding dimension. Here, we predict rating $\hat{r}_{uj}$ as follows:

$$\hat{r}_{uj} = \mu + b_u + b_j + (p_u + |\mathcal{I}_u|^{-\frac{1}{2}} \sum_{i \in \mathcal{I}_u} y_i)' q_j + \frac{\alpha}{|\mathcal{T}_u|} \sum_{k \in \mathcal{T}_u} w_u' E t_k + \frac{\beta}{|\mathcal{T}_j|} \sum_{k \in \mathcal{T}_j} x_j' F t_k, \tag{1}$$

where $w_u \in \mathbb{R}^d$ captures user u's preferences towards the tags and $x_j \in \mathbb{R}^d$ is item j's characteristics towards the tags. We obtain embeddings for tags from word2vec [12]. Thus, any two tags k and k' can be compared by its embeddings t_k and t'_k from the start of the training. Due to this fact, if two users share similar tag preferences, irrespective of their domain difference, we can obtain preference of users on tags from different domains. That is, p_u and q_j are learned from only the available ratings from the corresponding domains since users (items) from different domains do not share any item (user) in common. However, w_u and x_j are learned combinedly irrespective of its domain difference through the tag embeddings. Here, α and β control the influence of tags on predictions. In our model, to share a common embedding space and to have flexibility in choosing the dimension of t_k, we use projection matrices E and $F \in \mathbb{R}^{d \times c}$.

The number of occurrences of the tags associated with users and items plays an important role in characterising users and items. To adapt this knowledge we modify the definition $|\mathcal{T}_u|$ to be $\sum_{k \in \mathcal{T}_u} \eta_{uk}$, where η_{uk} denotes the frequency of the tag t_k associated with the user u. Similarly, we define $|\mathcal{T}_j|$ to be $\sum_{k \in \mathcal{T}_j} \eta_{jk}$. Therefore, Eq. (1) becomes:

$$\hat{r}_{uj} = \mu + b_u + b_j + (p_u + |\mathcal{I}_u|^{-\frac{1}{2}} \sum_{i \in \mathcal{I}_u} y_i)' q_j +$$

$$\frac{\alpha}{|\mathcal{T}_u|} \sum_{k \in \mathcal{T}_u} \eta_{uk} w_u' E t_k + \frac{\beta}{|\mathcal{T}_j|} \sum_{k \in \mathcal{T}_j} \eta_{jk} x_j' F t_k. \tag{2}$$

Additionally, we use a weighted-regularization technique [6,9] to control overfitting which arises due to the sparsity issue. Here, popular users and items are penalized less since more ratings are available, and users and items having less ratings are penalized more since only a few ratings are available for them. For instance, we regularize the user representation p_u by multiplying scalar $|\mathcal{I}_u|^{-\frac{1}{2}}$ instead of $|\mathcal{I}_u|$. Note that, the former penalizes less on users who rated more items as compared to the latter also the weighted-regularization does not drop any user or item. Let λ and λ_M be the positive hyperparameters to control over-fitting. We have the following optimization problem:

$$\min_{p_*,q_*,y_*,x_*,w_*,b_*} \mathcal{L} = \frac{1}{2} \sum_{(u,j)\in\Omega^S\cup\Omega^T} (\hat{r}_{uj} - r_{uj})^2 + \frac{\lambda}{2}(\sum_u |\mathcal{I}_u|^{-\frac{1}{2}} b_u^2 +$$

$$\sum_u |\mathcal{U}_j|^{-\frac{1}{2}} b_j^2) + \frac{\lambda}{2}\sum_u |\mathcal{I}_u|^{-\frac{1}{2}}(\|p_u\|^2 + \|w_u\|^2) + \quad (3)$$

$$\frac{\lambda}{2}\sum_j |\mathcal{U}_j|^{-\frac{1}{2}}(\|q_j\|^2 + \|x_j\|^2) + \frac{\lambda}{2}\sum_i |\mathcal{U}_i|^{-\frac{1}{2}}\|y_i\|^2 + \frac{\lambda_M}{2}(\|E\|_F^2 + \|F\|_F^2).$$

Complexity Analysis. Let a_I and a_T be the average number of rated items by users and average number of tags used by users (or tags assigned to items). Naive implementation of TagEmbedSVD takes $O(a_I|\Omega|d + a_T|\Omega|dc)$ to compute the objective value in Eq. (3) since we project tags into lower dimension using E and F. However, we can use hashtable to store and retrieve the sum of the embeddings of the tags corresponding to each user and item and this results in $O(a_I|\Omega|d + |\mathcal{T}^{uniq}|dc)$ time complexity, where $|\mathcal{T}^{uniq}|$ represents the number of unique tags in the system. Note that, in real time $|\mathcal{T}^{uniq}|dc << a_I|\Omega|d$. In addition, gradient computational effort of TagEmbedSVD requires $O(a_I|\Omega|d + d|\mathcal{T}^{uniq}|c + a_T|\Omega|dc)$ whereas SVD++ requires $O(a_I|\Omega|d)$. This is due to bottleneck in computing $\frac{\partial\mathcal{L}}{\partial E}, \frac{\partial\mathcal{L}}{\partial F}$. It can be reduced considerably by updating $\frac{\partial\mathcal{L}}{\partial E}, \frac{\partial\mathcal{L}}{\partial F}$ after some fixed number of intervals instead of every iteration. In our experiments, we observed that updating E and F after every ten iterations does not degrade the performance significantly.

Table 1. Dataset statistics

	#users	#items	#ratings	sparsity	#tag-assignments
MovieLens (ML)	5000	5000	2,68,092	98.93%	54,830
LibraryThing(LT)	5000	5000	1,05,171	99.58%	2,96,829

3 Experiments

Datasets and Evaluation Methodology. We used two publicly available datasets – MovieLens-10M[2] and LibraryThing[3] for cross-domain collaborative filtering setting. Statistics of the datasets are given in Table 1. For constructing training and validation split pairs, we follow the same procedure as used in [13,14]. From the target domain, we extract $K\%$ of the available ratings for the training set and the remaining $(100 - K)\%$ ratings are used for either validation or test purpose. We extract six such pairs. The first pair is used for tuning the hyperparameters, that is, the corresponding left out $(100 - K)\%$ ratings act as a

---

[2] https://grouplens.org/datasets/movielens/10m/.
[3] http://www.macle.nl/tud/LT/.

Table 2. Performance comparison for **setting 1**.

Setting 1	Metric	TagCDCF	GTagCDCF	TagGSVD++	TagEmbedSVD	Improvement (%)
ML (all)	MAE	0.6911	0.6905*	0.6926	**0.6524**	5.84
source (LT)	RMSE	0.8922	0.8915*	0.8920	**0.8439**	5.64
ML (cold-start users)	MAE	0.8652	0.7943*	0.7962	**0.7421**	7.03
source (LT)	RMSE	1.0786	1.0048*	1.0267	**0.9370**	7.24
ML (cold-start items)	MAE	0.6925	0.6895	0.6841*	**0.6526**	4.83
source (LT)	RMSE	0.8910	0.8901*	0.8914	**0.8434**	5.54
LT(all)	MAE	0.6658	0.6621	0.6543*	**0.6329**	3.38
source (ML)	RMSE	0.8618	0.8412*	0.8476	**0.8192**	2.69
LT (cold-start users)	MAE	0.8066	0.7503	0.7468*	**0.7141**	4.58
source (ML)	RMSE	1.0284	0.9652*	0.9928	**0.9321**	3.55
LT (cold-start items)	MAE	0.6697	0.6602	0.6556*	**0.6333**	3.52
source (ML)	RMSE	0.8646	0.8387*	0.8486	**0.8184**	2.48

Table 3. Performance comparison for **setting 2**.

Setting 2	Metric	TagCDCF	GTagCDCF	TagGSVD++	TagEmbedSVD	Improvement (%)
ML (all)	MAE	0.7117	0.7183	0.7095*	**0.6783**	4.60
source (LT)	RMSE	0.9146	0.9226	0.9135*	**0.8738**	4.54
ML (cold-start users)	MAE	0.8646	0.8009*	0.8282	**0.7767**	3.12
source (LT)	RMSE	1.0850	1.0241*	1.0839	**0.9949**	2.93
ML (cold-start items)	MAE	0.7121*	0.7163	0.7135	**0.6779**	5.04
source (LT)	RMSE	0.9150*	0.9197	0.9196	**0.8723**	4.90
LT(all)	MAE	0.7037	0.6828*	0.7053	**0.6556**	4.15
source (ML)	RMSE	0.9059	0.8635*	0.9027	**0.8411**	2.66
LT (cold-start users)	MAE	0.7971	0.7260*	0.7549	**0.6881**	5.49
source (ML)	RMSE	1.0051	0.9201*	0.9711	**0.8894**	3.45
LT (cold-start items)	MAE	0.7065	0.6830*	0.7017	**0.6554**	4.21
source (ML)	RMSE	0.9075	0.8632*	0.9009	**0.8411**	2.63

validation set. Once the hyperparameters values are obtained, we train the other five pairs with these values and obtain the test set performance. Here, in these five pairs, left out $(100 - K)\%$ ratings act as a test set. We report the average test error obtained from these five test sets as the final performance. During these extractions, we make sure that there exists at least one rating for all the users and items in the training set.

We conduct experiments under the following settings. In all the above settings, we add the source domain ratings (if multiple source domains are available we combine their ratings together) as a part of the training set.

1. **Setting 1**: We set $K = 80$.
2. **Setting 1, cold-start users (items)**: This is same as **setting 1**, but, results corresponding to only cold-start users (items) are reported.
3. **Setting 2**: Here, we set $K = 40$ to introduce more sparsity in the target domain part of the training set.
4. **Setting 2, cold-start users (items)**: This is same as **setting 2**, but, results corresponding to only cold-start users (items) are reported.

Further, **all** in Tables 2 and 3 indicates that all the users and items are included in test set, where as, **cold-start users** (**cold-start items**) indicates only users who rated less than five items (items which received ratings from less than five users) are included in test set. Similar definitions are followed in [6]. In Tables 2 and 3 bold-faced value indicates best performance, and 'Improvement' indicates the relative improvements that TagEmbedSVD achieves against the best performance among the comparison models highlighted by symbol *.

Comparison of Models. We compare our model with the following tag based CDCF models.

1. **TagCDCF** [13] extends matrix factorization and leverages tag information to improve the performance by understanding the similarities between users and items with the help of common tags.
2. **GTagCDCF** [14] connects the source and target domain by common tags. It additionally takes the frequency of the tag usage into account.
3. **TagGSVD++** [5] is an extension to SVD++ model. It uses tag information in place of implicit feedback in SVD++ to obtain user and item representations.

Since it has been demonstrated in [13,14] that the tag based CDCF models perform better than single domain models we do not include them for comparison.

Metrics. We employ two well-known metrics, Mean Absolute Error (MAE) and Root Mean Square Error (RMSE) for performance comparisons [6,9,10].

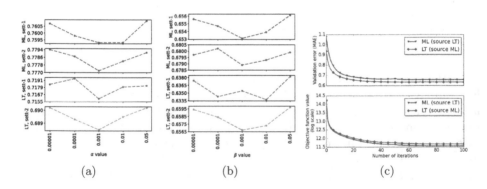

      (a)   (b)   (c)

Fig. 2. Impact of parameter (a) α and (b) β on cold-start users and items respectively. Here y-axis indicates MAE value.

Parameter Setting. We tune our hyperparameters using random hyperparameter search [1] and validation set with 100 trials for each model. For TagEmbedSVD, we tune λ from [0.01, 2], α from [0.00001, 0.05], β from [0.00001, 0.05] and latent dimension (d) from {5, 10, 20, 30, 40}. Ranges for comparison models were selected from the respective papers [5,13,14].

3.1   Results

Performance: Tables 2 and 3 compare the performance of TagEmbedSVD with the other tag-based CDCF models. We conduct a paired t-test and all the improvements are statistically significant for $p < 0.01$. The main findings from Tables 2 and 3 are summarized as follows:

1. TagEmbedSVD performs better than the other models by up to 5.84% (setting 1) and up to 4.60% (setting 2) when all the users and items are used. Similarly, it gives improvements up to 7.24% (setting 1) and 5.49% (setting 2) for cold-start users, and 5.54% (setting 1) and 5.04% (setting 2) for cold-start items respectively. This demonstrates the significance of using distributed representations for tags to improve the performance within and across domains.
2. One of the main reasons for TagEmbedSVD's better performance than that of TagCDCF and GTagCDCF is the utilization of all the available tags instead of just common tags. Further, despite both being extensions to SVD++, we gain improvement in TagEmbedSVD over TagGSVD++. The reason is that the former treats tags as tokens, hence *fiction* and *science-fiction* (or *hitler* and *ww2*) are two different tags. Whereas, the latter utilizes the pre-trained distributed representations for the tags, hence, they are very close to each other in the embedding space.

Impact of Parameters α and β: We investigate the effect of parameters α and β in Eq. (2), that control the influence of tag information to TagEmbedSVD. Note that, letting α and β to zero, it results in the SVD++ model. For cold-start users in datasets ML and LT, we fixed the other hyperparameter values and varied α in both setting 1, and setting 2. As we increase the value of α, the performance of the model improves as shown in Fig. 2(a). If α is set to the higher value, the performance decreases. It is because the information comes from tags dominates the rating values. We get similar behavior from the parameter β in cold-start item setting. This is illustrated in Fig. 2(b). Further, objective function value with respect to number of iterations are given in Fig. 2(c).

4   Conclusion

In this paper, we propose a simple and easy to train tag based cross-domain collaborative filtering model – TagEmbedSVD for leveraging tag information to cross-domain recommendations. TagEmbedSVD differs from the other models by employing distributed representations for tags to bridge the source and target domains. In this way, any two tags can be compared. Our experimental results show that our model performs better than other tag-based models in various sparse and cold-start settings. Although we use a single source and single target domain, TagEmbedSVD is general and any number of domains can be used without further modifications in the model.

References

1. Bergstra, J., Bengio, Y.: Random search for hyper-parameter optimization. J. Mach. Learn. Res. **13**(Feb), 281–305 (2012)
2. Bharti, R., Gupta, D.: Recommending top N movies using content-based filtering and collaborative filtering with hadoop and hive framework. In: Kalita, J., Balas, V.E., Borah, S., Pradhan, R. (eds.) Recent Developments in Machine Learning and Data Analytics. AISC, vol. 740, pp. 109–118. Springer, Singapore (2019). https://doi.org/10.1007/978-981-13-1280-9_10
3. Cantador, I., Fernández-Tobías, I., Berkovsky, S., Cremonesi, P.: Cross-domain recommender systems. In: Ricci, F., Rokach, L., Shapira, B. (eds.) Recommender Systems Handbook, pp. 919–959. Springer, Boston, MA (2015). https://doi.org/10.1007/978-1-4899-7637-6_27
4. Fang, Z., Gao, S., Li, B., Li, J., Liao, J.: Cross-domain recommendation via tag matrix transfer. In: ICDMW, pp. 1235–1240. IEEE (2015)
5. Fernández-Tobías, I., Cantador, I.: Exploiting social tags in matrix factorization models for cross-domain collaborative filtering. In: CBRecSys, pp. 34–41 (2014)
6. Guo, G., Zhang, J., Yorke-Smith, N.: TrustSVD: collaborative filtering with both the explicit and implicit influence of user trust and of item ratings. In: AAAI (2015)
7. He, M., Zhang, J., Yang, P., Yao, K.: Robust transfer learning for cross-domain collaborative filtering using multiple rating patterns approximation. In: WSDM, pp. 225–233 (2018)
8. Khan, M.M., Ibrahim, R., Ghani, I.: Cross domain recommender systems: a systematic literature review. ACM Comput. Surv. (CSUR) **50**(3), 36 (2017)
9. Koren, Y.: Factorization meets the neighborhood: a multifaceted collaborative filtering model. In: SIGKDD, pp. 426–434. ACM (2008)
10. Li, B., Yang, Q., Xue, X.: Can movies and books collaborate? cross-domain collaborative filtering for sparsity reduction. In: IJCAI, vol. 9, pp. 2052–2057 (2009)
11. Lops, P., Jannach, D., Musto, C., Bogers, T., Koolen, M.: Trends in content-based recommendation. User Model. User-Adapted Interact. **29**, 239–249 (2019)
12. Mikolov, T., Sutskever, I., Chen, K., Corrado, G.S., Dean, J.: Distributed representations of words and phrases and their compositionality. In: NIPS (2013)
13. Shi, Y., Larson, M., Hanjalic, A.: Tags as bridges between domains: improving recommendation with tag-induced cross-domain collaborative filtering. In: Konstan, J.A., Conejo, R., Marzo, J.L., Oliver, N. (eds.) UMAP 2011. LNCS, vol. 6787, pp. 305–316. Springer, Heidelberg (2011). https://doi.org/10.1007/978-3-642-22362-4_26
14. Shi, Y., Larson, M., Hanjalic, A.: Exploiting social tags for cross-domain collaborative filtering. arXiv preprint arXiv:1302.4888 (2013)
15. Shi, Y., Larson, M., Hanjalic, A.: Collaborative filtering beyond the user-item matrix: a survey of the state of the art and future challenges. ACM Comput. Surv. (CSUR) **47**(1), 3 (2014)
16. Wang, W., Chen, Z., Liu, J., Qi, Q., Zhao, Z.: User-based collaborative filtering on cross domain by tag transfer learning. In: Workshop on Cross Domain Knowledge Discovery in Web and Social Network Mining, pp. 10–17. ACM (2012)

On Applying Meta-path for Network Embedding in Mining Heterogeneous DBLP Network

Akash Anil[(✉)], Uppinder Chugh, and Sanasam Ranbir Singh

Department of Computer Science and Engineering,
Indian Institute of Technology Guwahati, Guwahati, India
a.anil.iitg@gmail.com, uppinderchugh@gmail.com, ranbir@iitg.ac.in

Abstract. Unsupervised network embedding using neural networks garnered considerable popularity in generating network features for solving various network-based problems such as link prediction, classification, clustering, etc. As majority of the information networks are heterogeneous in nature (consist of multiple types of nodes and edges), previous approaches for heterogeneous network embedding exploit predefined meta-paths. However, a meta-path guides the model towards a specific sub-structure of the underlying heterogeneous information network, it tends to lose other inherent characteristics. Further, different meta-paths capture proximities of different semantics and may affect the performance of underlying task differently. In this paper, we systematically study the effects of different meta-paths using recently proposed network embedding methods (`Metapath2vec`, `Node2vec`, and `VERSE`) over DBLP bibliographic network and evaluate the performance of embeddings on two applications, namely (i) Co-authorship prediction and (ii) Author's research area classification. From various experimental observations, it is evident that embeddings exploiting different meta-paths perform differently over different tasks. It shows that meta-path based network embedding is task-specific and can not be generalized for different tasks. We further observe that selecting particular node types in heterogeneous bibliographic network yields better quality of node embeddings in comparison to considering specific meta-path.

Keywords: Heterogeneous network · Meta-path · Heterogeneous network embedding · DBLP · Co-authorship prediction · Author classification

1 Introduction

Recently there is a surge in applying network embedding for addressing various tasks in network science such as classification, clustering, link prediction, community detection etc. [5,7,12,18]. Network embedding aims at learning low dimensional feature vector for a node capable of preserving its structural characteristics [4,7]. Majority of the network embedding models proposed previously

© Springer Nature Switzerland AG 2019
B. Deka et al. (Eds.): PReMI 2019, LNCS 11942, pp. 249–257, 2019.
https://doi.org/10.1007/978-3-030-34872-4_28

consider homogeneous networks, i.e. network consisting of singular type of nodes and relations [7,12,16,18]. However, majority of the real-world information networks and social networks are heterogeneous in nature i.e. networks consist of multiple types of nodes and relations [15]. For example, an academic bibliographic network may be represented using Author (A), Paper (P), Venue (V) (conference/journal) as nodes and different contextual relations such as Author-writes-Paper (AP), Author-publishes-at-Venue (AV), etc.

Majority of the previous studies on mining heterogeneous networks [3,14] exploit *meta-path* [8] which is a sequence of relations between different node types. Further, symmetric meta-paths are capable of preserving heterogeneous proximity between the underlying nodes. For example, in a bibliographic network, meta-path APA gives the proximity estimate between two authors collaborating on the same paper whereas AVA represents proximity between two authors publishing at the same venue. While exploring a network, a meta-path defines a specific path the explorer should follow. Recently, meta-paths have been used to generate network embedding [5,6] and reported to obtain promising results for various applications in network mining such as node classification, link prediction, clustering, etc. In this paper, we systematically analyze the effectiveness of considering meta-path for generating network embedding, specifically for bibliographic network. Since, meta-path guides to explore only the partial network defined by the meta-path, it may lose some of the inherent network properties. Motivated by this, this paper attempts to understand the following two important issues while considering meta-paths for generating network embeddings.

1. Does meta-path lose network information which can degrade the network embedding performance?
2. Are meta-path based embeddings independent to the end task?

To investigate the above-discussed problems, we evaluate embeddings generated using different types of meta-paths using three state-of-the-art embedding models, namely, (i) Metapath2vec [5], (ii) Node2vec [7], and (iii) VERSE [18] on Co-authorship prediction task and Author's research area classification in DBLP[1] heterogeneous bibliographic network. From various experimental observations, it is evident that meta-path based network embedding cannot be generalized for graph-based problems of diverse nature. Further, selecting suitable node types in the underlying heterogeneous network seems to be more important than considering different meta-paths for heterogeneous network embedding.

Rest of the paper is organized as follows. Section 2 presents some of the previous works on network embedding. Section 3 gives a brief description for heterogeneous network, meta-path, and network embedding. Section 4 describes the experimental setups and results. Finally, Sect. 5 concludes the paper.

---

[1] https://dblp.uni-trier.de/.

2   Literature Survey

For network embedding, a majority of the initial studies attempt to map the natural graph representations like normalized adjacency or Laplacian matrix to lower dimensions by using spectral graph theory [2,10] and various non-linear dimensionality reduction techniques [1,13,17]. However, these models are not scalable to large real-world networks as they exploit graph decomposition techniques at the core which requires the whole matrix beforehand.

To overcome the above limitations, many network embedding models exploit a framework which first generates a neighborhood sample using a random walk or proximity measure and then leverages it to learn the node embeddings using a skip-gram [9] based neural network model [7,12,16]. For example, Node2vec [7] uses a second order random walk to generate the neighborhood samples and learn the node embedding using skip-gram model, VERSE [18] preserves the vertex-to-vertex similarity using Personalized PageRank [11] and then exploits a single layer neural network to learn the embeddings.

All the above graph embedding models are proposed for homogeneous network. Recently, `Metapath2vec` [5] is proposed for heterogeneous network embedding which samples the node neighborhoods using a random walk guided through a meta-path. In a similar direction, study in [6] exploits the combined effect of different meta-path of predefined length to generate node embeddings in heterogeneous network.

3   Background Study

Definition 1 (Heterogeneous Network). *A Heterogeneous Network can be defined as six-tuple $<N, E, N^\tau, E^\tau, \phi, \psi>$ where N is a set of nodes, E is a set of edges, N^τ is a set of node types, E^τ is a set of edge types, $\phi : N \to N^\tau$ maps any node $n \in N$ to a node type $n^\tau \in N^\tau$, and $\psi : E \to E^\tau$ maps any edge $e \in E$ to an edge type $e^\tau \in E^\tau$. A homogeneous network is a special case of heterogeneous network where cardinalities of N^τ and E^τ are equal to one i.e. $|N^\tau| = |E^\tau| = 1$.*

Definition 2 (Meta-path). *Given a heterogeneous network G where $N^\tau = \{n_1^\tau, n_2^\tau, \cdots, n_l^\tau\}$ and $E^\tau = \{e_1^\tau, e_2^\tau, \cdots, e_{l-1}^\tau\}$, a meta-path $\mathcal{P}_{(n_1^\tau, n_l^\tau)}$ can be defined as an ordered sequence of edge types required to traverse for visiting a node type n_l^τ from node type n_1^τ, i.e. $\mathcal{P}_{(n_1^\tau, n_l^\tau)} = n_1^\tau \xrightarrow{e_1^\tau} n_2^\tau \xrightarrow{e_2^\tau} \cdots \xrightarrow{e_{l-1}^\tau} n_l^\tau$.*

3.1   Homogeneous Network Embedding

With the popularity of word2vec model using skip-gram proposed in [9] for generating word embedding from large sentence corpus, studies in [7,12,16] adapt skip-gram for network embedding. These network embedding frameworks exploit random walk based sampling strategy to generate node sequences capturing node's neighborhood characteristics similar to a sentence which captures contextual relation between two words. Formally, for a given network $G(N, E)$, network embedding using skip-gram model aims at maximizing neighborhood probability for a given node:

$$argmax_\theta \sum_{n \in N} \sum_{c \in \mathcal{N}(n)} log\ p(c|n; \theta) \qquad (1)$$

where $\mathcal{N}(n)$ gives the neighbors of n and $p(c|n; \theta)$ is the conditional probability of observing neighbor node c for the given node n.

3.2 Heterogeneous Network Embedding

For a given heterogeneous network $G(N, E, N^\tau, E^\tau)$, the skip-gram model defined in Eq. (1) can be transformed into heterogeneous skip-gram model as follows [5]:

$$argmax_\theta \sum_{n \in N} \sum_{\tau \in N^\tau} \sum_{c_\tau \in \mathcal{N}_\tau(n)} log\ p(c_\tau|n; \theta) \qquad (2)$$

where $\mathcal{N}_\tau(n)$ gives the neighbor nodes of n from τ^{th} type. Furthermore, $p(c_\tau|n; \theta)$ is defined using softmax function, i.e. $p(c_\tau|n; \theta) = \frac{exp(X_{c_\tau} \cdot X_n)}{\sum_{u \in N} exp(X_u \cdot X_n)}$, where X_n corresponds to the embedding vector of node n.

3.3 Meta-path Based Heterogeneous Network Embedding

The meta-path based heterogeneous network embedding model exploits heterogeneous skip-gram defined in Eq. (2). Further, random walks guided through meta-paths are used to generate neighborhood samples for all the nodes. In other words, random walker traverses partial heterogeneous network specific to underlying meta-path. For example, Metapath2vec exploits APVPA (or AVA) meta-path while generating random walk based node sequences [5].

While Metapath2vec has been proposed specifically for heterogeneous network embedding, the above-discussed meta-path based network embedding framework can be easily adapted by homogeneous network embedding methods through redefining the input network with specific meta-path. Therefore, this paper further exploits two homogeneous network embedding models namely Node2vec [7] and VERSE [18] for meta-path based heterogeneous network embedding.

4   Experimental Setups and Analysis

4.1 Experimental Dataset

This paper uses DBLP bibliographic dataset (reported in [19]) covering publication information for the period between years 1968 to 2011. To generate various

Table 1. Characteristics of different networks constructed over DBLP data

Dataset	DBLP 1968-2008							DBLP 2009-2011
	AA	APA		AVA		All		
Node types	Author	Author	Paper	Author	Venue	Author	Paper	Venue	Author
#Nodes	162298	162298	155189	162298	621	162298	155189	621	18457
#Edges	461722	475828		326602		957856			29677

network embeddings using different meta-paths and to evaluate the embedding performance over different applications, we further divide the dataset into two parts; (i) from 1968 to 2008 for generating network embedding, and (ii) from 2009 to 2011 for evaluating the embeddings over different applications. This paper considers three types of nodes, namely (i) Author (A), (ii) Paper (P), and (iii) Venue (V) for constructing various types of networks defined by different meta-paths. We construct the following four types of undirected networks from the DBLP 1968-2008 dataset.

- **AA**: It is a homogeneous unweighted co-authorship network considering only Author node type. Two nodes are connected if they co-author a paper.
- **APA**: It is a heterogeneous unweighted network considering Author and Paper node types. An author is connected to a paper if he/she is one of the authors of the paper.
- **AVA**: It is a heterogeneous unweighted network considering Author and Venue node types. An author is connected to a venue if he/she published a paper in that venue. This network structure is similar to the structure considered in Metapath2vec [5].
- **All**: It is a heterogeneous unweighted network considering all three types of nodes (Author, Paper, and Venue) and corresponding relationships between them.

Table 1 shows the characteristics of these experimental networks.

4.2 Experimental Setups

As mentioned above, three popular recently proposed network embedding models, namely (i) Metapath2vec [5], (ii) Node2vec [7], and (iii) VERSE [18] are considered to generate node embeddings. For all the models, we use the same hyper-parameter values as described in the original studies cited above. All the embedding results reported in this paper consider 100-dimensional vector[2]. To investigate the performance of different meta-paths and their associated embedding, we evaluate the embedding quality using the following two applications.

---

[2] While testing with different dimensions 100, 200, 300, we did not observe significant differences. We therefore consider 100-dimensional vector.

Co-authorship Prediction: Like the study [18], we also consider Co-authorship prediction task as a classification problem i.e., given a node pair, classify if the node pair has a co-author relation or not. To model it as a binary classification problem, we generate feature vectors representing node pairs using Hadamard operator [7,18]. To avoid possible bias with the embedding towards the target application, we consider the DBLP 2009-2011 (non-overlapping with the embedding dataset) for generating samples for the classification task. In this sample, there are 29,677 number of co-authorship relations and 18,457 authors. We use random 80-20 split as training and test samples subjected to four different classifiers namely Gaussian Naive Bayes (NB), Random Forest (RF), Decision Tree (DT), and Logistic Regression (LR). To avoid over-fitting, the above setup has been repeated 10 times.

Research Area Classification: We now investigate the quality of the embeddings for predicting author's research area. For each author in DBLP 2009-2011, we further identify (considering the `Field` attribute in [19]) the area in which author has maximum publication and consider it as the author's class label. Like Co-authorship prediction, we use similar random 80-20 split for all the classifiers and repeated 10 times.

4.3 Result and Discussion

Tables 2 and 3 present the Accuracy for Co-authorship prediction and Author's research area classification respectively using three network embedding models discussed above for all networks, i.e. `AA`, `AVA`, `APA`, and `All`. From Tables 2 and 3, it is observed that LR out-performs other classifiers in 93% times for Co-authorship prediction and 75% times for Author's research area classification task. Therefore, we select LR Accuracy for further analysis.

Table 2. Accuracy for co-authorship prediction by classifiers for different networks, (**Combine = Concat(Metapath2vec, Node2vec, VERSE)**)

Classifier	Metapath2vec				Node2vec				VERSE				Combine			
	AA	APA	AVA	All	AA	APA	AVA	All	AA	APA	AVA	All	AA	APA	AVA	All
NB	0.585	0.633	0.694	0.717	0.688	0.699	0.697	0.719	0.725	0.756	0.733	0.746	0.673	0.745	0.737	0.758
RF	0.761	0.724	0.698	0.720	0.749	0.731	0.698	0.730	0.760	0.754	0.707	0.744	0.772	0.753	0.714	0.748
DT	0.683	0.654	0.628	0.644	0.678	0.658	0.632	0.657	0.688	0.674	0.642	0.678	0.699	0.673	0.645	0.678
LR	0.736	0.739	0.738	**0.766**	0.773	0.766	0.75	**0.777**	0.788	0.784	0.764	**0.796**	0.799	0.795	0.778	**0.806**

We first investigate if meta-path based embedding loses information or not. It is evident from Tables 2 and 3 that almost all the models perform best by exploiting `All` network and show poor performance with `AA`, `APA`, and `AVA` networks for both tasks, i.e. Co-authorship prediction and area classification. Thus,

Table 3. Accuracy for author's research area classification by classifiers for different networks, (**Combine = Concat(Metapath2vec, Node2vec, VERSE)**)

Classifier	Metapath2vec				Node2vec				VERSE				Combine			
	AA	APA	AVA	All	AA	APA	AVA	All	AA	APA	AVA	All	AA	APA	AVA	All
NB	0.392	0.476	0.503	0.499	0.500	0.582	0.497	0.488	0.492	0.557	0.550	0.552	0.429	0.58	0.529	0.522
RF	0.484	0.486	0.491	0.482	0.488	0.536	0.518	0.509	0.495	0.499	0.530	0.545	0.499	0.529	0.527	0.53
DT	0.442	0.439	0.439	0.428	0.436	0.481	0.472	0.449	0.445	0.440	0.476	0.490	0.456	0.471	0.474	0.495
LR	0.504	0.539	0.565	**0.566**	0.486	0.544	**0.559**	0.555	0.536	0.531	0.605	**0.624**	0.552	0.592	0.612	**0.625**

it can be inferred that meta-path alone may be a weak representation for the network because it does not incorporate the impacts of other relational properties while capturing node neighborhood.

Secondly, we intend to investigate if the same embedding responds coherently to different problems. From Tables 2 and 3, it is clearly visible that APA performs better than AVA for Co-authorship prediction whereas AVA performs better than APA for classifying Author's research area. This observation is true for all the embedding techniques used in this study. Thus, meta-path based heterogeneous network embedding cannot be generalized for the tasks of different nature.

The homogeneous network AA and heterogeneous network APA, preserve similar proximity, i.e. co-authorship between underlying pair of authors. From Table 2, it is evident that AA performs better than APA for Co-authorship prediction in majority of the cases. However, for Author's research area classification in Table 3, APA performs better than AA in almost all the scenarios. Thus, it can be inferred that meta-path based heterogeneous network embedding may perform differently (poor or better) compared to homogeneous network embedding when subjected to tasks of diverse nature.

Among all the embedding models, VERSE consistently outperforms others for almost all the networks and classifiers for both Co-authorship prediction and research area classification tasks. It may be because unlike Metapath2vec and Node2vec, VERSE exploits a Personalized PageRank [11] capturing vertex-to-vertex similarity while generating the neighborhood sequences.

We further investigate combining all the three embeddings (Metapath2vec, Node2vec, VERSE) by concatenating the feature vectors. From Tables 2 and 3, it is observed that combined embedding always out-performs individual embedding for Co-authorship prediction and Author's research area classification over all the four networks.

5   Conclusion

In this paper, we investigate the applicability of meta-paths in heterogeneous network embedding for Co-authorship prediction and Author's research area classification problems in heterogeneous DBLP database. From various experimental results, we observe that by using appropriate node types, majority of the embedding methods out-perform their counter-parts exploiting meta-path based network for both of the above-mentioned tasks. Further, it is also evident that exploiting past co-authorship relation or APA meta-path yields better co-author prediction in comparison to AVA meta-path which exploits author's publication venue. On the other hand, AVA meta-path contributes positively to Author's research area classification problem and have superior performance than APA meta-path. Thus, for heterogeneous network embedding one should carefully choose the node types, relation types, and meta-paths which can capture better the network characteristics to address the underlying problem.

References

1. Ahmed, A., Shervashidze, N., Narayanamurthy, S., Josifovski, V., Smola, A.J.: Distributed large-scale natural graph factorization. In: WWW, pp. 37–48 (2013)
2. Belkin, M., Niyogi, P.: Laplacian eigenmaps and spectral techniques for embedding and clustering. In: NIPS, pp. 585–591 (2002)
3. Cao, B., Kong, X., Philip, S.Y.: Collective prediction of multiple types of links in heterogeneous information networks. In: ICDM, pp. 50–59 (2014)
4. Cao, S., Lu, W., Xu, Q.: Grarep: learning graph representations with global structural information. In: CIKM, pp. 891–900 (2015)
5. Dong, Y., Chawla, N.V., Swami, A.: metapath2vec: scalable representation learning for heterogeneous networks. In: SIGKDD, pp. 135–144 (2017)
6. Fu, T.y., Lee, W.C., Lei, Z.: Hin2vec: Explore meta-paths in heterogeneous information networks for representation learning. In: Proceedings of the 2017 ACM on Conference on Information and Knowledge Management CIKM 2017, pp. 1797–1806. ACM, New York (2017). https://doi.org/10.1145/3132847.3132953
7. Grover, A., Leskovec, J.: Node2vec: scalable feature learning for networks. In: SIGKDD, pp. 855–864. ACM (2016)
8. Kong, X., Yu, P.S., Ding, Y., Wild, D.J.: Meta path-based collective classification in heterogeneous information networks. In: CIKM, pp. 1567–1571 (2012)
9. Mikolov, T., Sutskever, I., Chen, K., Corrado, G.S., Dean, J.: Distributed representations of words and phrases and their compositionality. In: NIPS, pp. 3111–3119 (2013)
10. Ou, M., Cui, P., Pei, J., Zhang, Z., Zhu, W.: Asymmetric transitivity preserving graph embedding. In: SIGKDD, pp. 1105–1114 (2016)
11. Page, L., Brin, S., Motwani, R., Winograd, T.: The pagerank citation ranking: Bringing order to the web. In: Proceedings of the 7th International World Wide Web Conference,Brisbane, Australia, pp. 161–172 (1998). https://www.citeseer.nj.nec.com/page98pagerank.html
12. Perozzi, B., Al-Rfou, R., Skiena, S.: Deepwalk: online learning of social representations. In: SIGKDD, pp. 701–710 (2014)
13. Roweis, S.T., Saul, L.K.: Nonlinear dimensionality reduction by locally linear embedding. Science 290(5500), 2323–2326 (2000)
14. Sun, Y., Barber, R., Gupta, M., Aggarwal, C.C., Han, J.: Co-author relationship prediction in heterogeneous bibliographic networks. In: ASONAM, pp. 121–128. IEEE (2011)
15. Sun, Y., Han, J.: Mining heterogeneous information networks: principles and methodologies. Synth. Lect. Data Min. Knowl. Discovery 3(2), 1–159 (2012)
16. Tang, J., Qu, M., Wang, M., Zhang, M., Yan, J., Mei, Q.: Line: large-scale information network embedding. In: WWW, pp. 1067–1077 (2015)

17. Tenenbaum, J.B., De Silva, V., Langford, J.C.: A global geometric framework for nonlinear dimensionality reduction. Science **290**(5500), 2319–2323 (2000)
18. Tsitsulin, A., Mottin, D., Karras, P., Müller, E.: Verse: versatile graph embeddings from similarity measures. In: WWW, pp. 539–548 (2018)
19. Yang, D., Xiao, Y., Xu, B., Tong, H., Wang, W., Huang, S.: Which topic will you follow? In: Flach, P.A., De Bie, T., Cristianini, N. (eds.) ECML PKDD 2012. LNCS (LNAI), vol. 7524, pp. 597–612. Springer, Heidelberg (2012). https://doi.org/10. 1007/978-3-642-33486-3_38

An Improved Approach for Sarcasm Detection Avoiding Null Tweets

Santosh Kumar Bharti[1]([✉]) [iD], Korra Sathya Babu[2] [iD],
and Sambit Kumar Mishra[3] [iD]

[1] Pandit Deendayal Petroleum University,
Gandhinagar 382007, Gujarat, India
sbharti1984@gmail.com
[2] National Institute of Technology,
Rourkela 769008, Odisha, India
prof.ksb@gmail.com
[3] ITER, SOA University,
Bhubaneswar, Odisha, India
skmishra.nitrkl@gmail.com

Abstract. Among the plethora of social media, Twitter has emerged as the favorite destination for researchers in recent times. Many researchers are inclined to work on Twitter due to the availability of massive tweets and its unique features like hashtags and short messages. In recent times, various studies have preferred the hashtags (#sarcasm and #sarcastic) to collect Twitter dataset for sarcasm detection. However, hashtag-based distant supervision suffers from the problem of the inclusion of null tweets in the datasets which can be considered as a critical one for sarcasm detection. In this article, an algorithm is proposed for automatic detection and filtration of null tweets in the Twitter data. Additionally, an algorithm to identify sarcastic tweets using context within a tweet is also proposed. This approach use dictionaries of handpicked hashtag words, emoticons as the context within a tweet. Finally, we deployed a rule-based algorithm to analyse the performance of the proposed approach. The proposed approach attains the accuracy of 97.3% (after filtering null tweets) and 83.13% (without filtering null tweets) using a rule-based approach. The attained results conclude that after elimination of null tweets, the performance of the proposed system improved significantly.

Keywords: Context-based · Emoticons · Hashtag word · Negation word · Null tweets · Social media · Sarcastic · Sentiment

Adoption of business tactics based on the reviews and discussions in social media has gained much importance in recent times. The sentiments of the stakeholders are taken into consideration during planning and decision making. Sentiments can be classified as positive, negative, and neutral. The algorithms available for sentiment analysis focus mostly on availability of positive deviated words or phrases. Majority of sentiment analysis algorithms are likely to fail in case of the

© Springer Nature Switzerland AG 2019
B. Deka et al. (Eds.): PReMI 2019, LNCS 11942, pp. 258–266, 2019.
https://doi.org/10.1007/978-3-030-34872-4_29

presence of sarcasm in the text. The presence of sarcasm in the textual data such as tweets, reviews and discussions pose challenges to the automated systems for identifying actual sentiment [10]. In the textual data, detection of sarcasm is tough due to the lack of intonation and facial expressions. In fact, according to the BBC report, the U.S. Secret Service was looking for a software system that could detect sarcasm in social media data [2]. Therefore, an automated system is required for sarcasm detection in the text.

Due to the restriction on tweet's length (140 characters), users' often use symbolic notations such as smilies, emoticons, @User, *etc.* to accommodate more information. While posting a tweet, people often include videos, images, #hashtag, *etc.* along with text to indicate context of text in tweet. These context shows more visual information which cannot be demonstrated through text. To make the tweet as self-explanatory, few more features are added by the users, such as #trending, @User, RT, *etc.* These features of tweets make it unique over other social media text. According to Davidov *et al.* [7], around 20–25% of tweets are falls into one of the following three categories after downloading it from Twitter.

1. A tiny tweet having a length upto three or four words.
2. A tweet contains only handles, i.e., @, RT, URL, and #tag.
3. An indirect tweet which depends on either videos or images to conveys the theme of the tweet.

In this article, these tweets are referred to as null tweets which often miss the important features within a tweet in the context of sarcasm detection. The context within a tweet may be topical, historical, temporal, or situational [1,8, 9,13]. The conventional method of collecting sarcastic tweets is hashtag based distant supervision using #sarcasm and #sarcastic.

Some of the past studies on sarcasm detection in tweets are based on context. They have used features like relationships, the chain of conversations, intersentential incongruity and embedded multimedia posts [1,8,9,11,13]. To identify sarcasm based on this context, they require additional information such as the user's profile, chat history, the cohesion of sentences, *etc.* These context-based approaches are likely to fail in the identification of a sarcastic tweet when a single tweet is given for detecting the context. This article exploits the context within a tweet and proposed a rule-based approach using a list of manually collected hashtag words and emoticons as shown in Table 1 that play a role of the context within a tweet to identify sarcasm. These hashtag words and emoticons are usually appended by the user at the end of the tweet to indicate the topical, situational context.

In Table 1, the hashtag words and emoticons act as a guiding factor for polarizing the orientation of the tweet. It can be considered as a context of the tweet. For example: "Super easy to focus at work today #kidding". In this example, the tweet sentiment seems positive, but due to "#kidding" being appended at the end of the tweet, it acts as a context here for this particular tweet. Due to the hashtag appended at the end, the sentiment of the tweet flips to negative. It indicates that the user had written the tweet intentionally to make the tweet

Table 1. Dictionary used for sarcasm detection in tweets.

Negation words	Not, Never, No
Hashtag words	#Not, #NotAtAll, #NotReally, #Lying, #Kidding, #Joking, #Pretend, #Justkidding, #Justjoking
Emoticons	☺ , ☹

as sarcastic. Similarly, a sample list of such sarcastic tweets is given in Table 2. These sarcastic tweets are based on "negation words", "hashtag words" and "emoticons" dictionaries as shown in Table 1. In Table 2, the "negation words", "hashtag words" and "emoticons" are indicated in underlined italics.

Table 2. Hashtag and Negation word based Sarcastic Tweets

#	Sarcastic Tweets
1	I am _never_ late for office. _#joking_
2	@MrNiallMcGarry That's OK Niall. We actually only use 20% of our brainpower. The other 54% is dormant _#NotAtAll_
3	Finally got my Lagunitas mason jars delivered. They make your beer taste better. _#notreally_
4	Oh great...can't wait to see that then! _;-(_
5	I see the diet is going well! ☹

The contribution of this article is as follows:

1. Proposed an algorithm to detect and eliminate null tweets automatically.
2. Proposed an algorithm to detect sarcastic tweets using manually collected dictionary words includes hashtag words, negation words and emoticons as shown in Table 1.
3. Experimented the proposed approach and observe that after eliminating null tweets, the performance of sarcasm detection system improves significantly in some of the existing system as well.

The rest of this paper is organized as follows. Section 1 presents related work on sarcasm detection in Twitter data. The proposed scheme has been described in Sect. 2. Section 3 presents the performance analysis of the proposed schemes. Finally, the conclusions are drawn in Sect. 4.

1 Related Work

In the current era, the research on sarcasm detection in text is grown rapidly [1,3–9,11–13]. The objective of research emphasizes on analyzing sentiment in text data in the presence of sarcasm. The content of social media such as tweets often carries sarcasm. Sarcasm detection techniques in the past emphasize on several classification techniques. The complete classification of sarcasm detection techniques is shown in Fig. 1.

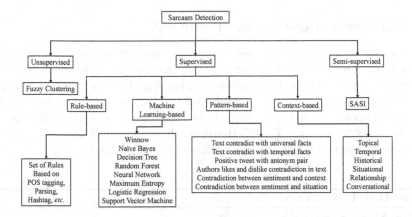

Fig. 1. Various classification approaches used for sarcasm detection in text.

This article focused on one the supervised method given in Fig. 1, *i.e.*, context-based sarcasm detection. This method is used in several text contexts in literature, such as topical, situational, relational, historical, *etc.*

1.1   Context-Based Approach

The relationship between an author and audience followed by the immediate communicative context can be helpful to improve the sarcasm prediction accuracy [1]. Message-level sarcasm detection on Twitter using a context-based model were used for sarcasm detection [13]. A framework based on the linguistic theory of context incongruity and an introduction of inter-sentential incongruity by considering the history of the posts in the discussion thread was considered for sarcasm detection [8]. A quantitative evidence of historical tweets of an author can provide additional context for sarcasm detection [9]. The author's past sentiment on the entities in a tweet was exploited to detect the sarcastic intent. Chains of tweets that work in a context were considered. They introduce a complex classification model that works over an entire tweet sequence and not on one tweet at a time. Integration between linguistic and contextual features extracted from the analysis of visuals embedded in multimodal posts was deployed for sarcasm detection [11].

2   Proposed Scheme

This section describes process of null tweets elimination as tweet filtration followed by sarcastic sentiment detection in filtered tweets using negation word, emoticons and hashtag word dictionaries.

2.1   Null Tweets Detection and Filtration

Preprocessing is an important step during the process of sarcastic sentiment detection in Twitter data. In the conventional method of preprocessing, one usually eliminates the trending information, *i.e.*, hashtag, URL of videos and images, re-tweets, uppercase word to lower case word conversion and @user information, *etc.* For example, a given tweet "yeah, right! #sarcasm!" will look after conventional preprocessing "yeah, right!". However, this article performs an additional preprocessing step, and the detail is discussed in Algorithm 1.

Algorithm 1: Null Tweets Detection

Data: *dataset* := Corpus of Raw tweets ($\mathbb{C}$)
Result: *Classification*: = Filtered Tweets
Notation: T: tweet, C: corpus, *tok*: token, LOT: list of tokens, $LOFT$: list of filtered tweets.
Initialization :
$handles = \{@, \#, RT, URL\}, LOFT = \{\Phi\}, LOT = \{\Phi\}, count = 0, flag = 0, flag1 = 1$
while T *in* $\mathbb{C}$ **do**
 while *tok in* T **do**
  | $LOT \leftarrow LOT \bigcup tok\ count + +$
 end
 if $count \leq 3$ **then**
  | Given tweet is null tweet.
 end
 else
  while *tok in* LOT **do**
   if *(tok[0] == http)* **then**
    | $flag1 = 0$ break
   end
   else if *(tok[0] ≠ @||#||RT||rt)* **then**
    | $flag1 = 0$ Break
   end
  end
 end
 if *(flag1 == 0)* **then**
  | Given tweet is null tweet.
 end
 else if *(flag1 == 1)&&(flag == 0)* **then**
  | Given tweet is null tweet.
 end
 else
  | Given tweet is valid for analysis. $LOFT \leftarrow LOFT \bigcup T$
 end
end

According to Algorithm 1, the tweet "yeah, right!" is considered as a null tweet. So, it is eliminated to enhance the accuracy of the proposed system. The Algorithm 1 shows the procedure for automatic detection and elimination of null tweets in the tweets corpus. Algorithm 1 takes tweet corpus (C) as input and tokenizes each tweet and stores in the list of tokens (LOT) file. It also counts the total number of tokens in each tweet. If the length count is less than or equal to three, then the given tweet is a null tweet and is discarded. Otherwise, if any token in LOT starts with HTTP://, then the given tweet is null tweet as it depends on some other source to conveys the meaning of the tweet (such tweets are called referred tweets). Similarly, if all the tokens in LOT contain only handles such as @, #tag, RT and no text, then the given tweet is a null tweet. If a tweet does not follow any of the three conditions as given in Algorithm 1, then the tweet is a valid tweet for sarcasm analysis and is stored in the list of filtered tweets (LOFT) file.

2.2   Proposed Algorithm for Sarcasm Detection

The proposed approach is based on the context within a tweet that is extracted from the three dictionaries namely, negation words, hashtag words, and emoticons as shown in Table 1. The negation words are capable of inverting the polarity of any word by appending it as a prefix whereas, the hashtag words dictionary is capable of flipping the polarity of the entire sentence by appending it as a suffix. For an instance of the negation word: "not happy", happy is known as a positive word. As we append 'not' as the prefix of happy, the polarity of happy becomes negative. Similarly, an instance of hashtag word: "This is a perfect solution for sarcasm detection #not". Here, without hashtag word "#not", the sentiment of this tweet is positive. However, after appending "#not" at the end of the tweet, the overall tweet's polarity inverted from positive to negative. Finally, emoticons are also capable of reversing the tweets polarity. For instance, "I see the diet is going well!". The tweet seems positive, but emoticons made it sarcastic. Therefore, according to the Macmillan English Dictionary for sarcasm, "#not" is capable of flipping the meaning of a text. Hence, it plays a role of context to make the tweet as sarcastic. These hashtag words are capable of making a sentence sarcastic under certain constraints.

The process of identifying a sarcastic tweet is given in Algorithm 2. It explains the step-wise procedure for sarcasm detection in a single filtered tweet based on the context of hashtags words and emoticons dictionaries. Algorithm 2 takes filtered tweets (LOFT) as input, and determines the sentiment value of each tweet and stored it in δ. Subsequently, the last bunch of hashtag words is stored in λ which is usually appended after the text part of the tweet. If any hashtag word appears within the text part, then the algorithm remove the hash symbol and treat as a word. Further, it tokenizes the given tweet and checks for the presence of negation words. If a negation word is found, then flip the sentiment value of the corresponding tweet and look for new sentiment value. If the new sentiment value is positive or neutral and any λ value is present in hashtag dictionary, then the given tweet is classified as sarcastic, otherwise, non-sarcastic. If negation word is not found in the tweet's text part and the sentiment value of the tweet is positive and any λ value present in hashtag dictionary, then the tweet is classified as sarcastic otherwise, it is considered as non-sarcastic.

Algorithm 2: Sarcasm Detection using hashtag context

Data: *dataset* := Corpus of filtered tweets (*LOFT*)
Result: *Classification*: = Classified tweets as sarcastic or not sarcastic.
Notation: T: tweet, C: corpus, *tok*: token, λ: Last bunch of hashtags in the tweet, δ:
Sentiment value of given tweet, $\hbar$: Opposite of original sentiment value.
Initialization: Negation Word (η) = {Not, Never, No}, Emoticons (ε) = {;-(, ☺, ☻ },
Hashtag word Dictionary (h) = {#Lying, #Kidding, #Joking, #Not, #Notreally,
#Notatall, #Pretend},
$handles$ = {@, #, RT, URL}, $LOFT$ = {Φ}, LOT = {Φ}, $count$ = 0, $flag$ = 0, $flag1$ = 1
while $T \in C$ **do**
    δ = find sentiment(T) **while** $tok \in T$ **do**
        λ = $find\_last\_bunch\_of\_tok(T)$ **if** *(tok $\in \eta$)* **then**
            Flip δ into negative to positive. **if** *($\delta \geq 0$)&& (any hashtag of $\lambda \in h$)* **then**
            | Tweet is sarcastic.
            end
            else
            | Tweet is not sarcastic.
            end
        else if *($\delta \geq 0$)&& ($\lambda \in h$)* **then**
        | Tweet is sarcastic.
        end
        else if *($\delta \geq 0$)&& ($\lambda \in \varepsilon$)* **then**
        | Tweet is sarcastic.
        end
        else if *(any h word present in tweet text without hashtag) && ($\lambda \in h$)* **then**
        | Tweet is not sarcastic.
        end
        else
        | Tweet is not sarcastic.
        end
    end
  end
end

Table 3. Manually annotated dataset for testing

Category of Tweets	Sarcastic	Non-sarcastic	Unpredictable
#Lying (582)	185	296	101
#Kidding (517)	178	260	79
#Joking (330)	84	204	42
#Not (1070)	467	474	129
#Notreally (58)	22	29	7
#Notatall (128)	47	55	26
#Pretend (84)	21	50	13
#Justjoking (126)	37	68	21
#Justkidding (105)	29	57	19
Total (3000)	1070	1493	437

3　Experimental Results

In this work, an experiment was evaluated with three statistical parameters
namely, *precision, recall*, and *F1score*. It starts with dataset annotation followed by performance analysis using confusion matrix.

3.1   Dataset Collection and Annotation

In this article, we collected and annotated 3000 tweets manually from Twitter using various hashtags, negation words, and emoticons given in Table 1. The manually annotated dataset (MADS) is shown in Table 3. We observed that 437 tweets are unpredictable during annotation. Most of the unpredictable tweets were missing the context of sarcasm meets the criteria of null tweet, which are treated as null tweets.

3.2   Performance Analysis

The performance of the proposed algorithm for sarcasm detection was analyzed using rule-based classification. The experimental results of the proposed algorithm is given in Tables 4 and 5 respectively. Table 4 describes the confusion matrix for error analysis and Table 5 shows the attained *precision*, *recall*, and *F1-score*.

Table 4. Confusion matrix of proposed approaches for error analysis

Approaches	Null Tweets	T_p	F_p	T_n	F_n
Rule-based	After Filtration	1033	32	1461	37
Rule-based	Without Filtration	1033	272	1461	234

Table 5. Compared *precision*, *recall*, and *F1-score* of proposed approaches with some of the existing work.

Approaches	Null Tweets	*Accuracy*	*Precision*	*Recall*	*F1 − score*
Tungthamthiti et al. [12]	Without filtration	80.4%	0.76	0.76	0.76
Bharti et al. [4]	Without filtration	79.4%	0.74	0.72	0.73
Rule-based	After filtration	97.3%	0.97	0.965	0.967
Rule-based	Without filtration	83.13%	0.815	0.791	0.807

4   Conclusion

This article deals with two things. First is the process to detect and eliminate null tweets which can be considered as a preprocessing step. Here, tweets that act as noisy in the dataset are eliminated. Secondly, a sarcasm detection algorithm based on context within a tweet was proposed. The properties of hashtags and emoticons were exploited as the context in a tweet to be identified as sarcastic. The proposed algorithm was implemented and evaluated with and without the presence of null tweets. Some of the state-of-arts algorithms for sarcasm detection were also evaluated in the same line. It is observed that the sarcasm detection algorithms perform better after filtration of null tweets in the dataset.

References

1. Bamman, D., Smith, N.A.: Contextualized sarcasm detection on twitter. In: Ninth International AAAI Conference on Web and Social Media (2015)
2. BBC: Us secret service seeks twitter sarcasm detector. http://www.bbc.com/news/technology-27711109/ (2014)
3. Bharti, S.K., Babu, K.S.: Sarcasm as a contradiction between a tweet and its temporal facts: a pattern-based approach. Int. J. Nat. Lang. Comput. (IJNLC) **7**, 67–79 (2018)
4. Bharti, S.K., Sathya Babu, K., Jena, S.K.: Harnessing online news for sarcasm detection in Hindi Tweets. In: Shankar, B.U., Ghosh, K., Mandal, D.P., Ray, S.S., Zhang, D., Pal, S.K. (eds.) PReMI 2017. LNCS, vol. 10597, pp. 679–686. Springer, Cham (2017). https://doi.org/10.1007/978-3-319-69900-4_86
5. Bharti, S.K., Babu, K.S., Raman, R.: Context-based sarcasm detection in Hindi Tweets. In: 2017 Ninth International Conference on Advances in Pattern Recognition (ICAPR), pp. 1–6. IEEE (2017)
6. Bharti, S., Vachha, B., Pradhan, R., Babu, K., Jena, S.: Sarcastic sentiment detection in tweets streamed in real time: a big data approach. Digit. Commun. Netw. **2**(3), 108–121 (2016)
7. Davidov, D., Tsur, O., Rappoport, A.: Semi-supervised recognition of sarcastic sentences in Twitter and Amazon. In: Proceedings of the Fourteenth Conference on Computational Natural Language Learning, pp. 107–116. Association for Computational Linguistics (2010)
8. Joshi, A., Sharma, V., Bhattacharyya, P.: Harnessing context incongruity for sarcasm detection. In: Proceedings of the 53rd Annual Meeting of the Association for Computational Linguistics and the 7th International Joint Conference on Natural Language Processing, vol. 2, pp. 757–762 (2015)
9. Khattri, A., Joshi, A., Bhattacharyya, P., Carman, M.J.: Your sentiment precedes you: using an author's historical tweets to predict sarcasm. In: 6th Workshop on Computation Approaches to Subjectivity, Sentiment and Social Media Analysis (WASSA) 2015, p. 25 (2015)
10. Maynard, D., Greenwood, M.A.: Who cares about sarcastic tweets? Investigating the impact of sarcasm on sentiment analysis. In: LREC 2014 Proceedings. ELRA (2014)
11. Schifanella, R., de Juan, P., Tetreault, J., Cao, L.: Detecting sarcasm in multimodal social platforms. In: Proceedings of the 24th ACM International Conference on Multimedia, pp. 1136–1145. ACM (2016)
12. Tungthamthiti, P., Kiyoaki, S., Mohd, M.: Recognition of sarcasms in tweets based on concept level sentiment analysis and supervised learning approaches. In: Proceedings of the 28th Pacific Asia Conference on Language, Information and Computing, pp. 404–413 (2014)
13. Wang, Z., Wu, Z., Wang, R., Ren, Y.: Twitter sarcasm detection exploiting a context-based model. In: Wang, J., et al. (eds.) WISE 2015. LNCS, vol. 9418, pp. 77–91. Springer, Cham (2015). https://doi.org/10.1007/978-3-319-26190-4_6

A Grammar-Driven Approach for Parsing Manipuri Language

Yumnam Nirmal[✉] and Utpal Sharma

Tezpur University, Tezpur 784028, Assam, India
{ynirmal,utpal}@tezu.ernet.in

Abstract. This paper presents attempts to parse Manipuri language using a context-free grammar (CFG) and Earley's parsing algorithm. For a low resource language like Manipuri where treebanks are not available, following a grammar-driven approach is the only viable solution as of now. The work covers simple, compound and complex Manipuri sentences. The system reports a Recall of 81.71%, Precision of 72.38%, and F-measure of 76.76%. The CFG and the parsing system we have created can be very useful in building a treebank for Manipuri.

Keywords: Parsing · Manipuri · Tibeto-Burman · Earley's parser

1 Introduction

Syntactic parsing is the process of formally analyzing a sentence or a string of words into its constituents, resulting in a parse tree, indicating the syntactic relationship between the constituents. A sentence that cannot be parsed have either grammatical errors or the sentence is too hard to read [8].

Manipuri (Meitei-lon) is one of the Tibeto-Burman [9] languages spoken in northeastern India, mainly in Manipur, and also in parts of Assam, Tripura, Bangladesh and Myanmar. It is a language that has been classified as vulnerable by UNESCO [12]. Being an extremely low resource language, developing a syntactic parser for Manipuri is challenging. This require grammar formalisms such as CFG, Tree-Adjoining Grammar (TAG), Dependency Grammar (DG), Head-Driven Phrase Structure Grammar (HPSG), etc. The grammar can be either learned from an existing treebank or hand-crafted. As treebanks are non-existent and parts-of-speech (POS) annotated data are scanty for Manipuri, learning its grammar from an existing treebank is not viable. Developing a treebank too requires a huge amount of time and manpower. This leaves us with only option to hand-craft syntax rules by extensively analyzing the language structure.

We have extensively analyzed the Manipuri language syntax and hand-crafted syntactic rules in the form of phrase structures, that in turn form a CFG. This CFG is then utilized to implement Earley's parsing algorithm to parse Manipuri sentences. In this paper we include only some of the rules of the grammar.

© Springer Nature Switzerland AG 2019
B. Deka et al. (Eds.): PReMI 2019, LNCS 11942, pp. 267–274, 2019.
https://doi.org/10.1007/978-3-030-34872-4_30

2 Related Work

Two broad approaches have been followed for parsing natural language texts: data-driven and grammar-driven. Data-driven approaches such as Maltparser [13], MSTParser [10] and Probabilistic CFG parser [6] rely on huge treebanks to learn the grammar structure and induce a parsing model. Grammar-driven approaches rely on hand-crafted grammar rules such as CFG and a lexicon to induce a parsing model.

Though the current scenario of syntactic parsing favors data-driven approach, grammar-driven approach tends to be the only option for low resource languages where a treebank is not available. Such approach can be a stepping stone to achieve parsing, which may eventually help in creating the desired treebanks.

Among grammar-driven approaches CFG has been a popular choice [2,7,11, 15]. CFGs can be utilized for parsing a sentence as well as for sentence generation.

A preliminary work on parsing of Manipuri language is described by Singh et al. [16]. They attempted parsing of Manipuri language using CFG and TAG, in a bottom-up approach. The proposed method has a low accuracy and does not handle issues such as ambiguity, multi-word expressions (MWEs), etc. In addition, the model handles limited structures of simple and compound sentences, but not complex sentences.

3 Parsing Manipuri Sentences

We have developed a CFG for Manipuri and implemented Earley's parsing algorithm in Python 3.6 using Natural Language Toolkit (NLTK) 3.4.1. For parts-of-speech (POS), we have used BIS tagset [1] for Indian languages. Parsing has been attempted for simple, compound and complex sentences.

3.1 Simple Sentences

Manipuri, a Tibeto-Burman language, follows subject-object-verb (SOV). It is a verb final language [3,5]. Simple sentences in Manipuri consist of a single or multiple noun phrases (NPs) followed by at least a verb phrase (VP) or a copula (COP). A COP acts as a verb connecting the NPs. Alternatively, an NP may not occur at all.

Example 1.

(a) ꯅꯤꯄꯤꯃꯆꯥ ꯑꯗꯨ ꯆꯍꯤ ꯱꯴ ꯆꯉꯂꯦ
[nipiməca ədu]$_{NP}$ [cəhi tərαməri]$_{NP}$ [cəŋle]$_{VP}$
small girl - that - year - 14 - approaching

(b) ꯇꯣꯝꯕꯒꯤ ꯃꯆꯥꯅꯤ
[tombə-gi]$_{NP}$ [məcα]$_{NP}$ [-ni]$_{COP}$
tomba's - son is

These can be explained by the CFG rules:

S → NP VP | NP COP
NP → NP NP

S is the start symbol.

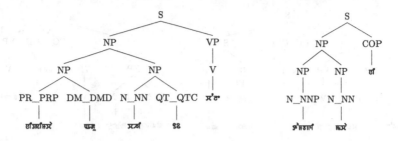

Fig. 1. Parse trees for Examples 1(a) and (b)

3.2 Compound Sentences

Compound sentences consist of more than one simple sentence conjoined together as one by coordinate conjunctions (CC_CCD).

$$S \rightarrow S \quad CC\_CCD \quad S$$

Example 2.

ꯑꯩꯅ ꯆꯥꯛ ꯆꯥꯔꯤ ꯑꯗꯨꯒꯥ ꯆꯥꯎꯕꯅ ꯉꯥꯏꯔꯤ (I - rice - eating - and - Chaoba - waiting)
[ei-nə cak cari]$_S$ [ədugə]$_{CC\_CCD}$ [caobə-nə ŋəiri]$_S$

There are cases where compound sentences are formed using co-referring verbs or nouns. In such a case one of the verbs or nouns is deleted (Figs. 2 and 3).

$$S \rightarrow NP \quad CC\_CCD \quad S \mid S \quad CC\_CCD \quad NP \mid S \quad CC\_CCD \quad VP$$

Example 3.

(a) ꯇꯣꯝꯕ ꯑꯃꯁꯨꯡ ꯆꯥꯎꯕ ꯆꯥꯛ ꯆꯥꯔꯤ (Tomba - and - Chaoba - rice - eating)
[tombə]$_{NP}$ [əməsuŋ]$_{CC\_CCD}$ [caobə cak cari]$_S$

(b) ꯆꯥꯎꯕ ꯆꯥꯛ ꯆꯥꯔꯤ ꯑꯗꯨꯒꯥ ꯑꯩꯁꯨ (Chaoba - rice - eating - and - me too)
[caobə cak cari]$_S$ [ədugə]$_{CC\_CCD}$ [eisu]$_{NP}$

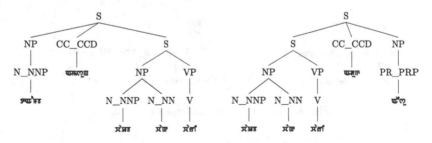

Fig. 2. Parse trees for Examples 3(a) and 3(b)

Example 4.

ꯑꯉꯥꯡꯗꯨ ꯀꯥꯔꯇꯨꯟ ꯌꯦꯡꯂꯤ ꯑꯗꯨꯒꯥ ꯅꯣꯛꯂꯤ (boy that - cartoon - watch - and - laughing)
[əŋaŋ-du kartun jeŋ-li]$_S$ [ədu-gə]$_{CC\_CCD}$ [nok-li]$_{VP}$

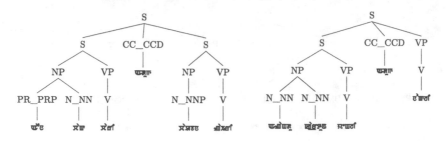

Fig. 3. Parse trees for Examples 2 and 4

3.3   Complex Sentences

Complex sentences in Manipuri are formed by joining two or more sentences with complex coordinate conjunction, or by embedding a sentence within another sentence [17]. Complex coordinate conjunction normally occurs in the form of a suffix.

Example 5. ꯇꯣꯝꯕꯒ ꯆꯥꯎꯕꯒ ꯅꯣꯛꯂꯤ   (Tomba   and   Chaoba   laughing)
[tombə-gə]$_{NP}$ [caobə-gə]$_{NP}$ [nok-li]$_{VP}$

Example 6.

S_1: ꯂꯩꯀꯣꯜꯗ ꯂꯩꯁꯤꯡ ꯁꯠꯂꯤ S_2: ꯂꯩꯁꯤꯡ ꯑꯗꯨ ꯐꯖꯩ
    ləykol-də ləysiŋ sat-li   ləy-siŋ ədu phəjəy
    (garden flowers bloom)   (flowers that beautiful)

$S_1 + S_2$: ꯂꯩꯀꯣꯜꯗ ꯁꯠꯂꯤꯕ ꯂꯩꯁꯤꯡ ꯑꯗꯨ ꯐꯖꯩ (garden bloom flowers that beautiful)
[ləykol-də sat-li-bə]$_{RelativeClause(RC)}$ [ləy-siŋ ədu]$_{NP}$ [phəjəy]$_V$

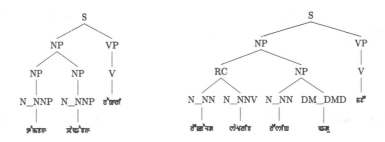

Fig. 4. Parse trees for Examples 5 and 6

In Example 5, the two simple sentences "ꯇꯣꯝꯕ ꯅꯣꯛꯂꯤ" (tombə nok-li) and "ꯆꯥꯎꯕ ꯅꯣꯛꯂꯤ" (caobə nok-li) have been combined into a single sentence using the complex coordinate conjunction "-ꯒ" (-gə). And, in Example 6, S_1 is embedded into S_2 (shown with the + sign) to form a single sentence using the verbal noun "ꯁꯠꯂꯤꯕ" (satlibə).

4   Issues with Parsing

Some issues encountered in parsing Manipuri sentences are mentioned below.

4.1   Ambiguity

Lexical and attachment ambiguities are two types of ambiguity that we encounter in Manipuri language. An example of lexical ambiguity can be seen in Example 7 where the word "ᯏᯟꯂ" (loynə) can be either a quantifier (QT) or an adverb (RB) (Fig. 5).

Example 7. ᯘꯥꯗꯢ ᯏᯟꯂ ꯊꯎꯥꯗꯢ (they all - know   or, they - know everything) (məkʰoy loynə kʰənhoui)

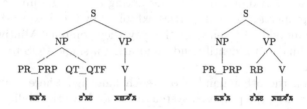

Fig. 5. Different parse trees for Example 7

Example 8 depicts attachment ambiguity where the constituents of the given sentence can be attached at different positions in the syntactic tree, giving rise to two trees. The first tree in Fig. 6 translates to "a shop that sells old books had broken", while the second tree translates to "an old shop that sells books had broken".

Example 8. ꯋꯦꯟꯗ ᯖꯁꯤꯟꯂ ꯁꯤꯕꯗ ꯀꯩꯅꯦ ꯋꯐꯪ ꯀꯦꯂ꯳꯳ꯢꯂꯛ

         (ancient - book - sell - shop - that - broken)
         əribə layrik yon-bə dukan ədu kintʰre

4.2   Copula

For syntactic parsing with CFGs for a highly agglutinative language like Manipuri, apart from POS tags of words, morphological features have to be considered. Such a case can be seen in Example 1(b), where the copula "ꯅꯤ" (-ni) acting as a verb has been attached to the noun "ꯃꯆꯥ" (məca) as a suffix. The copula had to be detached from the noun so that each of them could be assigned their respective POS tags.

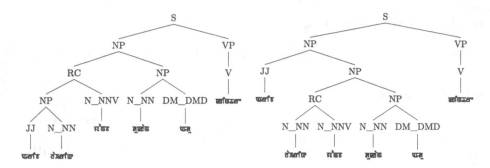

Fig. 6. Different parse trees for Example 8

4.3 Multiword Expressions

A MWE consists of two or more lexemes acting as a single unit and can have unpredictable properties. The properties exhibited by these expressions may not depend on the individual lexemes or the way they combine. Multiword expressions are also difficult to identify and annotating a single POS to MWE is not always possible.

Manipuri, on the other hand is one such language where multi-words are abundant. Currently, the proposed system does not have the ability, but may be refined further so as to be capable of identifying multi-words. The only viable solution as of now is to mark them as flat structure by hyphenating manually.

Example 9.

ꯃꯆꯤꯟ ꯃꯆꯦ꯫ ꯑꯗꯨ ꯂꯥꯛꯂꯦ (brothers - that - came)
[məcin mənao]$_{MWE}$ [ədu]$_{DM}$ [lakle]$_V$

The MWE in Example 9 can be hyphenated as "ꯃꯆꯤꯟ-ꯃꯆꯦ꯫" and will have PR_PRP as its POS tag.

4.4 Word Order

In Manipuri the position of the words are relatively free, provided they are sufficiently marked by case suffixes. The CFG rules need to cover all of these word order. Such a sentence is shown in Example 10. The sentence can have five more variants numbered 1–5 in the example. Excluding the verb, the remaining constituents can be arranged in any order without affecting the semantics.

Example 10.

ꯃꯍꯥꯛ - ꯉꯁꯤ - ꯏꯝꯐꯜꯗ - ꯆꯠꯂꯣꯌ (məhak ŋəsi ɪmˈfal-də ct-lə-roy)
he - today - Imphal-LOC - will not go

1. ꯃꯍꯥꯛ ꯏꯝꯐꯜꯗ ꯉꯁꯤ ꯆꯠꯂꯣꯌ (subject - object - adjunct - verb)
2. ꯏꯝꯐꯜꯗ ꯃꯍꯥꯛ ꯉꯁꯤ ꯆꯠꯂꯣꯌ (object - subject - adjunct - verb)
3. ꯏꯝꯐꯜꯗ ꯉꯁꯤ ꯃꯍꯥꯛ ꯆꯠꯂꯣꯌ (object - adjunct - subject - verb)
4. ꯉꯁꯤ ꯃꯍꯥꯛ ꯏꯝꯐꯜꯗ ꯆꯠꯂꯣꯌ (adjunct - subject - object - verb)
5. ꯉꯁꯤ ꯏꯝꯐꯜꯗ ꯃꯍꯥꯛ ꯆꯠꯂꯣꯌ (adjunct - object - subject - verb)

5   Evaluation

We have evaluated using Evalb [14], which is an open source implementation of PARSEVAL measures proposed by Black et al. [4]. It reports precision, recall, F-measure, non-crossing and tagging accuracy.

Labeled Precision (P): The average of how many brackets in the resulting parse tree match those in the gold standard.

$$P = \frac{\#\,\text{of correct constituents in candidate parse of sentence}}{\#\,\text{of total constituents in candidate parse of sentence}}$$

Labeled Recall (R): The average of how many brackets in the gold standard are in the resulting parse.

$$R = \frac{\#\,\text{of correct constituents in candidate parse of sentence}}{\#\,\text{of correct constituents in Treebank parse of sentence}}$$

F-measure: The weighted aggregation of precision and recall.

$$F = \frac{(\beta^2+1)PR}{\beta^2 P+R}, 0 \le \beta \ge +\infty,$$

when $\beta = 1$, P & R are weighted equally, $\beta > 1$, R is favored, $\beta < 1$, P is favored.

Cross-Brackets: The average of how many constituents in the resulting parse tree cross over the brackets in the gold standard.

cross-brackets = the number of crossed brackets

5.1   Evaluation

For evaluating the proposed parser we have prepared a gold standard treebank of 100 sentences. The sentences have been selected in such a way that it cover the overall structures of simple, compound and complex sentences. We have used the Peen treebank format with a modified tagset. The parse tree for Example 1(a) can be represented using Penn treebank format as follows:

```
(S
    (NP (NP (N_NN ঢ়সাতিদর্) (DM_DMD ঙর্))
        (NP (N_NN সর্) (QT_QTC ৪৪)))
    (VP (V স্ত্র্)))
```

The system reported a Recall (R) of 81.71%, Precision (P) of 72.38% and F-measure (F) of 76.76%.

6   Conclusion

To develop a parser for a language for which little computational work exists, a rule based approach is feasible and effective. We have created the rules for a CFG for the Manipuri language and developed a parser using the Earley's parsing method. We could identify the issues in this approach for the language and find ways to tackle some of them. The quality of parsing output is impressive and further improvement in future is possible. Such a parser can also be useful in creating a Treebank for the language, which will be a valuable resource for computational linguistic work in future.

References

1. Unified Parts of Speech (POS) Standard in Indian Languages - Draft Standard - Version 1.0. http://www.tdil-dc.in/tdildcMain/articles/134692Draft%20POS%20Tag%20standard.pdf. Accessed 20 Feb 2019
2. Bal, B.K., Shrestha, P., Pustakalaya, M.P., PatanDhoka, N.: Architectural and system design of the Nepali grammar checker. In: PAN Localization Working Paper (2007)
3. Bhat, D.S.: Grammatical Relations: The Evidence Against their Necessity and Universality. Routledge, London (2002)
4. Black, E., et al.: A procedure for quantitatively comparing the syntactic coverage of english grammars. In: Speech and Natural Language: Proceedings of a Workshop Held at Pacific Grove, California, 19–22 February 1991 (1991)
5. Ch Yashawanta, S.: Manipuri Grammar. Rajesh Publications, New Delhi (2000)
6. Collins, M.: Probabilistic Context-Free Grammars (PCFGs). http://www.cs.columbia.edu/~mcollins/courses/nlp2011/notes/pcfgs.pdf. Accessed 25 Apr 2019
7. Hasan, K., Mondal, A., Saha, A., et al.: Recognizing bangla grammar using predictive parser. arXiv preprint arXiv:1201.2010 (2012)
8. Jurafsky, D.: Speech & Language Processing. Pearson Education India, Bangalore (2000)
9. Matisoff, J.A., Baron, S.P., Lowe, J.B.: Languages and dialects of Tibeto-Burman (1996)
10. McDonald, R., Pereira, F., Ribarov, K., Hajič, J.: Non-projective dependency parsing using spanning tree algorithms. In: Proceedings of the conference on Human Language Technology and Empirical Methods in Natural Language Processing, pp. 523–530. Association for Computational Linguistics (2005)
11. Mehedy, L., Arifin, N., Kaykobad, M.: Bangla syntax analysis: a comprehensive approach. In: Proceedings of International Conference on Computer and Information Technology (ICCIT), Dhaka, Bangladesh, pp. 287–293 (2003)
12. Moseley, C.: Atlas of the World's Languages in Danger. UNESCO, Paris (2010)
13. Nivre, J., Hall, J., Nilsson, J.: Maltparser: A data-driven parser-generator for dependency parsing. In: LREC, vol. 6, pp. 2216–2219 (2006)
14. Sekine, S., Collins, M.: Evalb bracket scoring program (1997). http://www.cs.nyu.edu/cs/projects/proteus/evalb
15. Singh, G., Lobiyal, D.: A computational grammar for Hindi verb phrase. In: Proceedings of International Conference on Expert Systems for Development, pp. 244–249. IEEE (1994)
16. Singh, L.N., Sharma, U.: Modelling the syntax of Manipuri: a Tibeto-Burman language. M.Tech Dissertation form Tezpur University, Tezpur 784028, Assam, India (2012)
17. Thoudam, P.C.: A Grammatical Sketch of Meiteiron. Ph.D. thesis (1980)

Query Expansion Using Information Fusion for Effective Retrieval of News Articles

G. Savitha, Sriramkumar Thamizharasan, and Rajendra Prasath[(⊠)]

Indian Institute of Information Technology Sri City,
Chittoor Dt., Sri City 517 646, AP, India
{savitha.g15,sriram.t15,rajendra.prasath}@iiits.in
https://www.2power3.com/rajendra

Abstract. In this paper, we propose an information fusion based approach for query expansion for news articles retrieval. In this approach, we attempt to expand the user query using fusion of context oriented information collected from the underlying corpus statistics and topic diversity of news documents. The proposed methodology first collects graph based statistics and topic diversity to derive a list of additional terms and then uses fusion based approach to select a subset of terms from the derived list which are then used to expand the given query. Then the expanded query is used to perform news document retrieval. The proposed approach is tested on a specific collection of news articles collected from various news sources. The experimental results show that the proposed approach performs effective retrieval of news documents pertaining to the actual user intent.

Keywords: Query expansion · Understanding user query intent · Information fusion · News retrieval · Information retrieval evaluation

1 Introduction

Queries submitted to web search systems may contain a few keywords or under specified information which may lead to ambiguity in understanding user query intent. It is a challenging problem to understand the intent behind the actual user query and retrieve information accordingly in order to serve the user information needs. For instance, consider the query: "Bus Services in Java". This query is ambiguous as it may either point to information related to transportation in Java island (place) or bus architecture in JAVA programming language. But adding one extra term, for example, say "island" or "programming" would make the user intent clearer and reduce level of the ambiguity to larger extent. Thus, expanding the input query with addition terms would help understand the user query intent. This may result in better retrieval of information. A traditional IR system uses numerous representative queries to address the user intent. In order to achieve this, a huge amount of labelled examples are required which

© Springer Nature Switzerland AG 2019
B. Deka et al. (Eds.): PReMI 2019, LNCS 11942, pp. 275–284, 2019.
https://doi.org/10.1007/978-3-030-34872-4_31

276 G. Savitha et al.

is a laborious task. Thus, this problem needs automatic approaches that could assist in finding the additional terms for the query. We propose a fusion based approach that combines statistical information from the news corpus and topic diversity of news articles to expand the user query and therefore disambiguate the query.

The rest of the paper is organised as follows: Sect. 2 explains key related works. Section 3 describes the proposed query expansion approach using fusion of information collected from news articles. Section 4 shows the experimental results and its comparisons with different approaches. Finally Sect. 5 concludes the paper.

2 Related Work

In Information Retrieval systems, query expansion [4] has been well explored for better understanding of the information needs of users. Bai *et* al. [1] showed that the combination of 3 to 5 terms can capture the intention of user queries and improve the effectiveness of Web search. Zhao and Zhang [18] exploited global knowledge in knowledge graphs and local contexts in query logs to initialise and enrich the query representation and query intent understanding. Gonzalez-Caro and Baeza-Yates [6] proposed an approach to identify the query intent in a wide set of facets that are obtained from query intent classification model. He *et* al. [8] showed that proximity among query terms could be useful for improving the retrieval performance. This is achieved by improving the classical BM25 model using the term proximity evidence. Hu *et* al. [9] proposed an approach to discover intent concepts by leveraging Wikipedia in which each intent domain is represented as a set of Wikipedia articles and categories.

Jansen and Booth [10] classified web queries, (which can be of three types: informational, navigational, and transactional by topic and user intent. Most recently, Hashemi *et* al. [7] proposed a query classification approach to detect the intent of a user query using Convolutional Neural Networks (CNN). Queries are represented as vectors so that semantically similar queries can be captured by embedding them into a vector space. Kuzi *et* al. [11] proposed an approach for query expansion using word embeddings in which terms that are semantically related to the query are selected for query expansion.

3 Proposed Approach

The proposed approach expands the original query by augmenting it with additional terms. This results in the reformulation of the original query into multiple queries each pointing to a specific query intent. This would assist IR systems to understand the user search and their information needs. In the proposed approach, we first explore additional terms related to the actual user query terms and expand the query with these additional terms based on the fusion of information. Then we use the expanded query for the retrieval of news articles. Next we describe the components used in our proposed approach.

3.1   Construction of News Graph

The news articles are used to create a News Graph. Let the News Graph be $G = (V, E)$ where V is the set of Noun Phrases (NP) and E is set of edge weights measured as count of Verb Phrases between a specific pair of NPs. Stanford CoreNLP[1] is used to get the parsed tree of each sentence of a news article. A triplet of $NounPhrase_1$ - $VerbPhrase$ - $NounPhrase_2$ will indicate the relation between two Noun Phrases. With the generated triplets, the overall news graph has been constructed with Noun Phrases (NP) as nodes and Verb Phrases (VP) as the edges. Finally, the constructed news graph has 253,884 nodes in total.

The motivation behind the construction of the News Graph G is to construct a knowledge graph that captures association between any pair of entities. In the context of this work, entities represent persons, organizations, locations etc.

Moving forward, we use the terms entities and noun phrases interchangeably.

3.2   Topic Modeling

In this work, we focus on Latent Dirichlet Allocation (LDA) proposed by Blei *et* al. [2,3]. LDA is a generative probabilistic topic model. LDA performs hierarchical Bayesian model where each datum is modelled as a finite mixture over a set of topics. In turn, each topic is modelled as an infinite mixture over a set of topic probabilities. In text mining, the topic probabilities provide an explicit representation of a document.

As part of our proposed approach, the diversity of a term is measured using the possible topics covered by the term.

We used LDA with Gibb's sampling with the following parameters:

- Iterations $= 1000$
- Number of Topics $= 200$
- alpha $= 0.5$, beta $= 0.1$
- Number of Words in each Topic $= 200$

In the context of this work, more diverse a term, more prominent it is in the entire news domain across various news items. Moreover, the count across topics (where a word has) will help counteract highly connected but unimportant phrases to the query terms. From observation, it has shown to scale down the fusion score certain unimportant but highly connected phrases.

3.3   Vector Space Model

In this section, we describe the vector space as defined in [12,17]. Let $D = \{d_1, d_2, \cdots, d_N\}$ be the set of documents and $|D| = N$. Each document d_i ($1 \leq i \leq N$) is composed of one or more index terms t_j which may be weighted according to their importance. When each document d_i is represented using a k-dimensional vector and $d_i = (t_i^1, t_i^2, \cdots, t_i^k)$. Now the Term Frequency (TF)

---

[1] Stanford Core NLP - https://stanfordnlp.github.io/CoreNLP/.

of a term t_k of document d_i, denoted by $tf_{i,k}$, is defined as the number of occurrences of the term t_k in the document d_i. Document frequency of a term t, denoted by df_t is the number of documents that contain the term t. Inverse Document Frequency (IDF) of a term t, is denoted by $idf_t = log\frac{N}{df_t}$.

The TF-IDF weighting scheme [12] assigns a weight to a term t in a document d denoted by

$$tf - idf_{t,d} = tf_{t,d} \times idf_t \tag{1}$$

Average TF-IDF of a term t is the product of the average term frequency of t across the document space multiplied by IDF_t.

3.4   Information Fusion

Consider a query q with n terms and $q = w_1\ w_2\ \cdots\ w_n$. For each query, we generate r alternate queries by augmenting the selected terms with original query terms.

Normalised edge weight between 2 nodes n_i and n_j is given by:

$$normalized(e_{i,j}) = \frac{e_{i,j}}{max\{e_{i,1}, e_{i,2}, \cdots, e_{i,m}\}} \tag{2}$$

where $e_{i,j}$ is the edge weight between nodes n_i and n_j. $\{n_1, n_2, ..., n_m\}$ are the m neighbours of node n_i. Fusion Score of a phrase p is given by:

$$Fusion\_Score(p) = \frac{(normalized(e_{w,p}) * Average\ TF - IDF(p)) + num\_topics}{(rank(w))^k} \tag{3}$$

where $rank(w)$ is the rank of the word w in the user-defined *query* when ranked using Average TF-IDF and *num_topics* is the count of unique topics covered by words in phrase p. The topics are generated using LDA modelling on the documents. For our set of experiments, $k = 3$ provided the best outcome. For our experiments, we have considered 10 alternate queries. These alternate queries are used in place of the original user query to improve the effectiveness of news retrieval.

Firstly, terms in *query* are ranked according to the decreasing order of Average TF-IDF to scale the phrases generated from the terms in *query*. Secondly, for each term w_i in the *query* we consider the phrases that are directly connected to w_i in graph G. Score of each phrase is initially the normalised score of the edge weight (between phrase and term in the *query*) multiplied by the Average TF-IDF of the phrase. The number of unique topics covered by words in the phrase is also added to the score of the phrase to be taken into consideration. Score of each phrase is scaled by $rank(w_i)^k$ to incorporate the importance of the phrase with respect to terms in *query*. Finally, phrases are ranked according to their score and top r phrases are augmented to *query* to get alternate queries. The fusion in our approach is in fusing the information gained about a phrase from the news graph, topics and topic score derived from LDA and the importance of terms in query.

Algorithm 1. The proposed Query Expansion approach

Input: $query = w_1 \ w_2 \ \cdots \ w_n$ is an user query with n terms
Input: Graph $G(V, E)$ where V is set of Noun Phrases (NP) and E is set of counts of Verb Phrases between each pair of NPs
Input: m : Number of neighbouring nodes in the Graph G
Input: r : Number of alternate queries to be generated
 1: Rank terms in *query* based on their TF-IDF
 2: Initialise *candidate_phrases* ← NULL
 3: **for all** $w_i \in query$ **do**
 4:     $neighbours = \{p_j \mid 1 \leq j \leq m) \ \}$ where each p_j is a NP matching w_i in G
 5:     Add p_j to *candidate_phrases*
 6: **end for**
 7: Rank each $p \in$ *candidate_phrases* **according to** $Fusion\_Score(p)$
 8: **for** $i = 1 : r$ **do**
 9:     $alternate\_query_i = q + candidate\_phrase_i$
10: **end for**
Output: *alternate_queries*

---

4 Experimental Results

4.1 News Articles Collection

We have created a news articles corpus consisting of articles from various online sources like newspapers, web news and other news services. We used different news sources and meta search engines to collect the links leading to different news articles. We focused on the full coverage of the news item published by various sources at different periods of time. The document collection consists of 13,619 news articles in English Language. This news collection also consists of near duplicate news articles which could have been republished by different news agencies but with slightly updated content. We have identified and extracted various fields of the news articles like published date, source URL, title and full story (description) of the news.

The similarity between the query and news articles is calculated using BM25 [15]. We used DisMax Query Parser[2] for parsing the test queries. The collection covers a majority of articles in the following categories: India, Politics, Strike, Prime Minister, Cricket matches and players, Supreme court, and War.

Sources of top 10 news services (with their coverage given in brackets) are as follows: timesofindia.indiatimes.com (512), indianexpress.com (419), ndtv.com (413), news18.com (412), timesnownews.com (408), thehindu.com (334), indiatoday.in (330), hindustantimes.com (314), firstpost.com (304), and financialexpress.com (227)

---

[2] Lucene query parser - https://lucene.apache.org/core/3_0_3/api/core/org/apache/lucene/search/DisjunctionMaxQuery.html.

4.2   List of Queries

The selected list of 10 queries and different user intent for each query is illustrated in Table 1. We have explored the news corpus and identified multiple queries and their possible intents by reading through various articles. Queries are selected based on the news coverage and the number of news articles for each query is sufficiently larger in size.

Table 1. List of selected queries

QID	Actual query	User query intents
Q1	Settle Korean war	Settlement of war between North and South Korea, Settlement of Nuclear war between US and North Korea
Q2	Currency fall blame Jaitley	Jaitley blaming factors for currency fall Jaitley being blamed for currency fall
Q3	Cricket in Sri Lanka	Cricket tournament held in Sri Lanka Cricket matches played by Sri Lanka team
Q4	battle on court	Court case battle, a tennis match in the tennis court
Q5	BJP blame alliance	BJP coalition blames other parties People point out mistakes by alliance with BJP
Q6	Tesla problems	Issues with Tesla cars Tesla issues with leadership
Q7	India partnership	Run partnership by Indian cricket players, India partnership deal with countries/corporations
Q8	congress in court	Court case by Congress for Rafale deal Congress view on Sect. 377 court case
Q9	Attack Syria	Rebel attacks in Syria Syria attacking on rebels or embassies
Q10	Battle in Pakistan	Battles with Pakistan Crickct battle against Pakistan

4.3   Evaluation Methodology

Subjective evaluation of relevance in the content of news articles pertaining to the query intent is performed to measure the relevance of information retrieved. For each query, each evaluator evaluates the top 20 ranked news articles retrieved for that particular query, by analysing their context. The relevance of the news article is scored on a 3 point scale. The relevance of the news article is scored on a 3 point scale (0, 0.5,1). A document assigned a score 1 is completely relevant with respect to the query intent, 0.5 signifies the document is partially relevant, and 0 signifies that document is noisy/not relevant.

4.4   Comparison of Query Expansion Approaches

We have compared the performance of the proposed fusion based news articles retrieval with 3 different methods: (a) Word Embedding [13,14]; (b) RAKE [5,16]; (c) Modified Bo1 [14].

(a) **Word Embedding**: Word2Vec model is used to extract words that are semantically similar to the query.
(b) **RAKE**: RAKE extracts a collection of key phrases from a text based on the sum of 3 scores (degree of the keyword, frequency and ratio between them). The key phrases are extracted from the articles retrieved for the original user query.
(c) **Bo1**: This is a randomness weighting model used to find additional terms from the first k results for initial query.

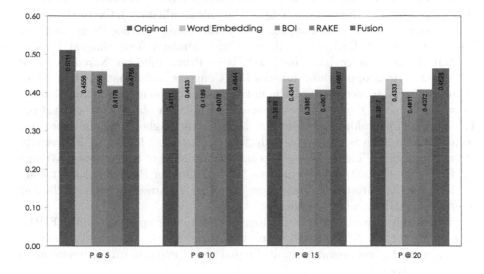

Fig. 1. Precision @ top k news documents

Figure 1 presents Precision @ top k documents for the user query in comparison with the various query expansion methods attempted. It can be observed that the proposed fusion approach outperforms the other methods at top 10, 15 and 20 documents. We performed statistical significance test with single-tailed paired t-test on Table 2. Our retrieved results obtained by the proposed fusion based approach are found to be statistically significant with 95% confidence level ($p < 0.05$). Table 3 shows the p values.

Table 2. P @ k values for various methods

Methods	P @ 5	P @ 10	P @ 15	P @ 20
WE	0.4555	0.4433	0.4340	0.4333
BO1	0.4555	0.4188	0.3985	0.40111
RAKE	0.4177	0.4077	0.4066	0.4072
Fusion	**0.4755**	**0.4644**	**0.4666**	**0.4627**

Table 3. Significance test Values

Method	P Values
WE vs. Fusion	0.001814176520
BO1 vs. Fusion	0.009916707469
RAKE vs. Fusion	0.000004949079

4.5 Discussions

In this section, we present our key observations related to an effective retrieval of news articles. For the user query Q5 ("BJP blame alliance"), the top 3 queries generated using fusion based approach: (1) BJP blame alliance Prime Minister Narendra Modi; (2) BJP blame alliance Union Minister Ravi Shankar Prasad; (3) BJP blame alliance India initiative. Here Prime minister Narendra Modi called out various opposition alliances like Congress and Janata Dal. A lot of blame-calling happened during India initiative and Union minister Ravi Shankar Prasad was one of the most documented persons who blamed the opposition. The phrase "Prime Minister Narendra Modi" ranks higher than the other two because of the high edge weight with "BJP" (562) and higher topic diversity (38 various topics). This helps to overcome the problem of having lower Average TF-IDF value (1.6602) as compared to "Union Minister Ravi Shankar Prasad" (2.1155). "India initiative" ranks below the other 2 phrases because of the low Average TF-IDF value of the phrase.

Experiments are currently being done for testing our approach on the FIRE[3] dataset for the English corpus. As part of future work, we intend to extend our approach to accommodate multi-lingual query expansion to improve news document retrieval.

5 Conclusion

In this work, we have attempted to identify the actual intent of the user query and use it to enhance the efficiency of information retrieval systems. The idea is to expand the query by adding additional terms from the corpus using graph

---

[3] FIRE stands for Forum for Information Retrieval and Evaluation - http://fire.irsi. res.in/fire/static/data.

and topic modelling. The additional terms are weighed using proposed fusion based approach. The preliminary experimental results show that this approach performs better retrieval compared to the base line approach. Graph properties and features of news documents such as the rate of decay of coverage of news items, importance of news source, named entity recognition can be also be used in the existing model to perform better news retrieval.

References

1. Bai, J., Chang, Y., Cui, H., Zheng, Z., Sun, G., Li, X.: Investigation of partial query proximity in web search. In: Proceedings of the 17th International Conference on World Wide Web WWW 2008, pp. 1183–1184. ACM, New York (2008)
2. Blei, D.M.: Probabilistic topic models. Commun. ACM **55**(4), 77–84 (2012)
3. Blei, D.M., Ng, A.Y., Jordan, M.I.: Latent dirichlet allocation. J. Mach. Learn. Res. **3**, 993–1022 (2003)
4. Carpineto, C., Romano, G.: A survey of automatic query expansion in information retrieval. ACM Comput. Surv. **44**(1), 1:1–1:50 (2012)
5. Garigliotti, D., Balog, K.: Generating query suggestions to support task-based search. In: Proceedings of the 40th International ACM SIGIR Conference on Research and Development in Information Retrieval SIGIR 2017, pp. 1153–1156. ACM, New York (2017)
6. González-Caro, C., Baeza-Yates, R.: A multi-faceted approach to query intent classification. In: Grossi, R., Sebastiani, F., Silvestri, F. (eds.) SPIRE 2011. LNCS, vol. 7024, pp. 368–379. Springer, Heidelberg (2011). https://doi.org/10.1007/978-3-642-24583-1_36
7. Hashemi, H.B., Asiaee, A.H., Kraft, R.: Query intent detection using convolutional neural networks. In: WSDM QRUMS 2016 Workshop, pp. 4285–4290. ACM, New York (2016)
8. He, B., Huang, J.X., Zhou, X.: Modeling term proximity for probabilistic information retrieval models. Inf. Sci. **181**(14), 3017–3031 (2011)
9. Hu, J., Wang, G., Lochovsky, F., Sun, J.T., Chen, Z.: Understanding user's query intent with Wikipedia. In: Proceedings of the 18th International Conference on World Wide Web WWW 2009, pp. 471–480. ACM, New York (2009)
10. Jansen, B.J., Booth, D.: Classifying web queries by topic and user intent. In: CHI 2010 Extended Abstracts on Human Factors in Computing Systems, pp. 4285–4290. ACM, New York (2010)
11. Kuzi, S., Shtok, A., Kurland, O.: Query expansion using word embeddings. In: Proceedings of the 25th ACM International on Conference on Information and Knowledge Management CIKM 2016, pp. 1929–1932. ACM, New York (2016)
12. Manning, C.D., Raghavan, P., Schütze, H.: Introduction to Information Retrieval. Cambridge University Press, New York (2008)
13. Mikolov, T., Chen, K., Corrado, G.S., Dean, J.: Efficient estimation of word representations in vector space. CoRR (2013)
14. Rattinger, A., Goff, J.L., Guetl, C.: Local word embeddings for query expansion based on co-authorship and citations. In: Proceedings of the 7th International Workshop on Bibliometric-Enhanced Information Retrieval (BIR 2018) co-located with the 40th European Conference on Information Retrieval (ECIR 2018) 26 March 2018, pp. 46–53 (2018)

15. Robertson, S., Zaragoza, H.: The probabilistic relevance framework: BM25 and beyond. Found. Trends Inf. Retr. **3**(4), 333–389 (2009)
16. Rose, S., Engel, D., Cramer, N., Cowley, W.: Automatic keyword extraction from individual documents, pp. 1–20, March 2010
17. Salton, G., Wong, A., Yang, C.S.: A vector space model for automatic indexing. Commun. ACM **18**(11), 613–620 (1975)
18. Zhao, S., Zhang, Y.: Tailor knowledge graph for query understanding: linking intent topics by propagation. In: Proceedings of the 2014 Conference on Empirical Methods in Natural Language Processing, EMNLP 2014, 25–29 October 2014, Doha, Qatar, A meeting of SIGDAT, a Special Interest Group of the ACL, pp. 1070–1080 (2014)

A Hybrid Framework for Improving Diversity and Long Tail Items in Recommendations

Pragati Agarwal, Rama Syamala Sreepada, and Bidyut Kr. Patra[✉]

National Institute of Technology Rourkela, Rourkela, Odisha, India
{217cs1089,515cs1002,patrabk}@nitrkl.ac.in

Abstract. In today's information overloaded era, recommender system is a necessity and it is widely used in most of the domains of e-commerce. Over the years, recommender system is improved to meet the main purpose of achieving better user experience, where accuracy is considered as one of the important aspects in its design. However, other aspects such as diversity, long tail item recommendation, novelty and serendipity are equally important while providing recommendations to the users. Research to improve above mentioned aspects is limited. In this paper, we propose an efficient approach to improve diversity and long tail item recommendations. The experiments are conducted on two real world movie rating datasets namely, MovieLens and Netflix. Experimental analysis shows that the proposed method outperforms the state-of-the art approaches in recommending diverse and long tail items.

Keywords: Recommender system · Diversity · Collaborative Filtering · Long tail items · Hybrid Reranking Framework

1 Introduction

With the advent of the World Wide Web, information has increased exponentially and so is the need of customized techniques to filter relevant and personalized information. Recommender system is an essential tool for providing customized information about products and services. It is widely used in various domains such as movies, music, books, research articles, electronic products, apparel, *etc.* Collaborative filtering (CF) approach is one of the popular approaches in the design of recommender systems [3]. It uses rating information of the items provided by the users in the past [2,3]. CF techniques are further classified as memory based CF and model based CF [6]. Memory based CF, also known as neighborhood based CF relies on a simple intuition that an item might be interesting to an active user if the item is appreciated by a set of similar users or if the user has appreciated similar items in the past [2,4,12]. On the other hand, model based CF learns the features or patterns from rating information using machine learning algorithms [3]. Once the model is trained, it is used to

© Springer Nature Switzerland AG 2019
B. Deka et al. (Eds.): PReMI 2019, LNCS 11942, pp. 285–293, 2019.
https://doi.org/10.1007/978-3-030-34872-4_32

predict the ratings of unrated items. Examples of model based collaborative filtering are matrix factorization [7,10], tendency based [3], *etc.*

All the aforementioned algorithms focus on improving the accuracy of recommender system. However, accuracy oriented approaches have two major limitations. These approaches recommend items that are very similar to the items consumed or rated by the user in the past, leading to lack of diversity in recommendations [1,8]. Further, these approaches are biased towards recommending popular items due to which non popular items (referred as long tail items) are ignored leading to loss in business [5]. Overcoming the above mentioned shortcomings result in an improvement in overall user experience and increase in business profits.

Over the last decade, there are a few approaches proposed in the literature that focus on improving diversity and long tail item recommendation. Adomavicius and Kwon proposed a ranking-based approach that improves diversity based on the statistics of recommended items such as reverse predicted rating, item popularity ranking, item absolute likeability, item relative likeability, *etc.* [1]. There are other variants of ranking-based approach proposed in the literature such as clustering-based approach (KRCF) [11] and graph-based reranking technique [9]. Clustering-based approach known as Knowledge Reuse Framework in Collaborative Filtering (KRCF) clusters predicted items and pick items from each cluster exploiting inter cluster dissimilarity [11]. Özge and Tevfik proposed two methods to improve diversity. First method is graph-based reranking technique which is applied after predicting the ratings of unrated items and the second method incorporates a diversity factor while training the model using matrix factorization approach [9]. Pareto-efficient multi objective ranking technique is proposed to maximize accuracy and diversity in order to recommend accurate and unpopular items [13]. Valcarce et al. proposed an approach referred as item-based relevance modelling (IRM) to improve long tail item recommendations [14]. IRM utilizes a probabilistic approach to build the relevance model on long tail items.

Though the above mentioned approaches improve diversity, these approaches do not integrate different aspects of diversity such as popularity of item, finding items dissimilar to users' past ratings *etc.* We address this issue by integrating these aspects and categorized them into different item classes in order to reap the maximum benefits. We propose a Hybrid Reranking framework in Collaborative Filtering (HyReCF) that utilizes various statistics of the recommended items to improve diversity and long tail item recommendations. Our contributions are summarized as follow.

- The predicted ratings obtained from collaborative filtering and the item classes are fused to generate $topN$ recommendations for each user.
- We identify personalized dissimilar items to improve diversity.
- The proposed technique, HyReCF is extensively tested on two datasets (MovieLens and Netflix) and exhaustive study shows that HyReCF outperforms the state-of-the-art in terms of diversity and long tail recommendations.

The rest of the paper is organized as follows. Section 2 discusses proposed methodology in detail. Experimental results and analysis are provided in Sect. 3. Finally, we conclude our work in Sect. 4.

2   HyReCF: Proposed Hybrid Reranking Framework in Collaborative Filtering

In this section, we explain the proposed approach termed as Hybrid Reranking Framework in Collaborative Filtering (HyReCF) in detail. The proposed approach is carried out in two phases. In phase I (referred as prediction phase), a list of predictions (I_u) is generated for each user applying a traditional collaborative filtering technique. Main idea of this approach is to deploy this framework with the existing collaborative recommender system to offer more diverse and long tail items to the users. Therefore, we can apply one of the popular traditional collaborative filtering approaches such as user-based CF, item-based CF, matrix factorization, *etc.* in the first phase to generate prediction list I_u for a user.

In phase II (referred as reranking phase), significant factors such as popularity of item, finding items dissimilar to users' past ratings and good predicted items need to be introspected to attain diverse and long tail item recommendations. The predicted items are categorized into different classes to achieve maximum diversity as well as to improve long tail item recommendations. Several aspects such as relevance, popularity, and similarity of the items are considered while defining the classes. For each user, the prediction items list (I_u) obtained from phase I is categorized into three classes: (1) Unpopular Items, (2) Personalized Dissimilar Items, (3) Good Predicted Items.

Unpopular Items: In order to improve long tail item recommendations, the prediction list I_u is ranked in the increasing order of their popularity. The popularity of an item is defined as number of users who rated that item. This approach ensures that less popular items or long tail items are prominent in the user recommendations.

Personalized Dissimilar Items: A predicted item is said to be personalized dissimilar if it is different from all the items rated by the user in the past. Let $I_R = \{i_1, i_2, \ldots, i_r\}$ be the set of items rated by user u in the past. For an unrated item i and each item j in I_R, the dissimilarity can be computed using Eq. 1.

$$dis(i, j) = 1 - sim(i, j) \tag{1}$$

where $sim(i, j)$ is similarity between item i and item j. Similarity measures such as Adjusted Cosine, Cosine, Pearson Correlation, *etc.* can be used to compute $sim(i, j)$. We compute aggregate dissimilarity between the item i and rated set I_R. Any one of the popular aggregate functions such as average, maximum or minimum can be used to compute aggregate dissimilarity. Once the aggregate dissimilarities of all the unrated items are computed, these items are ranked in the descending order of aggregate dissimilarities to obtain personalized dissimilarity items class.

Good Predicted Items: A predicted item is considered as a good predicted item, if its predicted rating is greater than or equal to a certain threshold value (Th).

$$I_u^{Rel} = \{i \in I | \forall r_{u,i}^* \geq Th\} \qquad (2)$$

where $r_{u,i}^*$ is the predicted rating of a user u for an item i and I is the set of items in the system. This class ensures accuracy in the recommendation list.

Most of the recommender systems provide $topN$ items, where $topN$ represents the number of items recommended to a user. HyReCF also recommends $topN$ items to each user. For each user, $\alpha \times topN/3$ items are selected from each of the above mentioned classes where α is an accuracy factor and holds an integer value. As the value of α increases accuracy also increases. Finally, standard ranking approach is applied on total of $\alpha \times topN$ items to generate a recommendation list I_u^{Rec} of topN items. This approach reranks the items in ascending order of their predicted rating values and picks the $topN$ items. The overview of the proposed approach is shown in Fig. 1.

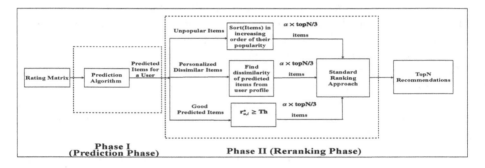

Fig. 1. Hybrid Reranking Framework in Collaborative Filtering (HyReCF).

3   Experimental Evaluation

The details of experiments and result analysis are described in this section. In phase I of HyReCF, for predicting users' ratings of unrated items, we use matrix factorization (MF) [7] and item based collaborative filtering techniques [12]. To evaluate performance of proposed HyReCF framework, we implemented the traditional matrix factorization approach [7], item based collaborative filtering [12], ranking-based approach [1] and clustering-based approach (KRCF)[11]. All variants of ranking based technique and KRCF are executed. Out of which Item Absolute Likeability provides the best results [1]. So, results are reported with "Item Absolute Likeability" ranking approach. We used Ward's Minimum distance clustering approach in our experiments as it provides best results. Throughout all the experiments $topN$ recommendations for all algorithms is considered, where the value of N is set to 8.

3.1 Dataset Description and Experimental Settings

All the implemented approaches are tested with two real world movie rating datasets namely, MovieLens and Netflix. Subsets of these datasets are created to ensure that each user has rated at least 20 items. Sparsity level of a dataset, denoted as K is the percentage of all missing ratings in a dataset. The statistics of working dataset are summarized in Table 1. In order to find long tail items, the item popularity graphs are plotted for MovieLens and Netflix datasets, as shown in Fig. 2(a) and (b), respectively. The items are sorted in descending order of their popularity. It can be noted that the items in the tail section of the plots received fewer ratings from the users, therefore considered as long tail items.

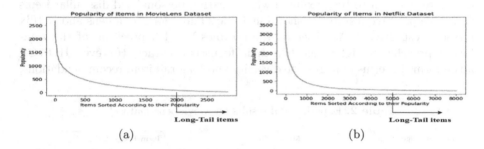

(a)   (b)

Fig. 2. Item popularity graph for (a) MovieLens Dataset (b) Netflix Dataset.

Table 1. Dataset description

S. no	Dataset	# Users ($|U|$)	# Head Items	# Tail Items	# Ratings (R)	K)
1	MovieLens	5,278	2,000	836	972,165	93.51
2	Netflix	8,139	5,000	3,086	2,005,997	96.96

3.2 Evaluation Metrics

The following evaluation metrics are used to measure the performance of the proposed methodology.

Precision-in-TopN: It is defined as the ratio of recommended items that are actually relevant to all the recommended items. Let $Rec(u)$ be the set of *topN* recommended items for user u, recommendation accuracy of a recommender system can be expressed as *precision-in-TopN* $= \sum_{u \in U} |Rec(u) \cap Rel(u)| / \sum_{u \in U} |Rec(u)|$, where $Rel(u)$ is the set of relevant items (ratings $\geq Th$) of user u in the test set.

Aggregate Diversity (AD): It is defined as number of distinct items that occur in the *topN* list of all the users [1]. The AD is computed as $|\cup_{u \in U} Rec(u)|$.

This metric gives an aggregate diversity achieved in the recommendations. However, it does not provide enough information regarding how many long tail items are recommended. To overcome this issue, we propose a new metric Long Tail denoted as LT.

Long Tail (LT): It is defined as number of times distinct long tail items occur in the *topN* list of all the users.

3.3   Experimental Results and Comparison

In order to compare improvement in diversity we keep precision of the proposed approach close to the standard approach. To achieve this we set $\alpha = 3$ and aggregate function to be maximum while creating personalized dissimilar items class. Comparative results are shown in Table 2 and 3 for MovieLens and Netflix datasets respectively. As observed from Tables 2 and 3 precision of the standard approaches is highest as their main focus is accuracy. However, HyReCF outperforms in terms of aggregate diversity and long tail item recommendations.

Table 2. Experimental results on MovieLens Dataset.

Prediction Algorithm →		Matrix Factorization			Item Based CF		
S. no	Recommendation Algorithm ↓	Precision (TopN)	AD	LT	Precision (TopN)	AD	LT
1	Standard Approach	**0.8326**	1,967	287	**0.8325**	1,660	202
2	KRCF Approach	0.7959	2,076	344	0.8290	1,698	246
3	Ranking Approach	0.8250	2,054	331	0.8304	1,708	222
4	HyReCF Approach	0.8225	**2,194**	**453**	0.8206	**1,958**	**418**

From Table 2, it can be noted that for prediction algorithm as matrix factorization approach, with 1% loss in precision, HyReCF approach achieves 227 items gain in aggregate diversity and 166 more long tail items are recommended with respect to traditional matrix factorization approach. In comparison to Ranking approach, with a little loss of 0.25% in precision, gain of 140 items in aggregate diversity and 122 items in long tail are obtained. Likewise, in item based CF, HyReCF earns a significant gain of 298 items in aggregate diversity and 216 in long tail with a loss of 1.1% in precision compared with traditional item based CF approach. A profit of 250 items in aggregate diversity and 196 items in long tail is attained with a loss of 0.9% in precision in comparison to Ranking approach.

Comparative results on Netflix dataset are reported in Table 3. For prediction algorithm as matrix factorization approach, we can observe that with a 1% loss in precision we attain 723 items gain in AD and 435 items in LT recommendation, while compared with standard matrix factorization. However, when compared to ranking approach, we achieve gain of 205 items in AD and 232 items in long tail recommendation with a loss of 0.41%. Similar observations can be drawn for item based CF approach.

Table 3. Experimental results on Netflix Dataset.

Prediction Algorithm →		Matrix Factorization			Item Based CF		
S. no	Recommendation Algorithm ↓	Precision (TopN)	AD	LT	Precision (TopN)	AD	LT
1	Standard Approach	**0.8323**	4,061	569	0.8188	4,023	730
2	KRCF Approach	0.7887	4,203	510	0.8137	3,901	595
3	Ranking Approach	0.8247	4,579	772	**0.8256**	4,173	918
4	HyReCF Approach	0.8206	**4,784**	**1,004**	0.8006	**4,919**	**1,229**

IRM is an approach that provides long tail item recommendation [14]. In order to observe the effectiveness of long tail item recommendations, we compare HyReCF with IRM. Table 4 and 5 report the results in terms of precision loss and gain in long tail item recommendation for IRM and HyReCF approach with respect to standard matrix factorization approach.

Table 4. Experimental results on MovieLens Dataset.

Precision (Standard)	Precision Loss	IRM		HyReCF Approach	
		AD Gain	LT Gain	AD Gain	LT Gain
0.8326	0.001	−3	0	4	2
	0.010	−31	5	42	27
	0.025	−79	12	106	68
	0.050	−158	25	212	136
	0.100	−316	51	424	273
Standard	0.0000	AD = 1,967		LT = 287	

From Table 4, it can be noted that with 0.100 precision loss, IRM gains 51 items whereas 273 LT items gain is achieved in HyReCF which is a significant improvement. IRM performs adversely in terms of aggregate diversity. With the loss in precision there is a loss in diversity instead of having gain whereas HyReCF shows a significant improvement in diversity with a small drop in accuracy. Table 5 reports result on Netflix dataset. With a loss of 5%, IRM gains 387 long tail items and HyReCF gains 590 long tail items. It can be observed that HyReCF outperforms IRM in both diversity and long tail item recommendation.

The proposed approach, HyReCF is further analyzed on varying precision using accuracy factor α. Figure 3(a) and (b) depict the performance of HyReCF on MovieLens and Netflix datasets respectively. From Fig. 3(a), in matrix factorization approach when $\alpha = 1$, precision is 0.6324 and aggregate diversity is 2817 items. At $\alpha = 2$, precision is 0.7776 and aggregate diversity is 2601. At $\alpha = 3$, precision is 0.8225 and aggregate diversity is 2194. It can be observed from the Fig. 3(a) and (b), as precision decreases, we obtain significant improvement in diversity.

Table 5. Experimental results on Netflix Dataset.

Precision (Standard)	Precision Loss	IRM		HyReCF Approach	
		AD Gain	LT Gain	AD Gain	LT Gain
0.8323	0.001	−3	7	19	11
	0.010	−30	77	192	118
	0.025	−77	193	481	295
	0.050	−154	387	963	590
	0.100	−308	775	1,926	1,181
Standard	0.0000	AD=4,061		LT=569	

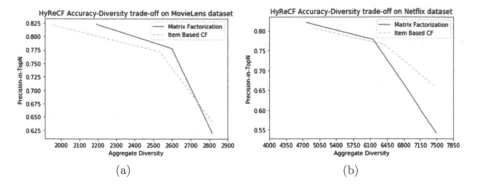

(a)   (b)

Fig. 3. Performance of HyReCF on (a) MovieLens Dataset (b) Netflix Dataset.

4 Conclusion

In this paper, we proposed Hybrid Reranking Framework in Collaborative Filtering (HyReCF), a technique to improve diversity and long tail item recommendations. Various statistics of the rating information and recommended items are integrated to improve diversity with a small drop in accuracy. The proposed approach provides a significant improvement in the diversity as compared to state-of-the-art. The long tail items are more prominent in the proposed framework than state-of-the-art which makes system interesting for the users as well as increases profits for the business organizations.

Acknowledgement. This work is partially funded by SERB, Department of Science and Technology, Govt. of India, Grant No. EMR/2017/004357, dated: 18th June 2018.

References

1. Adomavicius, G., Kwon, Y.: Improving aggregate recommendation diversity using ranking-based techniques. IEEE Trans. Knowl. Data Eng. **24**(5), 896–911 (2012)
2. Bobadilla, J., Ortega, F., Hernando, A., Gutiérrez, A.: Recommender systems survey. Knowl. Based Syst. **46**, 109–132 (2013). Kindly provide page range for the Ref. [3], if applicable.
3. Cacheda, F., Carneiro, V., Fernandez, D., Formoso, V.: Comparison of collaborative filtering algorithms: limitations of current techniques and proposals for scalable, high-performance recommender systems. ACM Trans. Web **5**(1) (2011). Article 2
4. Desrosiers, C., Karypis, G.: A comprehensive survey of neighborhood-based recommendation methods. In: Ricci, F., Rokach, L., Shapira, B., Kantor, P.B. (eds.) Recommender Systems Handbook, pp. 107–144. Springer, Boston, MA (2011). https://doi.org/10.1007/978-0-387-85820-3_4
5. Harald, S.: Item popularity and recommendation accuracy. In: Proceedings of the Fifth ACM Conference on Recommender Systems, pp. 125–132 (2011)
6. Koren, Y.: Factor in the neighbors: scalable and accurate collaborative filtering. ACM Trans. Knowl. Discov. Data **4**(1), 5–53 (2010)
7. Koren, Y., Bell, R., Volinsky, C.: Matrix factorization techniques for recommender systems. Computer **42**(8), 30–37 (2009)
8. Matevž, K., Tomaž, P.: Diversity in recommender systems-a survey. Knowl.-Based Syst. **123**, 154–162 (2017)
9. Özge, K.M., Tevfik, A.: Effective methods for increasing aggregate diversity in recommender systems. Knowl. Inf. Syst. **56**(2), 355–372 (2018)
10. Paterek, A.: Improving regularized singular value decomposition for collaborative filtering. In: Proceeding of KDD Cup Workshop at 13th ACM International Conference on Knowledge Discovery and Data Mining, pp. 39–42 (2007)
11. Pathak, A., Patra, B.K.: A knowledge reuse framework for improving novelty and diversity in recommendations. In: Proceedings of the Second ACM IKDD Conference on Data Sciences, pp. 11–19 (2015)
12. Sarwar, B., Karypis, G., Konstan, J., Riedl, J.: Item-based collaborative filtering recommendation algorithms. In: Proceedings of the 10th International Conference on World Wide Web, pp. 285–295 (2001)
13. Shanfeng, W., Maoguo, G., Haoliang, L., Junwei, Y.: A knowledge reuse framework for improving novelty and diversity in recommendations. In: Proceedings of the Second ACM IKDD Conference on Data Sciences, vol. 104, pp. 145–155 (2016)
14. Valcarce, D., Parapar, J., Barreiro, A.: Item-based relevance modelling of recommendations for getting rid of long tail products. Knowl.-Based Syst. **103**, 41–55 (2016)

Prioritized Named Entity Driven LDA for Document Clustering

Durgesh Kumar[✉] and Sanasam Ranbir Singh

Department of Computer Science and Engineering,
Indian Institute of Technology Guwahati, Guwahati, India
durgeshit@gmail.com, ranbir@iitg.ac.in

Abstract. Topic modeling methods like LSI, pLSI, and LDA have been widely studied in text mining domain for various applications like document representation, document clustering/classification, information retrieval, etc. However, such unsupervised methods are effective over corpus with well separable topics. In real-world applications, topics might be of highly overlapping in nature. For example, a news corpus of different terror attacks has highly overlapping keywords across reporting of different terror events. In this paper, we propose a variant of LDA, named as *Prioritized Named Entity driven LDA* (PNE-LDA), which can address the issue of overlapping topics by prioritizing named entities related to the topics. From various experimental setups, it is observed that the proposed method outperforms its counterparts in entity driven overlapping topics.

Keywords: Topic modeling · LDA · Entity-driven topics · PNE-LDA

1 Introduction

Topic modeling has been widely applied in various domains such as text mining [1,2], information retrieval [21] and image processing [5,12] etc. Latent Dirichlet Allocation (LDA) is one of the most studied topic modeling methods. This is a generative model based on the extension of probabilistic Latent Semantic Indexing (pLSI) [4]. LDA has been widely investigated in modeling text corpora with exchangeable discrete data (i.e. bag-of-words), where text documents are modeled as the mixture over topics drawn from a Dirichlet distribution, and topics are modeled as a multinomial distribution of words drawn from a finite set.

With the increase in the availability of text resources of various characteristics (temporal, tagged, multi-class, etc.) and different applications, researchers have proposed various LDA variants [14,16,18,19] in the literature. Traditional LDA does not perform effectively if text corpus has skewed topic distributions or highly overlapping keywords between topics. Further, many of the real world events are entity driven. For example, news reporting of different terror attacks share many important keywords such as blast, bomb, IED, terror, attack, etc.,

B. Deka et al. (Eds.): PReMI 2019, LNCS 11942, pp. 294–301, 2019.
https://doi.org/10.1007/978-3-030-34872-4_33

but differentiated by named entities like person name, locations, organizations, etc. Over such text collection, an unsupervised generative approach like LDA fails to separate documents related to different events effectively because of high overlapping keywords. In this paper, we address the above problem by assuming to have prior knowledge about the underlying topics and its representative named entities. With the increase in the availability of real-world datasets such as natural disaster, terror attack, political discussion, such assumption related to entity driven events becomes relevant in today's scenario. We refer to such representative named entities as *prioritized entities*. In the literature, SeededLDA [8] also considers similar assumptions, however for generalized keywords. It samples each word of a document as seed word or non-seed word for each topic and accordingly modifies the document-topic and topic word distributions. Whereas, in the case of *Prioritized Named Entity driven LDA* (PNE-LDA), we assume the information of each word to be treated as prioritized or non-prioritized as an external input and estimate the topic-word distribution.

From experimental observations over three different datasets of different nature (i.e. Bomb Blast, Reuters-21578, and 20-Newsgroup), it is evident that the proposed PNE-LDA outperforms its LDA counterparts for entity driven topics. The remainder of this paper is organized as follows. We discuss relevant related works in Sect. 2 which is followed by the proposed solution in Sect. 3. Section 4 describes the characteristics of datasets and insight into the results. Finally, we conclude our work with future work in Sect. 5.

2 Literature Review

Topic modeling has been used in past decades to mine the important hidden information in the large text corpus [1,2,21]. The application of text modeling varies from clustering the text corpus, mining the scientific articles, to the detection of the event on Twitter and news publications. LDA is one of the prominent topic modeling techniques that had been suitably modified to suit the need. Rosen et al. [16] proposed Author-Topic Model, a variant of LDA to mine the authors' research interest from the research paper collections by modeling the content of research paper. McCallum et al. [14] extend Author-Topic model to Author-Recipient-Topic model for mining topic, interaction relationship between sender and receiver, and people roles from Enron and academic email. Another variants of LDA Topics-over-Time LDA (*TOT-LDA*) [19], Dynamic LDA [3], and Temporal LDA [20] incorporate temporal factors to find granular topics and evolution of topics over time from the text documents like scientific publication and Twitter.

To include the supervised information into LDA, Labeled LDA (*L-LDA*) [15] establishes one to one relationship between LDA topics and user defined class labels. Thereafter several variants of LDA have been proposed to include the supervised information; for e.g, *sLDA* [13], *DiscLDA* [11] and *MedLDA* [23] for classification of documents with single label; while *DP-MRM* [10], and *Dep-LDA* [17] and Boost Multi class L-LDA [9] for classification of documents with

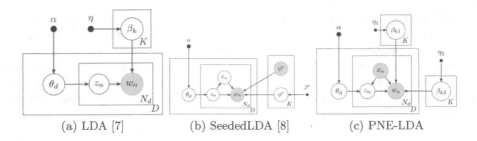

(a) LDA [7]   (b) SeededLDA [8]   (c) PNE-LDA

Fig. 1. Plate diagram of LDA, SeededLDA and the proposed PNE-LDA

multiple labels. Source-LDA [22] has been proposed to include external information from wikipedia to guide the topic word formation. Seed-LDA [8] is one of the variants of LDA which incorporates the user's understanding of the corpus and bias the topic formation process with the help of representative word of each topics which is more useful for the various extrinsic tasks such as document classification.

3 Methodology

In this section, we discuss the proposed prioritized named entity driven LDA *(PNE-LDA)*. LDA is a generic probabilistic graphical model capturing the latent semantic topic distribution across large document collections. PNE-LDA extends LDA for clustering the text corpus with the help of prior information in terms of prioritized named entities. By giving prior information about the cluster in terms of prioritized named entities, we can improve the quality of cluster, especially in case of highly overlapping clusters.

3.1 Prioritized Named Entity Driven LDA (PNE-LDA)

Despite having high overlapping of word distribution among clusters, every cluster has distinguishable named entities like persons, locations, and organizations, which we call as *prioritized named entity*. As the occurrence of such prioritized named entity is rare compared with general words, it is required to project these words separately onto topics from the general words.

Plate diagram of the LDA, SeededLDA and our proposed PNE-LDA, is shown in Fig. 1 for comparison. In PNE-LDA, we consider one observed random variable x_{nd} similar to SeededLDA, acting as a switch, which denotes about a word being a prioritized word or general word. The switch variable x_{nd} is observed in contrast to SeededLDA, where it is sampled for each topic based on beta distribution. In PNE-LDA, we use different Dirichlet distributions for prioritized and non-prioritized words to generate topic-word distributions as shown in Fig. 1(c). A list of prioritized named entity is provided to the model apriori to the start of PNE-LDA. The generative algorithm of PNE-LDA is given in Algorithm 1.

Algorithm 1. The generative algorithm of PNE-LDA. PNEWD represents Prioritized Named Entity Words Distribution and GWD represents General Words distribution

---

// Generating distribution of word in Topics
for *each topic k in [1,K] :* **do**
  | *Generate Topic-PNEWD:* β_{1k} using $\text{Dir}(\eta_1)$
  | *Generate Topic-GWD:* β_{2k} using $\text{Dir}(\eta_2)$
end
for *each document d in Corpus:* **do**
  | *// Document generation*
  | Sample a topic distribution θ_d using $Dir(\alpha)$
  | **for** *each of word w in document d :* **do**
  | | Sample a topic Z_{nd} using Multinomial(θ_d)
  | | **if** x_{nd} *equal to 0* **then**
  | | | *// Sample from Topic-PNEWD*
  | | | $w_{nd} \sim$ Multinomial(β_{1zdn})
  | | **end**
  | | **else if** x_{nd} *equal to 1* **then**
  | | | *// Sample from Topic-GWD*
  | | | $w_{nd} \sim$ Multinomial(β_{2zdn})
  | | **end**
  | **end**
end

We first sample prioritized named entity topic distributions (β_1) and general topic word distribution (β_2) using uniform Dirichlet parameter η_1 and η_2 respectively. For each document in corpus, we first sample the document topic distribution θ_d, using Dirichlet prior α. Further, for each word, we sample a word from either topic-prioritized named entity distribution or topic-general word distributions based on the values of x_{nd}. If x_{nd} is 0, we sample from topic-prioritized named entity distribution; else we sample from the topic-general word distribution.

PNE-LDA focuses on the study of emphasizing more weights on prioritized keywords. Varying the value of Dirichlet parameters i.e., prioritized named entity Dirichlet parameter (η_1) and general word Dirichlet parameter (η_2), we can put different weights on a prioritized named entity and general word respectively. Like SeededLDA, PNE-LDA divides topic-word relationship in two parts: (i) topic-prioritized named entity word relationship (η_1) and (ii) topic-general word relationship (η_2) respectively.

The conditional probability of assigning a topic j to a word w_{nd} of document d by PNE-LDA can be written as:

$$P(z_{nd} = j | z_{\neg nd}, w_{\neg nd}) \propto \begin{cases} (\alpha + n_{\neg nd,j}^{w_d}) \dfrac{\eta_1 + n_{\neg nd,j}^{(w_{nd})}}{V_1.\eta_1 + n_{\neg nd,j}^{(.)}}, & \text{if } x_{nd} = 0 \\[3ex] (\alpha + n_{\neg nd,j}^{w_d}) \dfrac{\eta_2 + n_{\neg nd,j}^{(w_{nd})}}{V_2.\eta_2 + n_{\neg nd,j}^{(.)}}, & \text{Otherwise} \end{cases} \qquad (1)$$

where $x_{nd} = 0$ indicates prioritized named entity and $x_{nd} = 1$ indicates general words. The various symbols used in Eq. 1 are as follows: (i) w_{nd} represents word at n^{th} index of document d (ii) z_{nd} represents topic of the word at n^{th} index of document d (iii) $z_{\neg nd}$ represents all topics-word assignment except the current word topic assignment (iv) $w_{\neg nd}$ represents all words in the vocabulary except the current word (v) $n_{\neg nd,j}^{w_d}$ represents number of words of current document assigned to current topic j except the current word w_{nd} (vi) $n_{\neg nd,j}^{(w_{nd})}$ represents number of words assigned to current topic j and similar to current word, except current word w_{nd} (vii) $n_{\neg nd,j}^{(.)} = \sum_{\forall w_{nd} \in V} n_{\neg nd,j}^{(w_{nd})}$ represents number of words assigned to current topic j except current word w_{nd}. (viii) $V1$ and $V2$ represent the number of vocabulary of prioritized named entity and non-prioritized named entity.

In Eq. 1, left hand side of equation $p(z_{nd} = j)$ resembles the probability of getting a topic j for word at n^{th} index of document d. The first term of right-hand side equation resembles the probability of choosing a topic j from a multinomial distribution of topics in d_{th} document. The second term of right-hand side of the equation refers to choosing a word w_{nd} from the topic j. If the word is prioritized word, we sample it from prioritized word-topic distribution parameterized by η_1; otherwise, we sample it from general word-topic distribution parameterized by η_2.

3.2   Prioritizing Named Entities

Ideally named entities representing the target topics can be given as external inputs. For a random dataset, identifying such entities is an expensive operation. In this study, we consider all the entities present in the document. For identifying the named entities, we have used Standford NER [6]. For the bomb blast dataset, since there is a need to identify Indian named entities, we have used and adapted Stanford NER, which is trained to recognize India named entities. Further, these named entities can be assigned different priority. However, for simplicity, this study assigns equal priority to all the named entities present in the documents. Once we identify representative named entities, the proposed PNE-LDA considers these entities as prioritized keywords and rest as the non-prioritized keywords.

Table 1. Characteristics of the experimental datasets. NE represent named entity

Dataset	#Doc	#class	Avg. #NE per doc	Avg. #doc per class
Bomb blast	855	53	225	16
Reuters-21578-R	5485	8	70	686
20-Newsgroups	11293	20	152	565

4   Experimental Setups

Datasets: We have experimented with three datasets namely Bomb Blast, Reuters-21578, and 20-Newsgroups respectively as described in Table 1. The Bomb Blast dataset is our locally collected and processed dataset, which consist of 855 news articles reporting 53 different bomb blast events occurred in different parts of India. The Reuters-21578 dataset consists of 5,485 documents spanning across 8 clusters and 20-Newsgroups dataset consists of 11,293 documents spanning across 20 clusters. Among the three datasets, Bomb Blast has most occurrences of named entities while Reuters dataset has least occurrences of named entities in the documents. Bomb blast dataset mostly has the person, location and organization name whereas 20-Newsgroups dataset has only person and organization names. Reuters-21578 dataset is mostly related to business articles and hence has a limited number of named entities.

Experimental Setup: We have experimented with different Document-Topic Dirichlet parameter α and set it as 0.1 for all LDA, SeededLDA and PNE-LDA as $\alpha = 0.1$ provides best result for all three datasets. For LDA, we have Topic-Word Dirichlet parameter η to 0.2 for Bomb blast Dataset, η to 0.3 for Reuters Dataset and η to 0.2 for 20-Newsgroup. For PNE-LDA and Seeded-LDA, to assign higher weights to named entities, we have assigned the weights of η_1 less than η_2 for all three datasets. For SeededLDA we have used the same value of η_1 and η_2 as PNE-LDA corresponding to three datasets. For Bomb blast dataset, we have set the topic-prioritized named entity Dirichlet parameter η_1 as 0.1 and topic-general word Dirichlet parameter η_2 as 0.2. For Reuters dataset, we have set η_1 as 0.2 and η_2 as 0.3 respectively. For 20-Newsgroup, we have set η_1 as 0.1 and η_2 as 0.2 respectively. Although we have experimented with other combination of η_1 (0.1, 0.2, 0.3, 0.4) and η_2 (0.1, 0.2, 0.3, 0.4), but we are reporting parameter with the best performance with respect to both F-measure and RandIndex. We have run 120 iterations of Gibbs sampling for all the LDA and PNE-LDA experiments.

In order to investigate the performance of different methods, we have used the document clustering task to measure the quality of topics given by LDA, SeededLDA and PNE-LDA. For all the experimental setups, we consider the number of topics as the number of clusters present in the respective datasets as shown in Table 1. LDA, SeededLDA and PNE-LDA returns the document-topic proportions. Each document is assigned to the cluster defined by the topic with maximum proportions. Once each document is assigned with a cluster id (which

Table 2. Topic evaluation of LDA, PNE-LDA and SeededLDA over Bomb Blast, Reuters-21578 and 20-Newsgroups datasets.

Model	Bomb blast		Reuters-21578		20-Newsgroups	
	F-measure	RandIndex	F-measure	RandIndex	F-measure	RandIndex
LDA	0.0802	0.7987	0.6052	0.7800	0.1382	0.6056
SeededLDA	0.0868	0.8319	**0.6234**	**0.7855**	0.3356	0.9126
PNE-LDA	**0.0887**	**0.8672**	0.5353	0.7603	**0.5003**	**0.9415**

is topic id with maximum proportions), we evaluate clustering performance using F-measure and RandIndex.

Observations: Table 2 represents comparative study of topic quality given given by three models LDA, PNE-LDA, and SeededLDA. As shown in the table, the proposed PNE-LDA outperforms both LDA and seeded LDA for both Bomb Blast and 20-Newsgroups. As mentioned above, topics in these two datasets are named entity driven. Whereas, in less named entity driven dataset (Reuter-21578), PNE-LDA under-performs both LDA and SeededLDA. It is evident from these observation that the proposed PNE-LDA is more effective to determine real world events which can be represented by named entities defining the topics.

5   Conclusion and Future Work

In this paper, we propose a novel entity driven Topic modeling based model namely PNE-LDA, for clustering documents with high overlapping word and named entities driven clusters. From the various experimental observation over three datasets namely Bomb Blast, Reuters-21578 and 20-Newsgroups, it is observed that the proposed methods outperforms its state of the art counterparts namely LDA and SeededLDA over dataset with high occurrences of named entity. Thus the proposed method is more suitable for detecting real world events like bomb blasts.

References

1. Aggarwal, C.C., Wang, H.: Text mining in social networks. In: Aggarwal, C. (ed.) Social Network Data Analytics, pp. 353–378. Springer, Boston (2011). https://doi.org/10.1007/978-1-4419-8462-3_13
2. AlSumait, L., Barbará, D., Domeniconi, C.: On-line LDA: adaptive topic models for mining text streams with applications to topic detection and tracking. In: Eighth IEEE ICDM 2008, pp. 3–12. IEEE (2008)
3. Blei, D.M., Lafferty, J.D.: Dynamic topic models. In: Proceedings of the 23rd ICML, pp. 113–120. ACM (2006)
4. Blei, D.M., Ng, A.Y., Jordan, M.I.: Latent Dirichlet allocation. J. Mach. Learn. Res. **3**(Jan), 993–1022 (2003)
5. Chong, W., Blei, D., Li, F.F.: Simultaneous image classification and annotation. In: 2009 IEEE Conference on CVPR, pp. 1903–1910. IEEE (2009)

6. Finkel, J.R., Grenager, T., Manning, C.: Incorporating non-local information into information extraction systems by Gibbs sampling. In: Proceedings of the 43rd Annual Meeting on ACL, pp. 363–370. ACL (2005)

7. Griffiths, T.L., Steyvers, M.: Finding scientific topics. Proc. Natl. Acad. Sci. **101**(suppl 1), 5228–5235 (2004)

8. Jagarlamudi, J., Daumé III, H., Udupa, R.: Incorporating lexical priors into topic models. In: Proceedings of the 13th Conference of the European Chapter of the ACL, pp. 204–213. ACL (2012)

9. Jankowski, M.: Boost multi-class sLDA model for text classification. In: Rutkowski, L., Scherer, R., Korytkowski, M., Pedrycz, W., Tadeusiewicz, R., Zurada, J.M. (eds.) ICAISC 2018. LNCS (LNAI), vol. 10841, pp. 633–644. Springer, Cham (2018). https://doi.org/10.1007/978-3-319-91253-0_59

10. Kim, D., Kim, S., Oh, A.: Dirichlet process with mixed random measures: a non-parametric topic model for labeled data. arXiv preprint arXiv:1206.4658 (2012)

11. Lacoste-Julien, S., Sha, F., Jordan, M.I.: DiscLDA: discriminative learning for dimensionality reduction and classification. In: Advances in NIPS, pp. 897–904 (2009)

12. Lienou, M., Maitre, H., Datcu, M.: Semantic annotation of satellite images using latent Dirichlet allocation. IEEE GRSL **7**(1), 28–32 (2010)

13. Mcauliffe, J.D., Blei, D.M.: Supervised topic models. In: Advances in NIPS, pp. 121–128 (2008)

14. McCallum, A., Wang, X., Corrada-Emmanuel, A.: Topic and role discovery in social networks with experiments on enron and academic email. J. Artif. Intell. Res. **30**, 249–272 (2007)

15. Ramage, D., Hall, D., Nallapati, R., Manning, C.D.: Labeled LDA: a supervised topic model for credit attribution in multi-labeled corpora. In: Proceedings of the 2009 Conference on EMNLP, vol. 1, pp. 248–256. ACL (2009)

16. Rosen-Zvi, M., Griffiths, T., Steyvers, M., Smyth, P.: The author-topic model for authors and documents. In: Proceedings of the 20th Conference on Uncertainty in Artificial Intelligence, pp. 487–494. AUAI Press (2004)

17. Rubin, T.N., Chambers, A., Smyth, P., Steyvers, M.: Statistical topic models for multi-label document classification. Mach. Learn. **88**(1–2), 157–208 (2012)

18. Tu, Y., Johri, N., Roth, D., Hockenmaier, J.: Citation author topic model in expert search. In: Proceedings of the 23rd International Conference on Computational Linguistics: Posters, pp. 1265–1273. ACL (2010)

19. Wang, X., McCallum, A.: Topics over time: a non-Markov continuous-time model of topical trends. In: Proceedings of the 12th ACM SIGKDD, pp. 424–433. ACM (2006)

20. Wang, Y., Agichtein, E., Benzi, M.: TM-LDA: efficient online modeling of latent topic transitions in social media. In: Proceedings of the 18th ACM SIGKDD, pp. 123–131. ACM (2012)

21. Wei, X., Croft, W.B.: LDA-based document models for ad-hoc retrieval. In: ACM SIGIR. ACM (2006)

22. Wood, J., Tan, P., Wang, W., Arnold, C.: Source-LDA: enhancing probabilistic topic models using prior knowledge sources. In: 2017 IEEE 33rd ICDE, pp. 411–422. IEEE (2017)

23. Zhu, J., Ahmed, A., Xing, E.P.: MedLDA: maximum margin supervised topic models. J. Mach. Learn. Res. **13**(Aug), 2237–2278 (2012)

Context-Aware Reasoning Framework for Multi-user Recommendations in Smart Home

Nisha Pahal[1(✉)], Parul Jain[1], Ruchika Saxena[2], Abhinesh Srivastava[2], Santanu Chaudhury[1], and Brejesh Lall[1]

[1] Indian Institute of Technology, New Delhi, India
nisha23june@gmail.com, paruljain@gmail.com, schaudhury@gmail.com, brejesh.lall@gmail.com
[2] Samsung Research Institute, Delhi, India
{r.saxena,s.abhinesh}@samsung.com

Abstract. This paper introduces a context-aware reasoning framework that adapts to the needs and preferences of inhabitants continuously to provide contextually relevant recommendations to the group of users in a smart home environment. User's activity and mobility plays a crucial role in defining various contexts in and around the home. The observation data acquired from disparate sensors, called user's context, is interpreted semantically to implicitly disambiguate the users that are being recommended to. The recommendations are provided based on the relationship that exist among multiple users and the decision is made as per the preference or priority. The proposed approach makes extensive use of multimedia ontology in the life cycle of situation recognition to explicitly model and represent user's context in smart home. Further, dynamic reasoning is exploited to facilitate context-aware situation tracking and intelligently recommending appropriate actions which suit the situation. We illustrate use of the proposed framework for Smart Home use-case.

Keywords: Context-aware · Multimedia ontology · Dynamic Bayesian Network (DBN) · Internet of Things (IoT) · Recommendations

1 Introduction

In today's world, almost every thing is on the verge of automation whether it is home, office, industry, health care, city or shopping to enable smart applications. To support such an environment, IoT allows for context to be available through devices and sensors which are ubiquitously present. The context-aware applications aim to provide personalized and customized recommendations by utilizing a variety of contexts obtained from these devices. The real-time data acquired by such devices is heterogeneous and need to be exploited using semantic web technologies. Ontology in IoT requires real-time modelling of concepts along with media features to minimize the semantic gap that exists between

© Springer Nature Switzerland AG 2019
B. Deka et al. (Eds.): PReMI 2019, LNCS 11942, pp. 302–310, 2019.
https://doi.org/10.1007/978-3-030-34872-4_34

sensor observation and the perceptual model. The shared devices or applications poses the challenge to recommendation systems as the suggestions suitable to one user may actually be more appropriate for another user. Thus, to design a generic context-aware service framework for multi-user recommendation the objective of this paper is three-fold:

- To acquire user context from disparate sensors and external inputs in IoT.
- To recognize the semantics of the acquired context with the help of ontological knowledge.
- To apply dynamic reasoning for situation tracking and to intelligently recommend actions which suit the situation.

The present work offers several significant contributions: First, we propose a context-aware Multi-user recommendation system for a smart home environment. It is designed to react to continuous changes in the environment; adapt the users behavior according to needs and situation; infer the group/ relationship dynamically in order to provide contextual recommendations. Second, the approach makes use of domain knowledge represented in the form of multimedia ontology encoded in MOWL. It explicitly models and represents the users context in smart home and exploits DBN reasoning [3] to track the situation and infer the relationship that exists between multiple users present at one location. Our approach makes use of DBN to predict the current situation based on previously inferred situation and current context data to make contextual recommendations to multiple users. Third, based on the tracked situation and inferred relationship/group, the relevant recommendations are provided. While making recommendations it is observed if there exists a priority among the identified user relationship if yes, then the recommendation is according to the highest priority. If there is no priority among the users present in a group then the recommendation involves preferences that are of interest to all the users in identified group. For instance, in a smart-tv situation if person A is Listening-toMusic in livingRoom and if a new user *say* person B joins then whether the new user will be a part of group or not is decided by considering the previous situation. If the last situation says he was in kitchen and has an intention to go back as the device which he has been working is turned ON. Then, person B will not be included in the current context and will not form the group and no new recommendations are provided. Consider another case, where it has been detected by sensors that the device which the user has been working on previously in the kitchen is turned OFF. Now, from the previous situation it has been identified that person B does not have any intention to go back. In such a scenario, person B will be considered as part of group and the new relationship is inferred. The new user (say person B) is identified based on the face recognition module using face-example images and reasoning is specified for the identification of user relationship. Depending on the underlying relationship as defined in the user ontology contextual recommendations are provided.

A great deal of work has focused on describing context-aware applications. Wilson and Atkeson in [8] proposed a simultaneous tracking and activity recognition to provide automatic health monitoring. They had used a particle filter approach to recognize the mobility of occupants. Nguyen et al. in [5,6] proposed a Hierarchical Hidden Markov Model (HHMM)-Joint Probabilistic Data Association Filter (JPDAF) to recognize behaviors of multiple people. The authors in [7] had presented a survey on context awareness from an IoT perspective. They had analyzed the past research work and discussed their applicability towards IoT. Paola et al. in [2] had proposed a multi-sensor data fusion system that makes use of context information to recognize users activity in a smart home environment. The authors in [3] had presented a context-aware situation tracking framework using DBN for smart mirror use-case.

Different from the above techniques, the proposed framework offers context-aware multi-user recommendations in a smart home environment. It utilizes multimedia ontology that exploits DBN based reasoning to track situations dynamically and to make contextual recommendations to multiple users present at same location.

2   Overview

The domain knowledge corresponding to Smart Home is encoded in several multimedia ontology's encoded in MOWL [1]. These ontologies have inter-dependencies and can have relations between concepts belonging to same or different ontology's. The snippet ontology is depicted in Fig. 1. Various ontology's that are required to model the smart home environment are discussed below:

- Group Ontology: It encodes the users' personal information such as name, age, gender, etc. and social context such as likes, preferences and daily schedule. The information related to groups are encoded based on the relationship among family members and their acquaintances.
- Device Ontology: This encodes the information related to the sensor devices and actuators installed in a smart home.
- Event Ontology: This encodes the users general life situations and events. These situations are linked with location, time and involved user relationship.
- Action Ontology: This encodes the actions and dynamic recommendations that are to be made based on identified users relationship and the situation tracked. It is linked to group ontology, device ontology and event Ontology.

MOWL infers a situation by exploiting Observation Model (OM) which is a bayesian tree wherein the root node specifies the situation. The OM describes the present universe of discourse that matches the inferred situation. MOWL realizes DBN [4] to model a dynamic world with changing situations. It differs from the standard Bayesian network as it maintains a spatial-temporal relationship with each node and handles real time series data along with the past activities. It represents the temporal properties and calculates the belief level by assessing current observations in combination with previously inferred situation to determine the likelihood of behaviours that represent the present situation. DBN incorporates probability distribution function on the series of T observable variables

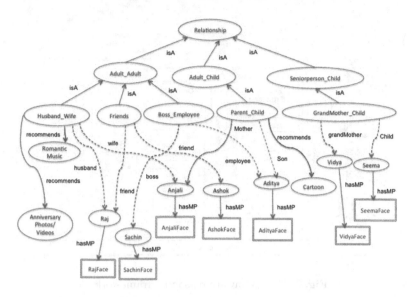

Fig. 1. Group ontology snippet

$V = v_0, v_1, \ldots, v_{T-1}$ and the series of T state variables $S = s_0, s_1, \ldots, s_{T-1}$. This is depicted below:

$$P(S \mid V) = \prod_{t=1}^{T-1} P(S_t \mid S_{t-1}) \prod_{t=0}^{T-1} P(V_t \mid S_t) P(S_0) \tag{1}$$

where, $P(S_t \mid S_{t-1})$ is the state transition probability distribution functions (PDFs), $P(V_t \mid S_t)$ is the observation PDFs and $P(S_0)$ is the initial state distribution.

3 Framework for Multi-user Recommendation in Smart Home Environment

The context-aware reasoning framework for multi-user recommendations is depicted in Fig. 2. It interprets the rich semantic data coming from the IoT and exploits multimedia ontology and dynamic reasoning capabilities to offer contextually relevant recommendations to multiple users in a smart home environment. It has Four Layers namely, the Physical Layer, Data Analysis and pre-processing Layer, Semantic Derivation and Reasoning Layer and an Application Layer.

– Physical Layer: Physical Layer, the first layer that senses the environment and the heterogeneous data generated is acquired and stored in the database through gateway/network. The acquired data is either textual or multimedia type. The general data about the weather, events or activities is gathered from the social media and open APIs.

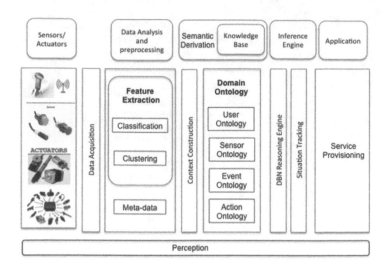

Fig. 2. Context-aware reasoning framework

- Data Analysis and Pre-processing Layer: The raw data obtained from the previous layer needs to be analyzed to fetch the important information describing the attributes present in them. For analysis, we have used various preprocessing techniques such as object detection and identification, classification, clustering, etc.
- Semantic Derivation and Reasoning Layer: In this layer, the semantics of the acquired context is recognized by means of ontology. The reasoning engine dynamically tracks the system as it moves from one situation to other in order to provide contextual recommendations.
- Application Layer: This layer deals with the domain-specific applications that seamlessly provide smart recommendations for enhancing the life style of inhabitants. In this paper, we have considered Smart Home as an application domain that interprets the rich semantic data coming from the IoT to provide contextually relevant recommendations to multiple users present at same location.

4   Multi-user Recommendations: A Context-Aware Service in Smart Home Environment

In a Smart Home context, if multiple users are accessing the system/device from different locations then the handling is done through threads where each user request is handled independently. If the multiple users are using the same device *say television* at the same time then the handling is done through group formation/ relationship identification. Group is identified based on the abstract information provided by the users like age, relationships with other users as specified in the group ontology. This is linked to device ontology, event ontology and action ontology and using the dynamic reasoning capabilities of MOWL smart recommendations are provided based on changing situations. Consider the following scenarios:

- Case 1: when a single user participates in the scenario then the recommendations are based on the tracked situation, user's preferences, likes and events in calender.
- Case 2: When more than one user participates in the scenario then:
  - if the multiple users are present at independent locations then the request from each user is handled separately by initiating a number of threads, one thread for each user. In such a case, recommendation's are made by considering user preferences for each independent user as specified in the knowledge base.
  - if multiple users are present at same location then they are grouped into conceptual classes based on the relationship that exists between them. In such a case the entire group is being recognized by the observed conceptual class and the same thread is initiated for the group. The actions are recommended by considering the preferences for each group as specified in the knowledge base.
    * If there is a priority among the users present in a group then the recommendation is according to the highest priority and in order. For instance, if the identified relationship is of *parent_child* then the resulting preferences list is overwritten by child's preferences in general.
    * If there is no priority among the users present in a group then the recommendation is made by considering the common user preferences specified in the ontology. For instance, if the identified relationship is *husband_wife* then the resulting preferences list is the intersection of their individual preferences say, listening to music.

To illustrate this support, let us consider the scenario of a smart home where the two users Raj and Anjali were tracked at same location. As depicted in Fig. 3 the users were identified through Face Recognition Module with probability $P(RajFace) = 0.68$ and $P(AnjaliFace) = 0.67$ and location sensor was also enabled for tracking the movement in home. Since the users were tracked at same location so the user ontology facilitates identifying the age category like *adult_adult, adult_child* or *adult_seniorperson* along with the relationship that exists between them like *husband_wife, parent_child, siblings, friends*, etc. Depending on these identifications they are grouped together and passed as a single concept to event ontology for situation tracking and making smart recommendations. For the given context, *adult_adult* has been recognized as a conceptual class/group and the following relationship probabilities were obtained, $P(adult\_adult) = 0.85$ and $P(husband\_wife) = 0.78$ with all relationships initialized to a prior probability of 0.4. Based on the identified relationship romantic songs were recommended. In other case, the users Raj and Anjali were identified at different locations. Since the users were present at diverse locations so no relationship was inferred and were passed to event ontology as different contexts.

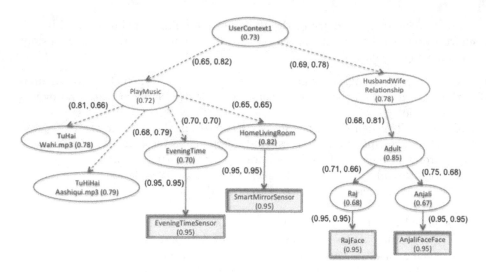

Fig. 3. Observation model: recommendations based on identified relationship *husband_wife*

The Observation Model (OM) for the same is depicted in Fig. 4. In this, the user Raj was tracked in livingroom and Anjali was tracked in kitchen. Based on this context the recommendations were provided.

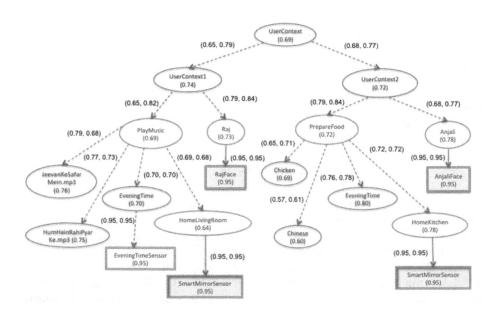

Fig. 4. Observation model: recommendations given when no group/relationship was identified

When a new user joins the existing group then it is checked whether the new user is going to stay at that location or leave in a short interval. If the users' intention is to leave the location within a short interval then the grouping is not done and no change in the context is suggested. The experiment was performed over the dataset obtained from MIT Home dataset available for tracking home situations and tested for accuracy of prediction. The resulted accuracy was measured to be 86.4% which marked an efficiency of our approach from earlier to be 72% with static approach.

5   Conclusion

This paper presents a novel context-aware recommendation framework that exploits the users context to provide smart suggestions to group of users in a smart home environment. As our domain is based on an IoT paradigm, the data obtained through various IoT devices acts as a contextual information for assessment of situation dynamically. The acquired context information is processed semantically to identify the relationship that exists among multiple users. It exploits multimedia ontology and associated DBN reasoning to intelligently recommend appropriate actions based on the recognized situation/group formed. The proposed approach facilitates resolving conflicts of multi-users sharing devices within a smart home environment.

References

1. Chaudhury, S., Mallik, A., Ghosh, H.: Multimedia Ontology: Representation and Applications. Chapman and Hall/CRC, Boca Raton (2015)
2. De Paola, A., Ferraro, P., Gaglio, S., Lo Re, G.: Context-awareness for multi-sensor data fusion in smart environments. In: Adorni, G., Cagnoni, S., Gori, M., Maratea, M. (eds.) AI*IA 2016. LNCS (LNAI), vol. 10037, pp. 377–391. Springer, Cham (2016). https://doi.org/10.1007/978-3-319-49130-1_28
3. Mallik, A., Tripathi, A., Kumar, R., Chaudhury, S., Sinha, K.: Ontology based context aware situation tracking. In: 2015 IEEE 2nd World Forum on Internet of Things (WF-IoT), pp. 687–692. IEEE (2015)
4. Mihajlovic, V., Petkovic, M.: Dynamic Bayesian networks: a state of the art. University of Twente Document Repository (2001)
5. Nguyen, N., Venkatesh, S., Bui, H.: Recognising behaviours of multiple people with hierarchical probabilistic model and statistical data association. In: BMVC 2006: Proceedings of the 17th British Machine Vision Conference, pp. 1239–1248. British Machine Vision Association (2006)
6. Nguyen, N.T., Phung, D.Q., Venkatesh, S., Bui, H.: Learning and detecting activities from movement trajectories using the hierarchical hidden Markov model. In: 2005 IEEE Computer Society Conference on Computer Vision and Pattern Recognition (CVPR 2005), vol. 2, pp. 955–960. IEEE (2005)

7. Perera, C., Zaslavsky, A., Christen, P., Georgakopoulos, D.: Context aware computing for the internet of things: a survey. IEEE Commun. Surv. Tutor. **16**(1), 414–454 (2013)
8. Wilson, D.H., Atkeson, C.: Simultaneous tracking and activity recognition (STAR) using many anonymous, binary sensors. In: Gellersen, H.-W., Want, R., Schmidt, A. (eds.) Pervasive 2005. LNCS, vol. 3468, pp. 62–79. Springer, Heidelberg (2005). https://doi.org/10.1007/11428572_5

Sentiment Analysis of Financial News Using Unsupervised and Supervised Approach

Anita Yadav[1](✉) [ID], C. K. Jha[1], Aditi Sharan[2], and Vikrant Vaish[2]

[1] Department of Computer Science, AIM and ACT, Banasthali Vidyapith,
Jaipur, Rajasthan, India
anitayadav_07@hotmail.com
[2] School of Computer and Systems Sciences, Jawaharlal Nehru University,
New Delhi, India

Abstract. Sentiment Analysis aims to extract sentiments from a piece of text. In addition to numeric data, sentiments are being increasingly favored as inputs to decision making process. However extracting meaning automatically from unstructured textual inputs involves a lot of complexities. These often depend on the domain from which the text was taken. Our work focuses particularly on extracting sentiments from financial news. We have proposed and implemented a framework using unsupervised and supervised techniques. We have proposed a hybrid approach of using seed sets for calculating the semantic orientation of news articles in a semi-automatic way. This approach produces better results than the standard techniques used in unsupervised sentiment analysis. Then we performed the experiment using the supervised approach with machine learning classifier Support Vector Machine (SVM). We compared the results with those produced from the standard unigram and unigram + bigram approaches and found that the proposed approach produces better precision.

Keywords: Financial news · Machine learning · Sentiment analysis · Supervised techniques · SVM · Unsupervised techniques

1 Introduction

Pang and Lee [1] defined sentiment analysis as "the computational treatment of opinion, sentiment, and subjectivity in text." Turney [2] classified reviews from four domains into recommended and not recommended based on the average Semantic Orientation (SO) of phrases used in the review. Pang et al. [3] used the machine learning methods Maximum Entropy (ME), Naïve Bayes (NB) and Support Vector Machine (SVM) to classify movie reviews as positive or negative. In this paper we have focused on another domain – Financial news.

While a lot of work has been done in sentiment analysis in areas like product or movie reviews and social media postings, not much work has been done in the field of financial news. Whatever work has been done has also not generally used unsupervised techniques. Our paper tries to address this research gap. The following are the research objectives for this paper:

© Springer Nature Switzerland AG 2019
B. Deka et al. (Eds.): PReMI 2019, LNCS 11942, pp. 311–319, 2019.
https://doi.org/10.1007/978-3-030-34872-4_35

- Tuning the existing Semantic orientation based unsupervised approach to better extract features and thereby sentiments from financial news.
- Applying multiple approaches with SVM to identify the best option for supervised learning from a financial text corpus.

2 Literature Review

Turney [2] classified reviews using sentiment indicator phrases which were extracted using an unsupervised approach. Parts of speech phrases, that were usually adjectives, were extracted to calculate semantic orientation (SO). This was defined as "the Point-wise Mutual Information-Information Retrieval (PMI-IR) between the given phrase and the word EXCELLENT minus the mutual information between the given phrase and the word POOR of all the extracted phrases of a document." The average SO of all the phrases extracted gives the overall SO of the document.

Fung et al. [5] suggested a model for mining multiple time series and textual content simultaneously. Their experimental results indicate that such a model is practically possible and can achieve a good level of accuracy. Schumaker and Chen [6] examined the role of financial news articles on stock trend prediction using three different feature representation techniques. Articles and stock quote data are processed by an SVM derivative, sequential minimal optimization style of regression which can handle discrete number analysis. Tang et al. [7] proposed an algorithm which analyses the financial news of stocks market and integrates it with the time series results in order to enhance it. Cheng [8] extracted core phrases by using rough sets theories after the unstructured information had been transformed into structured data. Then, a prediction model was established based on SVM classifier. Kaya and Karsligil [4] used SVM for classification and used bigrams based on noun-verb combinations instead of taking unigrams. Deng et al. [9] used the lexicon SentiwordNet (SWN) and then analyzed the whole dataset on five different models of multiple kernel learning and support vector regression along with the sentiment analysis and numerical dynamics. Siering [10] used a two stage method. Firstly every news article was analyzed to calculate sentiment measure and then news articles categorization was done.

Rout et al. [13] used both types of approaches on several datasets. They have used multinomial NB, ME and SVM and achieved a top accuracy of 80.68%. Zheng et al. [14] did feature selection of online reviews of mobile phones in Chinese language. Using an improved document frequency method and Boolean weighting method, they have obtained higher accuracy when using 4-POS-grams as features. Naik et al. [15] enhanced accuracy of sentiment classification in movie and university selection data-sets using tri-co-occurrences sentiment words. Rani et al. [16] used association clas-sification methods for building a semi-supervised model for POS tagging. Their tagger works with small datasets too.

3   Proposed System

3.1   Unsupervised Approaches

We have conducted the experiment on financial news using different unsupervised approaches. The classical approach proposed by Turney [2] has been used on domains other than financial news. In this approach, combinations of adjective-noun phrases are used. Then the PMI is calculated as

$$PMI(word_1, word_2) = \log_2[\frac{p(word_1 \& word_2)}{p(word_1)p(word_2)}] \tag{1}$$

Here,

$p(word_1 \& word_2)$ = Probability that both words occurs together.

$p(word_1)p(word_2)$ = Probability of co-occurrence of $word_1$ and $word_2$, if both words are independent.

$\log_2\left[\frac{p(word_1 \& word_2)}{p(word_1)p(word_2)}\right]$ = Degree of statistical dependence between words.

The Semantic Orientation (SO) of a phrase is calculated as:

$$SO(Phrase) = PMI(Phrase, "Appreciation") - PMI(Phrase, "Adversely") \tag{2}$$

When we asked a financial market expert to examine the bigrams, it was found that adjectives were not capturing the tone of financial news effectively. Instead, the expert advised using noun-verb combination phrases for better results. This was selected as the second approach. After extracting phrases using this approach, we found that phrases having significant sentiment were still not extracted from the corpus. So we proposed a third approach. We had the expert determine a seed set of representative phrases. Then we used POS tags to tag the seed set in the corpus. Learning the POS tags, we used machine learning techniques to extract an expanded seed set of phrases. Then we calculated the SO of this resulting set of phrases. Some representative phrases extracted using the three approaches are given in Table 1.

Table 1.   Representative phrases obtained using the three unsupervised approaches

Turney's phrases	Noun-verb phrases	Hybrid phrases
Stable rupee	Gaining strength	Largest market
Core banking	Value dipped	Struggled fourth-quarter
Ongoing business	Risk stemming	Brand leading
Record high	Head retired	Deal won
Low dispute	Selling dollar	Value increasing
Heavy volumes	Looking pressure	Finance beating
Initial gains	Surged valuation	Value plunged
Low investors	Technology growth	Pegged growth
Main losers	Hit suffered	Valuation rose

3.2   Supervised (SVM Based) Approach for Finding Semantic Orientation

SVM constructs a set of hyper planes in a high or infinite-dimensional space, separating the classes which are then used for classification. Input given to SVM classifier is the reduced feature matrix along with its class labels [11]. The Architecture of the proposed framework is shown in Fig. 1.

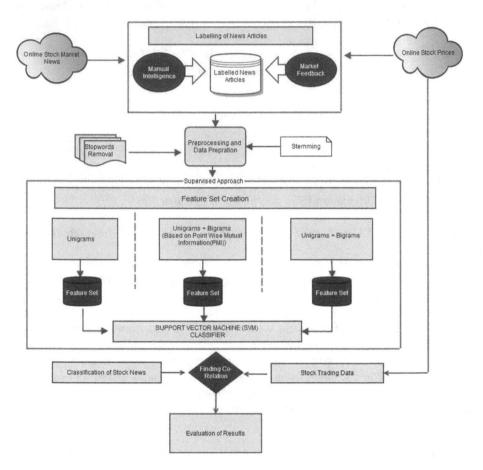

Fig. 1. General architecture of proposed framework using supervised techniques

Labeling of Documents

In order to do supervised learning, each financial news article must be labeled. In previous attempts generally two different methods were used for labeling the dataset as positive or negative. The first approach is to read the articles and label them manually. The chances of success are higher in this but the number of articles used in the dataset is very limited as it requires a lot of human effort. The second method is to label the articles automatically using some market feedback. So it is labeled on the basis of effects on their stock price. To label the financial news articles, delta price value (ΔP) for a particular day is calculated when the article was published.

$$\Delta P = P_t(\text{Open Price}:\text{current day}) - P_{t-1}(\text{Close Price}:\text{Previous day}) \quad (3)$$

If value of delta price of a day is greater than zero, the corresponding articles are labeled as positive. If value of delta price is less than zero all the corresponding articles are labeled as negative. To represent positive and negative class we used the notion of 1 for positive and −1 for negative.

Pre-processing and Extraction of Features from the Corpus
Pre-processing was done on the financial corpus i.e. removal of stop words and stemming in order to remove the redundant and noisy features so as to improve the document for further processing. Feature extraction is a major step in the classification process. Most of the work on classification is done using unigrams but as we see in the unsupervised technique, phrases or bigrams are also important in context of the financial domain. So the major contribution of this paper is to try to explore the use of bigrams in the feature extraction mechanism. Due to the combinatorial explosion in the number of bigrams and presence of noisy bigrams, we have used some standard techniques to reduce the number of bigrams. We used the well-known criteria of PMI to find the bigrams that form the phrases/features relevant for document classification. Based upon these criteria, we extracted features/phrases using three methods:

Method 1: Extracting Unigrams - In this we considered all the unigrams occurring in the corpus along with their frequencies and created a feature vector space model (VSM).
Method 2: Extracting Unigrams + Bigrams - In this we extracted all the unigrams along with all the bigrams occurring with their frequencies and eliminated those unigrams which occur in bigrams and create a feature VSM for that.
Method 3: Extracting Unigrams + Bigrams (filtered on PMI) - In this we extracted unigrams along with the bigrams filtered above certain threshold based upon their PMI scores and created feature VSM for that [12].

SVM Classification
All the features were taken individually along with the class labels given to the SVM classifier in order to build an SVM training model. We used n-fold cross validation i.e. dataset was divided into n subsets called training set and testing set of data. For every n subsets is used as the testing set and the other n − 1 subsets are used to as a training set and average error of all the n runs are calculated. The hypothesis function of SVM makes the approximation using the training set of data which is then asked to predict the output values for the data in the test set and calculate the parameters of the model from the training data and the class labels. The model was tested using the test data and the results of test data were compared with the actual class labels to calculate the accuracy and performance of the system.

4 Experimental Results and Analysis

There are no known publicly available Indian standard benchmark datasets which we could use for our testing so we have created our own dataset. As part of our experiments, we have downloaded financial news articles from online news sources like Reuters.com, Moneycontrol.com etc. for a company TCS (Tata Consultancy Limited) for a period of five months. This company is listed in the NSE (National Stock Exchange).

Tools Used

- Python version 2.7.
- NLTK Toolkit for Python.
- MATLAB version 2012.
- TMG version 6.0.
- Stanford POS tagger.

Results and experiments were evaluated using the common performance measures (Table 2).

Table 2. Results for unsupervised approach

	Turney's	Noun-verb	Hybrid
Accuracy	75.00	79.00	76.00
Precision	74.72	78.82	76.14
Recall	97.14	95.71	95.71
F-score	84.47	86.45	84.81

From the phrases in Table 1 for the three unsupervised approaches, we can see that noun-verb and hybrid phrases capture sentiment better than Turney's phrases. This can be observed in the results too where these approaches have yielded better results.

Next, experiments were conducted with an SVM using three different approaches. The detailed results are presented below (Tables 3, 4, 5 and 6).

Table 3. Confusion matrix for unigrams

	Classified positive	Classified negative
Actual positive	66	4
Actual negative	23	7

Table 4. Confusion matrix for unigrams + bigrams

	Classified positive	Classified negative
Actual positive	60	10
Actual negative	18	12

Table 5. Confusion matrix for unigrams + bigrams (PMI)

	Classified positive	Classified negative
Actual positive	61	9
Actual negative	15	15

Table 6. Results for supervised approach

	Unigrams	Unigrams + bigrams	Unigrams + bigrams (PMI)
Accuracy	73.00	72.00	76.00
Precision	74.16	76.92	80.26
Recall	94.29	85.71	87.14
F-score	83.02	81.08	83.56

As we move from unigrams to unigrams + bigrams, we notice that the precision improved but the accuracy, recall and F-score reduced. It could be due to presence of noisy bigrams in the dataset. We next used PMI to filter out such noisy bigrams. By experimentation we found that setting a threshold of 70% is effective in filtering noisy bigrams so we decided to keep that threshold. It can be observed from the results that careful filtering of bigrams has improved accuracy, precision, recall and F-score.

5 Conclusion

This paper has tried to investigate some techniques for sentiment analysis that have so far not been applied to financial domain. We faced a number of challenges in this task. Unstructured text has got a lot of noise that needs to be filtered out. The phrases containing tone are difficult to find in financial news.

In some domains like movie reviews, tone is captured mostly in adjectives but this is not found to be so in financial domain. First we have applied three different unsupervised techniques for feature selection and found that hybrid (seed word) approach gave the best results. Next we have explored the supervised approaches for text classification based on sentiments. We used three techniques – unigrams, unigrams + bigrams and unigrams + bigrams filtered on PMI. The results indicate that filtering based on PMI is effective in reducing noisy bigrams and improving the results.

Using both the techniques we finally found that the traditional way of finding the sentiments in the text does not apply to all types of dataset as the performance is dependent upon the type of dataset. We found that in unsupervised approach, the proposed technique performed better than the previous one and using supervised SVM approach features classified the data up to a good level of accuracy compared to the random model.

6 Future Work and Limitations

The work in this paper has been done on a limited dataset which was created for the purposes of this study. Labelling large data sets is a major challenge in financial domain in Indian stock markets. If this challenge can be addressed, larger scale studies can be carried out.

Use of hybrid approach of combining machine learning and lexicon may improve the results further. Cutting edge techniques like deep learning may also prove useful in this area.

References

1. Pang, B., Lee, L.: Opinion mining and sentiment analysis. Found. Trends Inf. Retr. **2**, 1–135 (2008)
2. Turney, P.D.: Thumbs up or thumbs down? Semantic orientation applied to unsupervised classification of reviews. In: Proceedings of the 40th Annual Meeting on Association for Computational Linguistics, Philadelphia, Pennsylvania, pp. 417–424. ACM (2002)
3. Pang, B., Lee, L., Vaithyanathan, S.K.: Thumbs up? Sentiment classification using machine learning techniques. In: Proceedings of the ACL-2002 Conference on Empirical Methods in Natural Language Processing, pp. 79–86. ACM, Stroudsburg (2002)
4. Kaya, M.I.Y., Karsligil, M.E.: Stock price prediction using financial news articles. In: 2nd IEEE International Conference on Information and Financial Engineering (ICIFE), pp. 478–482. IEEE, New York (2010)
5. Fung, G.P.C., Yu, J.X., Lam, W.: News sensitive stock trend prediction. In: Chen, M.-S., Yu, P.S., Liu, B. (eds.) PAKDD 2002. LNCS (LNAI), vol. 2336, pp. 481–493. Springer, Heidelberg (2002). https://doi.org/10.1007/3-540-47887-6_48
6. Schumaker, R.P., Chen, H.: Textual analysis of stock market prediction using breaking financial news: the AZFin text system. ACM Trans. Inf. Syst. (TOIS) **27**(2), 1–19 (2009)
7. Tang, X., Yang, C., Zhou, J.: Stock price forecasting by combining news mining and time series analysis. In: Proceedings of IEEE/WIC/ACM International Joint Conference on Web Intelligence and Intelligent Agent Technology, pp. 279–282. IEEE, New York (2009)
8. Cheng, S.H.: Forecasting the change of intraday stock price by using text mining news of stock. In: International Conference on Machine Learning and Cybernetics (ICMLC). IEEE, New York (2010)
9. Deng, S., Mitsubuchi, T., Shioda, K., Shimada, T., Sakurai, A.: Combining technical analysis with sentiment analysis for stock price prediction. In: Proceedings of the Ninth IEEE Ninth International Conference on Dependable, Autonomic and Secure Computing, pp. 800–807. IEEE, New York (2011)
10. Siering, M.: Boom or Ruin - does it make a difference? Using text mining and sentiment analysis to support intraday investment decisions. In: Proceedings of the 45th Hawaii International Conference on System Sciences, Waleia (Hawaii), pp. 1050–1059. IEEE, New York (2012)
11. Cristianini, N., Shawe-Taylor, J.: An Introduction to Support Vector Machines and Other Kernel-Based Learning Methods, 1st edn. Cambridge University Press, New York (2004)
12. Manning, C.D., Schutze, H.: Foundations of Statistical Natural Language Processing, 1st edn. MIT Press, Cambridge (1999)

13. Rout, J.K., Choo, K.K.R., Dash, A.K., Bakshi, S., Jena, S.K., Williams, K.L.: A model for sentiment and emotion analysis of unstructured social media text. Electron. Commer. Res. **18**(1), 181–199 (2018)

14. Zheng, L., Wang, H., Gao, S.: Sentimental feature selection for sentiment analysis of Chinese online reviews. Int. J. Mach. Learn. Cybern. **9**(1), 75–84 (2018)

15. Naik, M.V., Vasumathi, D., Siva Kumar, A.P.: An enhanced unsupervised learning approach for sentiment analysis using extraction of tri-co-occurrence words phrases. In: Bhateja, V., Tavares, J.M.R.S., Rani, B.P., Prasad, V.K., Raju, K.S. (eds.) Proceedings of the Second International Conference on Computational Intelligence and Informatics. AISC, vol. 712, pp. 17–26. Springer, Singapore (2018). https://doi.org/10.1007/978-981-10-8228-3_3

16. Rani, P., Pudi, V., Sharma, D.M.: A semi-supervised associative classification method for POS tagging. Int. J. Data Sci. Anal. **1**(2), 123–136 (2016)

Finding Active Experts for Question Routing in Community Question Answering Services

Dipankar Kundu[1]($\boxtimes$), Rajat Kumar Pal[2], and Deba Prasad Mandal[1]

[1] Machine Intelligence Unit, Indian Statistical Institute, Kolkata, India
d.kundu7681@gmail.com, dpmandal@isical.ac.in
[2] Department of CSE, University of Calcutta, Kolkata, India
pal.rajatk@gmail.com

Abstract. In this article, we propose a method for finding active experts for a new question in order to improve the effectiveness of a question routing process. By active expert for a given question, we mean those experts who are active during the time of its posting. The proposed method uses the query likelihood language model, and two new measures, activeness and answering intensity. We compare the performance of the proposed method with its baseline query likelihood language model. We use a real-world dataset, called History, downloaded from Yahoo! Answers web portal for this purpose. In every comparing scenario, the proposed method is found to outperform the corresponding baseline model.

Keywords: Active experts · Query likelihood language model · Question routing · Community question answering

1 Introduction

In community question answering (CQA) services, information seekers ask questions and other users share their knowledge by answering these questions. Due to their usefulness, these services have attracted a large number of users and have earned notable popularity. Some of trending CQA services are Yahoo! Answers[1], Answerbag[2], Wiki answer[3], Baidu Zhidao[4], and Stackoverflow[5].

In CQA services, question routing (QR) is the process of routing a new question to its potential answerers. Let, $\hat{q}$ be a new question represented by a sequence of terms (non-stop words) t_j's, $j \in \{1, 2, \cdots, |\hat{q}|\}$, i.e., $\hat{q} = \{t_1, t_2, \cdots, t_{|\hat{q}|}\}$. Let us also assume that A be a set of answerers. Then QR can be defined as the

---

[1] http://answers.yahoo.com.
[2] http://answerbag.com.
[3] http://answers.wikia.com.
[4] http://zhidao.baidu.com.
[5] http://stackoverflow.com.

© Springer Nature Switzerland AG 2019
B. Deka et al. (Eds.): PReMI 2019, LNCS 11942, pp. 320–327, 2019.
https://doi.org/10.1007/978-3-030-34872-4_36

procedure to select a set of suitable answerers $A_{\hat{q}}^* \subseteq A$, who have the expertise on $\hat{q}$. Generally, QR schemes follow three steps: (i) generation of answerers' performance profiles, (ii) estimation of answerers' expertise (based on their performance profiles) and (iii) routing of the question to the potential experts in the expert list. In this context, the performance profile of an answerer contains her answering history. Usually, it is considered to be the collection of all of the questions that she has answered previously. In our work, however, we maintain all questions with their posting time stamp and answering timestamp in the performance profile. In QR, expert finding (EF), is the process of identifying the possible experts corresponding to a given question. Let, $Exp(a_i, \hat{q})$ be the expertise score of the answerer $a_i \in A$ corresponding to $\hat{q}$. Depending upon expertise score, a system ranks all of the answerers, and finally, selects the top N ($N = |A_{\hat{q}}^*|$) ranked answerers, i.e., experts for routing $\hat{q}$.

The performance of CQA services may be affected if the referred experts are not currently active in the community. In this case, the average waiting time for an asker to obtain the first suitable answer may be too high to be useful. Therefore, in QR, it is desirable to route a particular question to the experts, who are available online. We call such users as *active experts*.

Here, we propose an active expert finding (AEF) method. It incorporates an answerer's activeness at the time of question posting and uses the query likelihood language (QLL) model to estimate the expertise of the answerer. We not only estimate an answerer's availability during question posting time but also consider the answering intensity during the same period. To show the effectiveness of the proposed method, we have compared it with the baseline system on a real-world dataset downloaded from Yahoo! Answers web site using three performance measures. We find that in every corresponding scenario, the proposed AEF system outperformed the baseline EF system.

2   Related Works

The literature of EF is quite affluent [1,2,4–6,8–10,14,18]. To keep this paper concise, however, we discuss some of the recent prominent works that deal with the availability of answerers [3,7,11,15,16].

In [7], the authors proposed a QR model that utilizes the language model to estimate a user's expertise. This model [7] integrates the quality of each answerer's previous answers along with her login availabilities. It [7] treats the subproblem of predicting a user's login availability as a time series forecasting problem. In [3], the authors proposed a recommendation system that takes into account the compatibility, topical expertise, and availability of users. To estimate the availability of the users, they applied three classification techniques on previous activities of the users. In another work [16], researchers proposed a dynamic modelling approach for QR. They used two temporal discounting functions to model the availability of the users. While modelling the EF problem, the authors of [11] considered the dynamic aspects of the problem and proposed a supervised learning framework. The authors of [15] proposed a QR technique to route questions to users who are the most suitable to answer them based on their past and recent activities. They [15] proposed a measure which divides users' activities into four parts depending upon the time to assess the activities of the users. They also assigned weights to each of the categories such that the recent categories attain comparatively higher importance.

3   Proposed Method

The present investigation is concerned with the formulation of a method for finding active experts for a new question in order to improve the effectiveness of question routing schemes. It consists of four phases: (i) estimation of expertise, (ii) estimation of activeness, (iii) estimation of answering intensity, and (iv) estimation of the active experts. We now discuss these phases in detail.

3.1   Estimation of Expertise

In information retrieval, QLL [12] model is used to estimate the similarity between a document and a given query. Treating answerers' profiles as documents, we apply it to estimate the expertise of each answerer for a given question.

Let us assume that θ_{a_i} denotes the language model associated with the performance profile of the answerer a_i, and θ_C denotes the language model associated with the entire collection of performance profiles. Then, the expertise score of a_i for a given question $\hat{q}$ is computed with the help of [12] as

$$Exp(a_i, \hat{q}) = P(\hat{q}|\theta_{a_i}) = \prod_{t \in \hat{q}} P(t|\theta_{a_i})^{n(t,\hat{q})}, \tag{1}$$

where $n(t, \hat{q})$ is the number of occurrence of t in $\hat{q}$. To avoid the zero probabilities, we use the Jelinek-Mercer's smoothing method [17] in (1) and obtain the following:

$$P(\hat{q}|\theta_{a_i}) = \prod_{t \in \hat{q}} \{\lambda p(t|\theta_{a_i}) + (1 - \lambda)p(t|\theta_C)\}^{n(t,\hat{q})} \tag{2}$$

where

$$p(t|\theta_{a_i}) = \frac{tf(t, \theta_{a_i})}{\sum_{t' \in \theta_{a_i}} tf(t', \theta_{a_i})}; \tag{3}$$

and

$$p(t|\theta_C) = \frac{tf(t, \theta_C)}{\sum_{t' \in \theta_C} tf(t', \theta_C)}. \tag{4}$$

Here, $tf(t, \theta)$ is the frequency of the term t in θ (θ_{a_i} or θ_C), and $\sum_{t' \in \theta} tf(t', \theta)$ is the total number of occurrences of all terms in θ. Moreover, λ $(0 < \lambda < 1)$ controls the influences of θ_{a_i} and θ_C.

3.2 Estimation of Activeness

To incorporate the activeness of an answerer in the proposed method, we define the activeness score of an answerer a_i as

$$AS(a_i) = exp\left(-\left(\texttt{time}_c - \texttt{time}_{a_i}\right)/(24 \times 7)\right), \tag{5}$$

where $\texttt{time}_c$ is the current system time and $\texttt{time}_{a_i}$ is the last answering time of a_i. Here, $(\texttt{time}_c - \texttt{time}_{a_i})$ is measured in hours. Note that, $AS(\cdot)$ is an exponentially decreasing function which ensures that an answerer who has answered during recently obtains a high activeness score.

3.3 Estimation of Answering Intensity

A suitable expert should not only be active (as we argued in the previous section), but also her participation in answering questions should be high. Accordingly, we introduce a new measure, called answering intensity that incorporates hourly answering activity and consistency of each answerer during each hour of a day.

In this regard, we construct answerer hourly answering activity matrix denoted by $\mathcal{HA}$, of size $|A| \times |H|$. Here, $\mathcal{HA}_{a_i,h}$ represents the total number of answerers submitted by the answerer a_i ($a_i \in A$) in the hth ($h \in H = \{0, 1, \cdots, 23\}$) hour. Then, similar to [3], a_i's answering activity during the hth hour is estimated as

$$Ans_{act}(a_i, h) = \frac{\mathcal{HA}_{a_i,h}}{\sum_{\hat{h} \in H} \mathcal{HA}_{a_i,\hat{h}}}. \tag{6}$$

Next, to take into account the consistency of an answerer we construct an hourly consistency matrix denotes as $\mathcal{HC}$, of size $|A| \times |H|$. $\mathcal{HC}_{a_i,h}$ indicates the number of days the answerer a_i answered during the hth hour of the days (under consideration). Now, we estimate a_i's answering consistency during the hth hour as

$$\mathcal{H}_{con}(a_i, h) = \frac{\mathcal{HC}_{a_i,h}}{\sum_{\hat{h} \in H} \mathcal{HC}_{a_i,\hat{h}}}. \tag{7}$$

Then, we calculate the answering intensity of a_i during the hth hour as

$$\mathcal{I}_{ans}(a_i, h) = Ans_{act}(a_i, h) \times \mathcal{H}_{con}(a_i, h). \tag{8}$$

Let, $h(\hat{q})$ be the hour when $\hat{q}$ is post and δ is the time window. We use $\delta = 2\,\mathrm{h}$, which has been chosen based on our ad-hoc experiments. Now, we calculate the answering intensity of an answerer a_i for a question $\hat{q}$ at the time of posting of the question i.e., for the time period $[h(\hat{q}) - \delta, h(\hat{q}) + \delta]$ as

$$\mathcal{I}_{ans}(u_i, h(\hat{q})) = \prod_{h=h(\hat{q})-\delta}^{h(\hat{q})+\delta} \mathcal{I}_{ans}(u_i, h). \tag{9}$$

Equation (9) attains the effect on the answering intensity during the hth hour.

Table 1. Summary of the history dataset

Property	Training	Test	Collection
The number of questions	72784	5458	78242
The number of answers	205828	19630	225458
The number of total answerers	36957	3839	39153
The number of total askers	45418	3902	48557

3.4   Estimation of Active Experts

When a new question $\hat{q}$ is posted at hour $h(\hat{q})$ of a day, the active expert score, $Exp_{act}(a_i, \hat{q}, h(\hat{q}))$, of answerer a_i for $\hat{q}$ is estimated as

$$Exp_{act}(a_i, \hat{q}, h(\hat{q})) = Exp(a_i, \hat{q}) \times \mathcal{AS}(a_i) \times \mathcal{I}_{ans}(a_i, h(\hat{q})). \qquad (10)$$

The answerers are finally ranked using active expert score.

4   Experiments and Results

We examine the performance of the proposed method on a real-world dataset, called history, which has been downloaded from Yahoo! Answers web site by us [6]. As the name of the dataset suggests, the questions in this dataset belong to the category history. It consists of 78,242 resolved questions and their answers posted from June 27, 2012 to September 06, 2013. We have marked 72,784 questions posted during June 27, 2012 to July 30, 2013, as the training set. Remaining 5,458 questions posted during July 31, 2013 to September 06, 2013 are marked as the test set. A summary about this dataset is provided in Table 1.

Initially, we remove stop words from the dataset. Then, we stem the questions with Porter stemmer [13]. This process removes the common or morphological, and inflection endings. Moreover, from the training dataset, we remove the answerers who answered less than ten questions. After this process, we obtain 1,901 answerers' profiles in the training dataset. To prepare the test dataset, furthermore, we remove the questions for which the best answerers do not exist in the filtered training set. Thus, we obtain 4,261 questions in the test data set.

To investigate the effect of *activeness*, we use two models: EF and AEF. In EF and AEF, we route the questions to the *experts* and *active experts*, respectively. Furthermore, to investigate the impact of the incorporation of answer quality in our method, we use two different configurations of the answerers' profiles. In the first configuration, we consider all questions that a user has answered as the profile of the user. In the second configuration, to constitute the profile of a user, we consider only the questions for which the user has given the best answer. We denote these two configurations as ALL and BEST, respectively. As discussed earlier, we have two models, namely, EF and AEF and two configurations of user profiles, namely, ALL and BEST. Thus, we have four cases: (i) EF-ALL, (ii) EF-BEST, (iii) AEF-ALL, and (iv) AEF-BEST.

Table 2. MMR of different cases with different values of λ

Model	Config.	$\lambda = 0.1$	$\lambda = 0.2$	$\lambda = 0.3$	$\lambda = 0.4$	$\lambda = 0.5$	$\lambda = 0.6$	$\lambda = 0.7$	$\lambda = 0.8$	$\lambda = 0.9$
EF	ALL	0.013	0.015	0.019	0.023	0.028	0.033	0.041	0.049	0.062
	BEST	0.012	0.015	0.020	0.026	0.034	0.041	0.049	0.062	0.076
AEF	ALL	0.063	0.070	0.077	0.084	0.092	0.098	0.107	0.114	0.121
	BEST	**0.093**	**0.099**	**0.105**	**0.113**	**0.121**	**0.127**	**0.134**	**0.141**	**0.151**

Table 3. C@N for different cases with $\lambda = 0.5$

Models	Configuration	$C@1$	$C@10$	$C@50$	$C@200$
EF	ALL	0.71	5.01	19.72	50.95
	BEST	1.11	6.26	23.44	58.25
AEF	ALL	3.20	20.99	**43.14**	**65.12**
	BEST	**6.80**	**21.53**	40.97	61.76

Table 4. S@N of different cases with $\lambda = 0.5$

Model	Configuration	$S@1$	$S@2$	$S@3$	$S@4$	$S@5$
EF	ALL	0.0071	0.0112	0.0124	0.0134	0.0143
	BEST	0.0106	0.0141	0.0158	0.0180	0.0192
AEF	ALL	0.0320	0.0488	0.0592	0.0651	0.0689
	BEST	**0.0680**	**0.0876**	**0.0947**	**0.0987**	**0.1017**

We use three performance measures here: mean reciprocal rank (MMR) [7], best answer coverage (C@N) [4], and success at the top N (S@N) [14]. When we use MRR, we examine with $\lambda \in \{0.1, 0.2, 0.3, 0.4, 0.5, 0.6, 0.7, 0.8, 0.9\}$ in Eq. (2) and provide the results in Table 2. However, when we use $C@N$ and $S@N$ for the comparison, we use $\lambda = 0.5$ in Eq. (2). In Tables 3 and 4, we provide the results using C@N with $N \in \{1, 10, 50, 200\}$ and S@N with $N \in \{1, 2, 3, 4, 5\}$, respectively.

From Tables 2, 3, and 4, we find that in two comparing scenarios (C@50 and C@200) AEF-ALL obtained the best performance, and in every other 16 comparing scenarios AEF-BEST obtained the best performance. We also observe that for any comparing scenario, the proposed AEF outperforms the baseline EF.

5   Conclusion

Here, we propose an active expert finding method for QR in CQA services. We use the QLL model to measure the expertise of an answerer for a given question. We propose two measures, called activeness score and answering intensity score. These two scores assess the activeness of an answerer and the intensity to which an answerer is consistent in answering questions during a particular hour of the day. Finally, aggregate the expertise score provided by the QLL model, the activeness score, and the answering intensity score to we define an active expert score.

We investigate the performance of the proposed method using a real-world dataset called History which has been downloaded from the Yahoo! Answers web site. We use three performance measures: MMR, C@N, and S@N. The proposed scheme is found to perform the best for every comparing scenario. However, we have made the choice of δ, in a crude way. A further investigation of the parameter δ and experiments on more datasets are required.

References

1. Aslay, Ç., O'Hare, N., Aiello, L.M., Jaimes, A.: Competition-based networks for expert finding. In: Proceedings of the 36th International ACM SIGIR Conference on Research and Development in Information Retrieval, pp. 1033–1036. ACM (2013)
2. Bouguessa, M., Dumoulin, B., Wang, S.: Identifying authoritative actors in question-answering forums: the case of Yahoo! answers. In: Proceedings of the 14th ACM SIGKDD International Conference on Knowledge Discovery and Data Mining, KDD 2008, pp. 866–874. ACM (2008)
3. Chang, S., Pal, A.: Routing questions for collaborative answering in community question answering. In: Proceedings of the 2013 IEEE/ACM International Conference on Advances in Social Networks Analysis and Mining, pp. 494–501. ACM (2013)
4. Fang, L., Huang, M., Zhu, X.: Question routing in community based QA: incorporating answer quality and answer content. In: Proceedings of the ACM SIGKDD Workshop on Mining Data Semantics, MDS 2012, pp. 5:1–5:8. ACM (2012)
5. Jiang, J., Lu, W.: IR-based expert finding using filtered collection. In: 2008 4th International Conference on Wireless Communications, Networking and Mobile Computing, pp. 1–5, October 2008
6. Kundu, D., Mandal, D.P.: Formulation of a hybrid expertise retrieval system in community question answering services. Appl. Intell. **49**(2), 463–477 (2019)
7. Li, B., King, I.: Routing questions to appropriate answerers in community question answering services. In: Proceedings of the 19th ACM International Conference on Information and Knowledge Management, pp. 1585–1588. ACM (2010)
8. Liu, D.R., Chen, Y.H., Kao, W.C., Wang, H.W.: Integrating expert profile, reputation and link analysis for expert finding in question-answering websites. Inf. Process. Manag. **49**(1), 312–329 (2013)
9. Liu, J., Song, Y.I., Lin, C.Y.: Competition-based user expertise score estimation. In: Proceedings of the 34th International ACM SIGIR Conference on Research and Development in Information Retrieval, SIGIR 2011, pp. 425–434. ACM (2011)
10. Mandal, D.P., Kundu, D., Maiti, S.: Finding experts in community question answering services: a theme based query likelihood language approach. In: 2015 International Conference on Advances in Computer Engineering and Applications, pp. 423–427, March 2015
11. Neshati, M., Fallahnejad, Z., Beigy, H.: On dynamicity of expert finding in community question answering. Inf. Process. Manag. **53**(5), 1026–1042 (2017)
12. Ponte, J.M., Croft, W.B.: A language modeling approach to information retrieval. In: Proceedings of the 21st Annual International ACM SIGIR Conference on Research and Development in Information Retrieval, pp. 275–281. ACM (1998)

13. Porter, M.F.: An algorithm for suffix stripping. Program **14**(3), 130–137 (1980)
14. Riahi, F., Zolaktaf, Z., Shafiei, M., Milios, E.: Finding expert users in community question answering. In: Proceedings of the 21st International Conference on World Wide Web, WWW 2012 Companion, pp. 791–798. ACM, New York (2012)
15. Roy, P.K., Singh, J.P., Nag, A.: Finding active expert users for question routing in community question answering sites. In: Perner, P. (ed.) MLDM 2018. LNCS (LNAI), vol. 10935, pp. 440–451. Springer, Cham (2018). https://doi.org/10.1007/978-3-319-96133-0_33
16. Yeniterzi, R., Callan, J.: Moving from static to dynamic modeling of expertise for question routing in CQA sites. In: International AAAI Conference on Web and Social Media (2015)
17. Zhai, C., Lafferty, J.: A study of smoothing methods for language models applied to information retrieval. ACM Trans. Inf. Syst. **22**(2), 179–214 (2004)
18. Zhang, J., Ackerman, M.S., Adamic, L.: Expertise networks in online communities: structure and algorithms. In: Proceedings of the 16th International Conference on World Wide Web, pp. 221–230. ACM (2007)

Remote Sensing

Stacked Autoencoder Based Feature Extraction and Superpixel Generation for Multifrequency PolSAR Image Classification

Tushar Gadhiya[✉], Sumanth Tangirala, and Anil K. Roy

Dhirubhai Ambani Institute of Information and Communication Technology, Gandhinagar, India
201621009@daiict.ac.in

Abstract. In this paper we are proposing classification algorithm for multifrequency Polarimetric Synthetic Aperture Radar (PolSAR) image. Using PolSAR decomposition algorithms 33 features are extracted from each frequency band of the given image. Then, a two-layer autoencoder is used to reduce the dimensionality of input feature vector while retaining useful features of the input. This reduced dimensional feature vector is then applied to generate superpixels using simple linear iterative clustering (SLIC) algorithm. Next, a robust feature representation is constructed using both pixel as well as superpixel information. Finally, softmax classifier is used to perform classification task. The advantage of using superpixels is that it preserves spatial information between neighboring PolSAR pixels and therefore minimizes the effect of speckle noise during classification. Experiments have been conducted on Flevoland dataset and the proposed method was found to be superior to other methods available in the literature.

Keywords: Polarimetric Synthetic Aperture Radar (PolSAR) ·
Multifrequency PolSAR image classification · Autoencoder ·
Superpixels · Simple linear iterative clustering (SLIC) · Optimized
Wishart Network (OWN)

1 Introduction

Synthetic Aperture Radar (SAR) has been popularized in recent years as a technique that captures high resolution microwave images of the earth surface. With the SAR technique, an image can be taken regardless of weather conditions or time of the day unlike optical sensors. The other major reason to use SAR is its operability over multiple frequency bands, viz, from the X band to P band. The penetrability of the L and the P bands allows SAR to capture data from even below ground level. In case of PolSAR it draws information of the target in four polarization states which makes it an information rich technique. These

© Springer Nature Switzerland AG 2019
B. Deka et al. (Eds.): PReMI 2019, LNCS 11942, pp. 331–339, 2019.
https://doi.org/10.1007/978-3-030-34872-4_37

are some of the reasons that establish the superiority of SAR over optical data capturing techniques.

One of the very early approaches of classification of multifrequency PolSAR data was attempted using DNN (Dynamic Neural Network) [1]. Deep learning based multiplayer autoencoder network was also proposed [2]. It uses Kronecker product of eigenvalues of coherency matrix to combine multiple bands information. Recently, an Optimized Wishart Network (OWN) for classification of multifrequency PolSAR data was reported [3].

Superpixel algorithms in conjunction with deep neural networks have gained popularity for capturing spatial information of a PolSAR image. Hou *et al.* presented [4] a way of using Pauli decomposition of PolSAR image to generate superpixels and autoencoder to extract features from the coherency matrix of each PolSAR pixel. Prediction of the network was then used to run a KNN (k nearest neighbors) algorithm in each superpixel to determine the class of the complete superpixel. Guo *et al.* introduced a method to apply Fuzzy clustering algorithm over PolSAR images to generate superpixels [5]. This method considered only those pixels which are similar to their neighbors, while pixels which are in all probability badly conditioned were ignored. In another work [6] Cloude decomposition features were used to generate superpixels and CNN (Convolutional Neural Network) to perform the classification. Adaptive nonlocal approach for extracting spatial information was also proposed [7]. It uses stacked sparse autoencoder to extract robust features.

In this paper we propose a classification algorithm for multifrequency PolSAR images. It is organized as follows: Sect. 2 discusses the proposed network architecture; Sect. 3 explains the experiments conducted on the Flevoland dataset; Sect. 4 concludes with our observations and the discussions based on the results.

2   Proposed Methodology

Phase information is very useful for PolSAR image classification. Since we are using a real valued neural network it may not be able to relate amplitude and phase information. It considers the real and imaginary components as separate features. This brings the need of decomposition features to the network instead of the raw features of coherency matrix. We have created 33 dimensional feature vector extracted from one frequency band of a PolSAR image as shown in Table 1. Hence combining information of all three bands we get 99 dimensional feature vector corresponding to each PolSAR pixel.

Freeman decomposition [8] has 3 features which describe the power of single-bounce(odd-bounce), double-bounce and volume scattering of the incident waves. The Huynen decomposition [9] aims to form a single scattering matrix to model the scattering mechanism of the surface. The Cloude decomposition [10] aims to model the surface scattering using the parameters such as Entropy, Anisotropy and other angles which describe the scattering mechanism. The Krogager decomposition [11] aims to factorize the scattering matrix as the combination of a sphere, a di-plane and a helix. The Yamaguchi decomposition

Table 1. List of extracted features

Features	Description
$\|T_{13}\|/\sqrt{T_{11}T_{33}}, \|T_{23}\|/\sqrt{T_{33}T_{22}}, T_{22}/S,$ $T_{33}/S, 10log_{10}(S), \|T_{12}\|/\sqrt{T_{11}T_{22}}$	6 features of Coherency matrix [13]
$F_{dbl}, F_{odd}, F_{vol}$	3 features of Freeman decomposition [8]
K_d, K_h, K_s, K_t	4 features of Krogager decomposition [11]
$P_{dbl}, P_{hlx}, P_{odd}, P_{vol}$	4 features of Yamaguchi decomposition [12]
$A, B_0, B, C, D, E, F, G, H$	9 features of Huynen decomposition [9]
$\alpha, \beta, \delta, \gamma, \lambda$, Entropy, Anisotropy	7 features of Cloude decomposition [10]

[12] adds a Helix scattering component to the Freeman decomposition to model complicated man-made structures.

As shown in Fig. 1 our proposed network architecture contains three modules. The first module is a two layer autoencoder network. The purpose of this module is to reduce the dimensionality of the input vector by learning efficient representation of combined PolSAR frequency bands information.

Let $\mathbf{X_i}$ be the input feature vector of ith PolSAR pixel. Let $\mathbf{W_{11}}, \mathbf{W_{12}}, \mathbf{b_{11}}$ and $\mathbf{b_{12}}$ be weights and biases of two encoder layers. A hidden representation of input feature vector $\mathbf{X_i}$ can be calculated as $\mathbf{h_i} = f(\mathbf{W_{12}}^T f(\mathbf{W_{11}}^T\mathbf{X_i} + \mathbf{b_{11}}) + \mathbf{b_{12}})$, where f is a $tanh$ activation function. Let $\mathbf{W_{21}}, \mathbf{W_{22}}, \mathbf{b_{21}}$ and $\mathbf{b_{22}}$ be weights and biases of two decoder layers. A reconstructed input vector can be calculated as $\mathbf{X_i}' = f(\mathbf{W_{22}}^T f(\mathbf{W_{21}}^T\mathbf{h_i}+\mathbf{b_{21}})+\mathbf{b_{22}})$. We have used mean square error to train the autoencoder. The cost function of first module of the proposed network is given as follows:

$$J_1 = \beta(\|\mathbf{W_{11}}\|_2^2 + \|\mathbf{W_{12}}\|_2^2 + \|\mathbf{W_{21}}\|_2^2 + \|\mathbf{W_{22}}\|_2^2)$$
$$+ \alpha\frac{1}{N}\sum_{i=1}^{N}\|\mathbf{X_i} - \mathbf{X_i}'\|^2 + \gamma\sum_{t=1}^{U_1}KL(\rho\|\rho_t). \tag{1}$$

Here β is a regularization parameter, α is the learning rate, γ is the sparsity parameter, N is total number of training samples, U_1 is a size of reduced dimensional feature vector $\mathbf{h_i}, \rho$ is a sparsity parameter, ρ_t is the average activation value of the tth hidden unit and $KL(\rho\|\rho_t)$ is a Kullback-Leibler divergence which encourages sparsity in the hidden representation $\mathbf{h_i}$. For our application we have set the value of $U_1 = 5$. Once the training of the first module is complete we disconnect the decoder layers.

Next, we feed the entire PolSAR image as an input to this network to obtain hidden representations of all pixels of the PolSAR image. We will use this hidden representation of all pixels of the PolSAR image to generate superpixels. This process has two advantages: (i) it contains feature information of all bands and (ii) its dimensionality is substantially reduced in comparison to the input feature vector. To generate superpixels we have used an algorithm similar to simple linear iterative clustering (SLIC) [14]. Instead of giving an RGB image as input, we

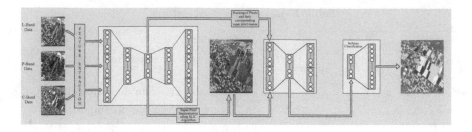

Fig. 1. Proposed network architecture.

will give the hidden representation of PolSAR image obtained from using first module of the proposed network as input. Using SLIC we measure the distance between any two pixels which is given by Eq. 2:

$$D = \frac{m}{s} D_s + D_h, \tag{2}$$

where $D_s = \sqrt{(x_i - x_j)^2 + (y_i - y_j)^2}$ and $D_h = \|\mathbf{h_i} - \mathbf{h_j}\|_2$. Here m is a parameter controlling the relative weight between D_s and D_h and s is a size of search space [14]. (x_i, y_i) and (x_j, y_j) are the positions of ith and j^{th} PolSAR pixels on Euclidean plane. This distance measure was finally used for superpixel generation.

The second module of our proposed architecture combines each pixel and corresponding superpixel information to construct robust feature vector. This is done by letting $\mathbf{S_j}$ be the jth superpixel and $\mathbf{h_i} \in \mathbf{S_j}$. Let $\mathbf{c_j}$ be the cluster center of $\mathbf{S_j}$. To extract robust feature using both pixel and superpixel information input for second autoencoder $\mathbf{H_i}$ can be constructed as $\mathbf{H_i} = [\mathbf{h_i}; \mathbf{c_j}]$ [7]. Since the dimensionality of the hidden representation h_i is low, a single layer autoencoder is sufficient for an effective reconstruction. Let $\mathbf{r_i} = f(\mathbf{W_{13}}^T \mathbf{H_i} + \mathbf{b_{13}})$ be the activation value obtained at hidden layer of second autoencoder. Let $\mathbf{H_i'} = f(\mathbf{W_{23}}^T \mathbf{r_i} + \mathbf{b_{23}})$ be the reconstructed input. The cost function for the second autoencoder can now readily be described by:

$$J_2 = \beta(\|\mathbf{W_{13}}\|_2^2 + \|\mathbf{W_{23}}\|_2^2) + \lambda \frac{1}{N} \sum_{i=1}^{N} \|\mathbf{H_i} - \mathbf{H_i'}\|^2 + \alpha \sum_{t=1}^{U_2} KL(\rho\|\rho_t). \tag{3}$$

Here U_2 is the size of feature vector $\mathbf{r_i}$. Once the training of the second module of the autoencoder is complete we again disconnect the decoder layer. Output of the second module of autoencoder contains both pixel and superpixel information. Finally we use softmax classifier to obtain predicted probability distribution.

3   Experiments

With this theoretical model we conduct the following experiments. We start with the details of dataset chosen for our experiments. After that we analyze the performance of the proposed network for different band combinations. Finally we will perform the analysis of performance for different feature decompositions. The complete experiment on the proposed network architecture was implemented using Python 3.6 and it was executed on a 1.60 GHz machine with 8 GB RAM for all the experiments.

Experiments have been conducted on a dataset of Flevoland [15], an agricultural tract in the Netherlands, whose data was captured by the NASA/Jet Propulsion Laboratory on 15 June, 1991. This dataset has often been viewed as the benchmark dataset for PolSAR applications. The intensities after Pauli decomposition of the dataset have been used to form an RGB image as shown in Fig. 2(a–c). The ground truth of the data set shown in Fig. 2(d) identifies a total of 15 classes of land cover.

Fig. 2. Pauli decomposition of (a) L band, (b) P band and (c) C band Flevoland dataset [15]. (d) Ground truth map of Flevoland dataset

We have evaluated the accuracy of each class with respect to all possible combinations of the data acquired in the three frequency bands. Please note that all the overall accuracy mentioned in the paper is the precision. Figure 3 shows classification maps obtained by the proposed method and Table 2 shows classwise accuracies for all possible band combinations. From Table 2 it can be observed that while the network with just the C band information recognizes Onions and Lucerne with high accuracy, it fails in the case of classes such as Beet or Oats. While in the case of the C band, the network fails to recognize Wheat accurately. It can be noted that the lack of information to recognize Wheat is compensated by the help of the L band. In the case of the P band, the network fails to accurately identify Onions and Peas but correctly identifies the majority of the Rapeseed and Fruit. It can also be observed that the L band performs better individually than the C or the P bands.

A total of six PolSAR decomposition methods have been applied to the PolSAR data and their features have been given as input to the proposed network. The significance of each decomposition technique is evident from the Table 3. It can be seen that the Krogager and Freeman decompositions fail to

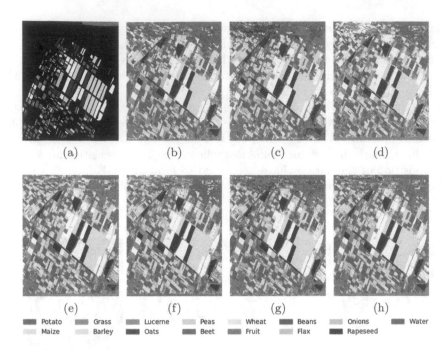

Fig. 3. (a) Ground truth map, Classification map of Flevoland dataset obtained using (b) L band, (c) P band, (d) C band, (e) P and C band, (f) L and C band, (g) L and P band and (h) L, P and C band.

Table 2. Class-wise accuracies for different band combinations

Class/bands	L	P	C	LP	LC	PC	LPC
Potato	0.9912	0.9636	0.9890	0.9977	0.9961	0.9931	0.9975
Maize	0.9859	0.9745	0.9292	0.9963	0.9932	0.9917	0.9979
Grass	0.9339	0.8282	0.8585	0.9741	0.9457	0.9608	0.9773
Barley	0.9757	0.9758	0.8848	0.9960	0.9930	0.9857	0.9979
Lucerne	0.9506	0.8148	0.9573	0.9798	0.9879	0.9638	0.9869
Oats	0.4641	0.1771	0.0000	0.8916	0.6275	0.7466	0.8718
Peas	0.9642	0.0024	0.8975	0.9952	0.9827	0.9807	0.9952
Beet	0.8024	0.9490	0.1667	0.9608	0.9545	0.9627	0.9772
Wheat	0.8756	0.6940	0.4792	0.9375	0.9328	0.8633	0.9538
Fruit	0.9505	0.9819	0.7591	0.9931	0.9674	0.9906	0.9944
Beans	0.8954	0.4990	0.8784	0.9410	0.9767	0.9534	0.9850
Flax	0.9856	0.9409	0.6749	0.9984	0.9860	0.9718	0.9966
Onions	0.2353	0.0000	0.8893	0.4706	0.9273	0.2907	0.9275
Rapeseed	0.9952	0.9951	0.9982	0.9986	0.9994	0.9988	0.9994
Water	0.9979	0.9964	0.9997	0.9975	0.9983	0.9983	0.9995
OA	**0.9784**	**0.9508**	**0.9211**	**0.9938**	**0.9900**	**0.9869**	**0.9959**

Table 3. Class-wise accuracies for different features

/	Yamaguchi	Coherency	Krogager	Freeman	Huynen	Cloude
Potato	0.9820	0.9860	0.9879	0.9721	0.9933	0.9941
Maize	0.9672	0.9696	0.9880	0.9465	0.9904	0.9884
Grass	0.9243	0.8898	0.8490	0.6757	0.9256	0.9555
Barley	0.9653	0.9516	0.9814	0.9554	0.9920	0.9846
Lucerne	0.6690	0.8568	0.4815	0.4820	0.9695	0.9522
Oats	0.0626	0.0031	0.2168	0.0000	0.3420	0.7252
Peas	0.9779	0.8686	0.7440	0.5245	0.9546	0.9904
Beet	0.1658	0.6794	0.0000	0.0000	0.8834	0.9244
Wheat	0.1460	0.7210	0.6841	0.0000	0.8916	0.9428
Fruit	0.9618	0.9661	0.9827	0.9027	0.9790	0.9874
Beans	0.7831	0.9265	0.8758	0.0160	0.9767	0.9167
Flax	0.9346	0.9636	0.7162	0.7470	0.9790	0.9688
Onions	0.0000	0.2561	0.0000	0.0000	0.8720	0.1246
R.seed	0.9921	0.9949	0.9976	0.9786	0.9976	0.9986
Water	0.9889	0.9957	0.9958	0.9926	0.9981	0.9968
OA	**0.9509**	**0.9609**	**0.9562**	**0.9105**	**0.9859**	**0.9856**

Table 4. Classification's overall accuracy comparison

Method	Number of classes	Accuracy
ANN [2]	7	98.23
OWN [3]	7	98.56
Stein-SRC [16]	14	99.00
Proposed method	7	99.93
Proposed method	14	99.69
Proposed method	15	99.59

provide enough information to recognize Beet and Onions. On the other hand Cloude decomposition and Huynen decomposition provide sufficient information for these crops respectively.

The proposed method for classification of multifrequency PolSAR image is also compared with other methods available in the literature. Comparison of overall accuracies is reported in Table 4. ANN [2] and OWN [3] have used small subset of Flevoland dataset containing 7 classes. On the other hand Stein-SRC [16] used ground truth with 14 classes. For parity in comparison with our results we have calculated the overall accuracy of the proposed method using ground truth of 7, 14 as well as 15 classes. We observe from Table 4 that the proposed method outperforms all the three methods [2,3,16].

4   Conclusion

In this paper a classification network for multifrequency PolSAR data is proposed. Proposed network involves three modules. First module reduces the dimensionality of the input feature vector. Output of the first module has been used to generate superpixels. The second module constructs robust feature vector using each pixel and its corresponding superpixel information. Finally the last module of the proposed network conducts classification using softmax classifier. It is observed that combining multiple frequency bands information improves overall classification accuracy. We validated our proposed network on Flevoland dataset resulted in 99.59% overall classification accuracy. Experimental result shows that this proposed network outperforms other reported methods available in the literature.

References

1. Chen, K.S., Huang, W.P., Tsay, D.H., Amar, F.: Classification of multifrequency polarimetric SAR imagery using a dynamic learning neural network. IEEE Trans. Geosci. Remote Sens. **34**, 814–820 (1996)
2. De, S., Ratha, D., Ratha, D., Bhattacharya, A., Chaudhuri, S.: Tensorization of multifrequency PolSAR data for classification using an autoencoder network. IEEE Geosci. Remote Sens. Lett. **15**, 542–546 (2018)
3. Gadhiya, T., Roy, A.K.: Optimized wishart network for an efficient classification of multifrequency PolSAR data. IEEE Geosci. Remote Sens. Lett. **15**, 1720–1724 (2018)
4. Hou, B., Kou, H., Jiao, L.: Classification of polarimetric SAR images using multi-layer autoencoders and superpixels. IEEE J. Sel. Top. Appl. Earth Obs. Remote Sens. **9**, 3072–3081 (2016)
5. Guo, Y., Jiao, L., Wang, S., Wang, S., Liu, F., Hua, W.: Fuzzy superpixels for polarimetric SAR images classification. IEEE Trans. Fuzzy Syst. **26**, 2846–2860 (2018)
6. Hou, B., Yang, C., Ren, B., Jiao, L.: Decomposition-feature-iterative-clustering-based superpixel segmentation for PolSAR image classification. IEEE Geosci. Remote Sens. Lett. **15**, 1239–1243 (2018)
7. Hu, Y., Fan, J., Wang, J.: Classification of PolSAR images based on adaptive nonlocal stacked sparse autoencoder. IEEE Geosci. Remote Sens. Lett. **15**, 1050–1054 (2018)
8. Freeman, A., Durden, S.L.: A three-component scattering model for polarimetric SAR data. IEEE Trans. Geosci. Remote Sens. **36**, 963–973 (1998)
9. Huynen, J.R.: Phenomenological theory of radar targets. Electronmagnetic Scattering (1970)
10. Cloude, S.R., Pottier, E.: An entropy based classification scheme for land applications of polarimetric SAR. IEEE Trans. Geosci. Remote Sens. **35**, 68–78 (1997)
11. Krogager, E.: New decomposition of the radar target scattering matrix. Electron. Lett. **26**, 1525–1527 (1990)
12. Yamaguchi, Y., Moriyama, T., Ishido, M., Yamada, H.: Four-component scattering model for polarimetric SAR image decomposition. IEEE Trans. Geosci. Remote Sens. **43**, 1699–1706 (2005)

13. Zhou, Y., Wang, H., Xu, F., Jin, Y.: Polarimetric SAR image classification using deep convolutional neural networks. IEEE Geosci. Remote Sens. Lett. **13**, 1935–1939 (2016)
14. Achanta, R., Shaji, A., Smith, K., Lucchi, A., Fua, P., Süsstrunk, S.: SLIC superpixels compared to state-of-the-art superpixel methods. IEEE Trans. Pattern Anal. Mach. Intell. **34**, 2274–2282 (2012)
15. Dataset: AIRSAR, NASA 1991. Retrieved from ASF DAAC on 7 December 2018
16. Yang, F., Gao, W., Xu, B., Yang, J.: Multi-frequency polarimetric SAR classification based on riemannian manifold and simultaneous sparse representation. Remote Sens. **7**(7), 8469–8488 (2015)

Land Cover Classification Using Ensemble Techniques

Hemani I. Parikh[1], Samir B. Patel[1]([⊠]), and Vibha D. Patel[2]

[1] Department of Computer Science and Engineering, School of Technology,
Pandit Deendayal Petroleum Univerisity, Gandhinagar, Gujarat, India
{hemani.pphd17,samir.patel}@sot.pdpu.ac.in
[2] Department of Information Technology,
Vishwakarma Government Engineering College,
Chandkheda, Ahmedabad, Gujarat, India
vibhadp@vgecg.ac.in

Abstract. This paper aims at a Land cover classification of compact polarimetric RISAT-1 data using pixel-based and patch-based input to an ensemble model. In the pixel-based approach, Support Vector Machine (SVM), Multi-Layer Perceptron (MLP) and Random Forest (RF) are ensembled to shape a single voting classifier using the soft voting and majority voting method. In the patch-based approach, Convolutional Neural Network (CNN), Neural Network (NN) and RF are ensembled through soft voting, majority voting and average voting methods. The experimental results show that the ensemble method on patch-based input provides better test accuracy than the pixel-based input. Visual evaluation of classified images indicates that water bodies and urban region are well classified with patch-based input using majority voting method, while the forest class is less misclassified in contrast to soft voting classification on pixel-based input.

Keywords: Ensemble model · Land cover classification · Compact polarimetry

1 Introduction

State of the art deep learning techniques has become popular in the field of remote sensing. Convolutional neural network (CNN) and Deep neural network (DNN) with multiple layers of perceptrons have proved to be highly efficient in image classification. Traditional machine learning techniques like Support vector machine (SVM), Random forest (RF) and Artificial neural network (ANN) have been applied in literature [1, 7, 8, 15, 19, 20] to classify compact polarimetric and full polarimetric data. In microwave imaging, hybrid or ensembled classifiers have been applied to assess the proficiency of various classifiers with majority voting, and soft voting approaches [10, 14]. Image classification task inculcates pixel-based or patch-based input. In patch-based approach, spatial information

© Springer Nature Switzerland AG 2019
B. Deka et al. (Eds.): PReMI 2019, LNCS 11942, pp. 340–349, 2019.
https://doi.org/10.1007/978-3-030-34872-4_38

of neighbouring pixels is utilized to classify a central pixel. In this paper, both the pixel-based and patch-based approaches have been used to classify RISAT-1 data. The main contribution of the paper includes:

1. Use of soft voting and majority voting classification on an ensemble of RF, SVM and ANN classifier on pixel-based values of RISAT-1 decomposed parameters.
2. Use of soft voting, majority voting, and average voting classification on an ensemble of RF, ANN, and CNN for patch-based values of RISAT-1 decomposed parameters.
3. Compare classification results of individual vs. ensembled approaches.

2   Related Work

Advantage of Compact Polarimetric (CP) data is broader coverage, low cost, low complexity and low data rate of the system. Due to such benefits, CP data have gained attraction for research [1]. State of the art work on CP data emphasized the SVM classifier [1,15,19] due to its benefit of high generalization capability.

Alban et al. [8] used an RF classifier for land cover classification and change detection of remotely sensed data. They preferred RF for classification due to its efficiency over high dimensional data [2], the capability to deal with overfitting and the discrimination capability among different plantations. Loosvelt et al. [13] used an RF classifier to assess the uncertainty of classification at pixel level using L-band and C-band data. It was observed that high dimensional data reduces uncertainty.

De et al. [7] used multi-layer perceptron to classify urban from non-urban areas. Turkar et al. [20] achieved high accuracy with ANN from multi-frequency data with different polarizations.

CNN has gained high popularity in image classification due to its inherent capability of feature extraction. It has been widely used for target recognition [5] and scene identification [21]. Polarimetric features have been provided to a CNN classifier and improved classification accuracy in [4,9].

3   Dataset Details

CP data of RISAT-1 in cFRS mode provided by SAC-ISRO has been used for experiments. Data covers Mumbai and surrounding regions. Its center is located at 19.220830° latitude and 72.932787° longitude. The resolution of the dataset is 3.332 m × 2.338 m. The major land cover types include forest, urban, water, wet-land, mangroves etc. Wet-land class integrates wet-land and salt panes due to their similar scattering behaviour. The data was preprocessed and CPR, SPAN and m-chi decomposed images were generated and land cover classification was done using SVM classifier in [19].

In this paper, initial processing has been performed using PolSARPro 5.0 (downloaded from https://earth.esa.int/web/polsarpro/). For experimental

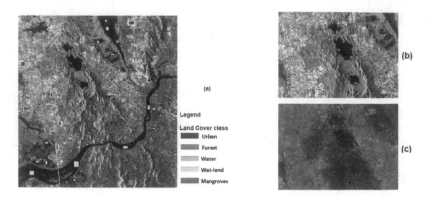

Fig. 1. a. Sample selection from 3×2 multilooked RISAT dataset and its legend b. Test image (subset of 3×2 multi-looked image) c. Google earth imagery of test image

work, the dataset is first multi-looked by 3 looks in azimuth and 2 looks in range direction followed by speckle noise removal using IDAN filter with 50 adaptive neighborhood size. Five decomposition parameters namely odd, even, double bounce, degree of polarization (m) and degree of circularity (χ) were generated by applying Raney's decomposition [16,17] on filtered data. Land feature like water body, bare surfaces etc. corresponds to odd bounce, human-made features like urban, bridges etc. corresponds to double bounce, while vegetation, forest etc. exhibits volume scattering. m and χ are useful for discrimination among similar targets. All these parameters are key indicators for SAR image interpretation for CP or dual polarimetric data and hence been used to train the algorithm. All the five settings were transformed using z-score normalization and then provided to the classifier. Google earth and Bhuvan's thematic map were used to label training and test pixels. The whole image was categorized into five classes – urban, forest, water bodies, mangroves, and wet-land. Among the labeled pixels, 45819 pixels were selected for training, 36654 pixels are selected for testing, and 4580 pixels are selected for validation. Pixel samples have been selected using the SNAP 5.0 tool (ESA-Sentinel Application Platform, http://step.esa. int). Figure 1(a) shows some distinct training and testing samples selected from each region on our dataset. A subset taken from original image is shown in Fig. 1(b) which is used as the test image and is visually assessed using Google Earth imagery.

4   Experimental Work and Results

4.1   Machine Learning Algorithms for Ensemble Classification

Support Vector Machine (SVM). Main aim of SVM [6] is to maximize margin between two hyperplanes or class boundaries. SAR data classification is generally non-linear due to the inclusion of noise. To deal with noisy data, A slack variable ψ is used to determine the error and a regularization parameter C is used, which defines the objective function Eq. 1 for non linear SVM.

$$Minimize \; \|w\|^2/2 + C\Sigma\Psi_i, \; where \; Y_i.(w.x_i + b) >= 1 - \psi_i \tag{1}$$

C is used to control the overfitting. Non-linear data can be efficiently classified using kernel trick that uses kernel function to replace inner product of feature vectors x_i and x_j. Kernel function on feature vectors resembles the dot product of two feature vectors in some high dimensional space. To identify the kernel function with optimal parameter values, grid search method was performed. Radial Basis Function (RBF) kernel resulted to be the best among the other kernels i.e linear, polynomial and sigmoid. RBF function is defined as:

$$k(x_i, x_j) = \frac{\exp{-\gamma\|x_i - x_j\|}}{2\sigma^2} \tag{2}$$

Through grid search, penalty parameter C was selected to be 1000 and γ as 1.0. Higher value of C may reduce margin width but to achieve acceptable accuracy on our data, value of C with 1000 is chosen. Support vector classifier used one vs rest approach.

Random Forest (RF). Random Forest [3] algorithm is applied using both pixel and patch-based approaches. Pixel-based input and patch-based input were applied for soft voting and majority voting classification. 100 random trees were chosen and number of variables for splitting was kept as default (i.e. $\sqrt{n}$; n = number of total features).

Multi Layer Perceptron (MLP). In this paper, Neural Network (NN) with multi-layer perceptron [18] has been applied on both pixel and patch-based inputs. Experimentation using a different number of nodes on single-layer model is shown in Fig. 2. The figure depicts MLP of one hidden layer with 12 nodes should be preferred but soft voting classifier with uniform weights for all three classifiers i.e RF, MLP and SVM, produce degraded result than MLP. So, a model with two hidden layers of 12 nodes in each layer was chosen. ReLU as an activation function and Adam as an optimizer with learning rate 0.001 were selected. It has been observed that neural network with pixel-based input requires more epochs for training as compared to epochs required with patch-based input. We also applied early stopping and L2 regularization with a penalty value of 0.0001.

Patch-based input was also provided to NN and incorporated into soft voting and majority voting classifier. NN is implemented with a single hidden layer of size 128 nodes with ReLU activation and one softmax layer. Each input pixel was evaluated using a 9×9 patch of surrounding pixels. Patch size of 9×9 was selected based on the experiments on 9×9, 12×12, and 15×15 patches on CNN, as shown in Fig. 3. Experimental analysis on single layer NN with 128 nodes, 256 nodes and 512 nodes on each patch size has been performed and

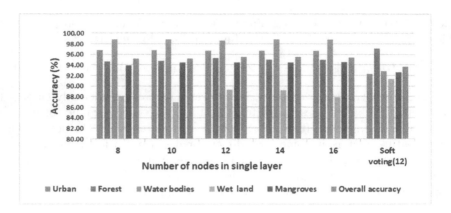

Fig. 2. Accuracy analysis of single layer neural network with different number of nodes on pixel-based input

based on the assessment of overall accuracy of all the classes, number of hidden nodes is selected.

To initialize weights, we have used he normal initializer [11]. Bias value was initialized to zero. Classification results were derived from average of 10 runs of the algorithm having each run with 100 epochs. To reduce overfitting, L2 regularization with penalty value of 0.0001, dropout of 0.8 and early stopping have been applied. Early stopping has been applied by assessing the validation accuracy of 10 consecutive epochs for a stable model. We have selected validation accuracy as a parameter for early stopping because when the model could not generalize any more, it reduces validation accuracy. A model with best validation score is used for prediction. RMSprop with learning rate 0.001 is used as an optimizer function.

Convolutional Neural Network (CNN). CNN [12] algorithm was applied with input size 9×9 for each pixel with 5 channels. Based on experiments on different patch sizes and different layers, as shown in Fig. 3, 9×9 patch size was selected with a single layer of 16 filters. Kernel size of 3×3 and pool size of 2×2 were adopted. To speed up the computations, batch normalization has been applied after convolution layer. Activation function, early stopping criteria, learning rate, and optimizer were the same as in NN with patch-based input. Classification results are derived from an average of 10 runs of algorithm, each with 80 epochs. Similar accuracy for both NN and CNN was observed using RMSprop and Adam. So, RMSprop optimizer have been selected for NN and CNN when patch-based input is provided.

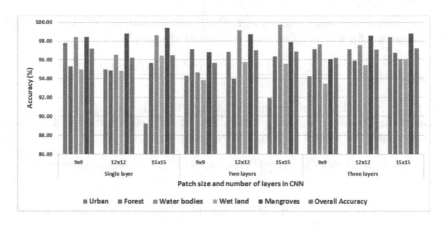

Fig. 3. Accuracy analysis of CNN with different layers and patch sizes

4.2 Pixel-Based Input: Soft Voting and Majority Voting Classification

Three classifiers RF, MLP, and SVM were incorporated in soft voting classifier with pixel values as input. Voting classifier performs soft classification based on predicted probabilities of each class by all three classifiers. It averages weighted probabilities for each class from all the classifiers and predicts class with the highest average probability. Similarly, majority voting was applied on three classifiers using a mode of three predictions. We have also investigated weighted soft voting classification where through grid search approach and 3-fold cross-validation, weights 3, 2 and 1 for RF, MLP and SVM respectively are identified as the best weights. We have restricted to only integer weights to simplify the approach. Weight parameters for three classifiers in grid search are selected as in Table 1. Equation 3 shows how voting classifier assigns a class to each pixel.

Table 1. Weight parameter used in grid search for soft voting approach

Classifier	Weight values						
RF	1	2	3	**3**	1	2	1
MLP	2	1	1	**2**	3	3	1
SVM	3	3	2	**1**	2	1	1

$$\hat{y} = argmax(\varSigma w_k p_k(i)) \tag{3}$$

Here, i represents class and k represents classifier. p_{ik} is the predicted probability of ith class by k^{th} classifier. w_{svm}, w_{RF} and w_{MLP} are the weights assigned to SVM, RF and MLP classifiers respectively and their predicted probabilities of class i as $p_{SVM}i$, $p_{RF}i$ and $p_{MLP}i$. Predicted probability of class i by soft voting classifier for a particular pixel is computed using Eq. 4.

$$p_{sv}(i) = \frac{w_{svm} * p_{svm}(i) + w_{RF} * p_{RF}(i) + w_{MLP} * p_{MLP}(i)}{3} \qquad (4)$$

Here, $p_{sv}(i)$ is predicted soft voting probability of ith class.

Soft voting classifier predicts probabilities of each class using Eq. 4 and predicts class with the highest probability by using Eq. 5.

$$\hat{y} = argmax(p_{sv}(urban), p_{sv}(forest), p_{sv}(water), p_{sv}(wetland), p_{sv}(mangroves))$$
$$(5)$$

$\hat{y}$ is the class predicted by soft voting classifier. Details of producer's accuracy for each class and overall accuracy are given in Table 2.

4.3   Patch-Based Input: Soft Voting and Majority Voting Classification

In this approach, a patch of pixels was provided to the classifier. Patch-based inputs uses spatial correlation among pixels to classify a pixel centered at the patch. In this paper, majority voting classification was performed on RF, NN,

Table 2. Classification accuracy with pixel-based input

Class	Accuracy (%)					
	SVM	RF	MLP	Soft voting (uniform weights)	Soft voting (variable weights)	Majority voting
Urban	96.98	97.17	96.7	96.98	97.02	97.16
Forest	94.14	94.86	95.34	95.7	95.73	96.03
Water bodies	80.85	90.23	98.33	94.36	95.03	92.3
Wet-land	85.89	89.70	87.61	91.02	90.8	89.14
Mangroves	93.38	92.94	94.17	93.72	93.69	93.58
Overall accuracy	90.10	93.01	95.23	94.59	94.74	93.94

Table 3. Classification accuracy with patch-based input

Class	Accuracy (%)					
	CNN	NN	RF	Average voting	Majority voting	Soft voting
Urban	97.49	94.23	99.33	93.72	97.92	97.54
Forest	96.75	94.41	97.17	96.49	97.41	97.04
Water bodies	97.80	98.72	98.08	96.93	99.31	99.42
Wet-land	92.67	93.74	90.33	85.28	94.58	94.86
Mangroves	98.29	98.22	98.38	97.27	98.51	98.72
Overall accuracy	97.07	96.33	97.29	95.24	97.94	97.89

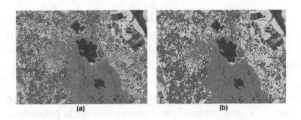

Fig. 4. Classification results of test image (a) Soft voting classified on pixel-based input (variable weights) (b) Majority voting classified on patch-based input

and CNN classifiers where a mode of three predictions is considered as the prediction of majority voting classifier. We have also assessed soft voting classifier on patch-based input. Average voting classification is also assessed where an average of three predictions is taken as the final prediction. The test image was predicted using the best model from each classifier and then the majority voting was performed. Majority voting gives visually better results than any individual classifier. Overall Accuracy and producer's accuracy computed on test data by all the classifiers with the patch-based approach is provided in Table 3.

Figure 4 shows classification result of test image (Fig. 1(b)) generated using soft voting classifier with variable (best) weights on pixel-based input and majority voting classifier on patch-based input.

5   Discussions on Results

In this paper, number of train and test samples were selected to oversee overfitting and misclassification. In pixel-based approach, with an individual classifier, MLP beats the other two i.e. RF and SVM in test accuracy. Thus, soft voting classification with uniform weights and with variable weights have minor contrast in accuracy, therefore gave comparative yield when applied on test image. Water class is satisfactorily classified using MLP as depicted in Table 2, however RF and SVM incorporated water with wet-land in some pixels because of their comparable scattering conduct and thus has reduced classification accuracy in water class. Due to comparative higher accuracy of urban and wet-land class, higher weightage was given to RF during grid search with 3-fold cross-validation. When variable weights 3, 2, and 1 were assigned to RF, MLP and SVM respectively, the accuracy of water class is enhanced slightly than the accuracy achieved using classifier with uniform weights. In RF, it has been resolved that in classification, volume scattering is the most significant component among the five parameters. Majority voting results on pixel-based input could not outperform soft voting results.

On patch-based input, soft and majority voting yield similar results with little improvement in majority voting results. From Fig. 4, it can be observed that majority voting classifier on patch-based input give improved classification of forest as contrast to soft voting classifier on pixel-based input. Misclassification in forest class occurred due to shadows in that region. Urban is also better identified using the majority voting classifier. RF classifies urban better than the other classifiers. Due to this reason, though CNN has a higher influence on majority voting predictions, its accuracy is slightly lesser than RF. From the accuracy assessment of individual classifiers in the ensemble approach, it has been observed that NN has distinctive behaviour with pixel-based and patch-based input. With patch-based input, the NN classifier misclassifies some pixels of urban class into mangroves because of the vegetation part inside urban. This is depicted by reduced accuracy of urban in Table 3. RF gives enhanced results when patch-based input is provided, but some pixels of wet-land class is misclassified into the water body. CNN gives great overall outcomes for all classes. The test image was also classified using individual classifiers participated in the ensemble model. Considering urban region, SVM gave the most astounding number of urban pixels with pixel-based. While, with patch-based input, CNN characterized the most noteworthy number of urban pixels.

6   Conclusion

In this paper, ensemble classification using soft voting, majority voting, and average voting approaches have been investigated. Soft voting and majority voting classifier have been evaluated using pixel-based as well as patch-based input while average voting classifier has been evaluated using only patch-based input. Enhancement in classification accuracy has been observed with patch-based input when contrasted with pixel-based input, due to the use of spatially related information by patch-based input. On patches, the majority voting classifier with equally strong classifiers resulted in improvement in classification accuracy. To assess the strength of these approaches, comparison with the unsupervised approach can be done in the future work.

Acknowledgement. Authors are thankful to SAC, ISRO Ahmedabad, for providing RISAT-1 data used in this paper. Authors also appreciate anonymous reviewers for their significant suggestions.

References

1. Aghabalaei, A., Maghsoudi, Y., Ebadi, H.: Forest classification using extracted polsar features from compact polarimetry data. Adv. Space Res. **57**(9), 1939–1950 (2016)
2. Belgiu, M., Drăguţ, L.: Random forest in remote sensing: a review of applications and future directions. ISPRS J. Photogramm. Remote Sens. **114**, 24–31 (2016)
3. Breiman, L.: Random forests. Mach. Learn. **45**(1), 5–32 (2001)
4. Chen, S.W., Tao, C.S.: PolSAR image classification using polarimetric-feature-driven deep convolutional neural network. IEEE Geosci. Remote Sens. Lett. **15**(4), 627–631 (2018)

5. Chen, S., Wang, H., Xu, F., Jin, Y.Q.: Target classification using the deep convolutional networks for SAR images. IEEE Trans. Geosci. Remote Sens. **54**(8), 4806–4817 (2016)
6. Cortes, C., Vapnik, V.: Support-vector networks. Mach. Learn. **20**(3), 273–297 (1995)
7. De, S., Bhattacharya, A.: Urban classification using PolSAR data and deep learning. In: 2015 IEEE International Geoscience and Remote Sensing Symposium (IGARSS), pp. 353–356. IEEE (2015)
8. De Alban, J., Connette, G., Oswald, P., Webb, E.: Combined landsat and L-band SAR data improves land cover classification and change detection in dynamic tropical landscapes. Remote Sens. **10**(2), 306 (2018)
9. Gao, F., Huang, T., Wang, J., Sun, J., Hussain, A., Yang, E.: Dual-branch deep convolution neural network for polarimetric SAR image classification. Appl. Sci. **7**(5), 447 (2017)
10. Hao, P., Wang, L., Niu, Z.: Comparison of hybrid classifiers for crop classification using normalized difference vegetation index time series: a case study for major crops in North Xinjiang, China. PLoS ONE **10**(9), e0137748 (2015)
11. He, K., Zhang, X., Ren, S., Sun, J.: Delving deep into rectifiers: surpassing human-level performance on imagenet classification. In: Proceedings of the IEEE International Conference on Computer Vision, pp. 1026–1034 (2015)
12. LeCun, Y., et al.: Backpropagation applied to handwritten zip code recognition. Neural Comput. **1**(4), 541–551 (1989)
13. Loosvelt, L., et al.: Random forests as a tool for estimating uncertainty at pixel-level in SAR image classification. Int. J. Appl. Earth Obs. Geoinf. **19**, 173–184 (2012)
14. Ma, X., Shen, H., Yang, J., Zhang, L., Li, P.: Polarimetric-spatial classification of SAR images based on the fusion of multiple classifiers. IEEE J. Sel. Top. Appl. Earth Obs. Remote Sens. **7**(3), 961–971 (2014)
15. Ohki, M., Shimada, M.: Large-area land use and land cover classification with quad, compact, and dual polarization SAR data by PALSAR-2. IEEE Trans. Geosci. Remote Sens. **56**(9), 5550–5557 (2018)
16. Raney, R.K.: Dual-polarized SAR and stokes parameters. IEEE Geosci. Remote Sens. Lett. **3**(3), 317–319 (2006)
17. Raney, R.K., Cahill, J.T., Patterson, G.W., Bussey, D.B.J.: The m-chi decomposition of hybrid dual-polarimetric radar data with application to lunar craters. J. Geophys. Res.: Planets **117**(E12) (2012)
18. Rumelhart, D.E., Hinton, G.E., Williams, R.J.: Learning internal representations by error propagation. Technical report, California Univ San Diego La Jolla Inst for Cognitive Science (1985)
19. Turkar, V., De, S., Ponnurangam, G., Deo, R., Rao, Y., Das, A.: Classification of RISAT-1 hybrid polarimetric data for various land features. In: Conference Proceedings of 2013 Asia-Pacific Conference on Synthetic Aperture Radar (APSAR), pp. 494–497. IEEE (2013)
20. Turkar, V., Deo, R., Rao, Y., Mohan, S., Das, A.: Classification accuracy of multi-frequency and multi-polarization SAR images for various land covers. IEEE J. Sel. Top. Appl. Earth Obs. Remote Sens. **5**(3), 936–941 (2012)
21. Wu, W., Li, H., Zhang, L., Li, X., Guo, H.: High-resolution polsar scene classification with pretrained deep convnets and manifold polarimetric parameters. IEEE Trans. Geosci. Remote Sens. **56**(10), 6159–6168 (2018)

A Supervised Band Selection Method for Hyperspectral Images Based on Information Gain Ratio and Clustering

Sonia Sarmah[1,2(✉)] and Sanjib K. Kalita[1]

[1] Gauhati University, Guwahati, Assam, India
sarmahsonia07@gmail.com, sanjib959@gauhati.ac.in
[2] Assam Don Bosco University, Guwahati, Assam, India

Abstract. The high spectral dimension of hyperspectral images increases the computational complexity and processing time requirement for analysing such images. As a consequence, dimension reduction is an essential step to be performed prior to the processing of hyperspectral images. In this work we have presented a dimension reduction technique for hyperspectral images using a supervised band selection method. Initially significance of each band were evaluated by calculating information gain ratio. Then, by applying clustering technique, similar bands were grouped into k clusters, from each of which the band with the maximum information gain ratio value was selected as the representative band. In a subsequent band pruning step, this set of representative bands were further reduced by eliminating the bands with low information gain ratio. The results of the experiment showed that adequate accuracy could be achieved with relatively low number of bands selected by the proposed method.

Keywords: Hyperspectral · Supervised band selection · Clustering · Information gain ratio · Band pruning

1 Introduction

Hyperspectral images are remotely sensed images containing hundreds of spectral bands. Though, this increased number of bands helps in identifying the objects uniquely by providing significant amount of information, at the same time it also increases the computational complexity for processing such images. Moreover, as the bands are almost continuous, the neighbouring bands are highly correlated which leads to redundancy [2]. Besides, due to the presence of only limited number of labelled samples, such images suffer from curse of dimensionality which leads to well-known Hughes phenomena [6]. To avoid these issues, dimensionality reduction is an essential pre-processing step for analysis of hyperspectral images. Dimension reduction can be achieved by performing either features extraction or feature selection techniques. Principal component analysis, independent component analysis (ICA) and autoencoder etc [3,4,11] are examples of

© Springer Nature Switzerland AG 2019
B. Deka et al. (Eds.): PReMI 2019, LNCS 11942, pp. 350–358, 2019.
https://doi.org/10.1007/978-3-030-34872-4_39

some widely used feature extraction techniques. Though feature extraction techniques result in discriminating set of features, the physical interpretation of the spectral bands get compromised in such techniques due to the linear or non-linear transformation applied over the original feature set. In contrast, feature selection methods select a sub set of highly discriminating features from the original set of features without using any transformation. Thus, band selection methods not only preserve the original interpretation of the spectral bands but also lessen the storage and processing time requirement for hyperspectral images. The feature selection methods can broadly be categorized into supervised and unsupervised methods. For supervised methods, additional information such as class signature, ground truth etc. is needed to be provided along with the dataset. Most of the supervised band-selection algorithms select the subset of bands, which results in the highest class separability. Regression tree, instance based methods are some of the techniques used for supervised band selection [2].

This work presents a supervised band selection algorithm which requires labelled samples as additional information. These labelled samples have been used for calculating information gain ratio of each spectral band of the input hyperspectral image. Initially, correlation among the spectral bands was utilized to cluster the highly correlated bands together and then form each cluster, the band with the highest information gain ratio was selected. Further reduction was carried out by applying band pruning method, which is a second level of filtering. The method is discussed in detail in Sect. 2. Experimental results have shown that the dataset with only the selected bands by the proposed supervised method gave comparable classification accuracy as that achieved by the original high dimensional dataset.

2   Proposed Methodology

This section provides a brief outline of the related techniques followed by a detail description of the proposed methodology.

2.1   Background

Information Gain: Let $A = (a_1, a_2, \cdots, a_m, C)$ be the feature vector where a_i represnts the ith feature and C stands for the class label. Let $X = \{x_1, x_2, \cdots, x_n\}$ be the set of training samples such that $\|X\| = n$. Each element $x_i \in X$ is an ordered tuple and can be expressed as $x_i = (x_{i1}, x_{i2}, \cdots, x_{im}, c_i)$, where x_{ij} is the value of the sample x_i for the feature $a_j \in A$ and c_i is the corresponding class label. The function $values(X, a_j)$ denotes the set of all possible values of the feature a_j in the given training set X. For each value $v \in values(X, a_j)$ we can define a subset S_v of X such that [9]-

$$S_v = \{x_i | x_i \in X \wedge x_{ij} = v \wedge v \in values(X, a_j)\} \tag{1}$$

Using Eq. 1 information gain for a feature a_j for the given training set X can be calculated as-

$$IG(X, a_j) = H(X) - \sum_{v \in values(X, a_j)} \frac{\|S_v\|}{n} \cdot H(S_v) \tag{2}$$

where, $H(X)$ is the entropy of the training set X and $H(S_v)$ is the entropy of the subset S_v extracted using Eq. 1 respectively.

Entropy of a given random variable measures the average information content of the variable [9]. Mathematically, entropy for a random variable Y with $p(y)$ as probability distribution can be defined using the following equation

$$H(Y) = -\sum_{y \in Y} p(y) log p(y) \tag{3}$$

Using Eq. 3 entropy for the training set X and the subset S_v may be calculated as

$$H(X) = -\sum_{c \in values(X, C)} p(c) log p(c) \tag{4}$$

$$H(S_v) = -\sum_{c \in values(S_v, C)} p(c) log p(c) \tag{5}$$

Information gain is widely used as a measure of relevance of a feature. Higher is the information gain of a feature more relevant it is. However, it is biased towards the features having many distinct values.

Intrinsic Gain: Intrinsic gain of a feature a_j for a training set X may be defined as [9]

$$IV(X, a_j) = \sum_{v \in values(X, a_j)} \frac{\|S_v\|}{n} \cdot log_2 \frac{\|S_v\|}{n} \tag{6}$$

Information Gain Ratio: For a given training set X, the ratio of information gain to intrinsic information of an a feature a_j is referred to as information gain ratio.

$$IGR(X, a_j) = \frac{IG(X, a_j)}{IV(X, a_j)} \tag{7}$$

Correlation: Correlation is the measure of statistical relationship between two variables. In statistics, Pearson's product moment correlation between two random samples X and Y can be calculated as [10]

$$\rho(X, Y) = \frac{Cov(X, Y)}{\sigma_X \sigma_Y} \tag{8}$$

where, $Cov(X, Y)$ is the co-variance of X and Y; σ_X and σ_Y are the standard deviations of X and Y respectively. The correlation coefficient can take values in the range $+1$ to -1. Positive value indicates linear relationship of the samples whereas a negative value specifies the inverse linear relationship between the samples. High magnitude of correlation coefficient implies that the variables are strongly related and low magnitude indicates that the variables are almost independent.

2.2 Description of the Proposed Method

In the proposed supervised band selection method, we have considered each labelled pixel of the input hyperspectral image as a training sample. All the spectral bands along with the class label represent the set of features. So, if the input hyperspectral image contains 'n' labelled pixels and 'd' spectral bands then the set of training samples X can be represented as-

$$X = \begin{bmatrix} x_{11} & x_{12} & \cdots & x_{1d} & c_1 \\ x_{21} & x_{22} & \cdots & x_{2d} & c_2 \\ \vdots & \vdots & \ddots & \vdots & \vdots \\ x_{n1} & x_{n2} & \cdots & x_{nd} & c_n \end{bmatrix} \tag{9}$$

where, x_{ij} represents the reflectance value of the pixel x_i at spectral band b_j and c_i is the corresponding class label.

Using Eqs. 1 to 8, we then calculated the information gain ratio of every spectral band in the input hyperspectral image with respect to X. The calculated information gain ratios of the spectral bands can be represented as a row vector using the following notation-

$$IGR_{bands} = [IGR_1, IGR_2 \ldots \ldots, IGR_d] \tag{10}$$

where, IGR_i is the information gain ratio of the ith spectral band with respect to X. In the next phase of the proposed method, the similar bands were grouped together using clustering technique. For our experiment we used two different clustering techniques namely hierarchical clustering and spectral clustering. The two methods were evaluated separately and obtained results are discussed in Sect. 4. The clustering phase resulted in k number of clusters such that $k <= d$. Then from each cluster we chose the spectral band having the highest information gain ratio and added it to the set of representative bands. Then, a classifier was trained and tested using samples consisting of only the k representative bands. Let's assume that the accuracy achieved was ACC_k. We further reduced the set of representative bands by applying band pruning method to get the final set of selected bands. In band pruning phase, first the set of representative bands were arranged in descending order of their information gain ratio. Then from this sorted list of 'k' bands only the leading 'm' bands were selected such that $ACC_m \geq ACC_k$.

3 Experimental Set Up

This section presents a brief overview of the experimental set up used for analysis of the proposed method.

3.1 Dataset Description

For carrying out the experiments, two datasets namely- Indian Pines and Kennedy Space Centre (KSC) were used. Brief description of the datasets is presented below.

Indian Pines Dataset. was acquired by AVIRIS (Airborne Visible/Infrared Imaging Spectrometer) over Indian Pines test center, North-western Indiana. This original scene consists of 224 bands from which a corrected image of 200 bands was constructed by eliminating 24 noisy and water absorption bands. The image comprises of 21025 pixels (10249 labelled and 10776 unlabelled) and 16 classes.

Kennedy Space Centre Dataset. was captured by AVIRIS Sensor over Kennedy Space Centre, Florida. The original dataset contained 224 bands from which only 176 bands were retained in the corrected dataset. The dataset contains total 314368 pixels out of which 5211 are labelled. The image contains total 13 classes.

3.2 Clustering Techniques Used

For clustering the similar bands of the input hyperspectral image, we have used two clustering techniques namely –**Agglomerative Hierarchical Clustering (HC)** and **Spectral Clustering (SC)**. In agglomerative hierarchical clustering technique, initially each individual band image is treated as a cluster then by applying "bottom-up" approach, in every iteration the pair of clusters exhibiting highest similarity is combined together unless there remains only one cluster [7]. For our experiment we have used correlation distance between the bands as the dissimilarity metric, which can be defines as $|1 - \rho(b_i, b_j)|$, where, b_i and b_j are two bands images. Average linkage has been used as the merging criterion for the clusters. The output of the hierarchical clustering can be depicted with the help of a dendogram tree. The dendogram may be cut either by specifying the desired height or desired number of clusters to obtain the clusters. For our experiment, desired number of clusters, k, has been used to cut the dendogram.

In spectral clustering, considering the bands as the set of vertices, first a similarity matrix among the vertices is generated. From this similarity matrix, in the subsequent step the graph Laplacian is calculated. Then Eigen values and Eigen vectors of the Laplacian matrix are calculated by performing Eigen analysis. If k number of features are to be selected then only the first k Eigen vectors are computed which are treated as the new features. Finally using k-mean clustering these points are grouped into k clusters [8].

3.3 Classifiers

For evaluating the worth of the selected bands by the proposed algorithm we have used two classifiers namely: K- Nearest Neighbour (KNN) and multi class Support Vector Machine (SVM) opting one-against-one scheme [1,5]. For SVM classifier we have used the polynomial kernel with parameters $C = 46$ and $d = 5$. For KNN classifier we have used the neighbourhood parameter equal to 101(square root of number of samples) and Euclidean distance as the distance metric. For training and testing the classifiers, the labelled pixels from the dataset were first extracted to be used as input set of samples. The reflectance values of a pixel across the selected bands were considered as the associated feature vector. Cross validation with 10-fold stratified partitions was performed to estimate the classification accuracy. Section 4 presents the detailed classification results achieved by the classifiers using different number of selected bands. The classification accuracy was presented as the mean accuracy of the 10 folds and standard deviation. Accuracy of the ith fold can be calculated using the following formula-

$$AC_i = \frac{TP_i}{Total_i} \tag{11}$$

where, TP_i and $Total_i$ designate the total number of true positives (i.e., the number of cases where both actual and predicted class labels are same) and total number of testing samples respectively in the ith fold. For 10 folds the mean accuracy and standard deviations are defined as-

$$ACC = \frac{\sum_{i=0}^{10} AC_i}{10} \tag{12}$$

$$stdv = \sqrt{\frac{\sum_i (AC_i - ACC)^2}{10}} \tag{13}$$

4   Results and Discussion

At first information gain ratio of each band of the input hyperspectral image was calculated. Figure 1a and b show the information gain ratio of each spectral band of Indian Pines and KSC dataset respectively. For each dataset experiments were performed with different number of clusters ranging from 5 to 40 in a step of 5. Tables 1 and 2 summarizes the results obtained by the classifiers for each clustering technique. The highest achieved accuracy is presented in bold.

For KSC dataset, the SVM classifier could achieve the highest accuracy using only 35 representative bands for both the clustering techniques. However, the accuracy obtained by using SC+SVM (spectral clustering and SVM) was 0.94 which is much superior to the highest achieved accuracy of 0.89 by HC+SVM(hierarchical clustering and SVM) technique. After the application of band pruning method, similar accuracy could be achieved using only 25 and 31 selected bands respectively for SC+SVM and HC+SVM techniques. Similarly, for HC+KNN (hierarchical clustering and KNN) and SC+KNN (spectral clustering and KNN) techniques highest accuracy of 0.87 and 0.93 respectively were achieved using only 30 representative bands initially, which were further refined to 27 and 24 bands respectively in the band pruning phase. The accuracy achieved by all the 176 bands in KSC dataset is 0.94 for the SVM classifier and 0.93 for KNN classifier. So, it can be claimed that using less than 18% of the total bands comparable accuracy could be achieved to that of using all bands. Table 3 present the final number of selected bands for different techniques. Here, we have shown the results of application of band pruning technique only to the set of representative bands which resulted in the highest accuracy for the respective techniques.

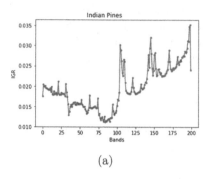
(a)

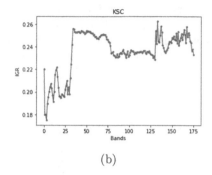
(b)

Fig. 1. Information gain ratios for the bands of -(a) Indian Pines dataset and (b) KSC dataset

For the Indian Pines dataset applying HC+SVM and SC+SVM techniques highest accuracy of 0.82 and 0.81 could be achieved using 40 representative bands. The band pruning technique further reduced the number of bands to 37 and 36 respectively by eliminating the insignificant bands. However, when the KNN classifier was used highest accuracy degraded. Accuracy score of only 0.76 and 0.80 were achieved using 35 and 40 initial representative bands respectively, which later were subsequently reduced to 25 and 34 selected bands. Using all the bands accuracy score of 0.87 and 0.79 were achieved.

Table 1. Experimental results for KSC dataset

Number of bands	Accuracy SVM ($ACC \pm stdv$)		Accuracy KNN ($ACC \pm stdv$)	
	HC	SC	HC	SC
5	$0.68 \pm .01$	$0.62 \pm .02$	$0.69 \pm .01$	$0.63 \pm .01$
10	$0.84 \pm .02$	$0.81 \pm .02$	$0.82 \pm .02$	$0.78 \pm .00$
15	$0.86 \pm .01$	$0.89 \pm .02$	$0.83 \pm .01$	$0.88 \pm .01$
20	$0.86 \pm .01$	$0.93 \pm .01$	$0.83 \pm .01$	$0.92 \pm .01$
25	$0.87 \pm .02$	$0.93 \pm .01$	$0.84 \pm .02$	$0.91 \pm .02$
30	$0.88 \pm .01$	$0.93 \pm .00$	$\mathbf{0.87 \pm .01}$	$\mathbf{0.93 \pm .02}$
35	$\mathbf{0.89 \pm .01}$	$\mathbf{0.94 \pm .01}$	$0.87 \pm .00$	$0.92 \pm .01$
40	$0.89 \pm .00$	$0.93 \pm .01$	$0.87 \pm .01$	$0.92 \pm .00$

Table 2. Experimental results of Indian Pines dataset

Number of bands	Accuracy SVM ($ACC \pm stdv$)		Accuracy KNN ($ACC \pm stdv$)	
	HC	SC	HC	SC
5	$0.57 \pm .02$	$0.54 \pm .02$	$0.71 \pm .01$	$0.58 \pm .01$
10	$0.65 \pm .01$	$0.56 \pm .02$	$0.73 \pm .02$	$0.65 \pm .00$
15	$0.72 \pm .02$	$0.61 \pm .02$	$0.73 \pm .01$	$0.72 \pm .01$
20	$0.74 \pm .01$	$0.68 \pm .01$	$0.73 \pm .01$	$0.74 \pm .01$
25	$0.78 \pm .01$	$0.71 \pm .01$	$0.73 \pm .02$	$0.74 \pm .02$
30	$0.80 \pm .01$	$0.75 \pm .00$	$0.75 \pm .01$	$0.77 \pm .02$
35	$0.81 \pm .01$	$0.79 \pm .01$	$\mathbf{0.76 \pm .00}$	$0.78 \pm .01$
40	$\mathbf{0.82 \pm .01}$	$\mathbf{0.81 \pm .01}$	$0.75 \pm .01$	$\mathbf{0.80 \pm .00}$

Table 3. Final number of selected bands after band pruning

Dataset	Technique	Highest accuracy achieved	#Representative bands	#Selected bands
KSC	HC+SVM	$0.89 \pm .01$	35	31
	SC+SVM	$0.94 \pm .01$	35	25
	HC+KNN	$0.87 \pm .01$	30	27
	SC+KNN	$0.93 \pm .02$	30	24
	SVM	$0.96 \pm .01$	176(all bands)	–
	KNN	$0.93 \pm .01$	176(all bands)	–
Indian Pines dataset	HC+SVM	$.82 \pm .01$	40	37
	SC+SVM	$0.81 \pm .01$	40	36
	HC+KNN	$0.76 \pm .00$	35	25
	SC+KNN	$0.80 \pm .00$	40	34
	SVM	$0.87 \pm .02$	200(all bands)	–
	KNN	$0.79 \pm .00$	200(all bands)	–

From obtained results, it may be observed that in most cases SC+SVM technique gave better performance than other techniques. Individually also, the Spectral clustering and the SVM classifier performed much better than their counterparts.

5   Conclusion

In this work, an information gain ratio based supervised band selection technique for hyperspectral images has been presented. Experimental analysis of the proposed algorithm over two hyperspectral images revealed that with a very small percentage of selected by the proposed algorithm satisfactory accuracy score could be achieved. In the first round, representative bands, one from each cluster of similar bands, were selected based on highest information gain ratio. A second round of reduction was applied to this set of representative bands to ensure the elimination of the insignificant bands which might have been selected in the first round. This step further reduced the size of the input image and made it suitable for processing and storing.

References

1. Bailey, T.: A note on distance-weighted k-nearest neighbor rules. Trans. Syst. Man Cybern. **8**, 311–313 (1978)
2. Bajcsy, P., Groves, P.: Methodology for hyperspectral band selection. Photogram. Eng. Remote Sens. **70**(7), 793–802 (2004)
3. Chang, C.I., Du, Q., Sun, T.L., Althouse, M.L.: A joint band prioritization and band-decorrelation approach to band selection for hyperspectral image classification. IEEE Trans. Geosci. Remote Sens. **37**(6), 2631–2641 (1999)
4. Du, H., Qi, H., Wang, X., Ramanath, R., Snyder, W.E.: Band selection using independent component analysis for hyperspectral image processing. In: 2003 Proceedings of 32nd Applied Imagery Pattern Recognition Workshop, pp. 93–98. IEEE (2003)
5. He, X., Wang, Z., Jin, C., Zheng, Y., Xue, X.: A simplified multi-class support vector machine with reduced dual optimization. Pattern Recogn. Lett. **33**(1), 71–82 (2012)
6. Hughes, G.: On the mean accuracy of statistical pattern recognizers. IEEE Trans. Inf. Theory **14**(1), 55–63 (1968)
7. Milligan, G.W., Cooper, M.C.: A study of the comparability of external criteria for hierarchical cluster analysis. Multivariate Behav. Res. **21**(4), 441–458 (1986)
8. Ng, A.Y., Jordan, M.I., Weiss, Y.: On spectral clustering: analysis and an algorithm. In: Advances in Neural Information Processing Systems, pp. 849–856 (2002)
9. Quinlan, J.R.: Induction of decision trees. Mach. Learn. **1**(1), 81–106 (1986)
10. Shen, D., Lu, Z., et al.: Computation of correlation coefficient and its confidence interval in SAS. SUGI: Paper, pp. 170–31 (2006)
11. Xing, C., Ma, L., Yang, X.: Stacked denoise autoencoder based feature extraction and classification for hyperspectral images. J. Sens. **2016**, 10 (2016)

Remote Sensing Image Retrieval via Symmetric Normal Inverse Gaussian Modeling of Nonsubsampled Shearlet Transform Coefficients

Hilly Gohain Baruah[✉], Vijay Kumar Nath, and Deepika Hazarika

Department of Electronics and Communication Engineering, Tezpur University, Tezpur, Sonitpur, Assam, India
hgbaruah1990@gmail.com, {vknath,deepika}@tezu.ernet.in

Abstract. This paper presents a nonsubsampled shearlet transform (NSST) based remote sensing (RS) image retrieval technique where the NSST subband coefficients are modeled with symmetric normal inverse Gaussian (SNIG) distribution. The NSST is popular for its very good directional selectivity and shift invariance. The SNIG is a four parameter distribution which can describe a wide range of heavy tailed distributions. Through Kolmogorov-Smirnov (KS) goodness of fit test, it is shown that the SNIG best approximates the shape of distribution of detail NSST coefficients. For Maximum Likelihood (ML) estimation of SNIG parameters, an Expectation-Maximization (EM) type of algorithm is used. The RS images are first decomposed with NSST and the feature vector is formed with the estimated SNIG parameters from detail NSST coefficients. Here, the UC Merced land use (UCM) and the High-resolution satellite scene (HRS) datasets are used for the experiments conducted. The results of the proposed method reveal superior performance, when compared with some well known techniques.

Keywords: Remote sensing · Image retrieval · Nonsubsampled shearlet transform · Statistical model · Symmetric normal inverse Gaussian

1 Introduction

The changes occurring in remote sensing (RS) imagery over the past several decades are dramatic. It has affected the image's spatial resolution and also the acquisition rate [3] which has resulted into large volume RS image datasets. The retrieval of the image of interest from these large databases is a challenging task. Earlier remote sensing image retrieval (RSIR) approaches were based on manual tags. These manual tag based RSIR approaches, while dealing with large databases faced trouble as the process of creating manual tags are quite difficult. To compensate this, content based RSIR techniques were developed. The

© Springer Nature Switzerland AG 2019
B. Deka et al. (Eds.): PReMI 2019, LNCS 11942, pp. 359–368, 2019.
https://doi.org/10.1007/978-3-030-34872-4_40

framework of content based image retrieval (CBIR) consists of two fundamental components, feature extraction and similarity measurement [6]. In feature extraction the important features from images are extracted and in similarity measurement the mathematical distance between query image and images of the database are calculated. The effectiveness of the features extracted has high impact on the performance of CBIR.

Traditional RSIR approaches involve extraction of local or global features from images using hand crafted feature descriptors. The global features are extracted from the whole image. Local binary pattern (LBP) [19,20], Gabor filters [4,5], histogram of oriented gradients (HoG) [12], morphological operator granulometry [9], these type of texture features can be used for RS image retrieval. Local hand crafted feature descriptors are extracted locally from the points of interest. The local features are robust as they are unaffected by lighting conditions and viewing angles. Vector of locally aggregated descriptors (VLAD) and fisher vector (FV) [15], bag of visual words (BOVW) [21] are few feature descriptors based on which different RSIR approaches have been proposed. RSIR system can also be designed with convolutional neural network (CNN) [25] and it is observed that CNN model based features fetch significant semantic information.

This paper presents a RS image retrieval technique by modeling the NSST coefficients with SNIG distribution. Through further investigation, it has been shown that the heavy tail non-Gaussian nature of distributions of detail NSST coefficients of RS images can be best approximated with SNIG. The estimated SNIG parameters from detail NSST coefficients constitute the texture feature.

There are a total of five sections in this paper. NSST is discussed in Sect. 2. In third section, proposed methodology is discussed. A brief discussion of experimental results is presented in Sect. 4. Conclusions are presented in the fifth section of the paper.

2   Nonsubsampled Shearlet Transform (NSST)

Wavelet transform is used for image representation widely as it can sparsely represent one dimensional signals. But the inability to capture the higher dimensional singularities effectively caused introduction of other multidimensional geometric tools. Shearlet transform (ST) [7] provides high directional sensitivity, spatial localization and optimal sparse representation of singularity of higher dimensional signals compared to other multigeometric analysis tool (MGA) such as wavelet, curvelet and contourlet [8]. Continuous ST of an signal p can be expressed as follows [8]

$$S_p(ast) = <p, \psi_{ast}>; \quad a \in R^+, t \in R^2, s \in R \tag{1}$$

a, t and s stands for scale, translation and shape/shearing factors respectively. ψ_{ast} is the shearlet function, represented as [8]

$$\psi_{ast}(f) = a^{-3/4}\psi(M_{as}^{-1}(f - t)) \tag{2}$$

where $M_{as} = \begin{pmatrix} a & \sqrt{as} \\ 0 & \sqrt{a} \end{pmatrix}$. By sampling the continuous ST with appropriate discrete set, the discrete ST can be obtained [7]. The discrete form of ST can be represented as

$$S_p(jlk) = \;<p, \psi_{jlk}>; \;\; j, l \in Z, k \in Z^2 \tag{3}$$

$$\psi_{jlk}(f) = |detA|^{j/2}\psi(B^j A^l f - k) \tag{4}$$

where $B = \begin{pmatrix} 1 & 1 \\ 0 & 1 \end{pmatrix}$, $A = \begin{pmatrix} 4 & 0 \\ 0 & 2 \end{pmatrix}$ are the shear and anisotropic dilation matrix respectively.

NSST is the multiscale, multidirectional and shift invariant version of ST proposed by Easley et al. [7]. NSST is the integration of several shearing filters and nonsubsampled laplacian pyramid filter. The non-existence of subsampling makes it shift invariant [8].

3   Proposed Methodology

The symmetric normal inverse Gaussian distribution (SNIG) has the ability to model the heavy tail nature of NSST coefficients [24]. This is a four parameter closed form flexible distribution [13].

Considering the normal distribution $N(\mu_m, \sigma^2)$ with variance σ^2, mean μ_m and Inverse Gaussian distribution $I_G(\kappa, \delta)$ with variance $V(y) = \dfrac{\delta}{\kappa^3}$ and $E(y) = \dfrac{\delta}{\kappa}$ as mean, the PDF of IG can be expressed as [13],

$$f_{IG}(y) = \frac{\delta}{\sqrt{2\pi}} e^{(\delta\kappa)} y^{-3/2} e^{-1/2(\frac{\delta^2}{y}+\kappa^2 y)} \tag{5}$$

If $N(\mu_m + \beta y, y)$ denotes the conditional distribution of c given y and if y obeys itself as an $I_G(\kappa, \delta)$, then the combined distribution that results is the $SNIG(\alpha, \beta, \mu_m, \delta)$ with $\alpha = \sqrt{\kappa^2 + \beta^2}$. The PDF of $SNIG$ as [13]

$$h(c; \alpha, \beta, \mu_m, \delta) = \frac{\alpha}{\pi} e^{(\delta\sqrt{\alpha^2-\beta^2}-\beta\mu_m)} \phi(c)^{-1/2} K_1(\delta\alpha\phi(c)^{1/2}) e^{(\beta c)} \tag{6}$$

with

$$\phi(c) = 1 + [\frac{c - \mu_m}{\delta}]^2 \tag{7}$$

The $K_b(c)$ represents the 3rd kind modified Bessel function of order b.

In this work we consider Kolmogorov-Smirnov (KS) goodness of fit test for justifying the use of SNIG in modeling the detail NSST coefficients. An EM type of algorithm is used for ML estimation of SNIG parameters [13]. KS test statistics is utilized to measure distance between empirical cumulative distribution function (ECDF) for a given dataset and the given model cumulative distribution (CDF) function [10]. The KS test is defined as [10,11,17,18]

$$KS_{res} = \max_{c \in N} |F(c) - \tilde{F}(c)| \tag{8}$$

where, $F(c)$ and $\tilde{F}(c)$ represents model CDF and ECDF respectively. The distribution that gives smallest value of KS_{res}, is considered as the best fit for the data. The KS test statistics for various subbands and various images are listed in Table 1. Table 1 and Fig. 1 clearly show the suitability of using SNIG for approximating the NSST coefficients statistics instead of other PDF such as Laplacian and BKF.

Table 1. Average KS statistics for both UCM and HRS dataset

Dataset	PDF	Level 1 Subband								Level 2 Subband				Level 3 Subband	
		1	2	3	4	5	6	7	8	1	2	3	4	1	2
UCM	Laplacian	0.0344	0.0447	0.0349	0.0370	0.0342	0.0417	0.0341	0.0289	0.0454	0.0416	0.0440	0.0390	0.0427	0.0409
	BKF	0.0722	0.0730	0.0677	0.0848	0.0579	**0.0116**	0.0956	0.0815	0.0543	0.0659	0.0635	0.0733	0.0362	0.0585
	SNIG	**0.0263**	**0.0233**	**0.0241**	**0.0272**	**0.0259**	0.0256	**0.0275**	**0.0266**	**0.0253**	**0.0247**	**0.0240**	**0.0220**	**0.0338**	**0.0277**
HRS	Laplacian	0.0417	0.0408	0.0381	0.0360	0.0406	0.0467	0.0418	0.0343	0.0489	0.0402	0.0458	0.0499	0.0514	0.0548
	BKF	0.0798	0.0844	0.0845	0.0986	0.0873	**0.140**	0.0836	0.0883	0.0625	0.0767	0.0643	0.0632	0.0619	0.0506
	SNIG	**0.0298**	**0.0260**	**0.0285**	**0.0282**	**0.0284**	0.0277	**0.0285**	**0.0286**	**0.0271**	**0.0296**	**0.0267**	**0.0280**	**0.0326**	**0.0314**

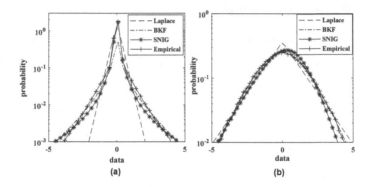

Fig. 1. Log histogram of one NSST subband coefficients (finest subband); Laplacian, BKF and SNIG pdf's are fitted to the empirical histogram in logarithmic domain (a) One image from HRS dataset (b) One image from UCM dataset

After the extraction of features from images, the main goal is to find the similar images related to the given query. Similarity measurement is done between the query and database image features. In the proposed framework, d_1 distance is found to give best results than other distance measures, for similarity measurement. d_1 distance between two vectors is defined as [22],

$$d_1(DB_t, A_q) = \sum_{p=1}^{L} \left| \frac{F_{DB_t}(p) - F_q(p)}{1 + F_{DB_t}(p) + F_q(p)} \right| \tag{9}$$

where L denotes the length of feature vector, F_q and F_{DB_t} are the feature vector of query A_q and t^{th} database image respectively and DB_t is the tth database image. The shortest distance stands for the best matched image present in the database.

The summary of the proposed approach is given below,

1. The RS images are transformed using NSST into different subbands.
2. The detail NSST coefficients are modeled using SNIG distribution and parameters of SNIG are estimated using ML scheme with an EM type of algorithm.
3. Feature vector is constructed by concatenating the extracted SNIG parameters from detail subbands and mean and standard deviation of approximation subband.
4. Similarity measurement is done using d_1 distance.

For a total of N_1 detail subbands and one approximation subband, the feature vector is constructed as follows:

$$FV = [\alpha_1, \beta_1, \mu_{m_1}, \delta_1, ...\alpha_{N_1}, \beta_{N_1}, \mu_{m_{N_1}}, \delta_{N_1}, \mu_A, \sigma_A] \qquad (10)$$

where α_1, β_1, μ_{m_1}, δ_1 are the parameters of SNIG distribution from the first detail subband; likewise α_{N_1}, β_{N_1}, $\mu_{m_{N_1}}$, δ_{N_1} are the parameters of SNIG distribution from the N_1^{th} detail subband. μ_A is the mean and σ_A is the standard deviation of the approximation subband.

4   Results and Discussions

For the experiments, three levels of decomposition with NSST are performed in 1, 2, 3 directions, yielding a total of 14 detail subbands and 1 approximation subband. Six well known local and global hand crafted descriptors are used for retrieval performance comparison with the proposed descriptor.

The UCM and the HRS datasets are considered to perform all the experiments. In this section, the datasets are introduced first and later, the experimental results are explained briefly.

The UCM dataset consists of a total of 2100, 256×256 RGB images where 100 images are present in a single class [23]. These images are perceived from the USGS National Map Urban Area Imagery collection [1]. In Fig. 2, sample image from all the image classes are shown. The HRS dataset contains satellite images of size 600×600 [2] in 19 different classes. Each class consists of 50 images. These images are extracted from Google Earth. Example of HRS dataset images from each class are shown in Fig. 3.

The proposed methodology's performance is evaluated with average normalized modified retrieval rank (ANMRR). Another metric, mean average precision (MAP) comparison is also examined. ANMRR is a commonly accepted measure of effectiveness of MPEG-7 retrieval. Range of ANMRR is $[0, 1]$. Small ANMRR value indicates better retrieval performance. For a query A_q, it is assumed that I_{th} ground truth image got retrieved at rank $R(I)$. $R(I)$ is defined as [16]

$$R(I) = \begin{cases} R(I), & \text{if } R(I) \leqslant P(A_q) \\ 1.25P(A_q), & \text{if } R(I) > P(A_q). \end{cases} \qquad (11)$$

Fig. 2. Example image from each class of UCM dataset

Fig. 3. Example image from each class of HRS dataset

where, a constant penalty, $P(A_q) = 2G(A_q)$ is commonly assigned to the items with higher rank and $G(A_q)$ is the ground truth set size for each query A_q. ANMRR is defined as [16]-

$$ANMRR = \frac{1}{Q} \sum_{A_q=1}^{Q} \frac{Ave(A_q) - 0.5[1 + G(A_q)]}{1.25K(A_q) - 0.5[1 + G(A_q)]} \tag{12}$$

where $Ave(A_q)$ is average rank for each query A_q and Q is the total query images considered.

MAP assembles the precision-recall (P-R) curves and results into a single value for evaluating all the ground truth rank positions [14]. The range of MAP value is [0,100]. Higher is the value of MAP, better is the retrieval performance. MAP is defined as [16]

$$M_{ap} = \frac{1}{Q} \sum_{A_q=1}^{Q} AvP \tag{13}$$

where, for a single query A_q, AvP is the average precision value.

Table 2 presents the comparison of performance of the proposed approach with LBP [19,20], HoG [12], Granulometry [9], Gabor L [5], Gabor RGB [4] and FV [15]. It is perceived that the proposed methodology outperforms the other methods for both the datasets in terms of all parameters which is shown in bold letters. The retrieval result of the proposed method for few classes of images for UCM and HRS datasets are shown in Figs. 4 and 5 respectively. Table 2 also presents the feature vector dimension of proposed method and all the other methods used for comparison in this work. It is seen that despite of having smaller dimensional feature vector size, proposed methodology is performing better than LBP, HoG, Granulometry, Gabor RGB, FV. Although the proposed method has feature vector slightly more than Gabor L, its performance is also much superior to Gabor L.

Table 2. Retrieval performance comparison of ANMRR and MAP in HRS and UCM datasets.

Dataset	Methods	ANMRR	MAP	Dataset	Methods	ANMRR	MAP	Feature dimension
UCM	LBP [19,20]	0.780	16.11	HRS	LBP [19,20]	0.674	23.55	256
	HoG [12]	0.751	17.85		HoG [12]	0.724	19.67	580
	Granulometry [9]	0.779	15.45		Granulometry [9]	0.717	21.41	78
	Gabor L [5]	0.740	17.98		Gabor L [5]	0.662	24.54	32
	Gabor RGB [4]	0.749	18.06		Gabor RGB [4]	0.649	27.00	96
	FV [15]	0.755	17.38		FV [15]	0.726	19.33	256
	Proposed	**0.725**	**19.81**		Proposed	**0.631**	**27.78**	58

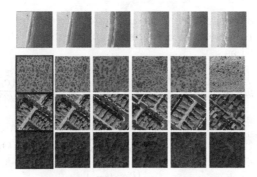

Fig. 4. Visual retrieval results for few classes of UCM Dataset (The extreme left images with black border are the query image and to it's right the top five corresponding retrieval results are presented)

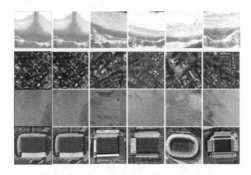

Fig. 5. Visual Retrieval results for few classes of HRS dataset (The extreme left images with black border are the query image and to it's right the top five corresponding retrieval results are presented)

5   Conclusions

This paper presents a statistical model based descriptor for RSIR. The multi-scale, multidirectional geometric analysis tool NSST with shift invariance property is used in the method leads to better representation of image details. With

KS goodness of fit test, we show that the NSST image coefficients are accurately modeled by SNIG distribution. The efficiency of the proposed descriptor has been examined in two datasets. The proposed technique, with SNIG as accurate statistical model achieves encouraging results with much less dimensions.

Acknowledgment. Authors would like to acknowledge the financial support given by Digital India Corporation, Ministry of Electronics and Information Technology (MeitY), Government of India through Visvesvaraya PhD scheme for Electronics & IT.

Appendix: EM algorithm

E step: Let the set of parameters to be estimated are $\theta = (\alpha, \beta, \mu_m, \delta)$. Suppose $\theta^{(k)}$ be the values of the parameters after the kth iteration, the pseudovalues S_i and w_i are calculated as

$$S_i = E(y_i|c_i, \theta^{(k)}) = \frac{\delta^k \phi^{(k)}(c_i)^{1/2} K_o(\delta^{(k)} \alpha^{(k)} \phi^{(k)}(c_i)^{1/2})}{\alpha^{(k)} K_1(\delta^{(k)} \alpha^{(k)} \phi^{(k)}(c_i)^{1/2})}$$

$$w_i = E(y_i^{-1}|c_i, \theta^{(k)}) = \frac{\alpha^{(k)} K_{-2}(\delta^{(k)} \alpha^{(k)} \phi^{(k)}(c_i)^{1/2})}{\delta^{(k)} \phi^{(k)}(c_i)^{1/2} K_{-1}(\delta^{(k)} \alpha^{(k)} \phi^{(k)}(c_i)^{1/2})}$$

for $i = 1, 2, ...n.$ where $\phi^{(k)}(c) = 1 + [\frac{(c - \mu_m^{(k)})}{\delta^{(k)}}]^2$

M step: The M-step updates the parameters using the pseudovalues calculated at the E-step. Compute $\hat{M} = \sum_{i=1}^{n} \frac{S_i}{n}$ and $\hat{\Lambda} = n(\sum_{i=1}^{n}(w_i - \hat{M}^{-1}))^{-1}$. The equations to update the parameters are

$$\delta^{(k+1)} = \hat{\Lambda}^{1/2}$$

$$\gamma^{(k+1)} = \frac{\delta^{(k+1)}}{\hat{M}}$$

$$\beta^{(k+1)} = \frac{\sum_{i=1}^{n} c_i w_i - \bar{c} \sum_{i=1}^{n} w_i}{n - \bar{S} \sum_{i=1}^{n} w_i}$$

$$\mu_m^{(k+1)} = \bar{c} - \beta^{(k+1)} \bar{S}$$

$$\alpha^{(k+1)} = [(\gamma^{(k+1)})^2 + (\beta^{(k+1)})^2]^{1/2}$$

where $\bar{S} = \sum_{i=1}^{n} \frac{S_i}{n}$.

The initial values in this algorithm are estimated using the moment based method [10] and EM steps are carried out in an iterative manner until the parameter converges.

References

1. UCMCVL. http://weegee.vision.ucmerced.edu/datasets/landuse. Accessed 2018
2. Yang wen. http://www.xinhua-fluid.com/people/yangwen/WHU-RS19. Accessed 2018
3. Aptoula, E.: Remote sensing image retrieval with global morphological texture descriptors. IEEE Trans. Geosci. Remote Sens. **52**(5), 3023–3034 (2013)
4. Bianconi, F., Fernández, A.: Evaluation of the effects of gabor filter parameters on texture classification. Pattern Recogn. **40**(12), 3325–3335 (2007)
5. Bianconi, F., Harvey, R.W., Southam, P., Fernández, A.: Theoretical and experimental comparison of different approaches for color texture classification. J. Electron. Imaging **20**(4), 043006 (2011)
6. Bosilj, P., Aptoula, E., Lefèvre, S., Kijak, E.: Retrieval of remote sensing images with pattern spectra descriptors. ISPRS Int. J. Geo-Inf. **5**(12), 228 (2016)
7. Easley, G., Labate, D., Lim, W.Q.: Sparse directional image representations using the discrete shearlet transform. Appl. Comput. Harmonic Anal. **25**(1), 25–46 (2008)
8. Farhangi, N., Ghofrani, S.: Using bayesshrink, bishrink, weighted bayesshrink, and weighted bishrink in NSST and SWT for despeckling SAR images. EURASIP J. Image Video Process. **2018**(1), 4 (2018)
9. Hanbury, A., Kandaswamy, U., Adjeroh, D.A.: Illumination-invariant morphological texture classification. In: Ronse, C., Najman, L., Decencière, E. (eds.) Mathematical Morphology: 40 Years On, pp. 377–386. Springer, Heidelberg (2005). https://doi.org/10.1007/1-4020-3443-1_34
10. Hazarika, D.: Despeckling of synthetic aperture radar (SAR) images in the lapped transform domain. Ph.D. thesis, Tezpur University (2017)
11. Hazarika, D., Nath, V.K., Bhuyan, M.: SAR image despeckling based on a mixture of Gaussian distributions with local parameters and multiscale edge detection in lapped transform domain. Sens. Imaging **17**(1), 15 (2016)
12. Junior, O.L., Delgado, D., Gonçalves, V., Nunes, U.: Trainable classifier-fusion schemes: an application to pedestrian detection. In: 2009 12th International IEEE Conference on Intelligent Transportation Systems, pp. 1–6. IEEE (2009)
13. Karlis, D.: An EM type algorithm for maximum likelihood estimation of the normal-inverse Gaussian distribution. Stat. Probab. Lett. **57**(1), 43–52 (2002)
14. Li, Y., Tao, C., Tan, Y., Shang, K., Tian, J.: Unsupervised multilayer feature learning for satellite image scene classification. IEEE Geosci. Remote Sens. Lett. **13**(2), 157–161 (2016)
15. Liu, Z., Wang, S., Tian, Q.: Fine-residual VLAD for image retrieval. Neurocomputing **173**, 1183–1191 (2016)
16. Napoletano, P.: Visual descriptors for content-based retrieval of remote-sensing images. Int. J. Remote Sens. **39**(5), 1343–1376 (2018)
17. Nath, V.K., Hazarika, D., Mahanta, A.: Lapped transform-based image denoising with the generalised Gaussian prior. Int. J. Comput. Vis. Robot. **4**(1–2), 55–74 (2014)
18. Nath, V.: Statistical modeling of lapped transform coefficients and its applications. Ph.D. Dissertation, Indian Institute of Technology Guwahati (IITG), Department of Electronics and Electrical Engineering (2011)
19. Ojala, T., Pietikäinen, M., Harwood, D.: A comparative study of texture measures with classification based on featured distributions. Pattern Recogn. **29**(1), 51–59 (1996)

368   H. G. Baruah et al.

20. Ojala, T., Pietikäinen, M., Mäenpää, T.: Multiresolution gray-scale and rotation invariant texture classification with local binary patterns. IEEE Trans. Pattern Anal. Mach. Intell. **7**, 971–987 (2002)
21. Özkan, S., Ateş, T., Tola, E., Soysal, M., Esen, E.: Performance analysis of state-of-the-art representation methods for geographical image retrieval and categorization. IEEE Geosci. Remote Sens. Lett. **11**(11), 1996–2000 (2014)
22. Verma, M., Raman, B.: Center symmetric local binary co-occurrence pattern for texture, face and bio-medical image retrieval. J. Vis. Commun. Image Representation **32**, 224–236 (2015)
23. Yang, Y., Newsam, S.: Bag-of-visual-words and spatial extensions for land-use classification. In: Proceedings of the 18th SIGSPATIAL International Conference on Advances in Geographic Information Systems, pp. 270–279. ACM (2010)
24. Zhang, X., Jing, X.: Image denoising in contourlet domain based on a normal inverse Gaussian prior. Digit. Sig. Process. **20**(5), 1439–1446 (2010)
25. Zhou, W., Newsam, S., Li, C., Shao, Z.: Learning low dimensional convolutional neural networks for high-resolution remote sensing image retrieval. Remote Sens. **9**(5), 489 (2017)

Perception Based Navigation
for Autonomous Ground Vehicles

Akshat Mandloi[(✉)], Hire Ronit Jaisingh, and Shyamanta M. Hazarika

Department of Mechanical Engineering, IIT Guwahati, Guwahati, India
akshat.2597@gmail.com

Abstract. Autonomous vehicles rely on their sensors specializing in specific tasks to understand their surroundings. Camera is perhaps one of the most versatile of them. Frames from a camera, when processed, form the basis for object detection in the surroundings, which is an embodiment of perception. The major tasks in perception are object tracking and localization. Neural Networks, especially Convolutional Neural Networks (CNN), exceed expectations in the tasks of detection and classification. This paper adds to the functionality of CNN by exploring the aggregation of additional knowledge from images, consisting of segmentation and localization information, as an attempt to predict safe driving zone for robust navigation. This aggregated knowledge can be visualized as a meta representation of the surroundings for the navigation algorithm.

Keywords: Autonomous vehicles · Computer vision · Navigation

1 Introduction

Perception based navigation for autonomous ground vehicles (AGVs) originated from the idea of developing a framework which mimic the human driving behaviour. Humans use vision as their main source of input for making key decisions while driving. Even though Convolutional Neural Networks (CNN) exceed expectations in the tasks of detection and classification, most of the frameworks previously developed for perception based navigation for AGVs usually lack the prediction knowledge. The ability to predict the activity of a moving object in near future enables a human driver to maintain complete control of the vehicle. Most of the end-to-end CNN frameworks developed for this task lack this ability. The performance of these neural network based frameworks for autonomous navigation have been criticized in literature, which drives the decision to keep the knowledge extraction part separate from the CNN architecture [9].

Tracking being one of the major tasks in perception has been studied extensively in literature. Redmon et al. [8] describes a object detection algorithm You Only Look Once (YOLO) that achieves real-time processing speeds. The performance of various tracking algorithms is compared in [5], suggesting Multiple Instance Learning (MIL) tracking algorithm [2] to be better for an application like navigation. Further, localization of ego vehicle is essential for successful navigation. [3] implements a brightness constancy based feature tracker to generate

© Springer Nature Switzerland AG 2019
B. Deka et al. (Eds.): PReMI 2019, LNCS 11942, pp. 369–376, 2019.
https://doi.org/10.1007/978-3-030-34872-4_41

motion vectors. The use of illumination invariant images for road segmentation is suggested in [1]. Tuohy et al. [10] uses inverse perspective mapping to establish a linear relationship between the distances in the real world and in the images. However, this depends heavily on camera orientation. This paper while incorporating the above ideas to develop a novel architecture for reliable navigation of autonomous vehicles in urban environment, performs multiple object detection and tracking with YOLO in association with MIL tracker over multiple threads.

Navigation is one of the most essential tasks for an AGV. Architectures such as the one suggested by Smith et al. [9] deploy learning where it excels and rely on model-based strategies where they perform the best. Pfeiffer et al. also demonstrate that the learned navigation model is directly transferable to unseen virtual and real world environments effectively [7]. They explore traditional machine learning framework for end-to-end navigation and demonstrate that global navigation approaches perform better, which involve exhaustive or highly sampled search. This paper develop an architecture for reliable navigation of autonomous vehicles in urban environment. The architecture uses computer vision and neural networks as primary tools to answer decisive questions which provides the AGV a local safe zone for driving. The decision making process takes into account results about localization and navigation.

2   Generating Qualitative Instructions for Navigation

Everyday reasoning is predominantly qualitative. Human route guidance is often in terms of qualitative abstractions rather than quantitative terms. In view of above, we have a novel architecture as shown in Fig. 1 which generates higher order qualitative instructions for navigation.

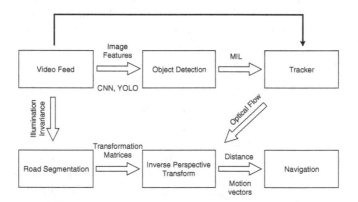

Fig. 1. Pipeline of the proposed architecture to generate higher order qualitative instructions for navigation. A CNN based object detection and MIL tracking is used to compute a representation of the dynamic environment. Aggregation of additional knowledge from images, consisting of segmentation and localization information, together with the tracking information predict safe driving zone for robust navigation.

The sequence of frames generated by the camera sensor are fed to the tracking paradigm. Tracking is achieved using YOLOv3 [8] in combination with a multi-instance MIL tracker. The knowledge about road surface is extracted using illumination invariant color space and likelihood based classifier. In order to localize the ego vehicle in the environment, the distance and driving lanes of detected vehicles are determined through inverse perspective transformation. The aggregated information incorporating localization and tracking is used to mark a safe and feasible driving zone for the ego vehicle, and, deliver high order qualitative navigation instructions. Probing into the above pipeline three pivot points are identified - a. Tracking, b. Knowledge Extraction and c. Navigation. MIL Tracking was limited to single instance which was upgraded to incorporate parallel tracking using threading. An algorithm encompassing all the elements of the knowledge vector is proposed which demarcates safe driving zones.

2.1   Tracking

The tracker takes object detection results from every 10th frame using YOLOv3 [8] and then tracks the objects in the following nine frames using multiple instances of MIL tracking algorithm [2].

Multiple Instance Tracking. In order to implement the MIL tracking algorithm for multiple objects, a hardware based scheme to utilize the multiple processors in a CPU to track several objects simultaneously has been proposed. A knowledge vector of objects received from the detection framework is used to initiate MIL trackers corresponding to each object. Each tracker is associated to an independent thread. Thus the MIL track algorithm runs on multiple threads simultaneously. The output from each thread corresponds to location of object in the next frame and is similar to the knowledge vector from detection. Tracking is as if detection is running on each frame, but faster. Figure 2 shows the graphical representation of multiple instance tacking using threading.

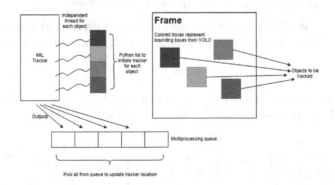

Fig. 2. Multiple instance tracking using threading in association with YOLO.

It may seem lucrative that tracking is enough to discard detection altogether after the first frame. But appearance model in a tracker loses properties of the original object after a few frames, the tracker loses track. New objects are often introduced in a dynamic environment. To counter these, we run the YOLO detection on every 10th frame and MIL tracker on every intermediate frame. Running the above scheme on a nvidia GeForce 920M GPU for 206 frames from the KITTI dataset [4] takes 37.48 s as opposed to 139.78 s taken by YOLO based pipeline. An IoU score of 75.083 is obtained for parallel tracking which further corroborates its accuracy.

2.2　Knowledge Extraction: Localization of Ego Vehicle

Localization of ego vehicle is necessary in order to predict the future action to be followed by the vehicle. This includes knowledge about road surface, lane, object distance and motion flow.

Road Surface Extraction: The usual likelihood based classifiers are susceptible to changes in lighting conditions and source illumination variations. The illumination invariant image is obtained using the following formula [6]

$$I = 0.5 + \log(R_2) - \alpha \log(R_1) - (1 - \alpha) \log(R_3) \tag{1}$$

For the camera sensors available commercially, a value of $\alpha = 0.4$ works well. The road surface is extracted from the image once the input image from the image sensor is processed to produce illumination invariant gray-scale image. The road segmentation is performed using a likelihood based classification. The algorithm classifies the pixels according to two labels, road surface and background using a similarity measure between each pixel and the road model and a threshold [1].

Distance Determination: The images captured are in perspective view, where the distance between objects in real world can not be mapped linearly in the image. We perform inverse perspective mapping to derive a linear map between number of pixels in the image and objects on road surface.

Inverse Perspective Mapping: A camera can be represented as a mathematical model, that maps world points X_w to image points x_i according to the equation

$$X_w = C x_i \tag{2}$$

where, C is the camera model. The points represented in a world coordinate frame are related to the required frame by $C = R * T * K$. where,

$$
R = \begin{bmatrix} 1 & 0 & 0 & 0 \\ 0 & \cos\theta & \sin\theta & 0 \\ 0 & \sin\theta & \cos\theta & 0 \\ 0 & 0 & 0 & 1 \end{bmatrix} \quad
T = \begin{bmatrix} 1 & 0 & 0 & 0 \\ 0 & 1 & 0 & 0 \\ 0 & 0 & 1 & -h/\sin\theta \\ 0 & 0 & 0 & 1 \end{bmatrix} \quad
K = \begin{bmatrix} f_{c,x} & 0 & u_{p,x} & 0 \\ 0 & f_{c,y} & v_{p,y} & 0 \\ 0 & 0 & 1 & 0 \end{bmatrix} \tag{3}
$$

$f_{c,x}$ and $f_{c,y}$ represent the focal length in x and y direction of the camera.

The above procedure is sensitive to change in orientation of the camera. The inverse perspective map (IPM) is made invariant to height variations using vanishing point as a measure of height change. The results shown in Fig. 3 depict the shift of vanishing point as the height of camera is lowered.

Fig. 3. Vanishing point shift as a measure of height change to make inverse perspective mapping invariant to height variations.

Lane Detection: Lane detection is essential to localize the ego and surrounding vehicles. This helps in predicting maneuvers such as lane departure, global constraints setting and overtaking. The lane detection consists of following steps:

Step-1: Process the frame to form illumination invariant image, perform road segmentation and derive inverse perspective transform.

Step-2: Perform thresholding to extract the lane markings and derive a histogram of the binary image thus formed.

Step-3: Find the argument for the peaks in histogram.

Step-4: In order to extract the pixels where the lane markings lie in the image, perform a window search in the vertical axis localized at that argument.

Step-5: Extract a list on nonzero pixels from the image, and using the fixated windows from the window search, extract the pixels of a given lane marking.

Step-6: Fit a quadratic polynomial to these pixels.

Once the pixels corresponding to the lane markings are achieved, localization of the vehicle can be performed. This requires calibration of IPM image across the width. The offset of the object center and ego vehicle center from the lane center and the lateral distance between the vehicles can be determined. For relation between the pixels and distance, calibration of the IPM image has to be performed along the vertical axis. Lane marking based calibration was performed.

Motion Flow. Using the techniques of optical flow a motion vector indicating the displacements of keypoints of detected objects is obtained to get a sense of direction of motion and possibility of interference in the path of the ego vehicle.

Iterative Lucas Kanade: We adopt the iterative Lucas-Kanade algorithm to generate motion vectors of keypoints.

1. Initialize a motion vector v
2. Compute the spatial gradient matrix

$$G = \sum_{x=p_x-w}^{p_x+w} \sum_{y=p_y-w}^{p_y+w} \begin{bmatrix} I_x^2(x,y) & I_x(x,y)I_y(x,y) \\ I_x(x,y)I_y(x,y) & I_y^2(x,y) \end{bmatrix} \tag{4}$$

3. For n number of iterations, compute the following:

$$\delta I_k(x,y) = I(x,y) - J(x+g_x+v_x, y+g_y+v_y) \tag{5}$$

$$b_k = \sum_{x=p_x-w}^{p_x+w} \sum_{y=p_y-w}^{p_y+w} \begin{bmatrix} \delta I_k(x,y)I_x(x,y) \\ \delta I_k(x,y)I_y(x,y) \end{bmatrix} \tag{6}$$

$$v^k = G^{-1}b_k \tag{7}$$

Pyramidal Lucas-Kanade: Large motions between keypoints cannot be accounted for. [3] implements a pyramidal Lucas-Kanade method - run an iterative Lucas-Kanade on different scales of the image, thereby large motion become small in a zoomed out image.

3   Navigation - Qualitative Higher-Order Instructions

As a proof of concept we implement and examine the effectiveness of the architecture over videos recorded from a moving vehicle. The navigation algorithm in its simplest form, as can be seen in Algorithm 1, is a group of if-else statements; trying to mimic the higher order decision making of a human driver. The algorithm takes the aggregated knowledge from localization and tracking and performs an iteration to produce the best feasible routine for navigation. This allows the ego vehicle to navigate and perform qualitative maneuvers like *follow, accelerate* or *decelerate* and *change lanes*.

Algorithm 1. Navigation Algorithm

---

set vehicle parameters **while** *True* **do**

    check distance to objects in current lane   check motion vectors of objects **if** *distance < safeZone* **then**

        if *distance < redZone* **then**

        | BRAKE

        else

            check side lanes **if** *not vacant* **then**

            | continue current state

            else

                $t_s = \frac{d_r \, \| \, d_l}{\gamma}$ $t_c = \frac{d}{v_s - v_o}$ **if** $t_s < t_c$ **then**

                | switch lane

                else

                    └ BRAKE;

    else

    └ continue current state

    └ Call subroutine: Interference

Algorithm 2. Subroutine: Interference

---

if *vector suggests interference* **then**

    check distance **if** *distance* $<$ *safeZone* **then**

        check opposite lane **if** *not vacant* **then**

            $t_e = \frac{d_r \;||\; d_l}{\gamma}$ $d_{fo} = t_e * v_o$ $d_s = t_e * v_s$ **if** $d_{fo} - d_s > d_{red}$ **then**

            | continue current state

            else

            | BRAKE

            end

        else

        | change lane

        end

    else

    | continue current state

    end

else

| continue current state

end

3.1   Experimental Results

Figure 4 depicts the output of the knowledge extraction. Frames show results for processes described in Sect. 2.2 - road segmentation, lane detection, distance determination and motion vector results in sequence.

Fig. 4. Knowledge extraction results.

The video frames in Fig. 5 depicts the output during two typical scenarios. The first two frames shows a vehicle on the left lane of the ego vehicle with a motion vector suggesting no interference in near future. There is no intersection of the vehicle lane with the red or danger zone around the vehicle, thus the ego vehicle can *accelerate* safely in its lane. In the second scenario depicted in the last two frames, the lane of the ego vehicle intersects with the red zone of a vehicle detected in its own lane. The ego vehicle is suggested to *decelerate* and *follow* the preceding vehicle at a safe distance. The proposed architecture predicts safe navigation zones with an accuracy greater than 85% when tested on our dataset. However, when in crowded scenarios the number of false positives from tracking increased affecting the overall performance of the proposed architecture.

Fig. 5. Navigation output during two typical instances. Frames on the left shows a vehicle detected; no interference and a free lane computed. Last two frames shows the lane of the ego vector detecting an interference and red zone of a tracked vehicle. (Color figure online)

4 Conclusion

The architecture presented here is aimed at providing qualitative higher-order instructions within a safe driving zone to navigate an AGV using perception knowledge. The architecture predicts safe navigation zones with an appreciable success rate on test videos. There is scope of improvement in terms of improving the tracker accuracy, road surface segmentation during occlusion and variance in IPM due to changes in orientation. This is part of ongoing research. The inability of tracking algorithms to perform in crowded scenarios is a ubiquitous major hindrance which needs to be addressed on a larger basis.

References

1. Alvarez, J.M., Lopez, A., Baldrich, R.: Illuminant-invariant model-based road segmentation. In: 2008 IEEE Intelligent Vehicles Symposium, pp. 1175–1180, June 2008
2. Babenko, B., Yang, M., Belongie, S.: Visual tracking with online multiple instance learning. In: 2009 IEEE Conference on Computer Vision and Pattern Recognition, pp. 983–990, June 2009
3. Bouguet, J.Y.: Pyramidal implementation of the lucas kanade feature tracker. Intel Corporation, Microprocessor Research Labs (2000)
4. Geiger, A., Lenz, P., Stiller, C., Urtasun, R.: Vision meets robotics: the KITTI dataset. Int. J. Rob. Res. **32**(11), 1231–1237 (2013)
5. Janku, P., Koplik, K., Dulik, T., Szabo, I.: Comparison of tracking algorithms implemented in OpenCV. MATEC Web Conf. **76**, 04031 (2016)
6. Maddern, W., Stewart, A., McManus, C., Upcroft, B., Churchill, W., Newman, P.: Illumination invariant imaging: applications in robust vision-based localisation, mapping and classification for autonomous vehicles. In: IEEE International Conference on Robotics and Automation (ICRA), Hong Kong, China (2014)
7. Pfeiffer, M., Schaeuble, M., Nieto, J.I., Siegwart, R., Cadena, C.: From perception to decision: A data-driven approach to end-to-end motion planning for autonomous ground robots. CoRR (2016)
8. Redmon, J., Farhadi, A.: YOLOv3: An incremental improvement. CoRR (2018)
9. Smith, J.S., Hwang, J., Chu, F., Vela, P.A.: Learning to navigate: Exploiting deep networks to inform sample-based planning during vision-based navigation. CoRR (2018)
10. Tuohy, S., O'Cualain, D., Jones, E., Glavin, M.: Distance determination for an automobile environment using inverse perspective mapping in OpenCV. In: IET Irish Signals and Systems Conference (ISSC 2010), pp. 100–105, June 2010

Automatic Target Recognition from SAR Images Using Capsule Networks

Rutvik Shah[1], Akshit Soni[1], Vinod Mall[2], Tushar Gadhiya[1(✉)],
and Anil K. Roy[1]

[1] Dhirubhai Ambani Institute of Information and Communication Technology,
Gandhinagar, India
201621009@daiict.ac.in
[2] ADG of Police, Government of Gujarat, Gandhinagar, India

Abstract. Synthetic Aperture Radar (SAR) imagery has become popular in the past few decades owing to its operability under difficult weather
conditions. In this paper, we introduce a deep learning architecture using
a Capsule Network (CapsNet) for automatic target recognition (ATR)
of images of targets captured using an X-band SAR sensor. The architecture consists of a single convolutional layer, followed by two Capsule
layers, and a decoder network at the end. Since, traditional Convolutional Neural Networks (CNNs) often require a significant number of
training images, their performance is limited for small number of training examples. Unlike CNNs, Capsule Networks encapsulate the instantiation parameters of an object within an image, thus, they do not require
a large number of training samples. In addition, Capsule Networks are
view-point invariant. For the evaluation of the proposed method, we have
used the MSTAR database, containing SAR images of 10-classes of military vehicles. We have achieved 98.14% overall classification accuracy on
this dataset.

Keywords: Capsule Network · Synthetic Aperture Radar ·
Automatic target recognition · Convolutional Neural Network (CNN)

1 Introduction

Synthetic Aperture Radar (SAR) has become widely popular in the area of
remote sensing due to its all-weather and day-and-night image capturing capability. This ability makes SAR suitable for wide range of civilian as well as military applications. One such important military application is automatic target
recognition (ATR) from SAR images.

In recent years Convolutional Neural Network (CNN) based architectures
have proven to be effective for ATR applications [3,6,10,12]. However, CNN
based architectures require a large number of training images which is often
too difficult to obtain. To compensate the smaller-sized datasets available, most
CNN based methods use data augmentation to increase the size of the training

© Springer Nature Switzerland AG 2019
B. Deka et al. (Eds.): PReMI 2019, LNCS 11942, pp. 377–386, 2019.
https://doi.org/10.1007/978-3-030-34872-4_42

dataset. Recently A-ConvNets [2] was proposed which extracts multiple 88×88 sized patches from a single SAR image of size 128×128. In another technique [4] the use of translation, speckle noise addition and pose synthesis is proposed for data augmentation. G-DCNN [6] uses a Gabor filter based data augmentation technique to increase the number of training images.

Another popular technique to improve the CNN classification accuracy introduces a pre-processing step for background clutter removal, where the input image is cropped to a smaller size, such that it only contains the object of interest. MFCNN [3] crops the original images of size 128×128 to a 40×40 image, containing only the target in the center, thus removing the background clutter from each image. In the method proposed by Dong et al. [5] the center 80×80 pixel patch of the raw image is cropped to exclude background clutter. In addition, the authors employed uniform down-sampling, concatenation and normalization of monogenic components to create an augmented feature vector. Zhou et al. [12] use a 60×60 cropped patch from the 128×128 image for mitigating the effects of background clutter.

To overcome these limitations, this paper proposes the use of a Capsule Network (CapsNet) [11] for SAR automatic target recognition. The advantage of a Capsule Network is that since it embeds the instantiation parameters of an object within an image, it eliminates the need of a larger training dataset, thus, eliminating the need of data augmentation or background clutter removal.

2   CapsNet

A neuron in a traditional Convolutional Neural Network uses a single scalar output to represent the activities of a local pool of replicated feature detectors. Capsules on the other hand, are a cluster of neurons which represents the instantiation parameters as well as the probability of presence of a particular entity over a limited domain of viewing conditions and deformations. This information is encapsulated in the vector output of a Capsule, where the length represents the presence of an entity and the orientation represents the instantiation parameters of the entity such as pose, lighting and deformation. Capsule Networks address the key drawback of CNNs, i.e., pooling and the subsequent loss of information due to it, by employing "Routing by Agreement". Outputs of Capsules in a lower layer, are routed to all the Capsules in the subsequent layer, however, the coupling coefficients are different. The output of a Parent Capsule is predicted by each Capsule in the lower layer, and subsequently the coupling coefficient between the two Capsules is altered based on how strongly the predicted and actual value agree. Multiple iterations of this procedure lead to the outputs of Capsules being "dynamically routed" [11] to specific Parent Capsules. The prediction of Capsule i for Parent Capsule j is given by the following equation, considering u_i is the output of Capsule i

$$\hat{u}_{j|i} = W_{ij} u_i. \tag{1}$$

Here $\hat{u}_{j|i}$ is the vector given by the ith Capsule for the predicted output of the jth Parent Capsule, and W_{ij} is the weight matrix learnt in the backward pass.

The following softmax function is used to calculate the coupling coefficient c_{ij} depending upon the degree of agreement between Capsules of the lower level and that in the higher level.

$$c_{ij} = \frac{exp(b_{ij})}{\sum_k exp(b_{ik})}, \tag{2}$$

where b_{ij} is initially assigned a value of 0 at the beginning of routing by agreement and represents the log probability of associative coupling between Capsule i and Parent Capsule j. The Parent Capsule j gets an input vector calculated by the following formula.

$$p_j = \sum_i c_{ij} \hat{u}_{j|i} \tag{3}$$

Since they represent probability of presence of an entity, to avoid output vector lengths of the Capsules exceeding one, the following squashing function is used to obtain the final output of each Capsule.

$$v_j = \frac{\|p_j\|^2}{1 + \|p_j\|^2} \frac{p_j}{\|p_j\|}, \tag{4}$$

Here the output of the jth Capsule is v_j, and p_j is its input value. The routing process checks agreement between the predicted output by Capsule i and v_j i.e. the output of Capsule j in the higher layer, by checking the inner product of v_j and $\hat{u}_{j|i}$. The inner product is large if the prediction is accurate and small if it is not accurate. Thus, to update log probability b_{ij} and coupling coefficients c_{ij}, agreement a_{ij} is calculated as follows

$$a_{ij} = v_j \cdot \hat{u}_{j|i}. \tag{5}$$

Agreement a_{ij} is added to b_{ij} to compute new values of c_{ij}, which determines the coupling of the Capsule i with all its Parent Capsules.

Since both, the SAR-target occurrence and the reconstructed image are estimated by the network, network optimization is carried out by minimizing the following loss function

$$L^{net} = L^{margin} + \alpha L^{reconstruction}, \tag{6}$$

where α determines the relative weightage accorded to each of the two losses.

All the p Capsules in the last layer, one for each of the p classes, have a margin-loss function L_p which sets a high loss value on Capsules wrongly predicting the existence of an entity, represented by a long output vector, when in fact, the entity is absent. The loss function L_p is given by

$$L_p = T_p max(0, m^+ - \|v_p\|)^2 + \lambda_p(1 - T_p)max(0, \|v_p\| - m^-)^2, \tag{7}$$

where $T_p = 1$ if the SAR target p is present in the image and $T_p = 0$ if the target p does not exist in the image. The effect of the presence or absence of a particular SAR target is specified by λ_p. m^+ and m^- are the margins specified, which signify the permissible error threshold in prediction. Consequently, the value of L_p, i.e. the margin loss, is 0 when $\|v_p\| > m^+$ and $T_p = 1$, i.e. target p is present and $\|v_p\| < m^-$ when $T_p = 0$, i.e. the target is absent.

The overall margin loss is given by the sum of margin losses for each of the p Capsules in the last layer.

$$L^{margin} = \sum_p L_p^{margin} \tag{8}$$

Finally, for the reconstructed-image loss, we use the sum of squared distances between the pixel-intensity of the original training image, and those obtained from the last fully-connected layer. A small value is chosen for α so that the margin loss is not dominated by the reconstruction loss for the calculation of the overall loss value.

3   Architecture

We experimented with different variants of the Capsule Network architecture, by changing the filter size in the Convolutional layer and the number of dynamic routing iterations. The filter sizes of 3×3, 5×5, 7×7, and 9×9 were evaluated, and we observed the highest classification accuracy for the filter size of 9×9. Further, two and three iterations of the Dynamic Routing algorithm were implemented, and we observed an increase in accuracy of 1.52% from 96.62% to 98.14% for the two experiments respectively. The specifics of the architecture depicted in Fig. 1 is described below.

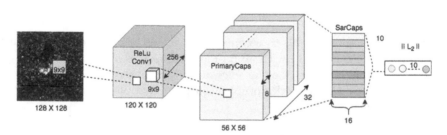

Fig. 1. Proposed Capsule Network architecture

Inputs to the proposed network are 128×128 SAR images from the MSTAR dataset. The input images are fed into a Convolutional-layer with 256 filters, with a 9×9 kernel size and a stride of 1, which leads to 256 feature maps of size 120×120 as output. This output is reshaped in the Primary Capsule layer into 32, 8-Dimensional Capsules with feature maps of size 56×56 using a convolution operation of kernel size 9×9 and step size of 2. This gives us, $32 \times 56 \times 56$ Capsule outputs, each of which is an 8D vector. The second and final Capsule layer includes 10 Capsules of 16 dimension, one for each class of MSTAR dataset. Each SAR-Capsule (SAR Caps) in this layer is connected with each of the Capsules in its previous layer.

The decoder at the end of the network contains three fully-connected layers of size 512, 1024 and 16384. For the first two layers, we employ ReLU activation function, whereas Sigmoid activation is used in the final layer. The predicted class is given by the output of the second Capsule layer, whereas the output of the decoder is used to generate the reconstructed images to calculate the reconstruction loss explained in Sect. 2.

The Adam optimizer [7] was used for optimizing the network parameters, with a learning rate of 0.001 and weight of decay learning rate 0.9. The values of margins m^+ and m^- for L^{margin} were assigned as $m^+ = 0.9$ and $m^- = 0.1$ as suggested in [11]. $\alpha = 0.0005$ was chosen for regularization of the loss value.

4 Dataset

The Moving and Stationary Target Acquisition and Recognition (MSTAR) [1] dataset has been used in this paper for evaluating the proposed architecture. It consists of ten different variants of ground-targets (tanks: T-62, T-72; bulldozers: D7; armored personnel carriers: BTR-60, BMP-2, BTR-70 and BRDM-2; air defense units: ZSU-234; rocket launchers: 2S1; trucks: ZIL-131). The number of images corresponding to all the target types is displayed in Table 1. SAR images with an angle of depression of 17° were utilized for training the CapsNet architecture and images corresponding to 15° were used for testing. SAR images and their respective optical images from the MSTAR dataset are depicted in Fig. 2.

Table 1. Number of images per class

Class	Serial No.	No. of train Img (17°)	No. of test Img (15°)
T-62	A51	299	273
ZSU-234	d08+	299	274
BTR-70	c71	233	196
T-72	132	232	196
2S1	b01	299	274
BTR-60	k10yt7532	256	195
ZIL-131	E12	299	274
BMP-2	9563	233	196
D7	92v13015	299	274
BRDM-2	E-71	298	274

SAR images suffer from speckle noise and deviations which hinders the effective extraction of information present in SAR images. Speckle noise thus, degrades the overall quality of SAR images and makes its interpretation difficult. To overcome the negative effects of speckle multiplicative noise, we apply the Lee filter [8] for removal of speckle noise, on the MSTAR dataset.

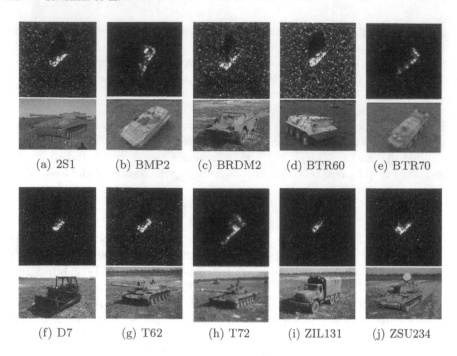

Fig. 2. Representative images of the MSTAR dataset

5   Experimental Setup

The proposed architecture was implemented in the Python using the Tensorflow. The experiments were carried out on a machine with Intel i7-7700 HQ, quad-core processor, with 32 GB RAM and the NVidia Titan X GPU with 12 GB VRAM.

6   Results

6.1   Interpretation of What the Dimensions of Capsules Represent

As explained in Sect. 3, a decoder network is used on the SAR Capsules, in order to reconstruct the original image and calculate the reconstruction loss. This causes the SAR Capsules to encapsulate the instantiation parameters of the original input SAR image. Since, we encode only the correct class, and mask out the reconstructions of other classes while training, the 16 dimensions of the SAR Capsule for a particular target should learn the space, pose and other variations of all images of that particular target. We can thus observe the property that each dimension of a particular SAR Capsule encapsulates, using the reconstruction images of that particular class.

It is also important to note that properties of SAR images are not easily distinguishable using the naked eye, and thus the changes in instantiation parameters encapsulated in the 16 dimensions of each SAR capsule become difficult to

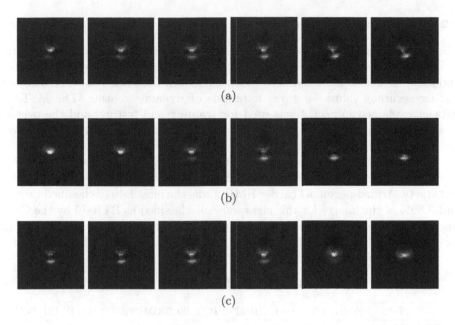

Fig. 3. Dimension perturbations. Each row displays the reconstructed image upon change in one of the sixteen dimensions of the SAR Caps by intervals of 0.2 in the range $[-0.5, 0.5]$.

interpret even while using the reconstruction images. However, since the input images of each class contain aerial images of a particular target, taken with the same depression angle but different azimuth angles (spanning from angles of $0°$ to $360°$), rotation becomes a key observable property of the input SAR images. A perturbed version of each activity vector of the correct SAR Capsule which was computed by the network, is fed to the decoder part of the architecture, to observe how each perturbation causes variations in the reconstructed image. Figure 3 contains few examples of reconstruction images constructed using these perturbations. We observed that for each of the SAR Capsules, at least one dimension from the sixteen, represented variation in the azimuth angle of the reconstruction. This is represented in row 1 of Fig. 3. The other notable observable variation was that of the vertical shift in concentration of the reflective surface (depicted by white portion) of the SAR target. As the azimuth angle of the SAR sensor changes, the reflective surface of the same target shifts slightly. This property is encapsulated in one of the perturbed dimensions as shown in row 2 of Fig. 3. The third row of Fig. 3 represents the concentration of the reflective surface towards the centre of the image as the dimension is perturbed. Similarly the other dimensions represent different instantiation parameters of the target images. Thus, we can successfully ascertain that the regularization of training using the reconstruction loss, is an effective method for our use-case.

6.2   Classification Accuracy

The Capsule Network architecture was implemented with the Dynamic Routing algorithm as discussed by Sabour *et al.* [11]. We experimented with both, two and three iterations of the dynamic routing between Capsules and observe higher testing accuracy values for three iterations of dynamic routing. The MSTAR dataset explained in Sect. 4 was used for training and testing, and the overall testing accuracy of 98.14% was observed for automatic target-classification on the 10-class MSTAR dataset.

Table 2 shows the class-wise classification accuracy for each of the ten classes. The highest classification accuracy of 100% was observed for ZSU-234 and BTR-70 targets. Armed-personnel carrier BMP-2 was the most falsely classified target, and 3.59% of the images for this class were misclassified as BTR-70 by the Capsule Network, which is also an armed-personnel carrier. However, there is no reciprocal misclassification for BTR-70. It is also noteworthy that the results

Table 2. Confusion matrix

	T-62	ZSU-234	D7	2S1	BMP-2	BTR-60	BRDM-2	BTR-70	ZIL-131	T-72
T-62	**96.34**	0.37	0.37	0.37	0.00	0.00	0.00	0.37	0.73	1.47
ZSU-234	0.00	**100.00**	0.00	0.00	0.00	0.00	0.00	0.00	0.00	0.00
D7	0.00	0.00	**98.18**	0.36	0.00	0.36	1.09	0.00	0.00	0.00
2S1	0.36	0.00	0.00	**96.35**	0.00	0.00	1.09	0.36	1.09	0.73
BMP-2	0.00	0.00	0.00	0.00	**94.36**	0.51	0.51	3.59	0.00	1.03
BTR-60	0.00	0.00	0.00	0.00	0.00	**99.49**	0.00	0.51	0.00	0.00
BRDM-2	0.00	0.00	0.00	0.73	0.00	0.00	**98.91**	0.00	0.36	0.00
BTR-70	0.00	0.00	0.00	0.00	0.00	0.00	0.00	**100.00**	0.00	0.00
ZIL-131	0.00	0.00	0.36	0.00	0.00	0.00	0.00	0.00	**99.64**	0.00
T-72	0.00	0.00	0.00	0.00	0.00	0.00	0.00	0.00	0.51	**99.49**

obtained were derived with no augmentation or cropping applied to the original images of the MSTAR dataset. This further corroborates the claim that CapsNets require less number of training images compared to traditional CNNs. This is encapsulated in the higher accuracy that our method has achieved compared to other Deep Learning based models [3,5,6,9,12] shown in Table 3. As discussed in the introduction section, these models utilize cropping for background clutter removal and augmentation of training data. In comparison, the proposed architecture uses the original images of size 128 × 128 as given in the MSTAR dataset, and does not employ any data augmentation to enhance the size of the training data.

Table 3. Comparison of accuracy with other methods

Methods	CDSPP [9]	G-DCNN [6]	MFCNN [3]	MSRC [5]	SVM	LM-BN-CNN [12]	Proposed
Avg. % accuracy	91.01	96.32	95.52	93.66	90.67	96.44	**98.14**

7 Conclusion

In this paper we proposed a unique Capsule Network model for automatic target recognition using SAR images. We have demonstrated that the proposed network works effectively with a smaller training dataset without using any augmentation or pre-processing techniques. Reconstruction as a regularization technique was also explored and was shown to be a competent technique to regularize the proposed Capsule Network. We evaluated our method using the MSTAR dataset containing 10 target classes and observed that we were able to outperform *state-of-the-art* CNN-based architectures.

References

1. The air force moving and stationary target recognition database. https://www.sdms.afrl.af.mil/index.php?collection=mstar. Accessed 01 June 2019
2. Chen, S., Wang, H., Xu, F., Jin, Y.: Target classification using the deep convolutional networks for SAR images. IEEE Trans. Geosci. Remote Sens. **54**(8), 4806–4817 (2016)
3. Cho, J.H., Park, C.G.: Multiple feature aggregation using convolutional neural networks for SAR image-based automatic target recognition. IEEE Geosci. Remote Sens. Lett. **15**(12), 1882–1886 (2018)
4. Ding, J., Chen, B., Liu, H., Huang, M.: Convolutional neural network with data augmentation for SAR target recognition. IEEE Geosci. Remote Sens. Lett. **13**(3), 364–368 (2016)
5. Dong, G., Wang, N., Kuang, G.: Sparse representation of monogenic signal: with application to target recognition in SAR images. IEEE Sig. Process. Lett. **21**(8), 952–956 (2014)
6. Jiang, T., Cui, Z., Zhou, Z., Cao, Z.: Data augmentation with gabor filter in deep convolutional neural networks for SAR target recognition. In: IGARSS 2018–2018 IEEE International Geoscience and Remote Sensing Symposium, pp. 689–692, July 2018
7. Kingma, D.P., Ba, J.: Adam: A method for stochastic optimization. arXiv preprint arXiv:1412.6980 (2014)
8. Lee, J.: A simple speckle smoothing algorithm for synthetic aperture radar images. IEEE Trans. Syst. Man Cybern. **SMC-13**(1), 85–89 (1983)
9. Liu, M., Chen, S., Wu, J., Lu, F., Wang, J., Yang, T.: Configuration recognition via class-dependent structure preserving projections with application to targets in SAR images. IEEE J. Sel. Topics Appl. Earth Observ. Remote Sens. **11**(6), 2134–2146 (2018)
10. Malmgren-Hansen, D., Kusk, A., Dall, J., Nielsen, A.A., Engholm, R., Skriver, H.: Improving SAR automatic target recognition models with transfer learning from simulated data. IEEE Geosci. Remote Sens. Lett. **14**(9), 1484–1488 (2017)

11. Sabour, S., Frosst, N., Hinton, G.E.: Dynamic routing between capsules. In: Advances in Neural Information Processing Systems, pp. 3856–3866 (2017)
12. Zhou, F., Wang, L., Bai, X., Hui, Y.: SAR ATR of ground vehicles based on LM-BN-CNN. IEEE Trans. Geosci. Remote Sens. **56**(12), 7282–7293 (2018)

Signal and Video Processing

Hiding Audio in Images: A Deep Learning Approach

Rohit Gandikota$^{(\boxtimes)}$ [iD] and Deepak Mishra [iD]

Indian Institute of Space Science and Technology, Trivandrum 695547, Kerala, India
grohit0@gmail.com, deepak.mishra@iist.ac.in

Abstract. In this work, we propose an end-to-end trainable model of Generative Adversarial Networks (GAN) which is engineered to hide audio data in images. Due to the non-stationary property of audio signals and lack of powerful tools, audio hiding in images was not explored well. We devised a deep generative model that consists of an auto-encoder as generator along with one discriminator that are trained to embed the message while, an exclusive extractor network with an audio discriminator is trained fundamentally to extract the hidden message from the encoded host signal. The encoded image is subjected to few common attacks and it is established that the message signal can not be hindered making the proposed method robust towards blurring, rotation, noise, and cropping. The one remarkable feature of our method is that it can be trained to recover against various attacks and hence can also be used for watermarking.

Keywords: Steganography · Generative Adversarial Network · Auto-encoder · Deep learning · Watermarking · Latent representations

1 Introduction

Audio is an essential mean of communication in the modern era. Through audio, one can communicate about things that can't be expressed in a visually representative way. It is quick and easy to initiate compared to typing a text message. One can argue that audio transmission is lighter compared to huge video data that use up the bandwidth. There could arise a need for secret communication without any adversary able to eavesdrop or a need to inject an audio message in every original content that we create. Our work attempts to study and understand the requirements for such scenarios and come up with a solution using the most popular architectures currently in deep learning that has the ability to generate data, Generative Adversarial Networks (GAN) [8].

Two common notions exist for hiding information in images. In steganography, the goal is secret communication between a sender and a receiver, while an adversary person cannot tell if the image contains any message. In digital watermarking, the goal is to encode information robustly. Even though an adversary

© Springer Nature Switzerland AG 2019
B. Deka et al. (Eds.): PReMI 2019, LNCS 11942, pp. 389–399, 2019.
https://doi.org/10.1007/978-3-030-34872-4_43

distorts the image, the receiver can extract the message. Watermarking is typically used to identify ownership of the copyright of data. Our method can be used in either case since the message is visually indistinguishable and the network is robust towards most of the attacks. Hiding data can be achieved through techniques that use domain transforms to embed message [4,5,7,9], and others using spatial domain [14]. Recently data hiding using deep neural networks were introduced [10,11,13]. Zhu et al. [10] has introduced the adversarial component in data hiding using an encoder-decoder model to embed and extract respectively.

Most of the works [10–13] deal with hiding images which are stationary signals and 3 dimensional. In this paper, we first discuss the possibilities and limitations on embedding a raw audio signal which is a one dimensional non-stationary signal inside an image using the existing deep learning methods. Further, we demonstrate that hiding the spectrogram of the raw audio and use its frequency domain information of audio as one of the feasible options in our work rather than directly dealing with the non-stationary audio signal. Therefore, we compute the spectrogram of the audio and use the frequency domain information in data hiding instead of dealing with the non-stationary time domain information. The idea of playing in frequency domain is inspired from work by Yang *et al.* [16], who used the adversarial networks for style transfer in audio by modifying spectrogram of the audio signal.

The main contributions of this work are

- Introducing the first generic end-to-end trainable GAN, **VoI-GAN**, for steganography and watermarking that can hide audio in images.
- Introducing latent representations in GAN training for data hiding which made the learning process more efficient, simpler and faster.
- A clean extraction of hidden audio from encoded images (even when distorted), preserving all the key features of the audio like frequency, pitch, speaking rate, and subjective features.

2   Method

2.1   Problem Formulation

Data hiding is to embed a message (MSG) into a cover image (CI) and then later to be able to extract the message from the encoded image (EI). So there is a need for two processes to accomplish a successful secret communication. In this paper, we propose a novel method for the two above mentioned tasks. We propose an autoencoder to reduce the messages to latent representations, so as to help the training of the embedder which is a multi-objective GAN. This base learning of simple latent representation makes the GAN training efficient. An exclusive extractor network with adversarial training to fundamentally extract the embedded messages is also required.

VoI-GAN consists of five nets; an embedder **Emb**, discriminator for embedder **D1**, an extractor **Ex**, a secondary discriminator for extractor **D2** and another autoencoder **AE**, for actual message to latent domain transformation.

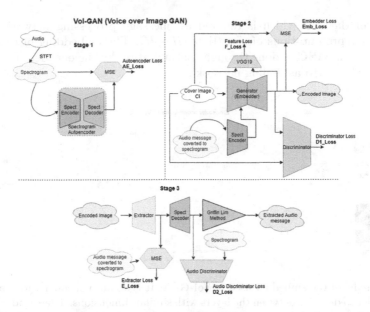

Fig. 1. Three stage explicit training stages of VoI-GAN, a network that can hide audio in images. It consists of a generator, a discriminator, an extractor, an autoencoder for spectrogram encoding and a pretrained VGG19 model for feature extraction.

It is important to note that, GANs are unstable during training [1] and may lead to distortions in output images. We propose a custom multi-objective framework with base training for VoI-GAN, which has resolved the issue of mode-collapse in GAN training.

2.2 Spectrogram Autoencoder

We believe it is since the GAN being a density estimation method, enforces the latent representation to follow simple distributions such as isotropic Gaussian as explored by Subakan *et al.* [15]. Since the autoencoder's latent space is much simpler than the base distribution, it enabled GAN in more accurate modeling over the complicated base distribution. We train an autoencoder to encode the spectrogram into the latent space. After the adversarial embedding is done, the embedded image is passed through an optional noisy channel and the message is recovered using a separately trained extractor. The output being in the latent space is converted back to the spectrogram using the decoder of **AE**. The whole process of embedding and extracting is shown in Fig. 1.

2.3 Embedder

A detailed representation of the embedder is depicted in Fig. 2. It is a convolution auto-encoder network with leaks from the encoder to decoder (U-Net structure). The message spectrogram of size $cW \times cH \times cC$, is first passed through the encoder of the **AE** which is referred as 'message processing network' to get the latent representations of size $lW \times lH \times lC$. The embedding of this latent representation **MSG** is done through concatenation by the encoder of the **Emb** which is referred to as 'hiding network'.

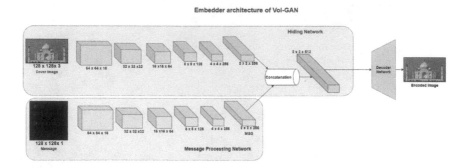

Fig. 2. Model of the embedder used in VoI-GAN. It is an auto-encoder with leaks from the encoder to decoder between the layers with similar dimensions. The encoder has two networks, message processing network and hiding network. Message processing network processes the message to concatenate with the cover image by the hiding network. It is also important to note that the message processing network is the encoder of the trained spectrogram autoencoder **AE**.

From the view of an embedder, the target is to embed the message in a visually indistinguishable way. Therefore, we design a discriminator that consists of stacks of convolutional layers with batch normalization and activation function, to evaluate whether each output of G is true or fake through lower level and semantic features. Through this discriminator network, we introduce the adversarial training into our method. To illustrate clearly, as shown in Fig. 1, the discriminator takes an image of size $cW \times cH \times cC$ as an input and outputs a single number to discriminate between the cover image and encoded image.

2.4   Extractor

To extract the hidden audio message from the encoded image, we propose an exclusive trainable CNN with U-net structure. The extractor **E**, takes EI with size $cW \times cH \times cC$ as input and is trained to extract the MSG of size $mW \times mH \times mC$ as output (as done by [3,10,12]). To get the spectrogram, **MSG** is passed through the decoder of **AE**. Optionally to improve the extraction quality, an adversarial loss can be added in addition to the pixel loss.

To preserve the features of audio like pitch, audio rate, etc., we design a secondary discriminator that distinguishes between the extracted spectrogram and the originally hidden spectrogram. For this end, we design a CNN that consists of stacks of convolutional layers with batch normalization and activation function, to evaluate the extracted spectrogram through lower level and semantic features.

Through this discriminator network, we introduce the adversarial training into the extractor. This discriminator preserves the features of the spectrogram which were ignored by the extractor when pixel loss alone was used.

2.5   Training of VoI-GAN to Encode Audio in Images

The training of VoI-GAN can be disintegrated into an implicit three-stage training process as shown in Algorithm 1. The first stage is to train the autoencoder **AE** to output an image that is similar to its input. As shown in Fig. 1, the autoencoder with parameters ϵ receives a spectrogram SPC^{in} as input and outputs a similar image SPC^{out} as output using Eq. 1.

$$AE_{Loss} = \frac{1}{cH \times cW} \sum_{i=1}^{cH} \sum_{j=1}^{cW} (SPC_{i,j}^{in} - SPC_{i,j}^{out})^2 \tag{1}$$

The second stage of training is to train the embedder to output an image that is similar to the cover image. As shown in Fig. 1, the generator with parameters θ receives a cover image CI and message image MSG as inputs and outputs an image EI as output. Using the adversarial loss, $D1_{Loss}$ in Eq. 2, alone to generate the encoded image was not sufficient as there was dilution effect near the boundary of the image. To overcome this dilution, we introduced a pixel loss between the encoded image and the cover image, referred as Emb_{loss} in Eq. 3.

$$D1_{Loss} = log\mathbf{D1}(CI) + log(1 - \mathbf{D1}(EI)) \tag{2}$$

$$Emb_{Loss} = \frac{1}{cH \times cW} \sum_{i=1}^{cH} \sum_{j=1}^{cW} (EI_{i,j} - CI_{i,j})^2 \tag{3}$$

Where $EI = Embedder_\theta(CI)$ and $\mathbf{D1}$ is Discriminator

Although this has improved the dilution effect, the images are still not sharp enough and had some color distortion. To achieve sharper images, we have introduced a feature loss. For this end, a pre-trained VGG19 network χ whose last pooling layer's output is considered as the features. We compared the features of encoded image EI with features of actual cover image CI and represented it as loss function F_{Loss} in Eq. 4. The feature loss has helped in restoring finer details in the encoded images. This also helped in maintaining similar texture and color information as that of the cover image making the watermark visually indistinguishable.

$$F_{Loss} = \frac{1}{fH \times fW} \sum_{i=1}^{fH} \sum_{j=1}^{fW} (\chi(EI)_{i,j} - \chi(CI)_{i,j})^2 \tag{4}$$

The third stage of training is to train the extractor, with parameters λ, to output the hidden message MSG given the current loop's EI as input. The pixel loss between the original message and the output of the encoder, E_{Loss} in Eq. 5, along with the adversarial loss, $D2_{Loss}$, for extractor output as are designed as loss functions for extractor.

$$E_{Loss} = \frac{1}{wH \times wW} \sum_{i=1}^{wH} \sum_{j=1}^{wW} (MSG_{i,j}^{extracted} - MSG_{i,j}^{actual})^2 \qquad (5)$$

The weights, $\beta_1, \beta_2, \beta_3, \beta_4, \beta_5$ were chosen in such a way that all the losses are scaled to have similar weights in the training. We have given a slightly higher weight for the F_{Loss} since it was experimentally proven to generate more appealing images. Once the hidden spectrogram is extracted, we use the Griffin-Lim's method [6] to recover the audio signal.

Algorithm 1. Three stage implicit training procedure for VoI-GAN

- Train the autoencoder with parameters ϵ such that:

$$\min_{\epsilon} AE_{Loss}$$

- Train the embedder with parameters θ such that:

$$\min_{\theta} \beta_1 Emb_{Loss} + \beta_2 F_{Loss} + \beta_3 D1_{Loss}$$

- Train the extractor with parameters λ such that:

$$\min_{\lambda} \beta_4 E_{Loss} + \beta_5 D2_{Loss}$$

2.6   Training on Attack Simulations

To handle situations like intermediate attacks to remove the message and to check the robustness, a few popular attacks like gaussian blurring, rotation, cropping, and various noises were simulated. The mentioned attacks were simulated on the encoded images and tested the extractor on the attacked samples to produce undistorted spectrogram. Experimentally, the network is inherently robust towards the mentioned attacks, but there is always a scope for a new attack that can break the robustness. The remarkable feature of the network is that it can be trained to recover the message under any attacks. The procedure is similar to that of Algorithm 1, except by inputting the attacked encoded images to the model and train the extractor to recover the message.

2.7   Implementation Details

For the sake of simplicity, we chose 800 audio files of 2 s length those are sampled at 8 kHz. For the embedder, stacks of convolution layers with filters 16, 32, 64, 128, 256 and stride 2 was used instead of max pooling to avoid the losses due to the pooling layer. This has shown an improvement in the generator's loss. Also the input dimensions were chosen as $128 \times 128 \times 3$. The **AE**'s encoder output is of size $2 \times 2 \times 256$ that is concatenated with the cover information of same size in the hiding network. The discriminators used were typical CNNs with an output layer of 1 dimension. In order to obtain a reduction of parameters, we avoided the extensive usage of fully connected layers. We had a total of 6 million parameters in the **Emb**, 2 million parameters in **AE**, 2.2 million parameters in **Ex** and, 1 million parameters in **D1** and **D2**. Adam optimizer with learning rate 0.0002 and momentum 0.5 was used as it is suitable for deep networks. The computational complexity in terms of training time for VoI-GAN with Div2K dataset is around 2 s for 1 epoch with GTX 1080Ti graphics processor.

3 Experiment

3.1 Dataset Details

For initial results, we have used the popular images in image processing society as referred in Table 1. We have used the DIV2K dataset [2] to train and evaluate our model. The dataset consists of 900 images, and we report the average PSNR, and SSIM on the test set. The training data comprises a total of 900 images where 800 are train set and the rest 100 are test images.

3.2 Without Attack Simulations

Apart from dataset analysis, we have also tested on various popular images as cover to check the generalization of the network and calculated the *PSNR* between the actual cover image and the encoded image. Shown in Table 1 are the comparison of different cover images their behavior in encoding the audio data and it is observed that the selection of cover image has not so significant role in encoding of data. *PSNR-CI* refers to the similarity between the encoded image and the cover image. As can be seen, there is no effect on the choice of the cover image. Our method works well for any cover image. On the Div2K dataset, we achieved embedding efficiency of **35.2 dB** PSNR. Extraction efficiency (measure of closeness between original and extracted spectrogram) of **31.44 dB** PSNR and finally structure similarity of **0.9828** between the cover and encoded image has been observed.

Table 1. Comparison on invisibility of message with different cover images.

Cover image	PSNR-CI (in dB)
Peppers	35.71
Mandrill	33.68
Tulips	36.08
Lena	33.95

3.3   After Attack Simulation

The model is quite efficient in hiding data in the cover image. For watermarking purposes, it should be capable of extracting the message from the encoded image even if it is attacked. Hence we test the efficiency of extraction on different audio files under different attacks. The extraction accuracy of the spectrogram of different files on attacks is shown in Table 2. As can be seen, the network is inherently robust to the considered attacks. We believe this inherent robustness comes from the deep architectures along with the autoencoder used to encode the spectrogram. From the extracted spectrogram, the audio is reconstructed using the Griffin-Lim method [6]. The mean opinion scores and *SSIM* of the reconstructed audio files are shown in Table 4. A common complaint from the M.O.S participants was the presence of a slight "hiss" sound in the background of the reconstructed audio. This can be avoided by adding a smoothing loss function in griffin-lim's method.

Table 2. Extraction efficiency of the spectrogram from attacked encoded Images in terms of PSNR (in dB). The last three columns corresponds to the types of noises.

Audio	Blur ($\sigma = 1$)	Rot90°	Rot10°	Crop 10%	S&P	Speckle	Gaussian
Female	55.26	26.86	51.89	58.17	54.80	54.32	54.12
Male	52.99	43.46	57.25	56.83	52.57	52.05	50.22
Car noise	52.76	24.95	51.62	53.90	52.96	53.24	52.92
White noise	51.99	20.81	57.73	52.04	51.33	51.79	51.40

To check the inherent robustness of the network, we gradually increased the blur variance, crop percentage, and rotation angles to plot *PSNR* and it is depicted in Fig. 3. We can further improve robustness by training the VoI-GAN with attacked samples as discussed in Sect. 2.4. However, the plot describes that the network is inherently robust to the popular attacks like blurring, crop, and rotation.

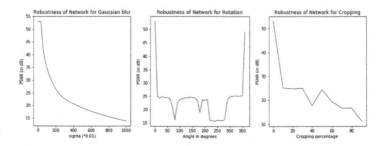

Fig. 3. Robustness of the models for the attacks before attack training. The extraction accuracy *PSNR* decreases with increase in blur, crop, and rotation till 180°. From 180° it is symmetry and hence the valley type of curve.

Since prior work in this area of hiding audio in images is a little scarce, we compared VoI-GAN with two other methods (Model 1 and Model 2) that we engineered over the time period in which we finally developed VoI-GAN. The architectural difference we had with Model 1 is that the extractor doesn't exist as a separate model. Instead, we trained the auto-encoder generator to embed and extract at the same time. This has reduced the number of parameters, however, this doesn't give the freedom to use multiple messages after training. The encoder takes the encoded image as input and is trained to output the message (spectrogram), while the entire generator takes the cover image as input and outputs similar image. Since encoder of the generator can extract the message, it is safe to say that the output of the generator has info about the message, making it "encoded" image. Hence the encoder of the generator is considered as the extractor. To prove that the existing image hiding techniques do not perform well on audio hiding in images, we propose Model 2 which is similarly built as Model 1, except we directly used the audio signal instead of the spectrogram and autoencoder base learning. Table 3 shows the *PSNR* and *SSIM* of the extracted audio files. We have ensured that Model 1 and Model 2 are similar to the prior work [10,12] by testing our Models on Div2K image dataset hiding. Model 2 directly hides the time-domain audio signal, while Model 1 hides the spectrogram. Even though Model 2 gave a decent *PSNR* and visually good extraction, the reconstructed audio was very bad. This is because of the fact that audio is highly non-stationary and can not be hidden directly by using prior methods.

Table 3. Comparison of audio extraction efficiency in terms of SSIM, PSNR (in db) and Mean Opinion Score between different models.

Model	SSIM	PSNR	Mean opinion score
Model 1	0.78	17.47	4.25
Model 2	0.81	21.42	0.00
VoI-GAN	0.998	53.45	4.53

Table 4. Audio Similarity in terms of SSIM and Mean opinion score between original and reconstructed audio file without any attack simulation on encoded Image

Audio file	SSIM	M.O.S
Female	0.998	4.53
Male	0.997	4.56
Car noise	0.992	4.49
White noise	0.989	4.43

4 Conclusion

The Audio over Image-GAN (VoI-GAN), to our knowledge, is the first end-to-end trainable adversarial model that can learn to embed audio signals into images.

The proposed method has shown robustness towards various attacks and hence can be used for watermarking images with audio as well. The network can be made more robust towards any attack by training, which is a remarkable feature. The proposed method is contributing to the field of Data Hiding by introducing a new and robust method to hide audio in images along with paving a research path for audio watermarking for an image. Due to inherent robustness, the model introduced in this work can be used to secretly transmit audio messages for strategic purposes. Methods to hide audio in images without deep learning would be very challenging and are hence not explored well. However, adopting deep learning ways may open new directions and find interesting applications in the field of data hiding. In the future, we would like to improve the architecture for multi-task training and replace the griffin-lim method with a deep network.

References

1. Radford, A., Metz, L., Chintala, S.: Unsupervised representation learning with deep convolutional generative adversarial networks. In: ICLR (2016)
2. Agustsson, E., Timofte, R.: Ntire 2017 challenge on single image super-resolution: dataset and study. In: 2017 IEEE Conference on Computer Vision and Pattern Recognition Workshops (CVPRW), pp. 1122–1131 (2017)
3. Baluja, S.: Hiding images in plain sight: deep steganography. In: Guyon, I., et al. (eds.) Advances in Neural Information Processing Systems 30, pp. 2069–2079. Curran Associates, Inc. (2017)
4. Chen, B., Coatrieux, G., Chen, G., Sun, X., Coatrieux, J.L., Shu, H.: Full 4-D quaternion discrete Fourier transform based watermarking for color images. Digit. Sig. Process **28**(1), 106–119 (2014)
5. Das, C., Panigrahi, S., Sharma, V.K.: A novel blind robust image watermarking in DCT domain using inter-block coefficient correlation. Int. J. Electron. Commun. **68**(3), 244–253 (2014)
6. Griffin, D., Lim, J.: Signal estimation from modified short-time Fourier transform. IEEE Trans. Acoust. Speech Sig. Process. **32**(2), 236–243 (1984)
7. Feng, L.P., Zheng, L.B., Cao, P.: A DWT-DCT based blind watermarking algorithm for copyright protection. In: Proceedings of IEEE ICCIST, vol. 7, pp. 455–458 (2010)
8. Goodfellow, I., et al.: Generative adversarial nets. NIPS **27**, 2672–2680 (2013)
9. Ouyang, J., Coatrieux, G., Chen, B., Shu, H.: Color image watermarking based on quaternion Fourier transform and improved uniform log-polar mapping. Comput. Electr. Eng. **46**, 419–432 (2015)
10. Zhu, J., Kaplan, R., Johnson, J., Fei-Fei, L.: HiDDeN: hiding data with deep networks. In: Ferrari, V., Hebert, M., Sminchisescu, C., Weiss, Y. (eds.) ECCV 2018. LNCS, vol. 11219, pp. 682–697. Springer, Cham (2018). https://doi.org/10.1007/978-3-030-01267-0_40
11. Kandi, H., Mishra, D., Gorthi, S.R.S.: Exploring the learning capabilities of convolutional neural networks for robust image watermarking. Comput. Secur. **65**, 2506–2510 (2017)
12. Zhang, K.A., Cuesta-Infante, A., Xu, L., Veeramachaneni, K.: SteganoGAN: High capacity image steganography with GANs. arXiv preprint arXiv:1901.03892, January 2019

13. Mun, S.M., Nam, S.H., Jang, H.U., Kim, D., Lee, H.K.: A robust blind watermarking using convolutional neural network. arXiv preprint arXiv:1704.03248 (2017)
14. Su, Q., Niu, Y., Wang, Q., Sheng, G.: A blind color image watermarking based on DC component in the spatial domain. Optik **124**(23), 6255–6260 (2013)
15. Subakan, C., Koyejo, O., Smaragdis, P.: Learning the base distribution in implicit generative models. arXiv:1803.04357 (2018)
16. Gao, Y., Singh, R., Raj, B.: Voice impersonation using generative adversarial networks. In: Proceedings of ICASSP, pp. 2506–2510, April 2018

Investigating Feature Reduction Strategies for Replay Antispoofing in Voice Biometrics

Sapan H. Mankad[✉], Sanjay Garg, Megh Patel, and Harshil Adalja

CSE Department, Institute of Technology, Nirma University, Ahmedabad, India
{sapanmankad,sgarg,16BIT020,16BIT063}@nirmauni.ac.in

Abstract. One of the biggest challenges for voice based biometric solutions is in handling replay spoofing attacks. These attacks pose enormous threat on speaker verification system wherein the recorded voice of a genuine user is played in front of the authentication system to attempt unauthorized access. The problem with such system is to distinguish between origin of the input signal whether it comes from a human (live signal) or a device (spoofed signal). In this work, we compare filterbank based features and attempt to choose prominent features by employing some dimensionality reduction strategies. Low level, short-term spectral features have been used to represent audio files. Three methods for feature selection and feature construction are implemented and tested on these features. Results obtained on ASVspoof 2017 version 2 corpus indicate that entropy based feature selection approach gains 9.98% relative improvement over other approaches for feature reduction studied in this work, and an overall performance gain of 13.2% in terms of equal error rate reduction.

Keywords: Replay attacks · ASVspoof 2017 · Feature selection · Entropy · Dimensionality reduction

1 Introduction

Feature reduction is an approach wherein a set of features is replaced with an alternative, lower dimensional representation without significant loss of information. This is done either by removing irrelevant features from original set (known as, feature selection), or by transforming the original input feature space to a new, reduced feature space (feature transformation) by deriving new features (feature construction). Apart from improving performance, feature reduction helps in decreasing training time for model, and computational cost.

Voice, as a biometric, is the easiest mode of communication, and suits the best for person recognition. It provides hands-free access, too. Automatic Speaker Verification (ASV) aims at identifying authenticity of a given voice signal by analysing its characteristics, determining the source of the audio signal, and comparing it with stored patterns of the claimant user. The claim is rejected

© Springer Nature Switzerland AG 2019
B. Deka et al. (Eds.): PReMI 2019, LNCS 11942, pp. 400–408, 2019.
https://doi.org/10.1007/978-3-030-34872-4_44

if the calculated score between the stored model and claimed model does not exceed above a predetermined threshold. An ASV system operates in two phases: (i) enrollment phase - during this phase, the system works by preparing models of all registered speakers, (ii) testing phase - in which the model of a test sample is compared with that of a claimant speaker from stored models, and decision is finally given based on some scoring mechanism. Replay attacks are carried out by presenting pre-recorded audio samples of a genuine speaker to the ASV system. Hence this kind of attacks are often referred to as presentation attacks.

The main contribution of this paper is two fold: (1) We attempt to explore the best set of features which can address the problem of replay spoofing, and (2) investigating the most suitable feature reduction strategy for this task.

The rest of the paper is organized as follows. Section 2 discusses about the existing literature related to this work. An audio representation using filterbank features is given in Sect. 3. Section 4 discusses our proposed work, experimental setup and results discussion. Finally, the paper is concluded in Sect. 5.

2 Related Work

The use of feature selection strategies for speaker identification task dates back in 1975 when [12] proposed a new probability-of-error criterion to rank 92 features derived from audio signals from a synthetic dataset. Only top 5 features were used in their experiment. Noisy and corrupted speech frame removal based feature selection was proposed in [14] on 2006 NIST SRE database. An Ant Colony Optimization (ACO) based feature selection for ASV system was proposed in [8] to optimize dimensionality of feature space on TIMIT dataset[1]. A review of ensemble for feature selection is provided in [1].

A combined approach of feature-feature and feature-class mutual information was proposed in [4] to find optimal features from a high dimensional feature set. Another study in [9] shows that minimal-redundancy-maximal-relevance criterion (mRMR) based feature selection yields promising scope on feature selection and classification accuracy over three classifiers and four datasets. A feature selection based approach using mutual information on incomplete data has been presented in [10], and found effective in terms of computational time and classification accuracy on incomplete data.

Some studies claim that high frequency components in a signal are better able to distinguish live speech from recorded speech. Inverted Mel-filtered cepstral coefficients (IMFCC) based features, which strongly emphasize high frequency regions of the signal have been found very successful in replay spoofing detection [3,6,13,15]. With this motivation, we investigate the best subset of filterbank features, and an alternative reduced representation for replay spoofing detection in this work. We explore entropy measures through decision tree feature selection. Further, we also evaluate the effectiveness of a variant of long-term spectral statistics (LTSS) based features, proposed in [7], and compare their performance with principal component analysis (PCA) based feature construction.

---

[1] https://catalog.ldc.upenn.edu/LDC93S1.

3   Audio Representation

An audio signal can be represented by several features which characterize the
inherent details inside the speech signal for a particular task. Often, in case of
this high dimensionality, it becomes difficult to manage or visualize the feature
space. Moreover, separability among classes is also not clear, feature selection
and dimensionality reduction plays key role in such cases.

3.1   Filterbank Based Features

We extracted four filterbank based features using Bob toolkit[2] as suggested in
[11], and computed their derivatives to obtain the dynamic information. Figure 1
depicts these filterbank structures. A typical process of extracting mel frequency
cepstral coefficients (MFCC) features is shown in Fig. 2. Rest of the features are
computed using the same process by replacing triangular mel-filter bank with
respective filterbank.

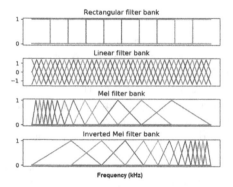

Fig. 1. Filterbanks used for computation [11]

Fig. 2. MFCC feature extraction

A feature vector is 156-dimensional vector consisting of all four features (each
with 39 coefficients). A feature set is 39-dimensional vector consisting of all
features of same kind. In nutshell, feature set is a subset of original feature
vector. For example, an MFCC feature set comprises 39 coefficients, whereas the
entire feature vector is comprised of 156 coefficients. This feature vector is made
up of concatenation of coefficients from all four feature sets.

---

[2] https://www.idiap.ch/software/bob/.

3.2   Entropy Based Feature Selection

Decision trees were primarily meant for classification and regression purpose, but they have also got popularity for feature selection task. A decision tree is typically built by arranging the most representative features near the root of the tree, and discarding the least important features from the dataset. This helps in feature elimination. Decision trees use measures like entropy (information gain) or gini index to decide the feature containing the maximum information with respect to class label information. Information Gain (Entropy) represents the discriminative power of a specific attribute. The foundation of decision trees is based on choosing the most important and meaningful attributes or features in a supervised manner. In this work, we exploit this phenomenun to reduce the dimensionality of our original feature vector.

3.3   Feature Construction

Feature construction is carried out by transforming the existing feature space into new feature space which has less dimensions, yet most of the information is preserved.

PCA transforms data points from existing feature space to a lower dimensional feature space. These lower dimensional features, termed as principal components, are linear combination of original features. These components are obtained in such a way that the first principal component (PC) captures the direction of the maximum variance, and subsequent components get the next possible highest variance. Every principal component is orthogonal to all other PCs.

Long-term spectral statistics (LTSS) features are obtained by computing the mean and variance of frame level short-term features. These features are not calculated exactly using the methodology used in [7], but motivated from there.

4   Proposed Approach

In this work, we compare two main approaches for reducing features: (i) feature construction through dimensionality reduction i.e. feature construction using (a) principal component analysis, and (b) LTSS based features, and (ii) feature subset selection using entropy.

4.1   Approach

Figure 3 describes the scenario of experiments. In one scenario, a PCA based reduction was carried out and impact of principal components over accuracy was computed, whereas in other scenario, entropy based feature selection was exploited. In this scenario, entropy values of all feature sets were computed. All the coefficients with entropy above 0.1 were selected, and rest were discarded. Entropy values for different coefficients of IMFCC feature set is shown in Fig. 4.

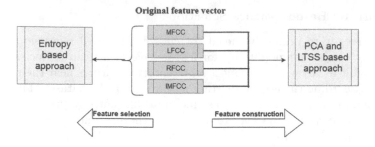

Fig. 3. Two scenarios for feature reduction used in this work.

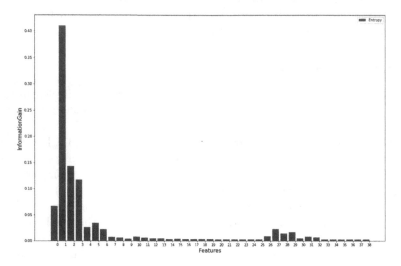

Fig. 4. Entropy values for IMFCC features

4.2   Implementation Scenario

In this section, we discuss scenario for implementation for our proposed work. Since the lower order coefficients contain more information pertaining to speaker-specific characteristics, we used the first 13 coefficients as the static features. After appending dynamic features, each audio file was represented by a feature of 39 coefficients.

A neural network classifier was trained with single hidden layer. Python Scikit-learn library was used for implementation. A Gaussian Mixture Model (GMM) with default parameters was also used for classification. A typical performance metric for speaker verification systems is equal error rate (EER). It is calculated based on threshold at which false acceptance rate (FAR) and false rejection rate (FRR) are equal. We have presented our results in terms of EER. We tested the proposed approach on ASVspoof 2017 version 2.0 corpus [5]. More details of the dataset can be found from [2].

Table 1. Results of four systems in terms of Equal Error Rate (EER) over features selected using entropy measure on two classifiers

Original feature set	No. of coefficients	Index	ANN	GMM
ALL156	156	—	0.4578	NA
BEST10	10	—	0.4567	NA
MFCC39	3	3,5,7	0.4391	0.4908
LFCC39	1	2	0.5156	0.5094
RFCC39	3	3,5,7	0.4445	0.51
IMFCC39	3	1,2,3	**0.397**	0.4926

4.3 Results and Discussion

Initially, a 156-dimensional feature vector was used to classify the dataset, and it achieved 0.4578 equal error rate (EER) on a multilayer perceptron classifier.

Table 2. EER performance of PCA based dimensionality reduction on Eval set

Feature	# of components	EER
MFCC	2	**0.4704**
LFCC	2	0.5502
RFCC	2	0.5180
IMFCC	2	0.5708
MFCC	3	0.5393
LFCC	3	0.5535
RFCC	3	0.5150
IMFCC	3	0.5660

Table 3. LTSS features performance (EER)

System	Using mean and variance	Mean only
IMFCC	0.5008	0.5433
MFCC	0.4986	0.4410
RFCC	0.4995	0.5
LFCC	0.5	0.4648

Table 1 lists the number of coefficients chosen from each feature set using entropy based feature selection. It is evident that all are static coefficients. A combination of ten best coefficients was used as a feature representation for audio which produced an EER value of 0.4567, stating that there is no significant improvement on performance if we combine the best coefficients from all four feature sets. As seen from Table 1, it is observed that using only three best coefficients of IMFCC feature yield considerable improvement of 13.2% on the performance (0.4578 to 0.397). GMM classifier also provided similar performance, hence we carried out further experiments with neural network. Using maximum mutual information (MMI) feature selection method, we obtained the same set of coefficients for each feature.

Table 2 summarizes the results on reduced data with two and three principal components. It can be seen that for 2-dimensional case, MFCCs show the best performance, whereas IMFCCs show the worst performance. This indicates that PCA is not able to capture high frequency characteristics of features while performing dimensionality reduction. Moreover, results with three PCs show either degradation or slight improvement in performance for all features. MFCC performance is highly affected due to addition of one more principal component. In no way, PCA based model is able to outperform the entropy based model.

We extracted a variant of long-term spectral statistics (LTSS) based features in following way. Mean and variance of all coefficients across all frames was computed to represent these long-term parameters. Thus, every audio file was represented by two dimensions for each feature. Table 3 shows the results of LTSS features on the evaluation set of the dataset. Each feature is represented by two coefficients i.e. mean and variance. It can be seen that in this case also, MFCCs give good results and they achieve the minimum EER. However, it is interesting to note that its performance is not better than the one shown in entropy based feature selection scenario. This shows that feature construction is related to human auditory perception and its representation (in lower dimension) goes alongwith auditory nerve behaviour. The best performance is given by a subset of static IMFCC features which is a relative improvement of roughly 10% compared to LTSS mean based classification.

Table 4. Performance of systems with combination of mean and variance coefficients.

# of coefficients	Features	EER	# of coefficients	Features	EER
4	MFCC+LFCC	0.4815	6	MFCC+LFCC+RFCC	0.4712
4	MFCC+RFCC	**0.4645**	6	MFCC+LFCC+IMFCC	0.4779
4	MFCC+IMFCC	0.4666	6	MFCC+RFCC+IMFCC	0.4728
4	LFCC+RFCC	0.476	6	LFCC+RFCC+IMFCC	0.4958
4	LFCC+IMFCC	0.5436	8	MFCC+LFCC+RFCC+IMFCC	0.5
4	RFCC+IMFCC	0.5047			

Results also show that RFCC features are not sensitive to standard deviation or variance coefficients as there is negligible change in their performance with or without variance. However, other three features show some sensitivity towards these coefficients. The equal error rate drops by almost 7.82% (0.5433 to 0.5008) when variance information is appended for IMFCCs. In contrast, MFCC and LFCC show good results in absence of variance information.

We tried different combination of these features to examine its effect on overall performance. From the results shown in Table 4, we can observe that minimum EER is achieved with combination of mean and variance of MFCC and RFCC features, yet they cannot outperform the standalone MFCC feature performance on the same case.

5   Conclusion and Future Work

We presented a comparative analysis of filterbank based short-term spectral features using feature selection and feature construction approach. A comparison was given among entropy based feature selection, and PCA and LTSS based feature construction. In this paper, we attempted to explore the possibility of finding best representative feature reduction strategies for the task of voice playback spoofing detection. Results indicate that entropy based feature selection gives promising results. Further, it can be seen that IMFCC features capture the replay attack information more significantly than any other feature. This shows that human perception (MFCCs) may not help much for detection of playback spoofing attacks, and this may also fail while performing subjective evaluation of such systems.

In future, recent techniques for feature visualization can be adopted to see class distribution among input data points. This information may be helpful for designing robust feature selection strategies.

References

1. Boln-Canedo, V., Alonso-Betanzos, A.: Ensembles for feature selection: a review and future trends. Inf. Fusion **52**, 1–12 (2019)
2. Delgado, H., et al.: ASVspoof 2017 version 2.0: meta-data analysis and baseline enhancements. In: Odyssey (2018)
3. Hanilci, C.: Features and classifiers for replay spoofing attack detection. In: 10th International Conference on Electrical and Electronics Engineering (ELECO), pp. 1187–1191, November 2017
4. Hoque, N., Bhattacharyya, D., Kalita, J.: MIFS-ND: a mutual information-based feature selection method. Expert Syst. Appl. **41**(14), 6371–6385 (2014)
5. Kinnunen, T., et al.: The ASVspoof 2017 challenge: assessing the limits of replay spoofing attack detection (2017)
6. Mankad, S.H., Shah, V., Garg, S.: Towards development of smart and reliable voice based personal assistants. In: TENCON 2018, pp. 2473–2478 (2018)
7. Muckenhirn, H., Korshunov, P., Magimai-Doss, M., Marcel, S.: Long-term spectral statistics for voice presentation attack detection. IEEE/ACM Trans. Audio Speech Lang. Process. **25**(11), 2098–2111 (2017)
8. Nemati, S., Basiri, M.E.: Text-independent speaker verification using ant colony optimization-based selected features. Expert Syst. Appl. **38**(1), 620–630 (2011)
9. Peng, H., Long, F., Ding, C.: Feature selection based on mutual information criteria of max-dependency, max-relevance, and min-redundancy. IEEE Trans. Pattern Anal. Mach. Intell. **27**(8), 1226–1238 (2005)
10. Qian, W., Shu, W.: Mutual information criterion for feature selection from incomplete data. Neurocomputing **168**, 210–220 (2015)
11. Sahidullah, M., Kinnunen, T., Hanilçi, C.: A comparison of features for synthetic speech detection. In: Interspeech (2015)
12. Sambur, M.R.: Selection of acoustic features for speaker identification. IEEE Trans. Acoust. Speech Sig. Process. **23**(2), 176–182 (1975)
13. Sriskandaraja, K., Suthokumar, G., Sethu, V., Ambikairajah, E.: Investigating the use of scattering coefficients for replay attack detection. In: APSIPA, pp. 1195–1198, December 2017

14. Sun, H., Ma, B., Li, H.: An efficient feature selection method for speaker recognition. In: 2008 6th International Symposium on Chinese Spoken Language Processing, pp. 1–4, December 2008
15. Witkowski, M., Kacprzak, S., Zelasko, P., Kowalczyk, K., Galka, J.: Audio replay attack detection using high-frequency features. In: Interspeech (2017)

Seismic Signal Interpretation for Reservoir Facies Classification

Pallabi Saikia[(✉)], Deepankar Nankani, and Rashmi Dutta Baruah

Indian Institute of Technology Guwahati, Guwahati, Assam, India
{pallabi.s,d.nankani,r.duttabaruah}@iitg.ac.in

Abstract. Understanding facies distribution in a hydrocarbon reservoir is an important aspect to characterise a hydrocarbon reservoir. Facies classes are basically based on rock characteristics that can indicate the locations of good quality of sand and presence of hydrocarbon, thereby helping the geologists to decide the location to drill a production well in a hydrocarbon field. However, due to the heterogeneous and nonlinear nature of earth subsurface, gathering facies information becomes a critical task. Researchers from different domains such as Machine learning, Geology, and Geophysics work towards the understanding of facies distribution. However, with the increased complexity of the reservoir, its interpretation becomes difficult. This work describes a case study that involves a framework to classify the facies categories in a reservoir using seismic data by employing different machine learning models. The framework is also capable to handle the data imbalance problem that occurs quite often while studying these kinds of datasets. Moreover, for gaining more confidence in the developed model, we used Local Interpretable Model-Agnostic Explanations to provide the interpretation of the model. The interpretations generated can be helpful for geologists to rate the applicability of our developed model in their domain.

Keywords: Seismic interpretation · Hydrocarbon reservoir characterisation · Neural network · Model interpretability

1 Introduction

Understanding a reservoir to determine where to drill a production well comes under Reservoir Characterisation (RC) domain. RC [11] aims to identify potential location of hydrocarbon presence by modelling earth's subsurface. To get most comprehensive understanding of a reservoir, geophysical data sources are utilized in RC. Seismic signals obtained from the seismic survey is the most common geophysical data source that is used to interpret the earth subsurface without a direct penetration into the earth crust. Seismic survey is performed by sending the acoustic signals to the earth crust, which are reflected back by the different layers of earth subsurface. The reflected signals are then recorded, analysed and processed to interpret the underlying subsurface characteristics. Interpretation of the earth subsurface in terms of facies classes is important to

© Springer Nature Switzerland AG 2019
B. Deka et al. (Eds.): PReMI 2019, LNCS 11942, pp. 409–417, 2019.
https://doi.org/10.1007/978-3-030-34872-4_45

distinctly define rocks of interest and to build a better understanding of the depositional environments of the earth subsurface [13]. With the emergence of Machine Learning (ML) in various fields of applications in recent years, many researchers suggested the use of data-driven methods in this domain [2,6,12,16–18,20,24]. Among the data driven models, the Artificial Neural Network (ANN) is one of the research hot spots in recent years. Many researchers are focusing on how to effectively apply ANN [10,21,22] in this field to improve the modelling accuracy. The researchers [8,10,14] suggests that neural network has the capability to solve complex calculations and can discover very complex relations within the data that the conventional algorithms often fail to discover. The model has also been successfully applied in petroleum engineering even with sparse data [4,5]. However, in the literature of facies classification [3,8,9,19], it is observed that facies classification is challenging when we try to interpret from seismic data compared to the interpretation from well logs. This is because of the complex relationship between geophysical data obtained from different sources. Moreover, the uncertainty from seismic data interpretation is higher compared to well logs interpretation as the seismic data cannot see the earth subsurface as the same resolution as well logs see. Hence, in the fields of having limited hydrocarbon wells understanding the relation between seismic and facies can be challenging. Moreover, due to the heterogeneous nature of subsurface, the data obtained can be imbalanced in terms of available classes and building a machine learning model over it can lead to a biased model towards the major classes.

In this work, we are considering a case study over a hydrocarbon field that consists of limited hydrocarbon wells (seven wells), having imbalanced data samples. The task here is to classify the three types of facies (shale, brine sand, and hydrocarbon bearing sand) available over the field from the acquired seismic data. To solve this problem, we developed a framework that consists of: Preprocessing to make the data suitable for the ML model applicability; Modelling using ANN to model the complex relationship between seismic and facies classes, Interpretation of the developed model to understand what made the model work as it is. The strategy of the framework used here can be used by any researcher in this field without the need of having much domain understanding and the end result of the framework can be provided to experts to analyse the model for its applicability in their field. It is worth to specify here that the interpretability of the ML model was never been performed in this field unlike other domains of ML. This is an initial work in this field to the best of our knowledge.

The organisation of this paper is as follows. Section 2 provides the data description. Section 3 provides the detailed description of the methodology followed. Section 4 provides details of experimentation performed with the results obtained. Section 5 provides the conclusions and the future work.

2 Data Description

The dataset we have is a confidential dataset provided by GEOPIC, ONGC, that consists of data from seven wells (Total 36000 instances with each well constituting around 5000 to 6000 instances) in a considered hydrocarbon field. The data consists of fourteen seismic attributes and three classes of facies. Class 0 corresponds to shale, Class 1 corresponds to Brine sand, Class 2 corresponds to Hydrocarbon bearing sand. Fourteen seismic attributes consists of: Amplitude Envelope(X1), Amplitude Weighted Cosine Phase (X2), Amplitude Weighted Frequency (X3), Amplitude Weighted Phase (X4), Derivative (X5), Dominant Frequency (X6), Instantaneous Frequency (X7), Instantaneous Phase (X8), Integrate (X9), Integrated Absolute Amplitude (X10), P-Impedance (X11), Quadrature Trace (X12), Seismic (X13), and VpVs (X14) that are extracted from the main reflected seismic signal recorded with respect to Time (X0). With this data, we need to model the relation between the seismic attributes and facies, to correctly identify the facies classes in the hydrocarbon field from seismic data.

3 Methodology

The proposed framework is provided in Fig. 1. The steps consists of Preprocessing, Modelling, and Interpretation. Preprocessing is performed to handle the imbalanced dataset by oversampling the minority classes and undersampling the majority class data, normalisation of the input (seismic) attributes using Z-score normalisation, and converting the output (facies classes) to the one hot encoding to make the data suitable for ML models. The modeling phase consists of applying ANN with different ML models for the comparisons in modelling the complex relationship between the input (seismic) and the output (facies class). The last step, interpretability is performed using Local Interpretable Model-Agnostic Explanations to understand the working of a model. The detailed description and motivation for each step of the framework is provided in Subsects. 3.1 to 3.3.

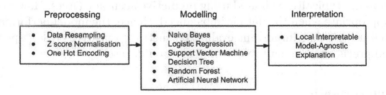

Fig. 1. Proposed framework

3.1 Preprocessing

It can be observed from Table 1 that the original dataset distribution depicts a high bias towards class 0 of around 85% whereas other two classes in combined contributed only 14% of the total data. If this dataset is used to build a model, the model will be highly biased towards majority class. This motivated us to use a resampling method. The resampling of the dataset is performed with Synthetic Minority Oversampling Technique (SMOTE) [7]. This is a popular technique in

Table 1. Description of classes before and after resampling

		Original database		Resampled database	
Class	Name	Instances	% of data	Instances	% of Data
0	Shale	30,720	85.33	15000	66.67
1	Brine Sand	4,152	11.53	6000	26.67
2	Hydrocarbon Sand	1,128	3.13	1500	6.67

literature to handle imbalance dataset. The technique is used to reduce the imbalance in the dataset by synthetically generating minority class instances based on the feature space similarities between existing minority instances. Table 1 shows the dataset distribution after resampling. SMOTE is used to generate more samples from Class 1 and Class 2. SMOTE alone is not sufficient to sample the data effectively as generating too many samples of a class will lead to dominance on noise in the dataset. Therefore, we also implement random under-sampling on Class 0 to reduce its samples and make a lesser biased dataset [1].

3.2 Modelling

For modelling, we applied various state of the art methods of classification and compared their results. However, due to the state of art performance of Neural Networks as provided in the literature, we preferred neural network over other ML models and compared it with Naive Bayes (NB), Support Vector Machine (SVM), Decision Tree (DT), and Random Forest (RF), to see how effectively they model the relation between seismic and facies. The performance of ML algorithms are typically evaluated using predictive accuracy (Acc). However, this is not appropriate when the data is imbalanced. Hence, the model performance is evaluated using other important evaluation metrics, sensitivity (Se), specificity (Sp) and precision (Pr), suitable for imbalanced dataset.

3.3 Interpretation

The last step of our framework explains the prediction of a model using Local Interpretable Model-Agnostic Explanations (LIME) [15]. LIME explains the model by perturbing the input data and correlating the effect of these changes on predictions made by the model. One advantage of using LIME is that it is model-agnostic technique and can be applied to understand the explanations of any model. Other model-specific methods try to understand the model by studying the internal components and their behavior. For example, investigation of activation units and linking of internal activations to inputs are performed in case of most deep learning models interpretability [23]. This requires deep knowledge of the model and is not applicable to other models. LIME generates a list of explanations, showing the effect of each feature on the output. The explanations are created by using interpretable models like decision trees and linear models. The interpretable models are trained in close vicinity of data instance i.e. understanding complex models by application of simpler models locally. With the use of LIME, we can provide the explanations of our model predictions to the geoscience experts to determine whether the model explanations satisfied their requirements for its deployment in the real world.

4   Experimentation and Results

To build and study the performance of models we first take one well aside (Well 1 with 5645 samples), to use it as test data to evaluate the performance of the models on unknown location. We then combine the data of the remaining six wells (31355 samples as train set) to train the models. To evaluate model performance we keep 30% of train data as validation set. Several runs of training, testing, and validation phases have been carried out in order to decide the best set of hyperparameters and parameters for every ML model. Figure 2 shows the performance on our dataset with resampling as well as without resampling with the considered ML models. NB performed poorly in validation dataset, however its performance is better in test dataset, in comparison to other models. On the contrary, DT and RF performed remarkably better (almost 100%) on the validation data but performed very poor on the test dataset. It looks a classic case of overfitting that the models could not generalise to the samples that belong to a different well from training wells. Coming to the ANN model performance, its performance is not as worse as NB in validation set, and does not overfit like DT and RF. Hence, the model is able to outperform all these models on the test dataset. We can also observe that ANN with more number of hidden layers (also called as deep neural network (DNN)) is able to improve accuracy on the test set but could not provide better Precision, Sensitivity and Specificity compared to its shallow version. This can be due to the more number of layers in DNN helping towards improvement of accuracy due to further improvement in loss function, but it is ignoring other important performance criteria of the model in imbalance dataset. However, in the scenario of balanced dataset, DNN has got the capability to improve performance in every aspects. Hence we can infer that ANN approximated a very good relation between seismic and facies classes as they can build more complex boundaries which is required for a real world dataset. ANNs generalises well, which can be seen as the improved accuracy of the test dataset. The above observations are true for both, with sampling as well as without sampling. However, better performance is obtained with the resampled dataset. The accuracy of the models remained almost similar, however, the sensitivity and specificity improved by almost 3% to 6%, which are the major evaluation criteria for imbalanced dataset. From the analysis, we came to a conclusion that ANN provided the best performance. However, being a black box model, the geoscientists find it difficult to gain trust in the model to make it deployable for the real world. So, we have used LIME interpretation as provided in Fig. 3 to explain about the predictions of the model. Figure 3 presents one explanation for each of the classes using LIME. The variables occurring in the right side of the axis represents positive contribution, whereas the variables occurring in the left side represents negative contribution. From the interpretation result of Class 0, it can be seen that features X11, X6, X0, X9, X1 and X2 contributes positively, whereas feature X13 and X10 contribute negatively for the prediction of Shale class. Other attributes have very negligible contribution or no contribution at all. Similarly, the result of model prediction belonging to Class 1, depicts that features X11, X6, and X10 contribute positively, whereas X0, X4, and X12 contribute negatively to the classification of this instance to Brine Sand Class. The result of model prediction belonging to Class 2 depicts that features X9, X11, and X8 contribute positively, whereas X12 and X4 contribute negatively to the classification of this instance to Hydrocarbon Sand Class. Moreover, Values of these features belonging to a certain range (as specified) is the sole reason behind classifying these instances to a particular class. This kind of interpretation can be helpful for geoscientists to validate if the reason for the classification of the instance should be trusted or not, and hence they can accordingly decide the models applicability for the real world.

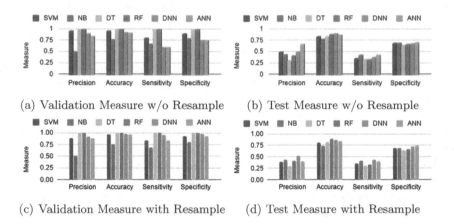

(a) Validation Measure w/o Resample   (b) Test Measure w/o Resample

(c) Validation Measure with Resample   (d) Test Measure with Resample

Fig. 2. Model evaluation using performance metrics: precision, accuracy, sensitivity, and specificity

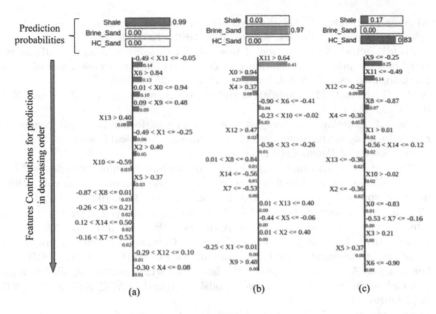

Fig. 3. LIME explanations for instances of (a) Class 0, (b) Class 1, (c) Class 2

5   Conclusion and Future Work

This paper presents a data driven based framework on facies classification. The challenge of having imbalanced classes in the data is successfully handled with an upsampling method, namely SMOTE. The method effectively deals with bias in the dataset towards specific classes. ANN is built for the purpose of modelling complex relationship within the data, provided the best performance measure on the test data that belongs to a completely new well. As important decisions on drilling a new well have to be taken based on model predictions, this paper also presented LIME results to interpret the predictions made by the model. It is observed from the interpretation that, not all seismic attributes are affected by facies characteristics. Some facies classes effect some seismic attributes more than the other. This is a very initial work on interpretation in this field. However, more analysis can be performed on interpretation of model that comprises the domain knowledge. Also, it's worth mentioning that the composition beneath the earth subsurface varies greatly from location to location that limits the applicability of a model from one field to another, that opens new avenues for more research on modelling the complex underlying relation in this field.

References

1. Agrawal, A., Viktor, H.L., Paquet, E.: Scut: multi-class imbalanced data classification using smote and cluster-based undersampling. In: International Joint Conference on Knowledge Discovery, Knowledge Engineering and Knowledge Management, vol. 1, pp. 226–234. IEEE (2015)

2. Al-Anazi, A., Gates, I.: On the capability of support vector machines to classify lithology from well logs. Nat. Resour. Res. **19**(2), 125–139 (2010)
3. Ashraf, U., et al.: Classification of reservoir facies using well log and 3D seismic attributes for prospect evaluation and field development: a case study of Sawan gas field, Pakistan. J. Petrol. Sci. Eng. **175**, 338–351 (2019)
4. Auda, G., Kamel, M.: Modular neural networks: a survey. Int. J. Neural Syst. **9**(02), 129–151 (1999)
5. Ayala, L.F., Ertekin, T.: Analysis of gas-cycling performance in gas/condensate reservoirs using neuro-simulation. In: SPE Annual Technical Conference and Exhibition. Society of Petroleum Engineers (2005)
6. Bhattacharya, S., Mishra, S.: Applications of machine learning for facies and fracture prediction using Bayesian network theory and random forest: case studies from the appalachian basin, USA. J. Petrol. Sci. Eng. **170**, 1005–1017 (2018)
7. Chawla, N.V., Bowyer, K.W., Hall, L.O., Kegelmeyer, W.P.: Smote: synthetic minority over-sampling technique. J. Artif. Intell. Res. **16**, 321–357 (2002)
8. Dubois, M.K., Bohling, G.C., Chakrabarti, S.: Comparison of four approaches to a rock facies classification problem. Comput. Geosci. **33**(5), 599–617 (2007)
9. Ferreira, D.J.A., et al.: Unsupervised seismic facies classification applied to a presalt carbonate reservoir, santos basin, offshore Brazil. AAPG Bull. **103**(4), 997–1012 (2019)
10. Imamverdiyev, Y., Sukhostat, L.: Lithological facies classification using deep convolutional neural network. J. Petrol. Sci. Eng. **174**, 216–228 (2019)
11. Lines, L.R., Newrick, R.T.: Fundamentals of geophysical interpretation. Society of Exploration Geophysicists (2004)
12. Mishra, S., Datta-Gupta, A.: Applied Statistical Modeling and Data Analytics: A Practical Guide for the Petroleum Geosciences. Elsevier, Amsterdam (2017)
13. Mur, A., Waters, K.: Play scale seismic characterisation-using basin models as an input to seismic characterisation in new and emerging PLA. In: 80th EAGE Conference and Exhibition (2018)
14. Nakutnyy, P., Asghari, K., Torn, A., et al.: Analysis of waterflooding through application of neural networks. In: Canadian International Petroleum Conference. Petroleum Society of Canada (2008)
15. Ribeiro, M.T., Singh, S., Guestrin, C.: Why should i trust you?: explaining the predictions of any classifier. In: Proceedings of the 22nd ACM SIGKDD International Conference on Knowledge Discovery and Data Mining, pp. 1135–1144. ACM (2016)
16. Salehi, S.M., Honarvar, B.: Automatic identification of formation lithology from well log data: a machine learning approach. J. Petrol. Sci. Res. **3**(2), 73–82 (2014)
17. Sebtosheikh, M.A., Salehi, A.: Lithology prediction by support vector classifiers using inverted seismic attributes data and petrophysical logs as a new approach and investigation of training data set size effect on its performance in a heterogeneous carbonate reservoir. J. Petrol. Sci. Eng. **134**, 143–149 (2015)
18. Silva, A.A., Neto, I.A.L., Misságia, R.M., Ceia, M.A., Carrasquilla, A.G., Archilha, N.L.: Artificial neural networks to support petrographic classification of carbonate-siliciclastic rocks using well logs and textural information. J. Appl. Geophys. **117**, 118–125 (2015)
19. Vashist, N., Dennis, R., Rajvanshi, A., Taneja, H., Walia, R., Sharma, P.: Reservoir facies and their distribution in a heterogeneous carbonate reservoir: an integrated approach. In: SPE Annual Technical Conference and Exhibition. Society of Petroleum Engineers (1993)

20. Wang, G., Carr, T.R., Ju, Y., Li, C.: Identifying organic-rich Marcellus Shale lithofacies by support vector machine classifier in the Appalachian basin. Comput. Geosci. **64**, 52–60 (2014)
21. Wei, Z., Hu, H., Zhou, H.W., Lau, A.: Characterizing rock facies using machine learning algorithm based on a convolutional neural network and data padding strategy. Pure Appl. Geophys. 1–13 (2019)
22. Zhang, D., Yuntian, C., Jin, M.: Synthetic well logs generation via recurrent neural networks. Petrol. Explor. Dev. **45**(4), 629–639 (2018)
23. Zhang, Q.S., Zhu, S.C.: Visual interpretability for deep learning: a survey. Front. Inf. Technol. Electron. Eng. **19**(1), 27–39 (2018)
24. Zhao, T., Jayaram, V., Roy, A., Marfurt, K.J.: A comparison of classification techniques for seismic facies recognition. Interpretation **3**(4), SAE29–SAE58 (2015)

Analysis of Speech Source Signals for Detection of Out-of-breath Condition

Sibasis Sahoo[⊠] and Samarendra Dandapat

Department of Electronics and Electrical Engineering,
Indian Institute of Technology Guwahati, Guwahati 781039, India
{sibasis2016,samaren}@iitg.ac.in

Abstract. This work focuses on analysing the changes in speech source signals under physical exercise induced out-of-breath conditions. A new database is recorded for sustained vowel phonations (SVPs). Electroglottogram (EGG) and speech signals are recorded in two simultaneous channels. Morphological changes in EGG signal are analysed using a set of five temporal features related to glottal opening and closing. Two source signals, zero frequency filtered (ZFF) and integrated linear prediction residual (ILPR) signals, are estimated from the speech signal. Their respective spectrums are analysed using a harmonic peak based feature. EGG based features indicate changes in vibration pattern of vocal folds. At the same time, analysis on ZFF and ILPR show that they too change from the normal condition. Classification performance of these features are evaluated by support vector machine (SVM), and k-nearest neighbour (KNN). The accuracies are obtained to be nearly 70% for both the classifiers in case of EGG and speech source signals.

Keywords: Out-of-breath · Glottal source · ZFF · ILPR · SVM · KNN

1 Introduction

While performing any physical exercise, we feel out-of-breath; our body demands more oxygen, which makes the rate of breathing faster and deeper. If we attempt to speak, the speech signal thus produced is perceptually different from that of normal condition; as the phenomenon of breathing provides the driving force behind speech communication [3]. A plot of a sample speech signal and its corresponding spectrogram is shown in Fig. 1 for both the normal and the out-of-breath conditions. For out-of-breath case it shows: reduction in signal duration, lessened pause duration, increase in signal amplitude as well as weakened harmonics at higher frequencies. According to the source-filter theory of speech production, the vocal tract (VT) system is driven by a source signal to produce sound. The source is characterized by air passing through the glottis: an opening formed by vibrating vocal folds situated inside the larynx. Considering the above delicate connection between the respiration process and the phonatory process, we can expect the source signal to get influenced by the out-of-breath condition.

© Springer Nature Switzerland AG 2019
B. Deka et al. (Eds.): PReMI 2019, LNCS 11942, pp. 418–426, 2019.
https://doi.org/10.1007/978-3-030-34872-4_46

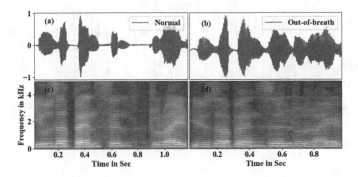

Fig. 1. Normal and out-of-breath speech signal in (a), (b) and their respective spectrograms in (c), (d) respectively.

There have been a few studies on the effect of out-of-breath condition on speech signal. Trouvain et al. [13] showed that under out-of-breath condition, there is an increase in subglottal pressure that leads to higher pitch frequency (F0). Godin et al. [8] analysed formant frequencies F1 and F2, and glottal open quotient using vowel and consonant-vowel utterances. Using both speech and EGG signals, they found that formats F1 and F2; and open quotient of glottis are effected differently for different speakers. Also in [9] they showed that the influence of out-of-breath condition on vocalfold vibration using six glottal features. Some authors studied the classification between normal and out-of-breath speech using speech extracted features like MFCC and teager energy based features [11]; Fourier model based harmonic features [5]. Few works have been reported which deal with the type of changes that occur on the vibrating pattern of vocal folds. It is also still unclear how much the source signal gets affected under out-of-breath condition. Hence, in this work, we have focussed on two tasks for analysing source signal. First, EGG signal has been analysed for finding out the changes in glottal vibration pattern. In the second task, two source signals, namely zero frequency filtered (ZFF) and integrated linear prediction residual (ILPR), are analysed. In literature, ZFF [10] and ILPR [12] are used as an approximation to the source for speech production. In this work, ZFF and ILPR are analysed to know whether they get affected under out-of-breath condition. Rest of the sections are divided as follows: in Sect. 2, the methodology of our analysis is presented. Results and discussions are described in Sect. 3 followed by conclusion in Sect. 4.

2 Methodology

Under physical exertion, the speaker tries to suppress its effect by adjusting breathing and speaking durations. In this work, we are considering vowel segments from SVPs to analyse the effect of out-of-breath condition. EGG and speech signals are both taken into consideration. The task of analysing these

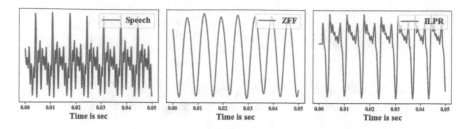

Fig. 2. Speech, ZFF and ILPR sample segment of /a/ vowel.

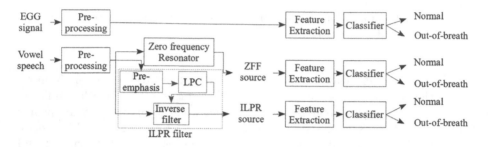

Fig. 3. Block diagram showing EGG and vowel speech processing and classification.

signals has three common sub-tasks: pre-processing, feature extraction and classification. The corresponding block diagram is shown in Fig. 3.

2.1   Pre-processing

The low-frequency trend in EGG signal is removed by filtering it by a Butterworth high-pass filter of order 3 with cut-off frequency $f_c = 50$ Hz set experimentally. The amplitude is normalised to the range -1 to $+1$. Along with EGG signal, the first difference of EGG (DEGG) is also considered, which indicates the instances as well as the rate of glottal opening and closing. Figure 4(a) shows a schematic EGG and DEGG signal. Similarly, the vowel speech signal is made to pass through mean removal and normalisation steps, followed by ZFF and ILPR extraction.

ZFF Source. Zero frequency resonator is applied on speech signal to remove the influence of vocal tract [10]. The signal thus obtained is an approximation of the source signal. It is sinusoidal like signal whose periodic property is inherited from the periodic vibration of vocal folds. It shows negative to positive zero crossing at the instances of glottal closure [1]. Hence, it is used for detecting glottal closure and voicing region in speech signals. Figure 2 shows sample segments of speech and its corresponding ZFF signal.

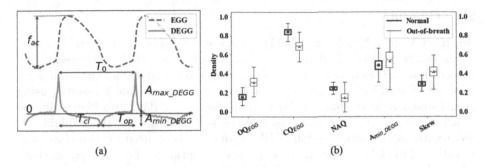

(a)   (b)

Fig. 4. (a) Sample EGG and DEGG signal with time and amplitude parameters, (b) Boxplots of five features for vowel /a/ for one speaker.

ILPR Source. It is obtained by performing inverse filtering on the speech signal. The inverse filter is derived from the pre-emphasized speech signal [12]. The choice of non-pre-emphasized speech makes the source signal to have more prominent peaks at glottal closures and smaller peaks at glottal opening instances. ILPR has a higher resemblance to DEGG signal as it is more or less unipolar in nature than that of the pre-emphasized source [1]. It is used for detecting voicing and glottal closing instances. Figure 2 shows sample segments of speech and its corresponding ILPR signal.

2.2   Feature Extraction

For analysis of EGG signal, a set of five EGG and DEGG based features are studied which indicate the changes in vocal fold vibration pattern. These are open quotient (OQ_{EGG}), close quotient (CQ_{EGG}), normalized amplitude quotient (NAQ), DEGG strength at glottal opening instance (A_{min_DEGG}) and skewness of the EGG waveform. At the same time, for analysis of ZFF and ILPR source signals, magnitude difference between the first two harmonics (H1-H2) is considered as a feature. Features are collected for every frame of duration 30 ms with overlapping of 20 ms.

Open quotient ($OQ_{EGG} = \frac{t_{op}}{T_0}$) and Close quotients ($CQ_{EGG} = \frac{t_{cl}}{T_0}$) define the fraction of time that the vocal folds remain open or closed with respect to a glottal cycle [14]. Where, t_{op} and t_{cl} are the duration of open phase and close phase, and T_0 is the time period. Figure 4(a) shows a schematic diagram of EGG and DEGG along with different timing intervals. Normalized amplitdue quotient given as $NAQ = \frac{f_{ac}}{A_{max\_DEGG}T_0}$ [2], is related to the closing phase of the glottis cycle. A_{min_DEGG} indicates the rate of opening of vocal folds [14]. Skewness is a statistical parameter that measure the asymmetry of a real-valued random variable about its mean. The harmonics based feature H1-H2 is the magnitude difference between the first two harmonics. It is given as $H1-H2 = 20log_{10}(\frac{H1}{H2})$.

2.3 Classifier

Support Vector Machine (SVM) Classifier. SVM is one of the widely used linear binary classifier. It determines a decision function that is maximally distanced from the training data [4]. Hence, it is also called the maximum margin classifier. The difficulty of getting a non-linear decision function is eased by using kernel functions that enable SVM to map the feature data into higher dimensional spaces where the optimal hyperplane is determined. In this experiment, SVM with radial basis function (RBF) kernel has been used for two-class classification. Five fold cross-validation is used to optimize the SVM parameters; where the training set is further divided into 5 sub-sets, four sub-sets are used for training the SVM model and the remaining one sub-set is used for testing.

K-Nearest Neighbour (KNN) Classifier. It is a non-parametric method of classification where membership of a test sample is computed by majority voting by K nearest neighbouring training samples [7]. In this work, K is set to 10 with distances computed using Euclidean measure.

3 Performance Analysis

The ability of a feature to indicate changes under out-of-breath condition is tested by Welch's t-test using speech and EGG signals recorded under constant vowel phonation.

3.1 Out-of-breath Data

A new database is created having speech, and EGG signals recorded simultaneously for SVP of sounds /i/, /a/ and /u/. It has two classes of signals, namely out-of-breath and normal. Out-of-breath class is recorded after performing two minutes of jump rope workout, whereas the normal signal is recorded right before the speaker undergoes the workout. Five male speakers, all are research scholars from Indian Institute of Technology Guwahati, participated in the recording process; they belong to the age group of 25–30 years. Total 191 number of SVPs of duration 1 sec each are collected. The normal class has 105 SVPs whereas the count is 86 for the out-of-breath class. All recordings are carried out using Tascam DR-100MK II linear PCM recorder and TechCadenza M2LU digital electroglottograph recorder for recording speech and EGG signals respectively. The sampling frequency of 48 kHz with 24-bit resolution has been used.

3.2 Result and Discussion

Table 1 shows the t-test values for the five glottal features. Boxplot for vowel sound /a/ is shown in Fig. 4(b) for representation. Under vowel phonations, it is observed that for OQ_{EGG} the interquartile range (IQR) is placed high for the out-of-breath condition. At the same time, the opposite behaviour is shown by

Table 1. Welch's t-test statistics for EGG.

Vowels	Statistics	Features				
		OQ_{EGG}	CQ_{EGG}	NAQ	A_{min_DEGG}	Skewness
/a/	t-value	48.79	48.79	21.18	13.12	36.67
	p-value	0.0	0.0	<0.0001	<0.0001	<0.0001
/i/	t-value	31.85	31.85	15.93	1.05	18.44
	p-value	0.0	0.0	<0.0001	0.29	<0.0001
/u/	t-value	22.54	22.54	18.61	30.85	10.88
	p-value	<0.0001	<0.0001	<0.0001	<0.0001	<0.0001
Combined	t-value	55.99	55.99	34.43	12.29	32.76
	p-value	0.0	0.0	<0.0001	<0.0001	<0.0001

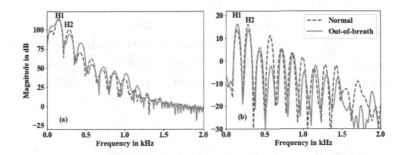

Fig. 5. Spectrum of a frame of vowel /u/ for source (a) ZFF, (b) ILPR.

CQ_{EGG} as expected. It indicates that the vocal folds do remain open for a longer period of a glottal cycle when a person is out-of-breath. NAQ shows higher t-value as well as a downward shift of IQR for the out-of-breath case than the normal case. It implies that the rate at which vocal folds close, increases for the out-of-breath condition. Such kind of behaviour is not observed in A_{max_DEGG}, which stands for strength of glottis closure. This may be due to the level of exertion under out-of-breath condition is different for different speakers. Similarly, a minor change is observed for glottal opening strength A_{min_DEGG}. Skewness has higher t-value with a lower mean for IQR in case of normal condition. This hints that the density function for EGG waveform is positively skewed under out-of-breath condition.

For ZFF and ILPR source signals, Fig. 5 shows spectrum of a frame of vowel /u/. For ILPR signal it is observed that, in majority of cases; the first harmonic peak magnitude (H1) is higher than that of the sencond harmonic peak (H2); which is opposite for that of normal condition. Thus the harmonic magnitude difference between H1 and H2 (H1-H2) is higher for out-of-breath condition. A similar trend is observed for H1-H2 in case of ZFF source signal. However for ZFF, H2 is more supressed in out-of-breath condition where as H1 remains high for both the conditions. This suggests that ZFF signal contains more low

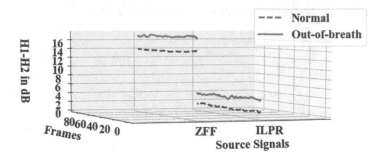

Fig. 6. Averaged H1-H2 over all frames of utterance /u/.

Table 2. Confusion matrix for classifiers SVM and KNN for the combined feature set of OQ_{EGG}, CQ_{EGG}, NAQ, A_{min_DEGG} and skewness.

	SVM		KNN	
	Out-of-breath	Normal	Out-of-breath	Normal
Out-of-breath	**74.41%**	27.61%	**55.81%**	12.38%
Normal	25.59%	**72.39%**	44.19%	**87.62%**
Average accuracy	**73.4%**		71.71%	

frequency components in case of out-of-breath condition. Figure 6 shows variation of averaged H1-H2 values for vowel sound /u/ uttered by all speakers. The Welch's t-test values appear high: for both the source signals as shown in Table 3. It hints that these approximated source signals can carry information about physical exertion.

Classification results have been obtained using leave-one-speaker out approach. Where, utterances of one speaker are considered for testing and others for training. Table 2 shows the confusion matrix for SVM and KNN classifiers for EGG based features. This shows an average binary classification rate of 73.40% and 71.24% for SVM and KNN respectively. Between the two approximated source signals, ZFF gives the best classification result with accuracies 70.40% and 71.0% for SVM and KNN classifiers respectively. ILPR source is not much far behind with accuracies of 63.60% and 68.60% for same set of classifiers. Table 4 shows these classification result. In literature, the highest classification rate of 91.90% is obtained by Suman et al. [5] on a regular speech corpus. They used a combination of harmonic features, teager energy based features, glottal features and mel frequency cepstral features for classification.

Table 3. Welch's t-test for ZFF and ILPR source signals

Source type	t-values	p-value
ZFF	54.23	0.0
ILPR	37.61	<0.0001

Table 4. H1-H2 feature classification for ZFF and ILPR source signals

Source	Classifier	
	SVM	KNN
ZFF	**70.40%**	**71.0%**
ILPR	63.60%	68.60 %
Combined	66.80%	67.80 %

4   Conclusion

In this work, we attempted to study the source characteristics of speech signal under out-of-breath condition. Different source signals like EGG, ZFF and ILPR are examined. Using sustained vowels, analysis of EGG showed that the glottal opening and closing pattern gets altered under out-of-breath case. It is expected under physical exertion as lungs require more air, and thus the breathing pattern becomes faster and deeper. As recording of EGG is not always possibile; we considered the sources extracted from speech signals like ZFF and ILPR. Spectral analysis of such sources showed alteration in the harmonic structure indicated by H1 and H2 harmonic peaks. Analysis by different statistical tools verifies that characteristics of source signals differ from normal to out-of-breath case.

References

1. Adiga, N., Prasanna, S.: Detection of glottal activity using different attributes of source information. IEEE Sig. Proc. Lett. **22**(11), 2107–2111 (2015)
2. Alku, P., Bäckström, T., Vilkman, E.: Normalized amplitude quotient for parametrization of the glottal flow. J. Acoust. Soc. Amer. **112**(2), 701–710 (2002)
3. Conrad, B., Schönle, P.: Speech and respiration. Archiv für Psychiatrie und Nervenkrankheiten **226**(4), 251–268 (1979)
4. Cortes, C., Vapnik, V.: Machine learning. Support-vector Netw. **20**(3), 273–297 (1995)
5. Deb, S., Dandapat, S.: Fourier model based features for analysis and classification of out-of-breath speech. Speech Comm. **90**, 1–14 (2017)
6. Drugman, T., Bozkurt, B., Dutoit, T.: A comparative study of glottal source estimation techniques. Comput. Speech Lang. **26**(1), 20–34 (2012)
7. Duda, R.O., Hart, P.E., Stork, D.G.: Pattern Classification. Wiley, Hoboken (2012)
8. Godin, K.W., Hansen, J.H.: Vowel context and speaker interactions influencing glottal open quotient and formant frequency shifts in physical task stress. In: Proceedings of INTERSPEECH, pp. 2945–2948 (2011)
9. Godin, K.W., Hasan, T., Hansen, J.H.: Glottal waveform analysis of physical task stress speech. In: INTERSPEECH (2012)
10. Murty, K.S.R., Yegnanarayana, B.: Epoch extraction from speech signals. IEEE Tran. Acoust. Sp. Lang. Proces. **16**(8), 1602–1613 (2008)

11. Patil, S.A., Hansen, J.H.: Detection of speech under physical stress: model development, sensor selection, and feature fusion. In: INTERSPEECH (2008)
12. Prathosh, A., Ananthapadmanabha, T., Ramakrishnan, A.: Epoch extraction based on integrated linear prediction residual using plosion index. IEEE Tran. Audio Speech Lang. Process. **21**(12), 2471–2480 (2013)
13. Trouvain, J., Truong, K.P.: Prosodic characteristics of read speech before and after treadmill running. In: INTERSPEECH (2015)
14. Yap, T.F., Epps, J., Ambikairajah, E., Choi, E.H.: Voice source under cognitive load: effects and classification. Speech Comm. **72**, 74–95 (2015)

Segregating Musical Chords
for Automatic Music Transcription:
A LSTM-RNN Approach

Himadri Mukherjee[1]([✉]), Ankita Dhar[1], Sk. Md. Obaidullah[2], K.C. Santosh[3],
Santanu Phadikar[4], and Kaushik Roy[1]

[1] Department of Computer Science, West Bengal State University, Kolkata, India
himadrim027@gmail.com, ankita.ankie@gmail.com, kaushik.mrg@gmail.com
[2] Department of Computer Science and Engineering, Aliah University, Kolkata, India
sk.obaidullah@gmail.com
[3] Department of Computer Science, The University of South Dakota,
Vermillion, SD, USA
santosh.kc@ieee.org
[4] Department of Computer Science and Engineering, Maulana Abul Kalam Azad
University of Technology, Kolkata, India
sphadikar@yahoo.com

Abstract. Notating or transcribing a music piece is very important for
musicians. It not only helps them to communicate among each other but
also helps in understanding a piece. This is very much essential for impro-
visations and performances. This makes automatic music transcription
systems extremely important. Every music piece can be broadly categorized into two parts namely the lead section and the accompaniment
section or background music (BGM). The BGM is very important in a
piece as it sets the mood and makes a piece complete. Thus it is very
much important to notate the BGM for properly understanding and performing a piece. One of the key components of BGM is known as chord
which is constituted of two or more musical notes. Every composition
is accompanied with a chord chart. In this paper, a long short term
memory-recurrent neural network (LSTM-RNN)- based approach is presented for segregating musical chords from clips of short durations which
can aid in automatic transcription. Experiments were performed on over
46800 clips and a highest accuracy of 99.91% has been obtained for the
proposed system.

Keywords: Chord identification · Music signal · LSTM-RNN

1 Introduction

A music piece is composed of musical notes. These notes occur in different combinations and timings which makes melodies different. In order to study such
music compositions it is very important to notate or transcribe them. This not

© Springer Nature Switzerland AG 2019
B. Deka et al. (Eds.): PReMI 2019, LNCS 11942, pp. 427–435, 2019.
https://doi.org/10.1007/978-3-030-34872-4_47

only helps in understanding them in a better way but also to communicate with other musicians. The BGM of a composition as important as the lead melody. It is the BGM which makes a piece sound complete and which goes on for almost the entire span of a piece. A change in the BGM can alter the mood of a composition and at times disrupt it completely. Thus it is very important to play the BGM flawlessly during performances to uphold the essence of a composition. One of the most important facets of BGM melody is known as a chord which is composed of two or more musical notes played simultaneously. Every composition has a chord chart associated with it whose transcription is essential.

Rajpurkar et al. [15] distinguished chords in real-time. They used hidden markov model (HMM) and Gaussian discriminant analysis in addition to chroma-based features and obtained an accuracy of 99.19%. Zhou and Lerch [18] used deep learning for distinguishing chords. They worked with 317 music pieces and obtained a recall value of 0.916 using max-pooling. Cheng et al. [4] distinguished chords for music classification and retrieval with the aid of N-gram technique and HMM. Different chord-based features like chord histogram and LCS were also involved in their experiments and a highest overall accuracy of 67.3% was obtained. Dylan Quenneville [14] has talked about multitudinous aspects of automatic music transcription. He has highlighted the basics of making music as well as that of transcription. He has talked about different techniques of pitch detection in the thick of fourier transform-based approaches, fundamental frequency-based approaches, harmonicity-based approaches to name a few.

Berket and Shi [3] presented a two phase model for music transcription. In the first phase, they used acoustic modelling to detect pitches and in the later phase it was transcribed. They worked with 138 MIDI files which were converted to audio. The train set consisted of 110 songs while the remaining were used for testing and reported results as high as 99.81%. Wats and Patra [17] used a non negative matrix factorization-based technique for automatic music transcription. They worked on the Disklavier dataset and obtained good results. Benetos et al. [1] presented an overview of automatic music transcription. They have touched on its various applications and challenges. They have also talked about several transcription techniques as well. Muludi et al. [12] frequency domain information and pitch class profile for chord identification. Their experiments involved 432 guitar chords and obtained an accuracy of 70.06%.

Osmalskyj et al. [13] used a neural network and pitch class profiles for guitar chord distinction. Their study involved other instruments in the thick of accordion, violin and piano. They performed instrument identification as well and obtained an error rate of 6.5% for chord identification. Benetos et al. [2] laid out different techniques and challenges which are involved in automatic music transcription. They have talked about various pitch tracking methods in the thick of feature-based approaches, statistical approaches, spectrogram factorization-based approaches and many more. They have also talked about several types of transcriptions including instrument and genre-based transcription as well as informed transcription. Kroher and Gomez [7] attempted to automatically transcribe flamenco singing from polyphonic tracks. They extracted predominant melody and eliminated contours of the accompaniments. Next the vocal contour was discretized into notes followed by assignment of a quantized pitch level. They experimented with three datasets totaling to more than 100 tracks and obtained results which was better than state of the art singing transcribers based on overall performance, onset detection and voicing accuracy. Costantini and Casali [5] used frequency analysis for chord identification. Experiments were performed with upto 4 note chords. Highest accuracies of 98%, 97% and 95% were obtained for the 2, 3 and 4 note chords.

Here, a system is proposed to distinguish chords from clips of very short duration. It works with LSTM-RNN based classification and has the potential of aiding in automatic music transcription for background music which is very vital. The system is illustrated in Fig. 1.

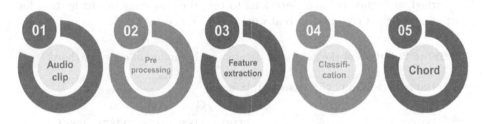

Fig. 1. Pictorial representation of the proposed system.

The rest of the paper consists of the details of dataset in Sect. 2. Sections 3 and 4 talk about the proposed method whose results respectively. Finally we have concluded in Sect. 5.

2   Dataset

Data is a very important aspect of any experiment. The quality of data plays a crucial part in development of robust systems as well. To the best of our knowledge, there is no publicly available dataset of chords and hence we put together a dataset of our own. In the present experiment, we consider two of the most popular chords from the major family (C and G) and two most popular chords from the minor family namely A minor (Am) and E minor (Em) [16]. The constituent notes of scales of the considered chords along with the notes of the chords is presented in Table 1. The chord pairs (G-Em) and (C-Am) have common notes which makes it difficult to distinguish them.

Volunteers were provided a Hertz acoustic guitar (HZR3801E) for playing the chords. They played different rhythm patterns and no metronome was used to allow relaxation with respect to tempo. Volunteers further used different type of plectrums which slightly change the sound thereby encompassing more variation. The audios were recorded with the primary line port of a computer having a motherboard (Gigabyte B150M-D3H). Further, studio ambience and use of

Table 1. Notes involved in the chords.

Scale	Notes	Chord	Similar notes
G	G,A,B,C,D,E,F#	G, B, D	G,B
Em	E,F#,G,A,B,C,D	E, G, B	G,B
C	C,D,E,F,G,A,B	C, E, G	C,E
Am	A,B,C,D,E,F,G	A, C, E	C,E

pre amplifiers was avoided to ensure real world scenario. The audio clips were recorded in .wav format at a bitrate of 1411 kbps.

Four datasets (D1-D4) having clips of lengths 0.25, 0.5, 1 and 2 s respectively were put together form the recorded data whose details are presented in Table 2. We worked with clips of such durations to test the efficiency of our system for short clips which is common in real world.

Table 2. Details of the generated datasets with number of clips per chord.

Datasets (length of clip in second)	Chords				
	C	G	Am	Em	Total
D1(0.25)	11613	11863	11537	11871	46884
D2(0.5)	5804	5928	5762	5930	23424
D3(1)	2899	2959	2876	2961	11695
D4(2)	1444	1476	1434	1475	5829

3   Proposed Method

3.1   Preprocessing

Framing. The clips were first subdivided into smaller segments called frames. This was mainly done to make the spectral contents stationary which otherwise show high deviations thereby making analysis a herculean task. The clips were divided into 256 point frames in overlapping mode with 100 common points (overlap) between two consecutive frames [11].

Windowing. Jitters are often observed in the frames due to loss of continuity at the boundaries. These disrupt frequency-based analysis in the form of spectral leakage. To tackle this, the frames are windowed with windowing function. Here we used hamming window [11] which is presented in Eq. (1).

$$w(n) = 0.54 - 0.46 \cos\left(\frac{2\pi n}{N-1}\right), \tag{1}$$

Feature extraction where n is a sample point within a N sized frame.

3.2   Feature Extraction

Each of the clips were used for extraction of the standard line spectral frequency (LSF) features at frame level. LSF [11] was chosen due to its higher quantization power [10]. Here, a sound signal is represented as the output of a filter H(z) whose inverse is G(z) where G $_{1....m}$ are the predictive coefficients

$$G(z) = 1 + g_1 z^{-1} + . + g_m z^{-n} \tag{2}$$

The LSF derived by decomposing G(z) into $G_x(z)$ and $G_y(z)$ which are detailed below

$$G_x(z) = G(z) + z^{-(m+1)}G(z^{-1}) \tag{3}$$

$$G_y(z) = G(z) - z^{-(m+1)}G(z^{-1}) \tag{4}$$

We had extracted 5, 10, 15, 20 and 25 dimensional features for the frames. Each of these dimensions correspond to bands that is 5 dimensional LSFs have 5 bands and so on. Next, these bands were graded in accordance with the total value of the coefficients. This band sequence was used as feature. It depicted the energy distribution pattern across the bands. Along with this, the mean and standard deviation of the spectral centroids per frame was also appended. When 5 dimensional LSF was extracted, a total of $5 \times 440 = 2200$ coefficients were obtained for a clip of only 1 s (1 s clip produced 440 frames). This dimension varied with disparate length of the clips. The band grades along with the mean and standard deviation of the centroids produced a $5 + 2 = 7$ dimensional feature when 5 dimensional LSFs were extracted. These were also independent of the clip lengths. So finally we obtained features of 7, 12, 17, 22 and 27 dimensions.

3.3   Long Short Term Memory-Recurrent Neural Network (LSTM-RNN) Based Classification

LSTM-RNN can preserve states as compared to standard neural networks [9] which makes them suitable for sequences. It further solves the vanishing gradient problem of simple RNNs [8]. A LSTM block comprises of a cell state and three gates namely forget gate, input gate and output gate. The input gate (i_n) helps to generate new state:

$$i_n = \sigma(Wt_i S_{n-1} + Wt_i X_n), \tag{5}$$

where Wt_i is the associated weight. The forget gate discards values form previous state to the present state:

$$f_n = \sigma(Wt_f S_{n-1} + Wt_f X_n), \tag{6}$$

where Wt_f is the associated weight. The output determines the next state as shown below:

$$o_n = \sigma(Wt_o S_{n-1} + Wt_o X_n), \tag{7}$$

where Wt_o is the associated weight. Our network comprised of a 100 dimensional LSTM layer. The output of this layer was passed through three fully connected layers of dimensions 100, 50 and 25 respectively. These layers had ReLU activation. The final layer was a 4 dimensional fully connected layer with softmax activation. We had initially used 5 fold cross validation with 100 epochs in our experiment and the network parameters were set after trials.

4   Result and Analysis

Each of the feature sets for the datasets D1-D4 were fed to the recurrent neural network as summarized in Table 3. It is observed that the best result was obtained for the 22 dimensional features on D3. To obtain better results, the training epochs were varied with 5 fold cross validation for 22 dimensional features of D3 as shown in Table 4. The best performance was obtained for 300 epochs. Increasing the training epochs even further led to over fitting and thus produced lower results. The confusions among the different classes for 300 iterations is presented in Table 5(a). It is observed that the highest confusion was among the minor chords. The clips were analyzed and it was found that the volunteers at times accidentally muted strings which interfered with the chord textures in the barred shapes. This could be one probable reason for such confusions.

Table 3. Results for different datasets with the disparate datasets.

Datasets	Feature dimensions				
	7	12	17	22	27
D1	62.03	74.36	79.29	75.41	77.74
D2	68.40	93.31	91.79	76.20	84.02
D3	71.01	85.16	93.64	**95.50**	92.70
D4	73.01	75.23	85.80	83.02	91.85

Table 4. Accuracy for different training epochs on D3 with 22 dimensional features.

Epochs	100	200	**300**	400	500
Accuracy (%)	95.50	95.45	**95.90**	94.50	94.24

In order to obtain further improvements, we varied the cross validation folds for 100 epochs for 22 dimensional features of D3. The obtained results are presented in Table 6. 20 folds produced the best result wherein the variation of the dataset was evenly distributed. The performance decreased on further increasing the folds of cross validation. The interclass confusions is presented in Table 5(b)

Table 5. (a) Confusion matrix for 300 epochs. (b) Confusion matrix for 20 fold cross validation. (c) Confusion matrix for 300 epochs 20 fold cross validation.

	C	G	Am	Em	C	G	Am	Em	C	G	Am	Em
C	2897	2	0	0	2899	0	0	0	2899	0	0	0
G	16	2903	40	0	9	2950	0	0	8	2951	0	0
Am	0	12	2840	24	0	3	2873	0	0	3	2873	0
Em	0	0	386	2575	0	0	0	2961	0	0	0	2961
		(a)				(b)				(c)		

wherein it is observed the chords C and Em were recognized with 100% accuracy. The confusions among the minor chords was also overcome in this setup. Finally the best fold value (20 fold) along with the best training epoch (300 epochs) were combined which produced an accuracy of 99.91 % (overall highest) whose confusions are presented in Table 5(c). It is observed that the confusions were exactly similar as compared to the 20 fold cross validation setup, only 1 more instance of G chord was identified correctly as compared to the former setup. Some of the other popular classifiers in the thick of bayesnet (BN), naïve bayes (NB), multi layer perceptron (MLP), random forest (RF), radial basis functional classifier (RBF) from [6] were also evaluated on D4 whose results are summarized in Table 7.

Table 6. Accuracy for different folds of cross validation on D3 with 22 dimensional features.

Folds	5	10	15	**20**	25
Accuracy (%)	95.50	99.63	99.48	**99.90**	99.85

Table 7. Performance of different classification techniques on D3 with 22 dimensional features.

Classifier	NB	BN	MLP	RF	RBF	**LSTM-RNN**
Accuracy (%)	42.26	80.86	93.63	98.97	80.47	**99.91**

5   Conclusion

Here, a system is presented to distinguish chords from clips of short durations. The system works with LSTM-RNN based classification technique and produced encouraging results. In future, we will experiment with a larger set of chords and involve other instruments as well. We will introduce other tracks along with the chords to observe the system's performance. We also plan to identify and discard silent sections in the clips to obtain better results. Finally, we will make use of other acoustic features coupled with different modern machine learning techniques to obtain further improvement in our results.

Acknowledgement. The authors would like to thank Mr. Soukhin Bhattacherjee, Mr. Debajyoti Bose of Department of Electrical, Power & Energy, University of Petroleum and Energy Studies for their help during the entire course of this work. They also thank WWW.PresentationGO.com for the block diagram template.

References

1. Benetos, E., Dixon, S., Duan, Z., Ewert, S.: Automatic music transcription: an overview. IEEE Sig. Process. Mag. **36**(1), 20–30 (2018)
2. Benetos, E., Dixon, S., Giannoulis, D., Kirchhoff, H., Klapuri, A.: Automatic music transcription: challenges and future directions. J. Intell. Inf. Syst. **41**(3), 407–434 (2013)
3. Bereket, M., Shi, K.: An AI approach to automatic natural music transcription (2017)
4. Cheng, H.T., Yang, Y.H., Lin, Y.C., Liao, I.B., Chen, H.H.: Automatic chord recognition for music classification and retrieval. In: 2008 IEEE International Conference on Multimedia and Expo, pp. 1505–1508. IEEE (2008)
5. Costantini, G., Casali, D.: Recognition of musical chord notes. WSEAS Trans. Acoust. Music **1**(1), 17–20 (2004)
6. Hall, M., Frank, E., Holmes, G., Pfahringer, B., Reutemann, P., Witten, I.H.: The weka data mining software: an update. ACM SIGKDD Explor. Newsl. **11**(1), 10–18 (2009)
7. Kroher, N., Gómez, E.: Automatic transcription of flamenco singing from polyphonic music recordings. IEEE/ACM Trans. Audio Speech Lang. Process. (TASLP) **24**(5), 901–913 (2016)
8. Li, J., Mohamed, A., Zweig, G., Gong, Y.: LSTM time and frequency recurrence for automatic speech recognition. In: 2015 IEEE Workshop on Automatic Speech Recognition and Understanding (ASRU), pp. 187–191. IEEE (2015)
9. Lipton, Z.C., Berkowitz, J., Elkan, C.: A critical review of recurrent neural networks for sequence learning. arXiv preprint arXiv:1506.00019 (2015)
10. Mukherjee, H., Dutta, M., Obaidullah, S.M., Santosh, K.C., Phadikar, S., Roy, K.: Lazy learning based segregation of Top-3 south indian languages with LSF-a feature. In: Santosh, K.C., Hegadi, R.S. (eds.) RTIP2R 2018. CCIS, vol. 1035, pp. 449–459. Springer, Singapore (2019). https://doi.org/10.1007/978-981-13-9181-1_40
11. Mukherjee, H., Obaidullah, S.M., Santosh, K., Phadikar, S., Roy, K.: Line spectral frequency-based features and extreme learning machine for voice activity detection from audio signal. Int. J. Speech Technol. **21**(4), 753–760 (2018)
12. Muludi, K., Loupatty, A.F.S., et al.: Chord identification using pitch class profile method with fast fourier transform feature extraction. Int. J. Comput. Sci. Issues (IJCSI) **11**(3), 139 (2014)
13. Osmalsky, J., Embrechts, J.J., Van Droogenbroeck, M., Pierard, S.: Neural networks for musical chords recognition. In: Journees d'informatique Musicale, pp. 39–46 (2012)
14. Quenneville, D.: Automatic Music Transcription. Ph.D. thesis, Middlebury College (2018)
15. Rajparkur, P., Girardeau, B., Migimatsu, T.: A supervised approach to musical chord recognition (2015)
16. Spotify, 6 Apr 2019. https://insights.spotify.com/us/2015/05/06/most-popular-keys-on-spotify/

17. Wats, N., Patra, S.: Automatic music transcription using accelerated multiplicative update for non-negative spectrogram factorization. In: 2017 International Conference on Intelligent Computing and Control (I2C2), pp. 1–5. IEEE (2017)
18. Zhou, X., Lerch, A.: Chord detection using deep learning. In: Proceedings of the 16th ISMIR Conference, vol. 53 (2015)

Novel Teager Energy Based Subband Features for Audio Acoustic Scene Detection and Classification

Madhu R. Kamble[1]([⊠])(iD), Maddala Venkata Siva Krishna[2]([⊠]),
Aditya Krishna Sai Pulikonda[2]([⊠]), and Hemant A. Patil[1]([⊠])(iD)

[1] Speech Research Lab, Dhirubhai Ambani Institute of Information and Communication Technology (DA-IICT), Gandhinagar, Gujarat, India
{madhu_kamble,hemant_patil}@daiict.ac.in
[2] Indian Institute of Information Technology (IIIT), Vadodara, Gujarat, India
{201551045,201551013}@iiitvadodara.ac.in

Abstract. Acoustic Scene Classification (ASC) is the task of assigning a semantic label for a given audio sample recorded in different acoustic environments. Sounds carry a significant information about everyday environment scenes, such as bus, tram, airport, concert hall, etc. Thus, extracting the sound signals of these acoustic scenes can be useful to detect and classify the audio signals. In this context, Detection and Classification of Acoustic Scenes and Events (DCASE) 2018 challenge provides a common framework for researchers to propose various approaches with an aim to extract this information present in different acoustical environments. In this paper, to capture the discriminative information between different acoustic scenes, Teager energies with both mel and linear scales are used. These are computed by applying Teager Energy Operator (TEO) on a narrowband filtered signal and are modeled with convolutional neural network (CNN) for detecting and classifying the acoustic scenes or events. The results obtained on the development set gave an overall accuracy of 67.3% using recommended cross-validation setup and thus, overcoming the performance of baseline by 6.3%.

Keywords: Acoustic Scene Classification (ASC) · Convolutional neural network (CNN) · Teager Energy Operator (TEO)

1 Introduction

Acoustic Scene Classification (ASC) is a challenging research problem which is seen as a subset of computational auditory scene analysis (CASA) [13]. It is the task of classifying acoustical scenes and events from the surrounding noise, silence, etc. where the environment can be a busy street, quite park, etc. Sounds carry significant information about our everyday environment, and events that happens around us. With recent advancements in technology especially in the field of machine learning, developing methods to capture this information can be

© Springer Nature Switzerland AG 2019
B. Deka et al. (Eds.): PReMI 2019, LNCS 11942, pp. 436–444, 2019.
https://doi.org/10.1007/978-3-030-34872-4_48

invaluable to number of applications, such as searching for multimedia-based on audio content [1], designing automated cars, robots that depends on the context [3], intelligent monitoring systems to recognize activities using acoustic information and many more. It may look trivial for humans classifying an acoustical scene after hearing the audio sample. However, it is challenging to develop artificial systems that classify the acoustical scenes especially the scenes with sound sources in real-life environments, where often multiple sounds are present.

The schematic representation of Acoustic Scene Classification (ASC) task is shown in Fig. 1. In ASC task, audio signals recorded in different acoustical environments are used for the training of models. This model training uses the front-end features which can be in cepstral-domain or the log-energy coefficients that are obtained from the filterbank energies. Depending on the features, the models are prepared on the training data with traditional or neural network-based classifiers. For the given test signal, the features are extracted and depending on the probabilities obtained from the trained models, the test signal is classified into corresponding acoustical scene.

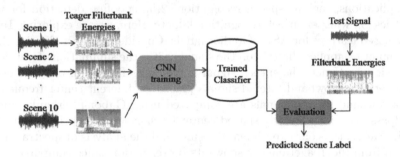

Fig. 1. Schematic representation of proposed Acoustic Scene Classification (ASC) system.

The first DCASE challenge was organized in 2013 to emphasize the problem of developing machines to perform ASC and to provide a publicly available database containing non-speech and non-music audio samples. This was followed by three more challenges in 2016, 2017, and 2018, where several researchers proposed various models using different classifiers, such as Gaussian Mixture Models (GMMs), Support Vector Machines (SVM), tree-bagger classifiers along with audio features. Convolutional Neural Networks (CNNs) proved to be successful for diverse audio-related tasks, such as speech recognition, environmental sound classification, robust audio event recognition and thus, motivated researchers to use these networks for acoustical scene classification as well [11].

In this paper, we explore the Teager energy-based subband features for the ASC task with predefined classes (subtask 1A) in DCASE 2018 challenge development dataset. In particular, Teager energy-based log-filterbank energies extracted from Mel and linearly-spaced Gabor filterbank are used to detect and classify different acoustical scenes with the help of CNN classifier. The use of linearly scale Gabor filterbank equally covers entire frequency range in low as well as in higher frequency regions compared to the Mel scale. Hence, with linear scale, we obtained better performance in classifying the acoustical scenes compared to the baseline system and the one using Mel scale accuracy.

2 Front-End Features

In this Section, we discuss the front-end features used for ASC task. The organizers of the DCASE 2018 challenge provided the baseline system that includes the state-of-the-art Mel filterbank energies along with CNN classifier. We are comparing the Teager energy-based log-energy coefficients with the baseline system. An algorithm derived by Teager uses a nonlinear energy tracking operator [6]. The Teager Energy Operator (TEO), $\Psi_d\{\cdot\}$, for a narrowband signal is defined as [6]:

$$E_n = \Psi_d\{x_i[n]\} = x_i^2[n] - x_i[n-1]x_i[n+1], \tag{1}$$

where E_n gives the running estimate of signal's energy. The TEO is used in several applications, such as speech recognition [2,5], spoofing detection for voice biometrics [9], whisper speech recognition [8], etc. Here, we are exploring Teager energy-based features for ASC of audio signals. Considering the audio signal, the TEO cannot be applied directly on the signal as it is the summation of multicomponent signals. Hence, the speech signal is bandpass filtered in order to obtain N number of narrowband filtered signals placed at different center frequencies. The narrowband filtered signals are computed using Gabor filterbank which is a bandpass filter with linearly-spaced center frequencies. We choose linear scale over the Mel scale as our experimental results and the analysis of spectral energy obtained from the Teager energy shows its better performance compared to the Mel scale.

Figure 2 shows the analysis of (a) linear, and (b) Mel frequency scale used in Gabor filterbank. With linear frequency scale, it can be observed that the subband filters are equally distributed across the frequency range. Whereas, with Mel scale, initial filters are compressed at the lower frequencies and has expanded bandwidth with a few number of filters in higher frequency regions. This can be also observed from the placing of center frequencies for linear and Mel scale, where the linear scale has the center frequency varying linearly from 0 to 8000 Hz. On the other hand, for Mel scale, approximately 20 to 25 subband filters are covered within 2000 Hz frequency while the remaining 20 subband filters are placed between the frequency range of 2000–4000 Hz. The spectral energy obtained from the linear scale with Teager energy-based approach shows the differences in lower as well as higher frequency regions compared to the Mel scale as shown in the last panel of Fig. 2.

The block diagram of log-energy coefficients-based on TEO is shown in Fig. 3. Here, the input audio signal is first given to the filterbank to obtain N = 40 number of subband filtered signals [7]. We have used linearly-spaced Gabor filterbank to have approximately equal bandwidth to cover the entire frequency range. The

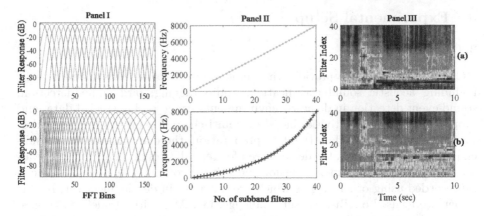

Fig. 2. Filterbank analysis of acoustical scene: (a) Linear scale *vs.* (b) Mel scale. Panel I: Filterbank response, Panel II: Frequency scale, and Panel III: Filterbank energies.

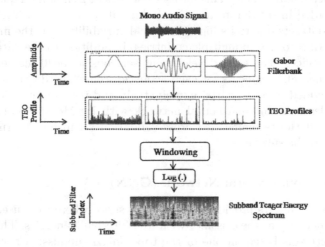

Fig. 3. Block diagram for the proposed feature extraction using subband Teager energy spectrum.

Gabor filterbank has *optimal* time-frequency resolution. Furthermore, these subband filtered signals are given to the TEO block to compute the energy profile of each subband filtered signals. These TEO profiles are passed through the frame blocking and averaging using a short window length of 40 ms with a shift of 20 ms followed by logarithm operation to compress the data. Finally, these log-energy coefficients are extracted from the audio signal to detect and classify the ASC task.

3   Experimental Setup

3.1   Database

DCASE 2018 challenge provides the audio signal data for five different tasks including acoustic sound classification, audio tagging, bird audio detection, sound event detection in domestic environments using weakly labeled data, and monitoring domestic activities based on multi-channel acoustics. In particular, for Task-1 (ASC), based on the data preparation, there are three different sub-challenges. For Task-1A, the devices used for recording of development and evaluation data are the same, while for the other two the evaluations, data can be recorded using other devices which are not used in development set. In this paper, we focused on Task-1A sub-challenge, i.e., ASC with pre-defined classes. The newly recorded TUT Urban Acoustic Scenes 2018 dataset is the largest freely available dataset that consists of ten different acoustic scenes, such as airport, park, metro station, etc. This dataset consists of 24 h of high quality audio that was recorded in six European cities making it relatively much harder than the previous datasets due to its high acoustical variability. It is the first dataset containing data recorded in multiple countries, in addition to the data recorded with mobile devices. The total development set consists 8640 segments of 10 s audio signal, i.e., 864 segments for each acoustic scene. The development set is further sub-divided into two parts, namely, train and test subsets. The baseline system includes convolutional neural network with log-Mel filterbank energies as features, and the recommended cross-validation setup for determining the performance on the sub-tasks.

3.2   Convolutional Neural Network (CNN)

The challenge organizers provided a baseline system, where they used convolutional neural network as the classifier with Mel filterbank energies. This baseline CNN architecture is based on one of the top ranked submission from DCASE 2016 [12], where changes are made to the regularizer used and the number of layers in the network. The baseline architecture consists of 2 convolutional layers, 1 Fully Connected (FC) layer. The input layer is of size 40×500, where 40 represents the number of subbands, and 500 denotes the number of frames. The first convolutional layer consists 32 subband filters, and 7×7 stride size, followed by batch normalization with Rectified Linear Unit (ReLU) as the activation function. The output obtained is passed through 2-D max-pool layer with stride size chosen as 5×5. The second CNN layer performs almost the same operation with changes made to the number of subband filters and pool size as 64, 4×100, respectively. Later, the output is flattened to give input to the dense layer with 100 neurons, and ReLU activation. Finally, to obtain the probabilities softmax is applied on the dense layer.

In this paper, for ASC task, we trained the filterbank Teager energies on the CNN classifier as shown in Fig. 4. The CNN architecture used consists of a series of Convolutional Block (ConvBlock), and pooling layers followed by FC layers. A

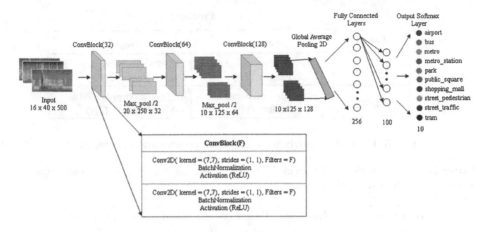

Fig. 4. Architecture of Convolutional Neural Network (CNN) used for ASC task.

ConvBlock is a combination of two convolutional layers with 7×7 kernel size, and ReLU activation. A total of three ConvBlock with 32, 64, and 128 subband filters, respectively, are used in the architecture. The max-pooling layers with 2×2 kernel and stride size are used in between the ConvBlocks to capture the important discriminative information. After the third ConvBlock, a global average pooling layer is used to pool the data across the filter maps. On the pooled data, two FC layers are used with 256 and 100 neurons, respectively. Finally, an output softmax layer with 10 neurons indicating the acoustical scenes is used to classify the input data. The batch normalization and dropout layers are used as the regularization parameters in the model. Batch normalization layers are used in between the convolutional layer, and its activation functions. The batch normalization across the filter maps helps to improve the generalization of the network [4]. A dropout of 0.3 was used after every pooling and FC layers. Adam optimizer with 0.001 learning rate is used to train the network with batch size of 16 for 200 epochs. The epoch with the best accuracy on test set is considered.

3.3   Performance Measures

Accuracy (ACC): It is defined as the number of all correct predictions divided by the total number of labels in the dataset. The best accuracy is (1.0), whereas the worst is (0.0). It can also be calculated by 1 - error rate (ER).

Confusion Matrix: It is used to describe the performance of a classification model or a classifier [10]. Assuming we are dealing with 2 classes. In Fig. 5, TP denotes the number of labels of Class 1 correctly predicted by our model, whereas FN gives us the number of labels incorrectly classified as Class 2. Similarly, FP denotes labels of Class 2 incorrectly predicted as Class 1, and TN shows the number of labels correctly classified as Class 2.

Predicted

Labels		Positive	Negative
Actual	Positive	#Positive predicted as Positive (TP)	#Positive predicted as Negative (FN)
	Negative	#Negative predicted as Positive (FP)	#Negative predicted as Negative (TN)

Fig. 5. Confusion matrix for a two-class classification problem.

Table 1. Individual Scene Accuracy (%) on test set for acoustic scene classification, subtask A in DCASE 2018 challenge

Scene label	Baseline	Teager-energy based subband features	
		Mel scale	Linear scale
Airport	72.9	49.4	52.5
Bus	62.9	51.2	57.4
Metro	51.2	49.4	64.8
Metro station	55.4	71.4	74.1
Park	79.1	78.5	83.9
Public square	40.4	44.0	54.6
Shopping mall	49.6	77.1	77.1
Street, pedestrian	50.0	61.5	51.4
Street, traffic	80.5	85.0	83.3
Tram	55.1	72.0	73.9
Average	59.7	64.0	**67.3**

4 Experimental Results

The performance of the Teager energy-based filterbank features is shown in Table 1 in terms of accuracy (in %) on the test set. In particular, we compare the performance of baseline and Teager filterbank energies obtained with linear and Mel frequency scales in Gabor filterbank. It can be observed that the accuracy obtained from the linear frequency scale gave better results compared to the Mel frequency scale (except for audio signals recorded in the street acoustic scene). The baseline system gave an average accuracy of 59.7%, with classwise results varying from 40.4% to 80.5%. Teager energy-based subband features that are obtained from Mel and linear scales gave 64% and 67.3%, respectively, have the classwise results from 44% to 85% (for Mel scale) and 51.4% to 83.9% (for linear scale). For individual acoustical scene classification, we have shown the confusion matrices in Fig. 6 for the proposed features with linear frequency scale used in Gabor filterbank. It can be observed from confusion matrix that with linear frequency scale, we obtained the better performance compared to the Mel frequency scale. For the street pedestrian scene, the Mel frequency scale performed better with less ambiguity than with linear scale. However, for rest of the acoustical scenes, linear frequency scale gave better performance. Comparing the system performance with linear scale, for most of the similar scenes gave same performance, such as for park and street traffic approximately 83% accuracy and metro station and tram gave 74% accuracy. The most difficult task is to detect the scenes of airport and street pedestrian giving lowest performance of 52.5% and 51.4%, respectively. This is also observed in terms of confusion matrix (as shown in Fig. 6), where we can observe more ambiguity for airport, metro and street pedestrian classes.

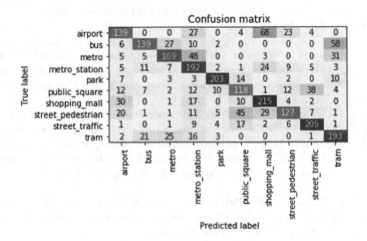

Fig. 6. Confusion matrix w.r.t performance of Teager energy-based subband filters with **linear** frequency scale for task 1 A performance.

5   Summary and Conclusions

In this paper, we proposed Teager energy-based log-filterbank energies for acoustical scene classification task. We analyzed the differences of the audio signals when recorded in different acoustical environments, such as airport, bus, tram, metro, etc. The filterbank Teager energies obtained from the linearly-spaced Gabor filterbank gave high spectral energy density compared to the traditional and Mel scale filterbank energies. In addition, we also observed that the performance of the similar scenes gave approximately same accuracy. For the acoustic scenes, such as airport and street pedestrians performance is low, which can be observed through the ambiguity in detection. Our future work will be directed towards more detailed analysis of Teager filterbank energies for the acoustic scenes, such as airport, bus, street which are relatively difficult to detect and classify.

References

1. Bugalho, M., Portelo, J., Trancoso, I., Pellegrini, T., Abad, A.: Detecting audio events for semantic video search. In: INTERSPEECH, pp. 1151–1154. Brighton, United Kingdom (2009)
2. Dimitrios, D., Petros, M., Alexandros, P.: Auditory Teager energy cepstrum coefficients for robust speech recognition. In: INTERSPEECH, pp. 3013–3016. Lisboa, Portugal (2005)
3. Eronen, A.J.: Audio-based context recognition. In: IEEE Transactions on Audio, Speech, and Language Processing, pp. 321–329 (2006)
4. Ioffe, S., Szegedy, C.: Batch normalization: accelerating deep network training by reducing internal covariate shift. In: International Conference on Machine Learning, pp. 448–456. Lille, France (2015)
5. Jabloun, F., Cetin, A.E., Erzin, E.: Teager energy based feature parameters for speech recognition in car noise. IEEE Sig. Process. Lett. **6**, 259–261 (1999)
6. Kaiser, J.F.: On a simple algorithm to calculate the energy of a signal. In: IEEE International Conference on Acoustics, Speech, and Signal Processing (ICASSP), pp. 381–384. Albuquerque, New Mexico, USA (1990)
7. Maragos, P., Quatieri, T.F., Kaiser, J.F.: Speech nonlinearities, modulations, and energy operators. In: IEEE International Conference on Acoustics, Speech, and Signal Processing (ICASSP), Toronto, Ontario, Canada, pp. 421–424 (1991)
8. Marković, B.R., Galić, J., Mijić, M.: Application of Teager energy operator on linear and mel scales for whispered speech recognition. Arch. Acoust. **43**, 3–9 (2018)
9. Patil, H.A., Kamble, M.R., Patel, T.B., Soni, M.: Novel variable length Teager energy separation based instantaneous frequency features for replay detection. In: INTERSPEECH, pp. 12–16. Stockholm, Sweden (2017)
10. Provost, F.J., Fawcett, T., Kohavi, R., et al.: The case against accuracy estimation for comparing induction algorithms. ICML **98**, 445–453 (1998)
11. Valenti, M., Diment, A., Parascandolo, G., Squartini, S., Virtanen, T.: DCASE 2016 acoustic scene classification using convolutional neural networks. In: Proceedings of Workshop on Detection and Classification of Acoustic Scenes and Events, pp. 95–99. Budapest, Hungary (2016)
12. Valenti, M., Squartini, S., Diment, A., Parascandolo, G., Virtanen, T.: A convolutional neural network approach for acoustic scene classification. In: 2017 International Joint Conference on Neural Networks (IJCNN), pp. 1547–1554. IEEE, Budapest, Hungary (2017)
13. Wang, D., Brown, G.J.: Computational Auditory Scene Analysis: Principles, Algorithms, and Applications. Wiley-IEEE Press, Hoboken (2006)

Audio Replay Attack Detection for Speaker Verification System Using Convolutional Neural Networks

P. J. Kemanth[✉], Sujata Supanekar[✉], and Shashidhar G. Koolagudi[✉]

Department of Computer Science and Engineering, National Institute of Technology
Karnataka, Surathkal, India
kemanth.raj@gmail.com, sujata.supanekar@gmail.com,
koolagudi@nitk.ac.in

Abstract. An audio replay attack is one of the most popular spoofing attacks on speaker verification systems because it is very economical and does not require much knowledge of signal processing. In this paper, we investigate the significance of non-voiced audio segments and deep learning models like Convolutional Neural Networks (CNN) for audio replay attack detection. The non-voiced segments of the audio can be used to detect reverberation and channel noise. FFT spectrograms are generated and given as input to CNN to classify the audio as genuine or replay. The advantage of the proposed approach is, because of the removal of the voiced speech, the feature vector size is reduced without compromising the necessary features. This leads to significant amount of reduction on training time of the networks. The ASVspoof 2017 dataset is used to train and evaluate the model. The Equal Error Rate (EER) is computed and used as a metric to evaluate model performance. The proposed system has achieved an EER of 5.62% on the development dataset and 12.47% on the evaluation dataset.

Keywords: Audio replay · Audio playback · Deep learning · CNN · GMM

1 Introduction

Over the past few years, biometric technology has evolved tremendously. Biometric authentication provides secure access to the system using human biological data such as DNA, facial features, fingerprint data, iris data, and voice data. Biometric data is unique for every individual and hence considered secure and easy to use. Face and fingerprint-based biometric systems became so advanced that they are widely used in various IoT devices and smartphones. Voice biometrics has a lot of potential because of convenience, low cost and readily available. Some of the applications of voice biometric would be smartphone user authentication, Interactive Voice Response (IVR) and voice authentication in banking.

© Springer Nature Switzerland AG 2019
B. Deka et al. (Eds.): PReMI 2019, LNCS 11942, pp. 445–453, 2019.
https://doi.org/10.1007/978-3-030-34872-4_49

Speech-based biometric systems are liable to a variety of attacks like impersonation, voice synthesis, voice conversion and replay attack [4]. In the impersonation attack, the imposter tries to mimic the voice of the authentic user and gain access to the system. Voice conversion systems are capable of converting the utterances spoken by person A to that of person B and they pose a grave threat to speaker verification systems. The voice synthesis system takes text as input and generates audio corresponding to the utterances of the text input. The replay attack is the most popular attack because it doesn't require any signal processing skills. The attacker has to capture the voice of the authentic user using a portable recording device and play it back to the speaker verification system.

The main purpose of this research is to investigate different ways to detect audio replay attack. ASVspoof 2017 challenge is solely focused on replay audio detection and as part of the challenge dataset containing genuine and replay audio is provided [6]. The audio replay attack detection problem reduces to a simple binary classification problem of classifying the audio as genuine or replay audio. In the ASVspoof 2015 challenge which was mainly focused on detecting voice conversion and speech synthesis CQT cepstral coefficients (CQCC) were used by many models as they provide high resolution in the lower frequency spectrum [14,17]. Hence in ASVspoof 2017 challenge, the baseline system used CQCC features and also gives good results in replay attack detection.

In this paper the proposed methodology consists of three major steps: removal of voiced segments, feature extraction and classification. The objective is to find discriminative features that will help the model to distinguish genuine and replay audio. The replay audio generally contains more noise as it comprises of environment noise from multiple channels and the effects of reverberation are also present due to reflection of sound. The removal of voiced segments results in much less feature vector size which makes it easier to train deep neural networks. In the next step, log magnitude FFT power spectrum is extracted and used as a feature to train a CNN classifier to mark the audio as genuine or replay. The ASVspoof 2017 dataset is used for training and evaluating the model [6].

The paper is organized as follows: Sect. 2 contains the related work done in reference to the task of replay attack detection. Section 3 provides detailed information about the proposed methodology and the CNN model which used to solve the problem. Section 4 gives the results and analysis of the proposed system. In Sect. 5 conclusion and further possible research direction is discussed.

2   Literature Survey

In past years, a lot of efforts was put to detect audio replay attacks and many features for audio replay attack detection have been identified. Features like spectral ratio, modulation index, and low frequency have been used to detect replay attacks [15]. After feature extraction, a feature vector is formed and Support Vector Machine (SVM) classifier was used to classify genuine and recorded voice. Till that moment there was no dataset publicly available for replay attack detection so the approach could not be validated properly.

In order to develop countermeasures for audio replay attack, ASVSpoof 2017 challenge was conducted [6]. The baseline system used constant Q cepstral transform coefficients (CQCC) for feature representation which uses constant Q transform (CQT) to represent the audio signal [14]. The CQCC features have proven to be a huge success in the ASVspoof 2015 challenge and hence used as baseline feature for ASVspoof 2017 challenge. The Gaussian Mixture Model (GMM) with expectation maximization (EM) algorithm was used as a classifier to classify whether the audio as genuine or replay. The idea was to find a discriminative feature in genuine and replay the audio. The replay audio goes through the analog to digital conversion process twice, as a result, channel noise and other recording device artifacts gets introduced in the audio. This distortion and noise can be used as a feature for the classification problem. The CQCC features increase the resolution of the lower frequency spectrum where most of the voice data and noise are present, which helps in identification of the artifacts and distortions in audio. The GMM by itself was unable to learn all the discriminative features from the CQCC feature, which was giving high equal error rate (EER) [13] on evaluation dataset. The point where the false acceptance and false rejection are minimal and optimal is called EER [13]. The system which gives lesser EER value is considered as good system.

Lavrentyeva et al. [7] had proposed a deep learning approach for audio replay attack detection. The authors make use of truncated normalized Fast Fourier transform (FFT) spectrograms as features to train a deep learning model architecture known as Light CNN (LCNN) [16]. LCNN uses max-feature-map (MFM) activation instead of ReLU, which takes a maximum between two convolution feature maps and also preserves the relavant information. This paper was considered as the state of the art since it gives the lowest EER of 4.53 on development dataset and 7.37 on the evaluation dataset. As the size of the feature vector is large in this approach, it takes more time to train the deep learning model and also requires high-end machines to train and evaluate such systems. Zhuxin Chen et al. [2] have investigated the use of recurrent neural networks which have gating mechanisms such as long short term memory (LSTM) unit and gated recurrent unit (GRU). Features such as MFCC, CQCC and Filter bank energy coefficients have been studied and the results show that the Filter bank energy coefficients provide better results for the task of audio replay attack detection. The problem with this approach is that when the training data is increased the model tends to overfit.

To avoid overfitting due to the presence of similar voiced segments in both genuine and replay audio MS Saranya et al. [12] had proposed the use of non-voiced segments of the audio to detect replay attack. Author suggests that it is easier to detect reverberation and channel condition using features extracted from non-voiced segments. Three different GMM's have been trained using features like CQCC, MFCC and Mel-Filterbank-Slope features (MFS) then voting was performed to decide whether the audio was replayed or not. This technique gives an EER of 2.99 on development dataset and 16.39 on the evaluation dataset. The stated approach takes a lot of time to train three GMM models.

3 Data Set

The ASVspoof 2017 database version 2.0 is used in all the experiments performed [6]. It is a subset of RedDots dataset [9]. The ASVspoof 2017 dataset contains three subsets namely training, development and evaluation. The train dataset is used for training and at the same time simultaneously, the development dataset is used for checking the model performance and adjustment of weights. The evaluation dataset contains new environment, recording device and playback device combination which is not present in train or development dataset. Therefore the evaluation dataset gives a better idea about model performance. The replayed utterances are recorded in 26 different environments labeled from E01 to E26. There are 26 different playback devices used for playing back the recorded utterances and labeled from P01 to P26. There are 25 recording devices used to record the replay audio and they are labeled from R01 to R25. The Sampling frequency of each record is 16 KHz. The training/train dataset contains 1508 genuine audio files and 1508 replay audio files. The development dataset contains 760 genuine audio files and 950 replay audio files. The evaluation dataset contains 1298 genuine audio files and 12922 replay audio files.

In the training dataset environments, E03 and E21 are used, playback devices P01 and P02 and recording device R01 is used. The development dataset is collected from environments E03, E04, E05, E06, E16 and E18 playback devices used are P07, P05, p09, P06, P01 and P08 and the following recording devices are used R04, R01, R02, R03, R07, R05 and R06. The replay attack detection systems are affected by the environmental conditions and the types of equipment used for playback and recording. High-quality recording and playback devices make it hard for the systems to detect replay attack and also if the environment is noisy, it becomes easier to detect replay attack. In the ASVspoof 2017 dataset the environments, playback devices, and recording devices are ranked using three colors green, yellow and red. The green ones pose less threat to the replay attack detection system whereas those marked with red color pose a greater threat.

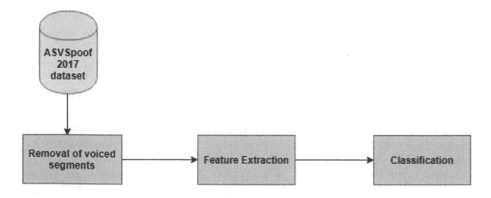

Fig. 1. Proposed methodology

4   Proposed Methodology

The proposed method of replay audio detection is displayed in Fig. 1. The approach mainly contains three steps. The first step consists of pre-processing of the audio signal where the silence, unvoiced and voiced regions are identified from the speech signal using voice activity detection algorithm [11]. The silence and unvoiced regions are concatenated together. The second step is feature extraction from the concatenated signal where log magnitude FFT power spectrum is extracted as shown in Fig. 3. In the last step, CNN model is used as a classifier to classify the audio as genuine or replay [8].

4.1   Pre-Processing

The silence and unvoiced regions are likely to contain more information about reverberation and channel noise which helps in distinguishing genuine and replay audio. Therefore the voiced regions are removed and silence and unvoiced regions are combined. Usually, the silence and unvoiced regions have low energy when compared to voiced regions which have higher energy. The energy threshold which separates the voiced region from other regions can be represented by Eq. 1.

$$Threshold = a \times average\ energy \tag{1}$$

where a is a constant which varies from 0 to 1. The energy values above the threshold are considered as voiced regions and hence removed. The threshold value is set to 0.15 after analysis.

The spectrogram of genuine and replay audio for the utterance "What sparked never boils" is shown in Figs. 2 and 3 respectively. In Fig. 3, the effect of reverberation on the speech signal can be seen, there is gradual degradation of energy around the voiced segments of the speech. Also there is more noise in the audio when compared to genuine audio. Therefore the voiced segments are removed using a voice activity detection algorithm (VAD) [11]. The rest of the audio segments are concatenated together and used for feature extraction process.

4.2   FFT Spectrogram Generation

In this phase framing is done to split the audio signal into frames and hamming window of size 256 is applied to each frame. A n - point Discrete Fourier Transform (DFT) of the speech signal is computed using a Fast Fourier Transform (FFT) algorithm. The audio signals with less frames are padded with zeros before computing FFT and at the end, FFT spectrograms of size 512*256*1 are obtained.

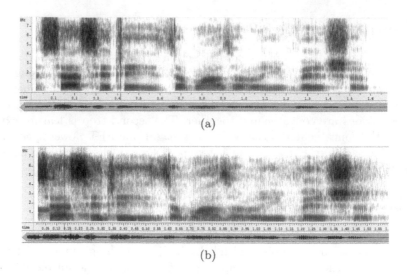

(a)

(b)

Fig. 2. (a) and (b) are the spectrograms of genuine and replay audio for the utterance "What sparked never boils"

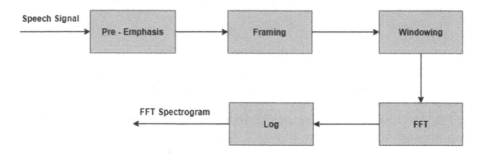

Fig. 3. Feature extraction

4.3   Convolutional Neural Network

Convolutional neural networks (CNN) [8] are used as a classifier to classify the audio as genuine or replay. The CNN is good at learning features for classification problem which give better performance compared to other traditional approaches where features have to be extracted manually. The model has 5 convolutional layers. The first convolutional layer has 32 neurons, and filters of size 7*7 are used. The second convolutional layer has 64 neurons, and filters of size 5*5 are chosen. For the remaining three convolutional layers, the neurons and filters are fixed at 96 and 3*3 respectively. Maxpool layers of size 2*2 is used and the output of which is flattened and passed through two fully connected layers with 4096 neurons each. The last layer of the network is a softmax layer which classifies the audio as genuine or replay.

To avoid overfitting, dropout with a probability of 0.2 is used during training. Dropout ignores random neurons with a certain probability p so that the neurons become less dependent on the other neurons with which they are connected. In general, the dropout helps the network to better generalize and increase the performance of the network. Adam optimizer is initialized with a learning rate of 0.0001 and momentum of 0.9 has been used for the optimizer [5]. The learning rate lowered by half if the training loss increases which makes the network converge faster and prevents overfitting. The initial weights of the network have been initialized with Xavier weight initialization method. The model is run for 20 epochs in general and early stop is used to stop the training if the training loss increases. The best model is saved every time better accuracy is obtained on the development dataset.

5   Results and Analysis

To evaluate the model performance, equal error rate (EER) [13] is used. EER is used as an evaluation measure in popular biometric systems it helps in equalizing false acceptance and rejections, the lower the EER the more accurate the system. For finding the EER of the development subset, the system is trained first on the training dataset and the posterior probabilities whether the audio is genuine or replay are obtained. The scores are generated by taking the difference between the two posterior probabilities. The BOASRIS Toolkit is used to estimate the EER from the scores [1]. Similarly, for finding the EER for the evaluation subset, the model is trained on both the training and development subset because it is seen that in the training set there are less environment, recording and playback device configurations.

In the evaluation dataset, there are many unknown environments, different playback, and recording devices and hence difficult to predict whether the audio is a replay or not. Hence the model is trained on both training and development subsets so that it can learn about more environments, recording and playback devices. Since the size of feature vector is small due to the removal of voiced speech segments, it is easier to train the network. In the experiments a simple three layer DNN [3] model was also considered. The model had 512 neurons in each layer and a softmax layer at the end. The model was trained on MFCC [10] features. The performance of the model was not good on the evaluation dataset mainly because of the problem of overfitting. The results are shown in Table 1,

Table 1. Results

System		Development dataset		Evaluation dataset	
		EER	Accuracy	EER	Accuracy
Baseline GMM	CQCC	10.83	80.16	28.65	45.62
DNN	MFCC	7.45	86.23	14.89	66.36
Proposed CNN	FFT	5.62	92.43	12.47	71.33

the proposed CNN model achieves an EER of 5.62% on development dataset and 12.47% on evaluation dataset.

6 Conclusion and Future Work

In this paper, we studied the applicability of the convolutional neural network (CNN) using features extracted from silence and unvoiced segments of speech signal for audio replay attack detection. The silent and unvoiced regions contain information about the channel and also reverberation of audio signal caused by the environment. The spectrograms of both genuine audio and replay audio are nearly identical because of the similar voiced regions which cause difficulty for CNN to learn the necessary features to discriminate genuine and replay audio. The removal of voiced speech segments also reduces the feature size which makes it easier to train CNN. The proposed approach is evaluated on ASVspoof 2017 dataset and it can be seen from the results that it outperforms the baseline system by 5.21% in development dataset and by 16.18% in the evaluation dataset. In the future, different features will be experimented with more sophisticated and newer deep learning architectures.

References

1. Brümmer, N., De Villiers, E.: The bosaris toolkit: theory, algorithms and code for surviving the new dcf. arXiv preprint arXiv:1304.2865 (2013)
2. Chen, Z., Zhang, W., Xie, Z., Xu, X., Chen, D.: Recurrent neural networks for automatic replay spoofing attack detection. In: 2018 IEEE International Conference on Acoustics, Speech and Signal Processing (ICASSP), pp. 2052–2056. IEEE (2018)
3. Deng, L., Hinton, G., Kingsbury, B.: New types of deep neural network learning for speech recognition and related applications: an overview. In: 2013 IEEE International Conference on Acoustics, Speech and Signal Processing, pp. 8599–8603. IEEE (2013)
4. Faundez-Zanuy, M.: On the vulnerability of biometric security systems. IEEE Aerosp. Electron. Syst. Mag. **19**(6), 3–8 (2004)
5. Kingma, D.P., Ba, J.: Adam: a method for stochastic optimization. arXiv preprint arXiv:1412.6980 (2014)
6. Kinnunen, T., et al..: The ASVspoof 2017 challenge: assessing the limits of replay spoofing attack detection (2017)
7. Lavrentyeva, G., Novoselov, S., Malykh, E., Kozlov, A., Kudashev, O., Shchemelinin, V.: Audio replay attack detection with deep learning frameworks. In: Interspeech, pp. 82–86 (2017)
8. LeCun, Y., Bengio, Y., et al.: Convolutional networks for images, speech, and time series. Handb. Brain Theory Neural Netw. **3361**(10), 1995 (1995)
9. Lee, K.A., et al.: The RedDots data collection for speaker recognition. In: Sixteenth Annual Conference of the International Speech Communication Association (2015)
10. Muda, L., Begam, M., Elamvazuthi, I.: Voice recognition algorithms using mel frequency cepstral coefficient (mfcc) and dynamic time warping (DTW) techniques. arXiv preprint arXiv:1003.4083 (2010)

11. Ramırez, J., Segura, J.C., Benıtez, C., De La Torre, A., Rubio, A.: Efficient voice activity detection algorithms using long-term speech information. Speech Commun. **42**(3–4), 271–287 (2004)
12. Saranya, M., Murthy, H.: Decision-level feature switching as a paradigm for replay attack detection. In: 19th Annual Conference of the International Speech Communication Association, pp. 686–690 (2018)
13. Soong, F.K., Rosenberg, A.E., Juang, B.H., Rabiner, L.R.: Report: a vector quantization approach to speaker recognition. AT&T Techn. J. **66**(2), 14–26 (1987)
14. Todisco, M., Delgado, H., Evans, N.: A new feature for automatic speaker verification anti-spoofing: Constant q cepstral coefficients. In: Speaker Odyssey Workshop, Bilbao, Spain, vol. 25, pp. 249–252 (2016)
15. Villalba, J., Lleida, E.: Preventing replay attacks on speaker verification systems. In: 2011 Carnahan Conference on Security Technology, pp. 1–8. IEEE (2011)
16. Wu, X., He, R., Sun, Z., Tan, T.: A light CNN for deep face representation with noisy labels. IEEE Trans. Inf. Forensics Secur. **13**(11), 2884–2896 (2018)
17. Wu, Z., et al.: ASVspoof 2015: the first automatic speaker verification spoofing and countermeasures challenge. In: Sixteenth Annual Conference of the International Speech Communication Association (2015)

Inverse Filtering Based Feature for Analysis of Vowel Nasalization

Debasish Jyotishi$^{(\boxtimes)}$ and Samarendra Dandapat

Department of Electronics and Electrical Engineering, Indian Institute of Technology
Guwahati, Guwahati 781039, India
{debasish.jyotishi07,samaren}@iitg.ac.in

Abstract. Vowel nasalization is present in almost every Indic languages. Detection of vowel nasalization can enhance the accuracy of Automatic Speech Recognition (ASR) systems designed for Indian languages. It also provides significant clinical information about the vocal tract. In pursuit of developing some acoustic parameters for detection of nasalized vowels, most researchers have extensively analyzed its spectral domain characteristics. In this work, we have used an inverse filtering based technique to develop a novel feature, which represents the amount of nasalization present in a vowel. The invariability of nasal filter for different nasalized vowels and addition of oral and nasal speech after radiation has been exploited to find out this feature. As the feature gives information about the amount of nasalization, this can be used for detection of vowel nasalization as well as for clinical purposes. Statistical analysis of the feature has been done in this work. The statistical analysis shows that the feature has good separability for oral vowels and nasalized vowels.

1 Introduction

Nasal sounds are produced when the glottal wave passes through the nasal cavity. The passage of glottal wave through the nasal cavity is controlled by the velum. When we intend to utter a nasal sound, the velum lowers and allows the glottal wave to pass through the nasal cavity [10]. The percentage of glottal wave passed through the nasal cavity determines the percentage of nasalization. The nasal sounds can be broadly categorized into two categories. First types are nasal murmur or nasal consonants, e.g. /m/ and /n/, which are produced by decoupling oral tract. And the second types are nasalized vowels or nasalized semi-vowels, which are produced by coupling of both oral tract and nasal tract. In Indian languages, one can deliberately utter nasalized vowels with the help of 'Matra', which are present in scripts. This is called phonemic nasalization. While in English language vowel nasalization is occurred mostly due to co-articulatory nasalization. Co-articulatory nasalization is the phenomenon in which the velum raises beforehand, in anticipation of nasal consonants and thus makes an oral vowel nasalized one. Sometimes due to the presence of nasal consonants before an oral vowel, the velum remains open for some moments. This

© Springer Nature Switzerland AG 2019
B. Deka et al. (Eds.): PReMI 2019, LNCS 11942, pp. 454–461, 2019.
https://doi.org/10.1007/978-3-030-34872-4_50

also contributes to the cause of Co-articulatory nasalization. Another kind of nasalization is functional nasalization, which occurs due to functional disorder of the velopharyngeal mechanism.

Nasalized vowels contribute to the vocabulary of almost every language. And there are words like dot and don't, which differed by the introduction of vowel nasalization. But difficulty in detecting vowel nasalization makes it a challenging task for ASR systems. Pruthi [9] has shown that accuracy of a Hidden Markov Model (HMM) based ASR system decreases if nasalized vowels are not detected. So detecting vowel nasalization is important for improving ASR system performance. Nasalized vowels can be considered as vowels having a higher degree of nasalization. And the degree of nasalization contains significant clinical information as well as information about speech intelligibility.

Many researchers have studied spectral domain properties and proposed acoustic parameters for nasalized vowels. Fant in his work showed that due to nasalization there is a decrease in amplitude of 1st formant and increase in its bandwidth [2]. House and Stevens [5] observed a spectral prominence around 1000 Hz and reduction of 2nd formant. Effects of nasalization on vowels /aaa/, /ooo/ and /uuu/ are studied by Fujimura and Lindqvist [3]. They observed the movement of 1st formant towards higher frequency and a pair of pole-zero are being introduced near 1st and 3rd formant. Glass and Zue had done extensive statistical analysis on spectral domain characteristics of nasalized vowels and proposed six features for automatic detection of nasalized vowels [4]. Chen had found out the difference of first formant and an extra peak to be a promising feature of nasality [1]. This property was exploited by Vijaylaxmi et al. for detection of hypernasality [11]. They have used a modified group delay based approach to resolve the first formant and extra resonance that manifest in hypernasal speech due to nasalization [8]. In their paper, they have reported that the proposed feature has limitation in case of nasalized vowel detection in healthy speakers' speech. Pruthi had analysed 37 acoustic parameters from the existing literature and selected nine knowledge based acoustic parameters for detection of nasalized vowels [9].

In this work, we have proposed an inverse filtering based feature which accounts for amount of nasalisation present in a vowel. The rest of the paper is organised as follows. Section 2 presents database description. In Sect. 3 inverse filtering based feature is proposed. In Sect. 4 analysis on different nasalised vowels is done. Section 5 summarises with the findings of the analysis.

2   Database Description

In this study, speech data from 15 speakers have been collected. Speech data are collected for vowels /e/, /u/, /i/, their nasalized counterparts i.e. /en/, /un/, /in/ respectively and the word 'summer'. The word 'summer' contains the nasal consonant /m/ [11]. The nasal consonant part is manually marked for all the speech files. All the recordings are done in a speech recording studio. So, the data are free from any background noises. Speech recordings are done by using Audacity software. All the speech data are collected at 48000 Hz sampling frequency. However, as information contained in speech signal above 5000 Hz is least, data in this work are resampled at 11025 Hz.

3   Inverse Filtering Based Feature

The nasalized vowels are the addition of oral sounds and nasal sounds. For different nasalized vowels, the oral filter changes its characteristics, while the nasal filter remains invariant [7]. And this invariant nasal filter has similar characteristics to the filter which produces nasal murmur [7]. So, for different nasalized vowels only one nasal filter can be modeled. The nasal filter can be estimated from nasal murmur sound. The coupled oral and nasal tract can be modeled as in Fig. 1.

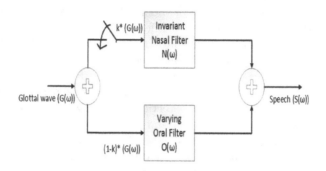

Fig. 1. Model of the speech production system

This understanding of our speech production system lets us model the speech sound as,

$$S(\omega) = k \times G(\omega) \times N(\omega) + (1 - k) \times G(\omega) \times O(\omega)$$
$$= G(\omega) \times N(\omega) \times (k + (1 - k) \times \frac{O(\omega)}{N(\omega)}) \tag{1}$$

In Eq. 1, 'k' represents fraction of glottal wave passed through the invariant nasal filter. $N(\omega)$ and $O(\omega)$ represents nasal filter and oral filter, respectively. And $G(\omega)$ and $S(\omega)$ represents glottal wave and speech signal, respectively.

Let's consider the nasal filter as,

$$N(\omega) = \frac{\prod_{z=1}^{Z} N(\omega - \omega_z)}{\prod_{p=1}^{P} N(\omega - \omega_j)} \tag{2}$$

Putting the above value in Eq. 1 we will get,

$$S(\omega) = G(\omega) \times N(\omega) \times (k + (1 - k)$$
$$\times O(\omega) \times \frac{\prod_{p=1}^{P} N(\omega - \omega_i)}{\prod_{z=1}^{Z} N(\omega - \omega_j)})$$
$$\implies S(\omega) \times N(\omega)^{-1} = G(\omega) \times (k + (1 - k)$$
$$\times O(\omega) \times \frac{\prod_{p=1}^{P} N(\omega - \omega_i)}{\prod_{z=1}^{Z} N(\omega - \omega_j)}) \tag{3}$$

Now, if we will evaluate Eq. 3 on a nasal pole, which doesn't have any oral pole nearby, then we will get,

$$S(\omega_p) \times N(\omega_p)^{-1} = G(\omega_p) \times (k)$$
$$\implies k = \frac{S(\omega_p) \times N(\omega_p)^{-1}}{G(\omega_p)} \tag{4}$$

In Eq. 4, 'k' represents the amount of nasalisation.

Equation 4 suggests that, if we have speech signal ($S(\omega)$), glottal wave ($G(\omega)$) and a mathematical model of nasal filter ($N(\omega)$) then we can find out amount of nasalisation present in the corresponding speech signal. However, in any speech application, we will be having $S(\omega)$ and hence $G(\omega)$. The only extra information needed is the person specific nasal filter.

In [6] we have shown that first three formants of nasal murmur occurs around 250 Hz, 1250 Hz and 2500 Hz. In [10] it is shown that /i/ and /e/ doesn't have any formant near 1250 Hz. So to find out the value of 'k' in case of /i/ and /e/, we will use the pole corresponding to formant location 1250 Hz. And in case of vowel /u/, we will be using the pole near 2500 Hz. The expression of 'k' is evaluated on the pole-zero circle rather than evaluating it on the pole location.

The assumptions taken in this study are,

i Effect of evaluating the value of 'k' on a unit circle instead of evaluating it on the pole location is negligible. This assumption is taken as the pole location of the nasal filter is very near to the unit circle.
ii There is no nasal zero present near the pole location chosen. Also the zeros of oral filters, if any, are also not taken into account in this study.

4 Analysis of Nasalised Vowels

Information needed to find out the value of 'k' are person specific nasal filter ($N(\omega)$), speech signal ($S(\omega)$) and glottal wave ($G(\omega)$). The person specific invariant nasal filter is estimated from nasal murmur, using LP analysis. LP coefficients of 12th order LP model are estimated, for each 20 ms windowed segment of nasal murmur. From the LP coefficients, poles for all frames are estimated and they are averaged to get a desired invariant nasal filter for each person. In this analysis $G(\omega)$ is taken as residual of an 12th order LP filter of the speech signal. In Fig. 2 pole-zero plot of nasal filter of a person is shown. The cross marks show the location of poles on the pole-zero plot. The pole locations marked in green color represents the locations corresponding to the estimated invariant nasal filter.

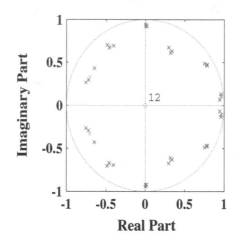

Fig. 2. Pole zero plot of the nasal filter of a person (Color figure online)

Value of 'k' is found out at five frequency locations and averaged to minimize any spurious peak that may arise due to division of residual signal. The five frequency locations are, the selected frequency and two lower and upper adjacent frequencies with differences of 5 Hz. Values of 'k' for vowel /i/, /e/ and /u/ are calculated and their box plots are also obtained. Figures 3, 4 and 5 correspond to the box plot of vowel /i/, /e/ and /u/ respectively.

From the box plots, it is observed that the value of 'k' for nasalised vowel is higher compared to their oral counterparts. It is also observed that the value of 'k' is within the range of 0 to 1 as desired. The box plot also shows that the proposed feature has high discriminatory capability, which is also validated using F-ratios. In Table 1 the median values of 'k' are tabulated. It is to be noted that for oral vowel case, the value of 'k' is non-zero. The possible reasoning for this can be the approximations that we have taken and also due to vibrations of velum during the utterance of oral vowels.

F-ratios and p values of 'k' are obtained using one-way ANOVA (Analysis Of Variance) for different vowels. ANOVA suggests whether different groups belong to the same distribution or they have come from different distributions. The small value of 'p' and high value of 'F-ratio' of two groups signify that the two groups have come from different distributions. From Table 2 it is observed that the F-ratios are high and p values are very low. This shows that oral vowels and nasalised vowels are highly discriminable for values of 'k'.

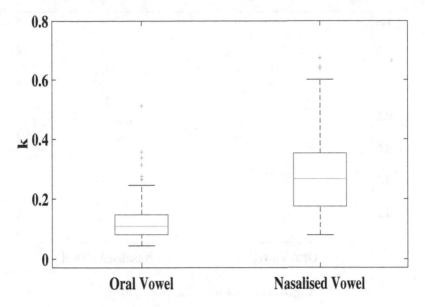

Fig. 3. Box plot of 'k' for vowel /i/

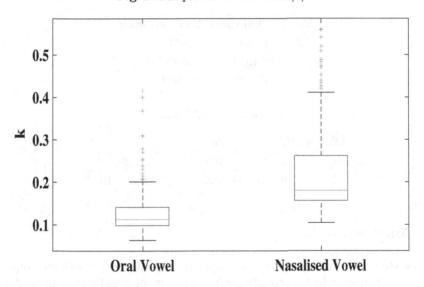

Fig. 4. Box plot of 'k' for vowel /e/

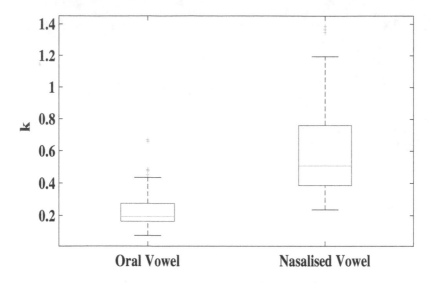

Fig. 5. Box plot of 'k' for vowel /u/

Table 1. Median values of 'k'

Vowels	Oral vowel	Nasalised vowel
/i/	0.1091	0.2671
/u/	0.1943	0.50804
/e/	0.1125	0.17972

Table 2. ANOVA values

ANOVA	/i/	/e/	/u/
F-ratio	203.74	109.19	288.69
p	1.9×10^{-38}	2.03×10^{-22}	1.42×10^{-47}

5   Conclusions

In this study, we have proposed a simple inverse filtering based technique to find out a feature which accounts for the amount of nasalisation present in a vowel. Nasalized vowels differ from oral vowels by containing more amount of nasalisation. So this feature becomes useful for detection of nasalised vowels. Statistical analysis of the feature has shown that this feature gives values which are well separable for nasalised vowels and oral vowels. As the mathematical basis of this feature is degree of nasalisation, a good correlation of this feature may be found out with the perceptual score.

References

1. Chen, M.Y.: Acoustic parameters of nasalized vowels in hearing-impaired and normal-hearing speakers. J. Acoust. Soc. Am. **98**(5), 2443–2453 (1995)
2. Fant, G.: Acoustic theory of speech production (1960)
3. Fujimura, O., Lindqvist, J.: Sweep-tone measurements of vocal-tract characteristics. J. Acoust. Soc. Am. **49**(2B), 541–558 (1971)
4. Glass, J., Zue, V.: Detection of nasalized vowels in American English. In: Acoustics, Speech, and Signal Processing, IEEE International Conference on, ICASSP 1985, vol. 10, pp. 1569–1572. IEEE (1985)
5. House, A.S., Stevens, K.N.: Analog studies of the nasalization of vowels. J. Speech Hear. Disord. **21**(2), 218–232 (1956)
6. Jyotishi, D., Deb, S., Abhishek, A., Dandapat, S.: Experimental analysis on effect of nasal tract on nasalised vowels. In: Tanveer, M., Pachori, R.B. (eds.) Machine Intelligence and Signal Analysis. AISC, vol. 748, pp. 727–737. Springer, Singapore (2019). https://doi.org/10.1007/978-981-13-0923-6_62
7. Jyotishi, D., Deb, S., Dandapat, S.: A novel feature for nasalised vowels and characteristic analysis of nasal filter. In: 2018 Twenty Fourth National Conference on Communications (NCC), pp. 1–5. IEEE (2018)
8. Murthy, H.A., Gadde, V.: The modified group delay function and its application to phoneme recognition. In: 2003 IEEE International Conference on Acoustics, Speech, and Signal Processing 2003, ICASSP 2003, vol. 1, pp. I-68. IEEE (2003)
9. Pruthi, T.: Analysis, vocal-tract modeling and automatic detection of vowel nasalization. Ph.D. thesis, University of Maryland, College Park (2007)
10. Rabiner, L.R., Schafer, R.W.: Digital Processing of Speech Signals, vol. 100. Prentice-hall Englewood Cliffs, New Jersey (1978)
11. Vijayalakshmi, P., Reddy, M.R., O'Shaughnessy, D.: Acoustic analysis and detection of hypernasality using a group delay function. IEEE Trans. Biomed. Eng. **54**(4), 621–629 (2007)

Iris Recognition by Learning Fragile Bits on Multi-patches using Monogenic Riesz Signals

B. H. Shekar[1]([⊠])[iD], Sharada S. Bhat[2][iD], and Leonid Mestetsky[3]

[1] Mangalore University, Mangalore, India
bhshekar@gmail.com
[2] Government First Grade College Ankola, Karnataka, India
sharadasbhat@gmail.com
[3] Lomonosov Moscow State University, Moscow, Russia
mestlm@mail.ru

Abstract. Unconstrained, at-a-distance iris recognition systems endure the problem of fragile bits. Existence of fragile bits in candidate iris results into low recognition rates. Proposed approach utilizes *fragile-bit* information for classification of iris bits as consistent or inconsistent. We divide the candidate iris into patches and propose monogenic signals of Riesz wavelets to learn fragile bits in each patch. We propose a new feature descriptor, Riesz signal based binary pattern (RSBP) to extract the features from these patches. Each patch is assigned with a weight pivoted on the fragile bits present in it. Dissimilarity score between the two irises is obtained by adopting weighted mean Euclidean distance (WMED). Experiments are conducted using both near infra red (NIR) and visible wavelength (VW) images, obtained from the benchmark databases IITD, MMU v-2, CASIA-IrisV4-Distance and UBIRIS v2. Results justify the applicability of proposed approach for iris recognition.

Keywords: Iris recognition · Fragile bits · Multi-patches · Monogenic signal

1 Introduction

Because of intricate textural pattern of the stroma in human iris, iris biometrics manifests low miss match rates compared to rest of the biometric traits [10]. Earlier works on iris biometrics deal with the iris images obtained under constrained scenario and have obtained promising results. However, when iris images are obtained from long stand-off distances under unconstrained imaging, acquired iris information will be very poor because of bad illumination and less cooperation by the subject and results into high intra-class variations [17]. Such degraded iris images are occluded by eye-lids, eye-lashes, eye-glasses and specular reflections. Occluded region of the candidate iris image leads to an iris code comprising of the bits which are not consistent (fragile) over irises of the same

© Springer Nature Switzerland AG 2019
B. Deka et al. (Eds.): PReMI 2019, LNCS 11942, pp. 462–471, 2019.
https://doi.org/10.1007/978-3-030-34872-4_51

person [6]. A fragile bit is defined as a bit in an iris code which is not consistent across the different iris images of the same subject. Probability of existence of noise at a specific region of iris varies from image to image and hence produces fragile bits in the corresponding iris code. Figure 1 illustrates presence of noise in sample eye images of the same subject, taken from UBIRIS v.2 database, and noise in their unwrapped irises which leads to fragile bits. We propose an iris recognition scheme in which, tracks and sectors are deployed to divide the unwrapped iris region into patches. Earlier researchers have adopted region-wise, multi-patch feature extraction approach for iris classification [11,13]. Recently Raja et al. [15] have proposed deep sparse filtering on multi patches of normalized mobile iris images. Histograms of each patch are used to represent iris in a collaborative sub space. In our earlier work [18], we have divided the unwrapped iris into M patches using p sectors and q tracks and have utilized Fuzzy c-means clustering algorithm to classify the patches into best iris region and noisy region. We have used probability distribution function in order to cluster the patches into iris or non-iris regions. However, due to varieties of noise present in unconstrained imaging, clustering the iris regions based on their statistical properties alone is difficult. We propose a learning technique which classifies the patches based on bit fragility in each patch using monogenic functions.

Monogenic signals are 2D extensions of 1D analytical signals. An Analytical signal, in its polar representation, gives the information about local phase and amplitude, hence, generalizing 1D analytical signal to its 2D counterpart, using Riesz transform, gives a deeper insight to low level image processing [5]. First order Riesz wavelets can extract 1D inherent signals such as lines and steps in an image and second order Riesz wavelets can capture 2D signals like corners and junctions [9]. In Fig. 2 we have presented example of an unwrapped iris from IITD database, its first order Riesz transformed output responses, $h_x(f)$ and $h_y(f)$ respectively. We can observe texture variations along horizontal and vertical directions. Further, Riesz transform allows us to extend Hilbert transform along any direction. This property is called steerable property. By the virtue of steerable property, a filter of arbitrary orientation can be created as a linear combination of a collection of basis filters. Steerable Riesz furnishes a substantial computational scheme to extract the local properties of an image.

A texture learning technique, exploiting local organizations of magnitude and orientations, using steerable Riesz wavelets, has been proposed by Depeursinge [4]. Zhang et al. [21] propose a *competitive coding scheme* (CompCode) for finger knuckle print (FKP) recognition based on Riesz functions and have obtained promising results for FKP. Two iris coding schemes, based on first and second order Riesz components and steerable Riesz are proposed in [19]. One of the coding scheme encodes responses of two components of first order Riesz and three components of 2nd order Riesz so that each input iris pixel is represented by five binary bits and another scheme generates three bits from steerable Riesz filters. Inspired by these works, we design a descriptor (RSBP) based on 1D and 2D Riesz signals which encodes each pixel into an 8-bit binary pattern. We present detailed explanation of RSBP descriptor and proposed iris recognition

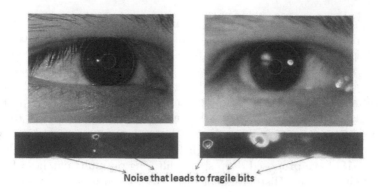

Fig. 1. Samples of two eye images of same subject and color maps of their unwrapped irises with noise that leads to fragile bits. (Color figure online)

Fig. 2. Example of an unwrapped iris from IITD database, its first order Riesz transformed output responses, $h_x(f)$ and $h_y(f)$ respectively.

scheme in Sect. 2. Result analysis of experiments is discussed in Sect. 3 and we conclude this paper in Sect. 4.

2   Proposed Iris Recognition Scheme

Outline of proposed scheme is demonstrated in Fig. 3. We create M patches of unwrapped iris using p tracks and q sectors. To extract predominant features from each patch, we deploy feature descriptor, Riesz signal based binary pattern (RSBP), which represents each pixel into 8-bit binary code. We adopt Fuzzy c-means clustering (FCM) to learn fragile bits and to cluster the iris patches into five classes, with labels 0 to 4, where 0 refers to non-iris region (with maximum number of fragile bits), and 4 refers to best iris region (with consistent bits). Based on these labels each patch is assigned with a weight. Further, a *magnitude weighted phase histogram*, proposed in [16], is adopted to represent each patch as 1D real valued feature vector. Dissimilarity between the two iris codes is computed using weighted mean Euclidean distance (WMED).

2.1   Riesz Functions

Riesz functions are the generalizations of Hilbert functions into d dimensional Euclidean space. If $\mathbf{x}$ is a d-tuple, $(x_1, x_2, \ldots, x_d)$ and f is L^2 measurable d-dimensional function, i.e. $f(\mathbf{x}) \in L^2(\mathbb{R}^d)$, then, Riesz transformation R^d of f is

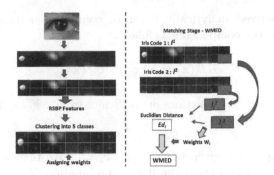

Fig. 3. Proposed iris recognition method

a d-dimensional vector signal. $R^d f(\mathbf{x})$ transforms a signal from $L^2(\mathbb{R}^d)$ space to $L^d(\mathbb{R}^d)$ space and is given by,

$$R^d f(\mathbf{x}) = (R_1^d f(\mathbf{x}), R_2^d f(\mathbf{x}), \ldots, R_d^d f(\mathbf{x})), \tag{1}$$

which can be simplified further as follows.

$$R^d f(\mathbf{x}) = ((h_1 * f)(\mathbf{x}), (h_2 * f)(\mathbf{x}), \ldots, (h_d * f)(\mathbf{x})), \tag{2}$$

where, $*$ represents convolutional operator and h_i is d dimensional Riesz kernel which is given by,

$$h_i = \frac{\Gamma^{\frac{(d+1)}{2}}}{\pi^{(d+1)/2}} \frac{x_i}{\|\mathbf{x}\|^{(d+1)}}. \tag{3}$$

Further, in 2-D space, taking $\mathbf{x} = (x, y)$, 2-D Riesz transform is given by,

$$R^2 f(\mathbf{x}) = ((h_x * f)(\mathbf{x}), (h_y * f)(\mathbf{x})) = (h_x f(\mathbf{x}), h_y f(\mathbf{x})), \tag{4}$$

where, h_x and h_y are 2-D Riesz kernels which are obtained by taking $d = 2$ in equation (3).

$$h_x = \frac{1}{2\pi} \frac{x}{\|\mathbf{x}\|^3}, \quad h_y = \frac{1}{2\pi} \frac{y}{\|\mathbf{x}\|^3}. \tag{5}$$

Components in the triplet $(f, h_x f, h_y f)$ are called first order monogenic components of the function f. Further, when $h_x f$ and $h_y f$ are convolved with the kernels, h_x and h_y, as the convolution operator is commutative, we get second order components of f as, $(h_{xx} f, h_{xy} f, h_{yy} f)$. One dimensional phase encoding method, based on zero crossings, is used to encode each of the five output signals obtained from first and second order components into binary iris code, so that each pixel in the input iris is represented by five binary bits. We call this iris code as Riesz filter based iris code, RFiC1.

2.2 Steerable Riesz

Because of the steerable property of Riesz filters, response of an ith Riesz component R_i^d of an oriented image f^θ of an image f, oriented by any arbitrary angle θ, can be derived analytically. It can be computed as linear combinations of responses of all the components as follows.

$$Rf^\theta(\mathbf{x}) = \sum_{i=0}^{d} g_i(\theta) R_i^d f(\mathbf{x}), \qquad (6)$$

where, $g_i(\theta)$ are coefficient functions or coefficient matrices. Equation (6) can be rewritten as,

$$Rf^\theta = \omega^T R^d, \qquad (7)$$

where ω is weight vector given by, $\omega = [\omega_1, \omega_2, \omega_3]$. We take a steering matrix M^θ proposed in [4] and obtain the linear combination of second order Riesz components h_{xx}, h_{xy} and h_{yy} of f as follows.

$$Rf^\theta(x, y) = \omega_1 M^\theta h_{xx} f(x, y) + \omega_2 M^\theta h_{xy} f(x, y) + \omega_3 M^\theta h_{yy} f(x, y) \qquad (8)$$

Based on the procedure given in [19], we obtain three more bits from the input pixel. If $\frac{k\pi}{n}$, $k = 0, 1, \ldots, n-1$ are n orientations of θ and $I(x, y)$ is the input unwrapped iris, then, at every pixel (x, y), linear sum of steerable Riesz responses $RI^\theta(x, y)$ is computed for each orientation θ using Eq. (8). Thus, each pixel will be having n representations for n orientations. Further, at every point (x, y) an integer value N which varies from 0, to $n-1$ is computed to represent the dominant orientation.

$$N(x, y) = argmax_\theta(RI^\theta(x, y)). \qquad (9)$$

In our experiments, we have taken $n = 6$, so that, each pixel has six orientation representations. Equation (9) gives the dominant orientation at (x, y) and generates a matrix consisting of integers from one to six, representing six orientations. These integers from 0 to 5 are represented by a corresponding three bit binary code in the set $\{000, 001, 011, 111, 110, 100\}$. These bits combined with $h_x f$ and $h_y f$, the horizontal and vertical responses, along with a Gabor bit, produce a 6-bit representation of a pixel and the iris code, thus obtained, is called RFiC2.

2.3 Riesz Signal Based Feature Descriptor (RSBP)

Design of RSBP is based on the method by Rajesh and Shekar [16], which uses complex wavelet transform for face recognition. However, we have developed this code on Riesz wavelet transforms. Proposed method is explained in Fig. 4. First and second order Riesz functions are operated on unwrapped iris using convolution operation, to obtain five real valued Riesz responses for each pixel. These responses are encoded into a binary bit using 1D phase informations of zero crossings. Further, steerable Riesz method produces three more bits based on the dominant orientation and thus, we obtain a binary pattern of 8-bits. While computing the decimal equivalent of this binary pattern, we have taken response of first order Riesz along horizontal direction as the most significant bit (MSB) and last bit in steerable Riesz output as LSB. Because, as we observe with experiments, responses of first order horizontal Reisz are more prominent. In Fig. 4 one can observe the distinguished features in RSBP image and in its color map.

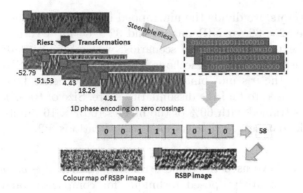

Fig. 4. Riesz signal based binary pattern (RSBP) (Color figure online)

2.4 Iris Code Matching

To explain representation and matching, let $\mathcal{I}^1$ and $\mathcal{I}^2$ be two unwrapped irises and $\mathcal{I}_i^1$ and $\mathcal{I}_i^2$ be the corresponding i^{th} patches, and ω_i^1 and ω_i^2 be the weights assigned to these patches where $i \in \{1, 2, \ldots, M\}$, M representing the total number of patches. In order to represent iris features, we adopt the approach in [16]. RSBP image is subdivided into P number of $p_1 \times q_1$ sized *sub-blocks*. In every sub-block, we compute k bin length histogram. Further, histograms of all of these sub blocks are concatenated to obtain 1D feature vector. We compute the dissimilarity score Ed_i between the feature vectors fv_i^1 and fv_i^2, corresponding to the patches $\mathcal{I}_i^1$ and $\mathcal{I}_i^2$, using Euclidean distance metric. Taking ω_i as the common weight for both $\mathcal{I}_i^1$ and $\mathcal{I}_i^2$, the weighted mean Euclidean distance (WMED) between the two irises is calculated by,

$$Ed(\mathcal{I}^1, \mathcal{I}^2) = \frac{\sum \omega_i Ed_i}{\sum \omega_i} \qquad (10)$$

When both $\mathcal{I}_i^1$ and $\mathcal{I}_i^2$ have non zero weights, ω_i will be the maximum of the weights ω_i^1 and ω_i^2 and when one of them has zero weight, then ω_i is set to zero.

3 Experimental Analysis

To justify the applicability of our scheme for iris recognition, we have experimented our approach on the benchmark NIR iris datasets IITD [8], MMU v-2 [2], CASIA-IrisV4-Distance [1] and VW dataset UBIRIS.v2 [14] and the results are compared with state-of-the-art publications which have worked on the same datasets. In this work, our primary objective is iris feature extraction and representation. Hence, iris segmentation is done using the approach given in [18] and for unwrapping (normalization), Daugman's rubber sheet model [3] is adopted. Regarding the parameter set up for our experiments, in case of NIR images, resolution for unwrapping is 64×256 and for UBIRIS v.2 it is 64×512. Using 2 tracks and 8 sectors, we divide the unwrapped iris into 16 patches so that, each patch is of size 32×32 for NIR images and 32×64 for VW images. We have conducted the experiments in two scenarios, without multi-patches and with multi-patches using the coding methods RFiC1, RFiC2 and RSBP. RFiC1 and RFiC2 represent the iris into binary-bit patterns and hence Daugman's Hamming distance is used to find the dissimilarity score. Size of Riesz kernel is set to 15 [19]. FCM is trained with 60% of the images (60% $\times$ 16 number of patches) from each of the dataset, so that training to test ratio is 3:2.

Discussion: We have used the equal error rate (EER), $d - prime$ values and ROC curve to evaluate proposed technique for comparison and analysis. In Table 1 we present the results of our experiments conducted without multi-patches scenario (S1) and with multi-patches scenario (S2). We observed average decrease of 7.9% in EER values and increase of 5.9% in $d - prime$ values from S1 to S2. With RSBP method there is 9.48% of decrease in average EER value and 6.7% of increase in $d - prime$ value. ROC curves of these experiments are presented in Fig. 5.

Table 1. EER and $d - prime$ values obtained by proposed methods with respect to different datasets without (S1) and with (S2) multi-patches.

Method	Scenario	IITD		MMU v-2		CASIA v-4 dist		UBIRIS v.2	
		EER	d'	EER	d'	EER	d'	EER	d'
RFiC1	S1	0.0228	5.6306	0.0575	3.3081	0.0952	2.5549	0.1120	2.178
	S2	0.0207	5.6321	0.0515	3.3891	0.0918	2.5651	0.1009	2.5592
RFiC2	S1	0.0206	5.8188	0.0636	3.3864	0.0879	2.7280	0.1033	2.176
	S2	0.0200	5.8811	0.0541	3.6956	0.0871	2.7358	0.0898	2.7113
RSBP	S1	0.0106	5.9106	0.0440	3.4166	0.0710	2.8160	0.09910	2.268
	S2	**0.0100**	**6.1167**	**0.0417**	**3.7783**	**0.0706**	**2.8203**	**0.0811**	**2.7290**

We compare the proposed technique with the state-of-art methods which work on fragile bits or multi-patches techniques. Results of comparison analysis are presented in the Table 2. In [12] author compare the usefulness of different regions of the iris for recognition using bit-discriminability. Kaur et al. [7] have computed discrete orthogonal moment-based features on the ROI divisions of the unwrapped iris. Vyas et al. [20] have extracted gray level co-occurrence matrix (GLCM) based features from multiple blocks of normalized iris templates and concatenated them to form the feature vector. Since, these authors have worked with multi-patches concept on the same datasets, we have compared their published results with our results and the figures displayed in Table 2 illustrate that our method compares favourably with existing approaches.

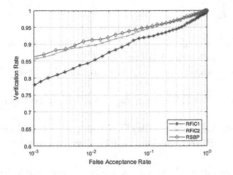 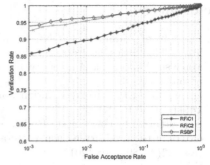

Fig. 5. ROC curves of the RFiC1, RFiC2 and RSBP obtained on MMU dataset without and with multi-patches approach respectively.

Table 2. EER and $d - prime$ values with respect to UBIRIS v.2 and IITD databases compared with the recent related publications. NA implies not available.

Database	Method	EER	$d - prime$
UBIRIS v.2	Proenca [12]	NA	1.8900
	Kaur et al. [7]	0.1000	2.8800
	Proposed	**0.0811**	2.7290
IITD	Kaur et al. [7]	0.0100	4.8000
	Vyas et al. [20]	0.0122	NA
	Proposed	**0.0100**	**6.1167**

4   Conclusion

Fragile bits present in an iris code mainly increase the intra-class variations, thereby increasing false reject rate. Proposed method uses the fragile bit information to rank the iris regions by assigning the weights which are further used in matching the iris code and hence, the intra-class variations are suppressed and at the same time inter-class variations are enhanced. We have also proposed a new iris feature extraction and representation approach using a descriptor RSBP, based on 1D and 2D Riesz transformations and experiments illustrate that it is well suited for iris recognition.

Acknowledgement. This work is supported jointly by the Department of Science and Technology, Government of India and Russian Foundation for Basic Research, Russian Federation under the grant No. INT/RUS/RFBR/P-248.

References

1. Institute of Automation, Chinese Academy of Sciences: CASIA Iris Database. http://biometrics.idealtest.org/, http://biometrics.idealtest.org/
2. Malaysia Multimedia University Iris Database. http://pesona.mmu.edu, http://pesona.mmu.edu
3. Daugman, J.: How iris recognition works. IEEE Trans. Circ. Syst. Video Technol. 14(1), 21–30 (2004)
4. Depeursinge, A., Foncubierta-Rodriguez, A., Van de Ville, D., Muller, H.: Rotation-covariant texture learning using steerable Riesz wavelets. IEEE Trans. Image Process. 23(2), 898–908 (2014)
5. Felsberg, M., Sommer, G.: The monogenic scale-space: a unifying approach to phase-based image processing in scale-space. J. Math. Imaging Vis. 21(1–2), 5–26 (2004)
6. Hollingsworth, K.P., Bowyer, K.W., Flynn, P.J.: The best bits in an iris code. IEEE Trans. Pattern Anal. Mach. Intell. 31(6), 964–973 (2009)
7. Kaur, B., Singh, S., Kumar, J.: Robust iris recognition using moment invariants. Wirel. Pers. Commun. 99(2), 799–828 (2018)
8. Kumar, A., Passi, A.: Comparison and combination of iris matchers for reliable personal authentication. Pattern Recogn. 43(3), 1016–1026 (2010)
9. Marchant, R., Jackway, P.: Local feature analysis using a sinusoidal signal model derived from higher-order Riesz transforms. In: 2013 20th IEEE International Conference on Image Processing (ICIP), pp. 3489–3493. IEEE (2013)
10. Nguyen, K., Fookes, C., Jillela, R., Sridharan, S., Ross, A.: Long range iris recognition: a survey. Pattern Recogn. 72, 123–143 (2017)
11. Pillai, J.K., Patel, V.M., Chellappa, R., Ratha, N.K.: Secure and robust iris recognition using random projections and sparse representations. IEEE Trans. Pattern Anal. Mach. Intell. 33(9), 1877–1893 (2011)
12. Proença, H.: Iris recognition: what is beyond bit fragility? IEEE Trans. Inf. Forensics Secur. 10(2), 321–332 (2015)
13. Proenca, H., Alexandre, L.A.: Toward noncooperative iris recognition: a classification approach using multiple signatures. IEEE Trans. Pattern Anal. Mach. Intell. 29(4), 607–612 (2007)
14. Proenca, H., Filipe, S., Santos, R., Oliveira, J., Alexandre, L.A.: The ubiris. v2: a database of visible wavelength iris images captured on-the-move and at-a-distance. IEEE Trans. Pattern Anal. Mach. Intell. 32(8), 1529–1535 (2010)
15. Raja, K.B., Raghavendra, R., Venkatesh, S., Busch, C.: Multi-patch deep sparse histograms for iris recognition in visible spectrum using collaborative subspace for robust verification. Pattern Recogn. Lett. 91, 27–36 (2017)
16. Rajesh, D., Shekar, B.: Undecimated dual tree complex wavelet transform based face recognition. In: 2016 International Conference on Advances in Computing, Communications and Informatics (ICACCI), pp. 720–726. IEEE (2016)
17. Rattani, A., Derakhshani, R.: Ocular biometrics in the visible spectrum: a survey. Image Vis. Comput. 59, 1–16 (2017)
18. Shekar, B.H., Bhat, S.S.: Multi-patches iris based person authentication system using particle swarm optimization and fuzzy C-means clustering. In: International Archives of the Photogrammetry, Remote Sensing & Spatial Information Sciences, vol. 42 (2017)
19. Shekar, B., Bhat, S.S.: Steerable Riesz wavelet based approach for iris recognition. In: 2015 3rd IAPR Asian Conference on Pattern Recognition (ACPR), pp. 431–436. IEEE (2015)

20. Vyas, R., Kanumuri, T., Sheoran, G., Dubey, P.: Co-occurrence features and neural network classification approach for iris recognition. In: 2017 Fourth International Conference on Image Information Processing (ICIIP), pp. 1–6. IEEE (2017)
21. Zhang, L., Li, H.: Encoding local image patterns using Riesz transforms: with applications to palmprint and finger-knuckle-print recognition. Image Vis. Comput. **30**(12), 1043–1051 (2012)

Shouted and Normal Speech Classification Using 1D CNN

Shikha Baghel[1(✉)], Mrinmoy Bhattacharjee[1], S. R. M. Prasanna[1,2], and Prithwijit Guha[1]

[1] Department of Electronics and Electrical Engineering,
Indian Institute of Technology Guwahati, Guwahati 781039, Assam, India
{shikha.baghel,mrinmoy.bhattacharjee,prasanna,pguha}@iitg.ac.in
[2] Department of Electrical Engineering, Indian Institute of Technology Dharwad,
Dharwad 580011, India

Abstract. Automatic shouted speech detection systems usually model its spectral characteristics to differentiate it from normal speech. Mostly hand-crafted features have been explored for shouted speech detection. However, many works on audio processing suggest that approaches based on automatic feature learning are more robust than hand-crafted feature engineering. This work re-demonstrates this notion by proposing a 1D-CNN architecture for shouted and normal speech classification task. The CNN learns features from the magnitude spectrum of speech frames. Classification is performed by fully connected layers at later stages of the network. Performance of the proposed architecture is evaluated on three datasets and validated against three existing approaches. As an additional contribution, a discussion of features learned by the CNN kernels is provided with relevant visualizations.

Keywords: Shouted and normal speech classification · Shouted speech detection · 1D CNN · Convolution filter visualization

1 Introduction

Automatic shouted speech detection has application in areas like health-care, security and home-care [12]. Moreover, it is also required as a preprocessing step in applications like ASR and speaker recognition systems. Performance of such systems that are mostly trained on normally phonated speech degrades when test utterances include shouted content [12,14]. Thus, this work focusses on efficient segregation of normal and shouted speech.

Production of shouted speech may be attributed to following situations – charged emotions while speaking, communicating over long-distances, or calling in distress. In such situations, speech production characteristics deviate from normal. The effect of change in production characteristics is reflected in vocal tract characteristics. Air pressure from lungs increases during shouting, leading to comparatively fast vibration of vocal folds [9]. This changes the fundamental

© Springer Nature Switzerland AG 2019
B. Deka et al. (Eds.): PReMI 2019, LNCS 11942, pp. 472–480, 2019.
https://doi.org/10.1007/978-3-030-34872-4_52

frequency (F_0) of produced speech. Properties of F_0 has been extensively studied in the context of shouted and normal speech. In literature, shouted speech is considered as a high vocal effort speech. Authors in [9] have studied deviation in vocal efforts for five vocal modes – whisper, soft, neutral (normal), loud and shout. Spectral tilt and log-linear predictive coding were examined in [14] to study the effect of different vocal efforts on automatic speech recognition. The classification performances were tested using three variants – (a) Bayesian classifier based on static GMMs with diagonal covariance matrices; (b) GMMs with full covariance matrices; (c) a multi-class Support Vector Machine (SVM) classifier with Radial Basis Function (RBF) kernel. To capture the variability in vocal tract characteristics due to different vocal modes, Mel-Frequency Cepstral Coefficient (MFCC) has been largely studied along with different combination of features [12]. Mittal et al. in [9] used features derived from the Hilbert envelope of double differenced Numerator of the Group Delay (HNGD) spectrum to analyze the differences between shouted and normal speech. Strength of Excitation (SoE), F_0 and dominant frequency (F_D) derived from the LP spectrum of vowel-like regions were examined to discriminate shouted and normal speech. Shifts in formant positions are also used in literature for detection of shouted speech. Recently, the standard deviation in frequency and energy of the first three formants have been explored for analyzing and characterizing shouted speech signals [8].

Existing works have mostly focused on using excitation or spectral features for detection of shouted speech [10]. Recent works in deep learning have shown that effective architectures can learn robust and discriminative patterns from data and provide promising classification performance [2]. Such representations are learned with a framework of many affine transformations followed by a nonlinearity [4]. These automatically learned features are comparatively more robust to local variations in data. Convolutional Neural Networks (CNN, henceforth) have been used for feature learning from audio data for several audio processing applications like speech-based emotion recognition [3], speech recognition [11], hate speech detection [1] etc. CNNs are capable of capturing patterns present in data. This motivated us to explore CNNs for shouted and normal speech classification. Motivated by the success of CNNs in various image processing applications [7], authors have deployed $2D$ CNNs on spectrograms (or other time-frequency variants) of audio data [2,3]. However, a drawback of this approach is the requirement of large datasets for learning the networks. To the best of our knowledge, most standard datasets (available in public domain) for shouted speech classification are small in size. Other works have reported results on private datasets (remains undisclosed). Hence, it is difficult to get a sufficiently large dataset for employing $2D$-CNNs for shouted and normal speech classification. This motivated us to explore $1D$-CNNs consisting of one-dimensional convolutional kernels. We propose a $1D$-CNN with two convolution layers, one max-pooling stage and three dense layers (Fig. 1). The proposed CNN is benchmarked on three standard datasets with respect to three baseline algorithms. To summarize, this work has the following contributions. First, a proposal of a

1D CNN architecture for shouted and normal speech classification. Additionally, CNN filters are analyzed in an attempt to interpret the learned features. Second, construction of a dataset of 20 speakers in both shouted and normal vocal mode.

The rest of the paper is organized in the following manner. The proposed approach is described in Sect. 2. The experimental results on standard datasets are presented and discussed in Sect. 3. Learned filters are visualized in Sect. 4. We conclude in Sect. 5 and sketch the future scope of present work.

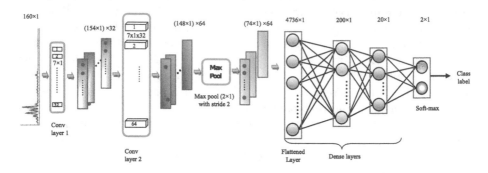

Fig. 1. Proposed 1D CNN architecture.

2   Proposed Approach

Existing works in shouted and normal speech classification have mostly used spectral features, like spectral tilt [14], formant locations [8], MFCCs [14] etc. This motivated us to first transform the input speech signal to frequency domain. Due to quasi-periodic nature of speech, it is segmented in small overlapping frames. Each frame $s[n]$ $(n = 0, 1, \ldots 2N - 1)$ of length $2N$ is transformed to frequency domain as $S[k] = \sum_{n=0}^{2N-1} s[n]e^{\frac{-j2\pi kn}{2N}}; k = 0, 1, \ldots, 2N - 1$, where $S[k]$ represents spectrum of $s[n]$. Since, $s[n]$ is real, only half the magnitude spectrum $(S[k]; k = 0, 1, \ldots N - 1)$ is used as input to 1D CNN. This work uses a frame size of 20 ms with a shift of 10 ms. We have used audio signals sampled at 16 KHz. Thus, each frame of 20 ms interval contains $2N = 320$ samples. Hence, input to the 1D CNN has $N = 160$ dimensions. The proposed architecture of the 1D CNN is described next.

A conventional architecture of 1D CNN comprises sets of convolutional and pooling layers followed by fully connected dense layers. Small-sized 1D filters are used in each convolutional layer to capture the local feature of input data. The output of convolutions is passed through an activation function for non-linear transformation of data. Pooling is performed to obtain the hierarchical representation in data. All tunable parameters of the architecture are learned through a feed-forward and back-propagation approach, which minimizes a cost function. The rest of the paper is organized in the following manner. The proposed approach is described in Sect. 2. The experimental results on standard datasets are presented and discussed in Sect. 3. Learned filters are visualized in Sect. 4. We conclude in Sect. 5 and sketch the future scope of present work. The $1D$ CNN is illustrated in Fig. 1.

Table 1. Classification performance in terms of F-score

	SNE-Speech		FIN-SN		IIIT-H VLSD	
	F_{nor}	F_{sh}	F_{nor}	F_{sh}	F_{nor}	F_{sh}
Raitio-FS [13]	0.79 ± 0.01	0.79 ± 0.00	0.9 ± 0.01	0.89 ± 0.00	0.72 ± 0.01	0.74 ± 0.02
Mittal-FS [9]	0.83 ± 0.01	0.79 ± 0.01	0.92 ± 0.01	0.90 ± 0.00	0.76 ± 0.02	0.79 ± 0.02
Zelinka-FS [14]	0.95 ± 0.01	0.93 ± 0.00	0.99 ± 0.00	0.99 ± 0.00	0.87 ± 0.01	0.86 ± 0.01
$1D$-CNN	0.97 ± 0.003	0.96 ± 0.003	0.99 ± 0.002	0.99 ± 0.001	0.9 ± 0.01	0.9 ± 0.01

Convolutional Layer. Convolutional operations are performed on local regions of input data through different kernels. Each convolutional kernel extracts certain patterns from the input data. The mathematical representation of convolutional operation in layer l is given by $y_i^l = f\left(x^l \star K_i^l + b_i^l\right)$ where, K denotes the convolutional kernel, i corresponds to kernel index and b_i^l is the bias of ith kernel in l^{th} layer. Here, x^l and y_i^l represent the input and output of lth layer respectively. The convolution operator is represented by $\star$ and $f(.)$ corresponds to activation function. The proposed architecture has two convolutional layers with 32 and 64 kernels respectively. Kernel size in both the layers is 7×1. ReLU $(max(0, x))$ is used as an activation function to address the vanishing gradient problem [6]. The 160×1 input is processed by these two convolutional layers to produce a $(148 \times 1) \times 64$ tensor.

Max-Pooling. Pooling operation is used to downsample data for reducing model complexity of CNN. Pooling retains necessary representational information for further layers. A single 2×1 max-pooling stage (with stride 2) is used in our proposal. This max-pooling stage downsamples the $(148 \times 1) \times 64$ tensor to $(74 \times 1) \times 64$.

Dense Layers. Fully connected dense layers are used at the later stages of the CNN. Each node in such a layer is connected to all nodes in the previous layer. The output of a dense layer node is mathematically represented as $d^l = f(W^l d^{l-1} + b^l)$, where, W and b represent weight and bias respectively. The $(74 \times 1) \times 64$ tensor is flattened to obtain a layer containing 4736 nodes. This is followed by two hidden layers with 200 and 20 nodes respectively till last (output) layer. Each dense layer uses a ReLU activation function. The last layer hosts soft-max activation function and has 2 nodes corresponding to shouted and normal speech.

Model Training. The predicted and actual output are used to estimate the prediction error of the CNN. The CNN parameters are optimized using cross-entropy loss function with ADAM optimizer [5]. Parameters of the model are

optimized by minimizing the error with a learning rate of 0.0001. We have trained
the CNN for a maximum number of 150 epochs with a mini-batch size of 512.

Table 2. Cross-dataset classification performance in terms of F-score for $1D$-CNN.

	SNE-Speech		FIN-SN		IIIT-H VLSD	
	F_{nor}	F_{sh}	F_{nor}	F_{sh}	F_{nor}	F_{sh}
SNE-Speech	–	–	0.54 ± 0.044	0.78 ± 0.01	0.65 ± 0.03	0.73 ± 0.002
FIN-SN [12]	0.86 ± 0.01	0.82 ± 0.01	–	–	0.82 ± 0.01	0.78 ± 0.01
IIIT-H VLSD [9]	0.80 ± 0.02	0.80 ± 0.02	0.79 ± 0.03	0.86 ± 0.01	–	–

3   Experiments and Results

The proposed approach is evaluated on three datasets. The first dataset (FIN-
SN, henceforth) contains 1024 Finnish sentences uttered by 22 (11 males and
11 females) native speakers of Finnish language [12][1]. The speech signal was
recorded at 16 kHz for both normal and shouted vocal modes. Second, the IIIT-
H Volume Level Study Database (IIIT-H VLSD, henceforth) [9] (See footnote 1)
comprises normal and shouted data of 17 (10 males and 7 females) non-native
English speakers. A total of 102 English sentences are recorded at 48 kHz sampling rate. Third, a dataset contributed by the authors (SNE-Speech, henceforth)
consists of 20 (10 female and 10 male) non-native English speakers. Speakers
were asked to utter 30 English sentences in normal and shouted vocal modes.
Speakers belong to different states of India. Therefore, the accent may vary
from speaker to speaker. Out of 30 sentences, 15 are in imperative mood that
people might use to threaten someone. While remaining 15 sentences have a
neutral mood. All recordings were done in a controlled environment with TAS-
CAM DR-100mkII 2-channel portable digital recorder, electroglottograph (with
two electrodes) and Praat software. Speech and corresponding Electroglottogram
(EGG) signals were sampled at 44.1 kHz. The dataset comprises a total of 1200
sentences in both normal and shouted vocal mode. All audio signals of all three
datasets are resampled at 16 kHz for our experiments.

Our proposal is validated against three baseline approaches. Features used
to analyze shouted speech in [13] are considered as baseline Raitio-FS. These
features are F_0, Normalized Amplitude Quotient (NAQ), Sound Pressure Level
(SPL) and the difference between first and second harmonics (H1-H2). The effect
of varying the glottal dynamics on speech production due to different vocal
modes are analyzed in [9]. The F_0, alpha (ratio of closed phase to the glottal
cycle), beta (ratio of low-frequency energy to the high-frequency energy in normalized HNGD spectrum) and standard-deviation of low-frequency energy have
been reported to have different characteristics in different vocal modes. This

---

[1] Authors of this work would like to thank Pohjalainen et al. [12] and Mittal et al. [9] for sharing
their dataset.

work is considered as baseline Mittal-FS [9]. Baseline Zelinka-FS [14] reported classification of five vocal modes using 20-MFCCs using 40 filters. Authors in [9,13] presented an analysis of shouted and normal speech. They did not report any classification result. For the comparison purpose, classification is done using SVM classifier. Classification performance for shouted (F_{sh}) and normal (F_{nor}) speech are reported in terms of F1-scores of individual classes.

We have used SVM (with RBF kernel) for the classification task of baseline approaches. Grid search is performed to find the optimal classifier parameters. All experiments are carried out with train-test split of 80:20. For each approach, the classification task is performed five times on randomly drawn instances of training and testing. Each reported result in Table 1 represents the mean and standard-deviation of F-scores for all five experiments. Classification performance of 20-MFCCs is higher than the other two baselines in all three datasets. The MFCC gives almost similar performance as of $1D$ CNN for FIN-SN dataset. The $1D$ CNN provides comparatively better classification results in all three datasets. All the approaches show lower results in IIIT-H VLSD dataset. This dataset contains few shouted recordings which are perceptually similar to normal speech. This may be the reason for lower performance in this dataset. Classification results using $1D$-CNN validate its ability of efficient representation learning.

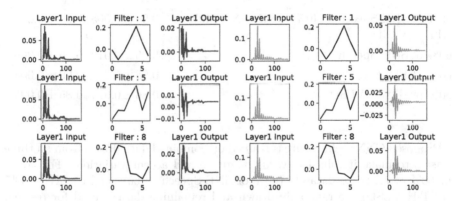

Fig. 2. Visualization of responses of few selected filters from the first convolution layer of the proposed CNN architecture as shown in Fig. 1. **First** and **fourth** columns correspond to inputs to the filters for normal and shouted speech respectively. Similarly, **third** and sixth columns correspond to outputs of the filters for normal and shouted speech respectively.

4 Analysis and Filter Visualization

The convolution filters of CNNs learn some specific properties of input data. The proposed $1D$-CNN is trained on the DFT spectrum of speech frames. Thus, filters of the proposed CNN might learn spectral characteristics of normal and shouted speech. We attempt to develop some intuitive understanding of the features learned by these filters. Figure 2 shows the shapes of some selected filters from the first convolution layer (CL, henceforth) of the trained CNN. We observe that each filter in first CL learns highly varying harmonic content of the input spectrum. The learned harmonic patterns seem to be quite different for normal and shouted speech (Fig. 2). We believe that this representation learning leads to better discrimination between the two classes.

A cross-dataset performance analysis is performed to establish the generalization performance of our proposal. The CNN is trained on one dataset and tested on the other two datasets. Table 2 shows the cross dataset performance. The results indicate the dependency of the CNN model on the training data up to some extent. Datasets FIN-SN and SNE-Speech have different languages. Dataset IIIT-H VLSD has a comparatively smaller corpus and accent of speakers is quite different from that of SNE-Speech. These might be the reasons of lower classification performance for cross datasets. This motivated us to train the CNN on combined dataset.

Table 3. Classification performance in terms of F-score for combined dataset.

Training data	1D-CNN	
	F_{nor}	F_{sh}
40% of SNE-Speech + 40% of FIN-SN + 40% of IIIT-H VLSD	0.96 ± 0.005	0.97 ± 0.009
60% of SNE-Speech	0.95 ± 0.012	0.96 ± 0.006
60% of FIN-SN	0.98 ± 0.003	0.99 ± 0.004
60% of IIIT-H VLSD	0.86 ± 0.029	0.85 ± 0.025

Proposed CNN architecture is trained on combined data drawn from all three datasets to make it robust towards language and accent variations. For training, 60% of all three datasets (60% of SNE-Speech + 60% of FIN-SN + 60% of IIIT-H VLSD) are randomly drawn and remaining data is used for testing. Table 3 illustrates the performance of the 1D CNN model trained on combined data. The results show that the trained model gives almost similar performance as mentioned in Table 1 except for IIIT-H VLSD dataset. IIIT-H VLSD is comparatively smaller dataset than the other two. Thus, the trained model might be biased towards the other two datasets. This conveys that the trained model is relatively robust towards language and accent than the one learned from individual datasets.

5   Conclusion and Future Scope

This work proposed a 1D CNN architecture for shouted and normal speech classification. The proposal is evaluated on three datasets and validated against three baseline methods. The CNN was able to efficiently learn discriminating features from the magnitude spectrum of speech frames. CNN convolution layer visualization reveals that filters in the first convolution layer learn the harmonic structure of speech spectra. Efficacy of the learned features is validated by decent generalization performance obtained with cross-data and combined data performance analysis. The CNN performance indicates that the proposed approach is effective for shouted and normal speech classification.

In the future, further exploration of the effect of different kernel and convolution layer sizes can be explored. Slight degradation in performance of cross-dataset experiments indicates the dependency of the model on language. Thus, further explorations are required to design language-independent CNN based methods. The performance of the proposed CNN based approach for different noisy speech conditions also requires further investigations.

References

1. Badjatiya, P., Gupta, S., Gupta, M., Varma, V.: Deep learning for hate speech detection in tweets. In: Proceedings of the 26th International Conference on World Wide Web Companion, WWW 2017 Companion, pp. 759–760. International World Wide Web Conferences Steering Committee, Republic and Canton of Geneva (2017)
2. Badshah, A.M., Ahmad, J., Rahim, N., Baik, S.W.: Speech emotion recognition from spectrograms with deep convolutional neural network. In: Proceedings of 2017 International Conference on Platform Technology and Service (PlatCon), pp. 1–5, February 2017
3. Huang, Z., Dong, M., Mao, Q., Zhan, Y.: Speech emotion recognition using CNN. In: Proceedings of the 22nd ACM International Conference on Multimedia, MM 2014, pp. 801–804. ACM, New York (2014)
4. Kim, J., Englebienne, G., Truong, K.P., Evers, V.: Deep temporal models using identity skip-connections for speech emotion recognition. In: Proceedings of the 25th ACM International Conference on Multimedia, MM 2017, pp. 1006–1013. ACM, New York (2017)
5. Kingma, D.P., Ba, J.: Adam: a method for stochastic optimization. arXiv preprint arXiv:1412.6980 (2014)
6. Krizhevsky, A., Sutskever, I., Hinton, G.E.: Imagenet classification with deep convolutional neural networks. Commun. ACM 60(6), 84–90 (2017)
7. LeCun, Y., Bengio, Y., Hinton, G.: Deep learning. Nature 521(7553), 436 (2015)
8. Mesbahi, L., Sodoyer, D., Ambellouis, S.: Shout analysis and characterisation. Int. J. Speech Technol. 22(2), 295–304 (2019)
9. Mittal, V.K., Yegnanarayana, B.: Effect of glottal dynamics in the production of shouted speech. J. Acoust. Soc. Am. 133(5), 3050–3061 (2013)
10. Mittal, V.K., Yegnanarayana, B.: An automatic shout detection system using speech production features. In: Böck, R., Bonin, F., Campbell, N., Poppe, R. (eds.) MA3HMI 2014. LNCS (LNAI), vol. 8757, pp. 88–98. Springer, Cham (2015). https://doi.org/10.1007/978-3-319-15557-9_9
11. Palaz, D., Magimai-Doss, M., Collobert, R.: Analysis of CNN-based speech recognition system using raw speech as input. In: Proceedings of Sixteenth Annual Conference of the International Speech Communication Association, pp. 11–15 (2015)

12. Pohjalainen, J., Raitio, T., Yrttiaho, S., Alku, P.: Detection of shouted speech in noise: human and machine. J. Acoust. Soc. Am. **133**(4), 2377–2389 (2013)
13. Raitio, T., Suni, A., Pohjalainen, J., Airaksinen, M., Vainio, M., Alku, P.: Analysis and synthesis of shouted speech. In: Proceedings of Fourteenth Annual Conference of the International Speech Communication Association (INTERSPEECH), pp. 1544–1548 (2013)
14. Zelinka, P., Sigmund, M., Schimmel, J.: Impact of vocal effort variability on automatic speech recognition. Speech Commun. **54**(6), 732–742 (2012)

Quantitative Analysis of Cognitive Load Test While Driving in a VR vs Non-VR Environment

Zeeshan Qadir$^{(\boxtimes)}$, Eashita Chowdhury, Lidia Ghosh, and Amit Konar

Artificial Intelligence Laboratory,
Department of Electronics and Telecommunication Engineering,
Jadavpur University, Kolkata, India
zeeshanqadir95@gmail.com, eashita06@gmail.com,
lidiaghosh.bits@gmail.com, amit.konar@jadavpuruniversity.in

Abstract. Brain-Computer Interface (BCI) based cognitive load assessment has been a topic of significant research for quite some time, especially that considering a driving scenario. And as methods for human-computer interactions has evolved, with virtual-reality (VR) coming into play, it soon found its applications in BCI as well. However, to date, very few research documents have considered the evaluation of cognitive load while driving in a VR environment. Moreover, none of them provide a qualitative or quantitative performance analysis in comparison to a non-VR scenario. This paper aims to provide a quantitative analysis of electroencephalography (EEG)-based cognitive load while driving in a very interactive VR environment compared to that in a traditional fixed non-VR environment, based on source localization through e-LORETA, cognitive load assessment, and that on the performance of our proposed Closed Interval Type-2 Fuzzy Set (CIT2FS)-induced pattern classifier.

Keywords: BCI · EEG · VR · Cognitive load · Driving simulation · CIT2FS

1 Introduction

BCI-based evaluation of cognitive load using devices like EEG, functional near-infrared spectroscopy (fNIRS), and other similar devices has been a popular research study. Cognitive failure while driving due to a lapse of visual alertness, cognitive planning, and motor execution was studied in [10]. Here, Recurrent Neural Networks (RNN) classifier based on Lyapunov energy surface was used for the processing of EEG signals. The work was further extended in [13], employing a novel two-stage motor intention classifier. A differential evolution (DE)-induced fuzzy neural classifier was used for the decoding and classification of motor imagery potentials while driving in [11]. As an improvement, authors in [12], proposed a general and an interval type-2-fuzzy set (GT2FS & IT2FS) classifier for cognitive load test while driving and was shown to outperform the

© Springer Nature Switzerland AG 2019
B. Deka et al. (Eds.): PReMI 2019, LNCS 11942, pp. 481–489, 2019.
https://doi.org/10.1007/978-3-030-34872-4_53

existing classifiers. This work considered event-related desynchronization (ERD) and event-related synchronization (ERS) stimuli responses and was extended in [2] using P-300 and N-400 stimuli responses. Different from the existing work, authors in [1], performed the analysis of cognitive load test while driving using a functional near-infrared spectroscopy (fNIRS) for a better spatial response of the stimuli.

Simultaneously, the research in the domain of VR based human-computer interactions has also been on the rise since the previous decade. In 2002, authors in [17] provided a glimpse of possible VR applications into different contexts of user-interface. Later, the VIRART team from the University of Nottingham, presented in [19] and [4] some more exciting and essential scenarios of VR modeling. In [16], a detailed study on a driver's working memory load assessment with the help of fNIRS was performed using a very realistic VR driving simulator setup at German Aerospace Centre. Similarly, in [18], a study on multitasking while driving in a visually interactive environment was done, involving participants of different age groups. However, both the above research considered either multiple displays or a huge display screen, which is not very cost-effective and also a fixed representation of a VR environment. Contrary to this, authors in [14] considered a head-mounted display (HMD)-based VR, which provides a 360° view of the environment depending on the subject's head movement, and studied the efficiency of the HMD-VR over traditional fixed simulations, based on a set of questionnaires. Since subjective questionnaires can't be considered as optimal human-machine interface performance indicator [9], we tried to evaluate the cognitive load test on the basis of EEG responses acquired for different instances of HMD-VR based driving simulations.

It can hence be observed that researchers nowadays prefer the evaluation of cognitive load test in VR environments over traditional fixed simulations, accounting for the better interactive capability of VR, which results in a more natural human response. However, to date, and to the best of the authors' knowledge, no study has been done to compare and evaluate the performance of the two techniques, qualitatively or quantitatively. Hence, based on source localization (brain-activation regions) obtained through e-LORETA, statistical measurement of cognitive load, as well as the performance evaluation of the proposed CIT2FS classifier, this paper tries to provide a quantitative comparative analysis of the two simulation techniques, keeping the qualitative performance analysis of human interaction with the environment a study for future research.

2   Principles and Methodologies

A quantitative analysis of cognitive load from the brain-signal response of a human subject while driving a car in VR and non-VR environments requires analysis of the acquired brain responses in subsequent stages involving data acquisition, pre-processing and artifact removal, feature extraction, and classification of cognitive load into three classes, viz., High, Medium and Low. This section aims at describing these principles and methodologies adopted for the analysis of the acquired EEG responses, as shown in Fig. 1.

2.1   Data Acquisition

Each session required a participant to perform a driving task using a setup mentioned in Sect. 3.1, for a duration of 15 min and every participant completed the session twice (first without VR, and then with VR). The EEG device collected the brain samples corresponding to the 19 electrodes placed on the scalp in synchronous with the driving simulation. The recorded data for each session was then stored in a MATLAB (.m) file for further processing.

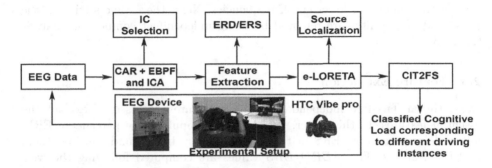

Fig. 1. Block diagram of our study

2.2   Preprocessing and Artifact Removal

The primary objective of the present study was to analyze the cognitive load for driving corresponding to seven driving instances, viz., acceleration (Acc), steering right (R), steering left (L), applying breaks (B), accident/collision (Acd), chase (Ch), and slow-driving (Sl). Hence, the data samples corresponding to only these instances were extracted from the EEG signal acquired for each session, and then each of them was stored as separate files. As the acquired EEG signals always remain contaminated with various types of noise like stray noise, ocular, and muscle artifacts, it was necessary to adopt some suitable artifact-removal approaches to make the raw EEG signals artifact-free. Common Average Referencing (CAR) was then performed to eliminate stray noise and ocular artifacts, based on the following equation:

$$EP_{CAR(i)} = EP_{(i)} - \frac{1}{N} \sum_{n=1}^{N} EP_{(n)} \tag{1}$$

where N is the total number of electrodes used for the recording of the EEG response (here 19) and $EP_{(n)}$ is the potential between the recording electrode i and the reference electrode n.

The CAR-filtered EEG signals were then band-pass filtered using an Elliptical band-pass filter (BPF) to remove the artifacts present due to un-volunteered muscle movements. The EEG signals were then filtered within four different frequency bands: theta (4–8 Hz), lower-alpha (8–10 Hz), upper-alpha (10–12 Hz) and beta (12–30 Hz) - independently by using the respective frequency ranges as the pass-bands of the BPF since strong evidence of cognitive load assessment tasks being associated with these frequency bands are available in the existing literature [1,2,10,12]. The center of each pass-band selection is guided by the location of the peaks in the Fourier spectra, and the bandwidth is determined by the upper and lower 3 dB (decibel) frequencies around the peaks. The Elliptical filter is selected from its competitors for its sharpest roll-off characteristics around the upper and lower cutoff frequencies. Next, the artifacts lying in the pass-band of the filter are removed by using independent component analysis (ICA) [3].

2.3   Feature Extraction

Next, the artifact-free independent components are processed to extract the essential features. In this paper, we focus on computing the percentage ERD, followed by percentage ERS in the four frequency bands, and consider them as EEG features. The $\%ERD/ERS$ values are computed following the well-established definition proposed in [8], as given below:

$$\%ERD/ERS = \frac{P_R - P_t}{P_R} \times 100\% \qquad (2)$$

where, P_R is the mean power of the base-line period or reference interval and P_t is the mean band power of task interval. Here, base-line period refers to the pre-stimulus duration without any task-demands where a subject has to concentrate on a displayed fixation cross for a duration of 2 s, and the task interval refers to the time-period while performing any experimental task, here car driving. The $\%ERD/ERS$ in each frequency band per EEG channel is evaluated for each of the seven driving instances.

2.4   Classifier Design

In the present cognitive load classification problem in VR and non-VR environments, we propose a CIT2FS [5] induced pattern classifier to classify into three cognitive load classes: High, Medium and Low.

Let $x_1, x_2, ..., x_n$ be the n features extracted for a given subject in a single experimental trial. Also, let x_i has r experimental instances, $x_i^1, x_i^2, ..., x_i^r$, obtained from r measurements undertaken on the same subject in the same experimental trial. Presuming that the instances of x_i support Gaussian distribution, we represent them by a Gaussian membership function (GMF) with mean x_i and variance σ_i^2, equal to the mean and variance of $[x_i^1, x_i^2, ..., x_i^r]$, respectively $\forall\ i \in [1, n]$.

Now, for the 11 subjects considered in our experiments, we would have 11 GMFs. Hence, to accommodate the effect of inter-subject variations, we take the minimum and maximum of these type-1 MFs to construct the lower membership function (LMF), given by $\underline{\mu}_{\tilde{A}}(x_i)$ and the upper membership function (UMF), given by $\overline{\mu}_A(x_i)$ of a CIT2FS, which is more complete than its type-1 counterpart. Thus for n features, we have n CIT2FS given by $[\overline{\mu}_{\tilde{A}_i}(x_i), \underline{\mu}_{\tilde{A}_i}(x_i)] \; \forall \; i \in [1, n]$.

Next, we adopt the following type-2 fuzzy rule j for the present multi-class classification problem:

If x_1 is $\tilde{A}_{j,1}$ and x_2 is $\tilde{A}_{j,2}$ and ... and x_n is $\tilde{A}_{j,n}$, then class is C_k where x_i is $\tilde{A}_{j,i} \; \forall \; i \in [1, n]$, are the n CIT2FS-induced propositions, and C_k is the kth class label for $k = [1, 3]$. This implies that, C_1, C_2, and C_3 refer to the High, Medium, and Low cognitive load classes respectively.

Let $[x_1 = x'_1, x_2 = x'_2, ..., x_n = x'_n]$ together be the a measurement point. We then obtain the lower firing strength (LFS_j) and the upper firing strength (UFS_j) at the measurement point by employing

$$LFS_j = \min_{\forall i \in [1,n]} (\underline{\mu}_{\tilde{A}_{j,i}}(x'_i)) \tag{3}$$

$$UFS_j = \min_{\forall i \in [1,n]} (\overline{\mu}_{\tilde{A}_{j,i}}(x'_i)) \tag{4}$$

We then use the Nie-Tan type de-fuzzification [6] to obtain the firing strength of the fuzzy inference (Fig. 2), here representing the degree of cognitive load. The Nie-Tan type de-fuzzification provides the firing strength of rule j as:

$$FS_j = \lambda.UFS_j + (1 - \lambda).LFS_j \tag{5}$$

for a suitable value of λ in [0, 1]. In case L rules are instantiated by the measurement point, all the rules are fired and the firing strength of the jth rule is computed. Let the firing strength of the jth rule be FS_j. Then the rule j is selected, if $FS_j > FS_l$, where $l \in [1, L]$ and $l \neq j$. We declare FS_j as the effective firing strength of the most promising rule. Thus from L competitive rules, each describing different classes, only one class corresponding to the jth rule is selected.

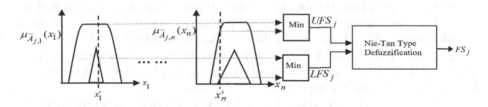

Fig. 2. Computation of firing strength of the proposed CIT2FS

3   Experimental Setup and Results

This section first describes, in detail, the experimental setup, and then details on the experimental results obtained through both the simulation techniques and hence, performs a comparative analysis to evaluate the performance variation of the two.

3.1   Experimental Setup

Eleven healthy right handed-subjects (5 females and 6 males, mean age: 25.2 years, range: 23–30 years) participated in this study given their written informed consents. The study was performed at the Artificial Intelligence Laboratory of Jadavpur University, Kolkata. All of these subjects were casual drivers while none of them had prior training on the experimental procedure of the present study.

EEG experiments were performed in a well insulated room. Subjects were seated comfortably in a chair similar to a car-driver seat. The EEG signals were recorded from a 32-channel Nihon Kohden EEG device and were acquired using 19 electrodes from different brain regions [1]. The data sampling rate was set at 500 Hz. The driving simulations were carried out using a standard LOGITECH driving simulator comprising a steering wheel, pedal foot, break foot, and a gear-box. Project CARS 2 ®© motorsport racing simulator having VR support along with HTC Vibe pro head-mount VR system was used for creating an immersive VR-driving environment.

3.2   Performance Analysis Based on Cognitive Load

The acquired EEG signals were then passed through e-LORETA software [7] for source localization, i.e., to detect the active brain regions responsible for the cognitive tasks corresponding to different driving instances of both the simulation environments. Figure 3a shows the e-LORETA response for accident (Acd) in non-VR driving simulation, while Fig. 3b corresponds to that for VR simulation. It can be observed from the two figures that the pre-frontal, frontal, as well as, parietal lobe activation for VR is higher as compared to that for non-VR under a complex situation (here accident/collision), thus signifying their involvement in cognitive load assessment. A slightly higher occipital lobe activation for the non-VR case suggests that the cognitive load during non-VR based driving simulation is only limited to that corresponding to our vision.

This can be further ascertained from Fig. 4, which plots the mean cognitive load corresponding to various driving instances in the two simulation environments. It can be clearly observed that the cognitive load corresponding to all the driving instances, especially accident/collision, for the case of VR is much higher. It can also be observed that the cognitive load while slow driving is comparable for both the cases, which can be accounted for a comparatively relaxed attitude of the subject corresponding to this task.

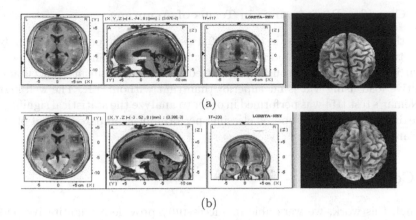

(a)

(b)

Fig. 3. e-LORETA based brain activation regions for accident while driving in (a) non-VR (b) VR experimental setup. Here the intensity of red color indicates the degree of brain activation. (Color figure online)

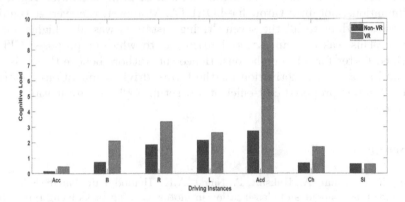

Fig. 4. Comparative analysis of mean cognitive load corresponding to various driving instances

Table 1. Classification accuracy with and without VR

Driving action	Non-VR (%)	VR (%)
Acceleration	79.3	**84.4**
Break	83.9	**87.5**
Slow driving	71.5	**79.5**
Right	76.5	**82.8**
Left	75.2	**82.2**
Chase	80.8	**83.7**
Accident	85.1	**92.6**
Average	78.9	**86.1**

3.3 Classifier Performance in Terms of Accuracy

Using the classifier designed in the Sect. 2.4 we evaluate the performance analysis of the two simulation methodologies in terms of their classification accuracy, as shown in Table 1, and it could be observed that the classification accuracy of cognitive load using VR is far superior than that without VR. The well-known Mc Nemar's test [15] was performed in order to analyze the statistical significance of the differences in our proposed classifier performance compared to that of other standard classifiers.

4   Conclusion

Through this work, we were able to successfully provide a quantitative performance analysis of a VR-based driving simulation compared to that of a non-VR case, based on source localization through e-LORETA, cognitive load assessment, and lastly based on the performance of our proposed CIT2FS-induced pattern classifier. We observed a higher number of brain activation regions for VR-simulations, obtained through e-LORETA. Moreover, the average cognitive load corresponding to all the seven driving instances was also higher for the VR case. This was also an essential reason as to why our proposed CIT2FS performed better for VR simulations. Hence the authors believe that this work can stand as a strong motivation for the future driving-simulation-research to be based on our proposed cost-efficient VR setup for better evaluation of their simulation results.

References

1. Ghosh, L., Konar, A., Rakshit, P., Nagar, A.K.: Hemodynamic analysis for cognitive load assessment and classification in motor learning tasks using type-2 fuzzy sets. IEEE Trans. Emerg. Top. Comput. Intell. **3**, 245–260 (2018)
2. Ghosh, L., et al.: P-300 and N-400 induced decoding of learning-skill of driving learners using type-2 fuzzy sets. In: 2018 IEEE International Conference on Fuzzy Systems, pp. 1–8. IEEE (2018)
3. Hyvärinen, A., Karhunen, J., Oja, E.: Independent Component Analysis, vol. 46. Wiley, Hoboken (2004)
4. Karaseitanidis, I., et al.: Evaluation of virtual reality products and applications from individual, organizational and societal perspectives-the "VIEW" case study. Int. J. Hum.-Comput. Stud. **64**(3), 251–266 (2006)
5. Mendel, J.M., Rajati, M.R., Sussner, P.: On clarifying some definitions and notations used for type-2 fuzzy sets as well as some recommended changes. Inf. Sci. **340**, 337–345 (2016)
6. Nie, M., Tan, W.W.: Towards an efficient type-reduction method for interval type-2 fuzzy logic systems. In: 2008 IEEE Fuzzy-Systems, pp. 1425–1432. IEEE (2008)
7. Pascual-Marqui, R.D., et al.: Assessing interactions in the brain with exact low-resolution electromagnetic tomography. Philos. Trans. Roy. Soc. A: Math. Phys. Eng. Sci. **369**(1952), 3768–3784 (2011)

8. Pfurtscheller, G., Aranibar, A.: Event-related cortical desynchronization detected by power measurements of scalp EEG. Electroencephalogr. Clin. Neurophysiol. **42**(6), 817–826 (1977)
9. Rimbert, S., et al.: Can a subjective questionnaire be used as brain-computer interface performance predictor? Front. Hum. Neurosci. **12**, 529 (2018)
10. Saha, A., Konar, A., Burman, R., Nagar, A.K.: EEG analysis for cognitive failure detection in driving using neuro-evolutionary synergism. In: 2014 International Joint Conference on Neural Networks (IJCNN), pp. 2108–2115. IEEE (2014)
11. Saha, A., Konar, A., Dan, M., Ghosh, S.: Decoding of motor imagery potentials in driving using DE-induced fuzzy-neural classifier. In: 2015 IEEE 2nd International Conference on Recent Trends in Information Systems (ReTIS), pp. 416–421. IEEE (2015)
12. Saha, A., Konar, A., Nagar, A.K.: EEG analysis for cognitive failure detection in driving using type-2 fuzzy classifiers. IEEE Trans. Emerg. Top. Comput. Intell. **1**(6), 437–453 (2017)
13. Saha, A., Roy, S.B., Konar, A., Janarthanan, R.: An EEG-based cognitive failure detection in driving using two-stage motor intension classifier. In: Proceedings of the 2014 International Conference on CIEC, pp. 227–231. IEEE (2014)
14. Sportillo, D., Paljic, A., Ojeda, L.: Get ready for automated driving using virtual reality. Accid. Anal. Prev. **118**, 102–113 (2018)
15. Sun, X., Yang, Z.: Generalized McNemar's test for homogeneity of the marginal distributions. SAS Global Forum. **382**, 1–10 (2008)
16. Unni, A., Ihme, K., Jipp, M., Rieger, J.W.: Assessing the driver's current level of working memory load with high density functional near-infrared spectroscopy: a realistic driving simulator study. Front. Hum. Neurosci. **11**, 167 (2017)
17. Van Dam, A., Laidlaw, D.H., Simpson, R.M.: Experiments in immersive virtual reality for scientific visualization. Comput. Graph. **26**(4), 535–555 (2002)
18. Wechsler, K., et al.: Multitasking during simulated car driving: a comparison of young and older persons. Front. Psychol. **9**, 910 (2018)
19. Wilson, J.R., D'Cruz, M.: Virtual and interactive environments for work of the future. Int. J. Hum.-Comput. Stud. **64**(3), 158–169 (2006)

Multipath Based Correlation Filter for Visual Object Tracking

Himadri Sekhar Bhunia[1]($\boxtimes$), Alok Kanti Deb[2]($\boxtimes$),
and Jayanta Mukhopadhyay[3]($\boxtimes$)

[1] Advanced Technology Development Centre, Indian Institute of Technology,
Kharagpur, India
hb.pku08@gmail.com
[2] Department of Electrical Engineering, Indian Institute of Technology,
Kharagpur, India
alokkanti@ee.iitkgp.ernet.in
[3] Department of Computer Science and Engineering, Indian Institute of Technology,
Kharagpur, India
jay@cse.iitkgp.ac.in

Abstract. This paper presents a new correlation filter based visual object tracking method to improve the accuracy and robustness of trackers. Most of the current correlation filter based tracking methods often suffer in situations such as fast object motion, the presence of similar objects, partial or full occlusion. One of the reasons for that is that object localization is performed by selecting only a single location at each frame (greedy search technique). Instead of choosing a single position, the multipath based tracking method considers multiple locations in each frame to localize object position accurately. In this paper, the multipath based tracking method is applied to improve the performance of the efficient convolution operator with handcrafted features (ECO-HC), which is a top performing tracker in many visual tracking datasets. We have performed comprehensive experiments using our efficient convolution operator with multipath (ECO-MPT) tracker on UAV123@10fps and UAV20L datasets. We have shown that our tracker outperforms most of the state-of-art trackers in all those benchmark datasets.

Keywords: Visual object tracking · Single path tracking · Multipath based tracking · Correlation filter

1 Introduction

Visual object tracking is one of the classic research problems in computer vision. It has numerous applications like surveillance, autonomous navigation, traffic monitoring, etc. to name a few. In recent years, discriminative correlation filters (DCF) [2–4,9,10] have shown excellent performance in visual tracking in terms of speed, accuracy, and robustness as the computation is done in the frequency domain. As mentioned in [6,8], these greedy search based methods often suffer in

© Springer Nature Switzerland AG 2019
B. Deka et al. (Eds.): PReMI 2019, LNCS 11942, pp. 490–498, 2019.
https://doi.org/10.1007/978-3-030-34872-4_54

many challenging situations such as the presence of similar objects, objects with the same color or texture, fast camera motion, partial or full occlusion. In this paper, a novel multipath based tracking algorithm using correlation filter has been proposed. This method generalizes any greedy search based tracker toward improving its performance. This paper is organized as follows; Sect. 3 represents our methodology. Performance analysis is presented in Sect. 4 and conclusions are drawn in Sect. 5.

1.1 Our Approach and Contributions

The contributions of this paper are summarized as follows. First, we have proposed a modified non-greedy search based multipath tracking scheme. Next, we enhance the performance of a correlation filter-based tracker (ECO-HC) by utilizing multipath tracking scheme in terms of improving accuracy and robustness. Extensive experiments have been carried out on various UAV datasets to analyze the performance of our method against the state-of-the-art trackers. Our tracker betters ECO-HC by 10.9% in precision and 6.04% in success rate on one pass evaluation on UAV20L dataset and 2.02% in precision and 2.12% in success rate on UAV123@10fps dataset.

2 Baseline Approach

2.1 Efficient Convolution Operator (ECO-HC)

The efficient convolution operator (ECO-HC) is the baseline of our approach. In this method, an interpolation model is used to transform the discrete spatial variable into a continuous spatial domain. For each feature channel d, the interpolation operator I_d is defined as:

$$I_d \left\{ x^d \right\} (t) = \sum_{n=0}^{P_d - 1} x^d [n] \beta_d \left(t - \frac{T}{P_d} n \right) \tag{1}$$

where, $[0, T)$ is the spatial support of the feature map, β_d is the interpolation function with period $T > 0$, P_d is the resolution of each feature layer x^d and $I\{x\}(t) \in \mathbb{R}^M$ is the entire continuous feature map. ECO-HC uses a factorized convolution operator to reduce optimized model parameters. It is defined as,

$$J_{Pf}\{x\} = Pf * I\{x\} = \sum_{c,d} p_{d,c} f^c * I_d\{x_d\} = f * P^T I\{x\} \tag{2}$$

For each feature channel d, a small subset of basis filters $f^1, \ldots, f^N$ are used, where $N < M$. Filter for feature layer d is derived as, $\sum_{c=1}^{N} p_{d,c} f^c$. Here, $p_{d,c}$ are the coefficients of matrix P, which is a linear dimensionality reduction matrix or projection matrix of size $M \times N$. The filter f and projection matrix P can be learned jointly by minimizing the following loss function,

$$L(f, P) = \left\| \hat{z}^T P \hat{f} - \hat{y} \right\|^2 + \sum_{c=1}^{N} \left\| \hat{w} * \hat{f}^c \right\|^2 + \lambda \left\| P \right\|_F^2 \tag{3}$$

Here, $\|P\|_F^2$ is the Frobenius norm of P, which is controlled by the weight parameter λ and $\hat{z}^d[k] = \widehat{I_d}\{x^d\} = X^d[k]\,\hat{\beta}_d[k]$. X^d is the discrete Fourier transform (DFT) of x^d. The top hat denotes DFT of the function. Predefined target detection score is denoted by y, which is a periodically repeated Gaussian function, w is the spatial penalty function and $f = (f^1...f^N)$. As Eq. 3 is a non-linear least square problem; the Gauss-Newton method is used to optimize it.

3   Proposed Approach

In this section, we present the multipath object tracking method for object localization step. We have divided it into two parts. First, we state the single path tracking method and some of its limitation. Then we discuss how these limitations can be overcome by using the multipath method.

3.1   Single Path Object Tracking

Single path object tracking algorithm is also known as a greedy search-based method. In this scheme, a patch containing the object to be detected is extracted from the previous image frame. Then a feature map is generated from that patch. Desired object position is predicted in the next frame by estimating the position with the maximum correlation score with the previous frame. In this paper, we have chosen the efficient convolution operator with handcrafted features (ECO-HC) as our single path based tracking algorithm. Section 2.1 depicts the brief description of ECO-HC tracker. As addressed in [5], single path tracking methods often fail to accurately detect object position in situations like the presence of similar color or texture objects nearby, the fast motion of object position, presence of occlusion, etc. In these cases, the correlation map contains multiple local maxima with comparable values. The tracker may pick a wrong peak. Selecting the maximum correlation score may not always guarantee accurate object localization.

3.2   Multipath Based Tracking

Multipath based methods can address the limitations of single path methods by selecting multiple local maxima. In the next frame, the tracker searches around all possible locations of previously selected position from single path mode and this process continues in successive frames. As selection of multiple positions in all frames improves computational overhead, we choose a threshold T_{MPT}

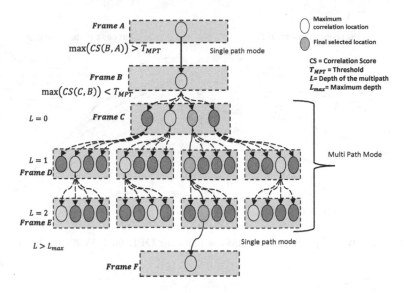

Fig. 1. Multipath based tracking method example (Color figure online)

which switches the algorithm between single path mode and multipath mode. An example describes this method in Fig. 1. From frame A to B maximum correlation score (CS) between frame A and B is greater than the selected threshold value and single mode continues and estimated position is the position of the maximum in the detection map shown in white. In between frame B to C, the condition fails, and the algorithm enters into the multipath mode. In multipath mode, four positions chosen by finding four local maxima in the detection map, which are shown in green and white. Here the white location has the maximum correlation score in this frame. In the next frame D, four local maxima are selected from each of the positions in C. So, a total of sixteen positions are considered at D. Out of those sixteen positions; the best four blocks are selected to propagate further in the next frame E, which are shown in white colors in Fig. 1. The same process is repeated in E. From E to F, current depth L crosses maximum allowable depth limit $L_{\max}$ and the algorithm automatically switched to single path mode. Final positions are selected (shown in yellow) by the shortest path Algorithm based on total correlation score from B to E. The detailed computational steps are described in Algorithm 1.

4   Experimental Analysis

4.1   Setup

The proposed ECO-MPT is implemented in MATLAB on a PC with i5 3.2GHz Intel CPU with 16GB of RAM. We have taken Histogram of Oriented Gradients (HOG) [1] and color names (CN) [13] features with the same combination as

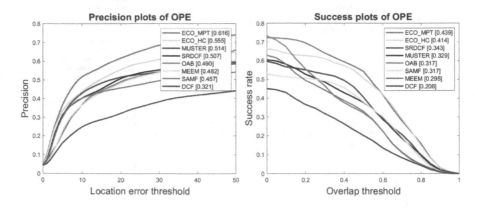

Fig. 2. Overall precision and success plots for OPE on UAV20L Dataset

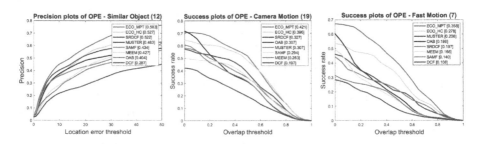

Fig. 3. Precision and success plots with different attributes for OPE on UAV20L dataset

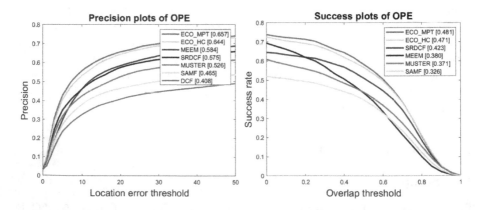

Fig. 4. Overall precision and success plots for OPE on UAV123@10fps dataset

---

Algorithm 1. Multipath based tracking approach (ECO-MPT)

Input: Previous target position p_{t-1} and scale s_{t-1}
Output: Estimated target positions $p_t, \ldots, p_{(t+L_{\max}-1)}$, and scales
$\quad\quad s_t, \ldots, s_{(t+L_{\max}-1)}$

1 **Initialization:** Threshold T_{MPT}, maximum depth $L_{\max}$.
2 **Single path Mode:** Get current frame target position p_t and correlation score
$\quad$ (CS_t) using ECO-HC tracker.
3 **if** $\max(CS_t) > T_{MPT}$ **then**
4 $\quad|\quad$ Repeat step 2
5 **else**
6 $\quad|\quad$ **MPT initialization:** Choose best k possible nodes at $p_t = p_1, p_2, \ldots, p_k$,
$\quad\quad\quad$ current depth $L = 0$.
7 $\quad|\quad$ **MPT recursion:** In the next frame, search k possible locations around
$\quad\quad\quad$ each p_t forms a set,

$$O_i^j = \{\{O_1^1, \ldots, O_1^k\}, \ldots, \{O_k^1, \ldots, O_k^k\}\} \tag{4}$$

$\quad\quad\quad$ where, $i, j = 1, \ldots, k$
8 $\quad|\quad$ **if** $CS\left(p_i, O_i^j\right) > T_{MPT}$, *for any* $i, j = 1, \ldots, k$ **then**
9 $\quad|\quad\quad|\quad$ Select O_i^j with maximum correlation score CS along with its parent p_i.
10 $\quad|\quad\quad|\quad$ Go to step 2 (single path mode)
11 $\quad|\quad$ **else**
12 $\quad|\quad\quad|\quad$ **if** $L < L_{\max}$ **then**
13 $\quad|\quad\quad|\quad\quad|\quad$ Select best k nodes with maximum CS from O_i^j.
14 $\quad|\quad\quad|\quad\quad|\quad$ $L = L + 1$
15 $\quad|\quad\quad|\quad\quad|\quad$ Repeat step MPT initialization
16 $\quad|\quad\quad|\quad$ **else**
17 $\quad|\quad\quad|\quad\quad|\quad$ Select the shortest path from beginning of current MPT
$\quad\quad\quad\quad\quad\quad$ initialization to O_i^j.
18 $\quad|\quad\quad|\quad\quad|\quad$ Go to single path mode.
19 $\quad|\quad\quad|\quad$ **end**
20 $\quad|\quad$ **end**
21 **end**

Table 1. A comparison of success rates with top 4 trackers for OPE on UAV20L dataset

Trackers	Camara motion	Fast motion	Out of view	Scale variation	Similar object	Low resolution	Viewpoint change	Overall
ECO-MPT	**0.421**	**0.358**	**0.343**	**0.430**	**0.476**	**0.325**	**0.386**	**0.439**
ECO-HC [2]	0.396	0.278	0.385	0.499	0.440	0.301	0.318	0.414
SRDCF [4]	0.327	0.197	0.329	0.332	0.397	0.228	0.303	0.343
MUSTER [11]	0.307	0.206	0.309	0.314	0.342	0.278	0.318	0.329
OAB [7]	0.307	0.198	0.308	0.301	0.310	0.261	0.303	0.317

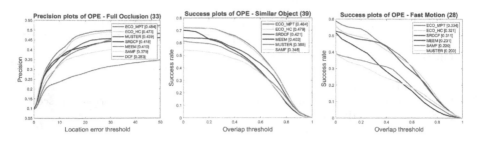

Fig. 5. Precision and success plots with different attributes for OPE on UAV123@10fps dataset

Table 2. Success rates comparison for OPE on UAV123@10fps dataset

Trackers	Camera motion	Similar object	Fast motion	Out of View	Scale variation	Low reso- lution	Viewpoint change	Overall
ECO-MPT	**0.463**	**0.484**	**0.334**	**0.405**	**0.446**	**0.323**	**0.417**	**0.481**
ECO-HC [2]	0.449	0.479	0.321	0.388	0.434	0.313	0.404	0.471
SRDCF [4]	0.399	0.421	0.311	0.368	0.390	0.237	0.356	0.423
MEEM [15]	0.361	0.403	0.231	0.325	0.340	0.241	0.348	0.380

used in ECO-HC tracker. Learning rate is taken as 0.010. Other parameters are the same as in [2]. The average correlation score (average of correlation score upto previous frame) is taken as the threshold value of ECO-MPT (T_{MPT}) and it is different for each dataset. We have taken maximum depth ($L_{\max}$) of the MPT as 3. The proposed tracker runs at 7 FPS. Increasing depth may increase accuracy but it also costs imposes extra computational burden.

4.2   Evaluation Metrics

Our tracker is evaluated with standard success and precision plots as suggested by Wu et al. in [14]. We have also used results in the one pass evaluation or OPE, which demonstrates average success rate or precision. We rank the trackers based on area under the curve (AUC). Our tracker is evaluated with top 5 trackers on two popular benchmark data sets.

4.3   Results on UAV123@10fps and UAV20L Datasets

We have evaluated the proposed tracker on two benchmark datasets, UAV20L and UAV123@10fps [12]. Our ECO-MPT outperforms all the state of art track- ers including ECO-HC by 2.02% for precision and 2.12% for success rate on UAV123@10fps dataset and by 10.9% in precision and 6.04% in success rate on one pass evaluation or OPE on UAV20L dataset respectively. It is observed that UAV20L dataset has 20 long duration sequences only and the improve- ment of the performance of the proposed technique is much better compared to that achieved for the UAV123@10fps which has 123 videos. Overall precision and success plots are shown in Figs. 4 and 2. Figures 5 and 3 show the precision and success plots of various attributes against the location error threshold and overlap threshold. Success rates are shown in Tables 2 and 1. The proposed ECO-MPT demonstrates superior performance for most of the attributes.

5 Conclusions

In this paper, we propose a novel correlation based multipath tracking method. We have availed the high performer ECO-HC tracker as a greedy search-based single path tracking method. Our tracking algorithm redresses some challenges associated with single path method by exploring multiple trajectories on a few looks ahead frames. Comprehensive experiments on significantly challenging UAV123@10fps and UAV20L datasets showed that our tracker has outperformed many of the state-of-art tracking methods including its base tracker ECO-HC.

Acknowledgement. The first author was financially supported for carrying out part of the work from project IMPRINT, MHRD, Government of India.

References

1. Dalal, N., Triggs, B.: Histograms of oriented gradients for human detection. In: International Conference on computer vision & Pattern Recognition (CVPR 2005), vol. 1, pp. 886–893. IEEE Computer Society (2005)
2. Danelljan, M., Bhat, G., Shahbaz Khan, F., Felsberg, M.: Eco: efficient convolution operators for tracking. In: Proceedings of the IEEE Conference on Computer Vision and Pattern Recognition, pp. 6638–6646 (2017)
3. Danelljan, M., Hager, G., Khan, F.S., Felsberg, M.: Discriminative scale spacetracking. IEEE Trans. Pattern Anal. Mach. Intell. **39**(8), 1561–1575 (2017). https://doi.org/10.1109/tpami.2016.2609928
4. Danelljan, M., Hager, G., Shahbaz Khan, F., Felsberg, M.: Learning spatially regularized correlation filters for visual tracking. In: Proceedings of the IEEE International Conference on Computer Vision, pp. 4310–4318 (2015)
5. Dogra, D.P., et al.: Video analysis of Hammersmith lateral tilting examination using Kalman filter guided multi-path tracking. Med. Biol. Eng. Comput. **52**(9), 759–772 (2014)
6. Gao, T., Li, G., Lian, S., Zhang, J.: Tracking video objects with feature points based particle filtering. Multimed. Tools Appl. **58**(1), 1–21 (2012)
7. Grabner, H., Grabner, M., Bischof, H.: Real-time tracking via on-line boosting. In: BMVC, vol. 1, p. 6 (2006)
8. Han, B., Roberts, W., Wu, D., Li, J.: Robust feature-based object tracking. In: Algorithms for Synthetic Aperture Radar Imagery XIV, vol. 6568, p. 65680U. International Society for Optics and Photonics (2007)
9. Henriques, J.F., Caseiro, R., Martins, P., Batista, J.: Exploiting the circulant structure of tracking-by-detection with kernels. In: Fitzgibbon, A., Lazebnik, S., Perona, P., Sato, Y., Schmid, C. (eds.) ECCV 2012. LNCS, vol. 7575, pp. 702–715. Springer, Heidelberg (2012). https://doi.org/10.1007/978-3-642-33765-9_50

10. Henriques, J.F., Caseiro, R., Martins, P., Batista, J.: High-speed tracking with kernelized correlation filters. IEEE Trans. Pattern Anal. Mach. Intell. **37**(3), 583–596 (2015)
11. Hong, Z., Chen, Z., Wang, C., Mei, X., Prokhorov, D., Tao, D.: Multi-store tracker (muster): a cognitive psychology inspired approach to object tracking. In: Proceedings of the IEEE Conference on Computer Vision and Pattern Recognition, pp. 749–758 (2015)
12. Mueller, M., Smith, N., Ghanem, B.: A benchmark and simulator for UAV tracking. In: Leibe, B., Matas, J., Sebe, N., Welling, M. (eds.) ECCV 2016. LNCS, vol. 9905, pp. 445–461. Springer, Cham (2016). https://doi.org/10.1007/978-3-319-46448-0_27
13. Van De Weijer, J., Schmid, C., Verbeek, J., Larlus, D.: Learning color names for real-world applications. IEEE Trans. Image Process. **18**(7), 1512–1523 (2009)
14. Wu, Y., Lim, J., Yang, M.H.: Online object tracking: a benchmark. In: Proceedings of the IEEE Conference on Computer Vision and Pattern Recognition, pp. 2411–2418 (2013)
15. Zhang, J., Ma, S., Sclaroff, S.: MEEM: robust tracking via multiple experts using entropy minimization. In: Fleet, D., Pajdla, T., Schiele, B., Tuytelaars, T. (eds.) ECCV 2014. LNCS, vol. 8694, pp. 188–203. Springer, Cham (2014). https://doi.org/10.1007/978-3-319-10599-4_13

Multiple Object Detection in 360° Videos for Robust Tracking

V. Vineeth Kumar[1]([☒]), Shanthika Naik[1], Polisetty L. Sarvani[1],
Shreya M. Pattanshetti[1], Uma Mudenagudi[1], Meena Maralappanavar[1],
Priyadarshini Patil[1], Ramesh A. Tabib[1], and Basavaraja S. Vandrotti[2]

[1] KLE Technological University, Hubballi, India
vellalavineethkumar@gmail.com, uma@kletech.ac.in
[2] Samsung R&D Institute, Bangalore, India

Abstract. In this paper, we propose an efficient way to detect objects
in 360° videos in order to boost the performance of tracking on the same.
Though extensive work has been done in the field of 2D video processing,
the domain of 360° video processing has not been explored much yet, as
it poses difficulties such as (1) unavailability of the annotated dataset
(2) severe geometric distortions at panoramic poles of the image and (3)
high resolution of the media which requires high computation capable
machinery. The State-of-the-art detection algorithm involves the use of
CNN (Convolution Neural Networks) trained on a large dataset. Faster
RCNN, SSD, YOLO, YOLO9000, YOLOv3 *etc.* are some of the detection
algorithms that use CNN. Among these, though YOLOv3 might not be
the most accurate, it is the fastest, and this trade-off between speed and
accuracy is acceptable. We improvise upon this algorithm, to make it
suitable for the 360° dataset. We propose YOLO360, a CNN network to
detect objects in 360° videos and thus increase the tracking precision and
accuracy. This is achieved by performing transfer learning on YOLOv3
with the manually annotated dataset.

Keywords: Computer vision · Equirectangular frames · 360° images ·
Detection · Tracking · Transfer Learning

1 Introduction

In this paper we address the problem of object tracking for multiple object detection and tracking in 360° videos, focusing on 'Person' and 'Car' class. Multiple object detection and tracking is a well-explored domain with respect to normal videos. However, the same is not true in case of 360° videos. Hence, our work focuses on fulfilling this gap in the domain of 360° videos. Multiple object detection and tracking find its application in the fields of security and surveillance [1–3], the interaction between humans and computer and navigation of UAVs (Unmanned Ariel Vehicles) and robots.

In what follows, we provide a literature survey of different object detection methods, object tracking methods, and applications of object tracking.

© Springer Nature Switzerland AG 2019
B. Deka et al. (Eds.): PReMI 2019, LNCS 11942, pp. 499–506, 2019.
https://doi.org/10.1007/978-3-030-34872-4_55

Object detection involves object localization [4,5] and object classification [6,7]. Object localization refers to locating an object(s) in an image by providing their coordinates. Object classification is categorizing these objects into their respective classes such as 'person', 'cat', 'dog', 'car' *etc.* Object detection finds its application in various fields such as face recognition, crowd counting, security, tracking *etc.*

Object tracking refers to keeping track of an object throughout a video or in an actual environment. There are two main approaches to object tracking: (1) Tracking by detection [8,9]. (2) Kernel-based tracking [10]. In tracking by detection objects are detected in every frame and then are assigned to their respective tracklets by data association algorithms. We have a set of tracks, representing the position of each object in each frame.

The primary challenge in object detection in 360° videos is severe distortion at projection poles. Objects in the images are deformed due to these distortions and are beyond recognition, even by humans. Classical detection algorithms, trained for normal video dataset cannot handle these distortions. We use tracking by detection based algorithm for our experimentation. The bottleneck here is that the efficiency of the tracking algorithm highly depends on the robustness and efficiency of the detection algorithm.

To address these issues, we propose a robust tracker for multiple objects in 360° videos. Towards this, our contributions are:

- We propose a YOLO360, an architecture trained to detect multiple objects in 360° videos, in-particular
  - We construct dataset with ground truth object positions in ER frames extracted from the 360° videos.
  - We propose YOLO360 to detect objects in 360° videos using transfer learning on YOLOv3, Which is used for training Dataset generation: We perform annotations for 'Person' and 'Car' classes on ER frames extracted from the 360° videos to generate ground truth for our dataset. We use "Yolomark" GUI for annotation of images.
  - We train YOLO360 using these annotated images.
- Ground truth for tracking algorithm: We generate ground truth of total 4000 frames from 9 different videos, which is further used to evaluate the performance of the tracker.

In Sect. 2, we provide the proposed framework of detection and tracking algorithm using YOLO360. We also provide details training YOLO360 using transfer learning of YOLOv3. We demonstrate our results in Sect. 3. We provide conclusions in Sect. 4.

2   Framework for Multiple Object Detection and Tracking

In this section, we provide the proposed framework of multiple object tracking, as shown in Fig. 1. The framework contains 3 major phases *viz,* (1) Creation of Dataset and Training, (2) Object Detection, (3) Object Tracking.

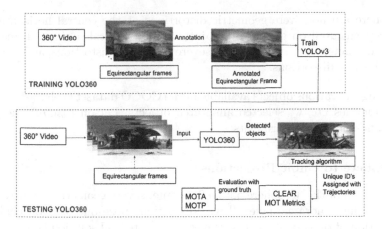

Fig. 1. Proposed pipeline of Multi object tracking in 360° videos.

We explain the first phase in Sect. 2.1. The second phase in every tracking framework is an object detection module. To propose a learning model for detection of objects in 360° videos, there are very few dataset available. This necessitates the creation of training data. We convert 360° videos into ER frames and annotate objects. Once the model is ready with adjusted weights we feed ER frames or 360° video as input to the model. The output is an image with the bounding box coordinates of the detected objects for each video. These coordinates are further given as an input to the tracking algorithms SORT [11] and DEEP SORT [12]. The output of Tracking algorithm comprises output video with the unique ID's assigned and trajectories traced along all the frames of the video. The output of Tracking is given to CLEAR MOT [13] metrics to evaluate the MOTA (Multiple Object Tracking Accuracy) and MOTP (Multiple Object Tracking Precision) score against the ground truth generated.

2.1 360° video dataset

360° images and video frames are the 3D representation of the real world. If the real world is perceived as a sphere, with the camera capturing the image as its center, then each point of the said sphere corresponds to a pixel in the 360° image. 2D projection of this 3D sphere is used to store these images. Equirectangular panorama (ER) is one such projection popularly used. The projection maps each point on the sphere represented by two angles: latitude $\psi \epsilon$ [90°, +90°] and longitude $\lambda \epsilon$ [180°, +180°] to an x,y coordinate of the 2D plane. In this work, we use ER frames for further processing.

YOLOv3 is originally trained on 'Microsoft COCO' [14] dataset. The significant differences between the COCO dataset and ours are as follows:

- In our dataset, there are a number of small objects because the YOLOv3 trained on COCO dataset did not produce accurate detection on 360° videos.

– As there are no severe geometric distortions at the central horizontal line, we concentrated on annotating objects at non central horizontal line (non-equator plane) where there are the severe geometric distortions, and the model learns these distortions.

Among the various objects available in the COCO dataset such as signboards, trucks, bicycles *etc*, we selected annotated objects: Person (3300 frames), Car (700 frames) for our experiments.

2.2 Sphere to Cubic Projections

On projection of the 3D sphere onto a 2D plane, severe geometric distortions are caused at the panoramic poles. These distortions pose one of main challenges for detection of objects in the ER frames. One potential solution is to map sub-windows of the 360° sphere to cubic map as shown in Fig. 2. The traditional detection algorithm work on these cubic projections just as well as they do on normal images.

This spherical to cubic mapping can be pictured as follows: The cube onto which the sphere is to be mapped contains the sphere. The center of both sphere and cube coincide. Consider a pyramid with its apex at the center of the sphere and a face of the cube as its base. The sector of the sphere enclosed within this pyramid is projected on to the base. Thus the entire sphere is divided into 6 sectors projected onto six faces of the cube.

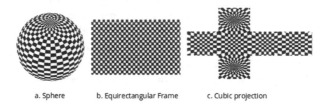

a. Sphere   b. Equirectangular Frame   c. Cubic projection

Fig. 2. Sphere to Equirectangular and Cubic mapping

This gives us a picture as to how the box is constructed. We employ the same technique for the ER frames extracted from the 360° videos. We convert the ER frames to cubic and run the detection algorithm YOLOv3 trained on "Microsoft COCO" dataset. The coordinates thus obtained are saved and further fed to SORT and DEEP SORT algorithm. Our experiments show that the detection in cubic frames using YOLOv3 had higher accuracy (4–5%) as compared to the detection on ER frames.

However this method is computationally expensive as it involves conversion from ER frames to cubic mapping, run detection methods trained on conventional images, get the coordinates and then use these coordinates for tracking in ER representation. By following the above method, we come to two conclusions:

(1) The detection accuracy was high in Method 1 when compared to Method 2.
(2) The time taken for Method 1 is twice the time taken for method 2 as it involved the conversion of ER frames to cubic projections. The drawbacks of this approach are overcome by our proposed approach 'YOLO360'.

2.3 YOLO360

The scarcity of the labeled data is the main bottleneck to train deep learning architecture for detection in 360° videos. Authors in [15, 16] show that the results of models trained on different tasks reduce the bottleneck of learning a new network for different tasks. The study also shows that learning the new architecture with pre-trained features improved the generalization even after fine tuning with the new dataset.

Hence we propose YOLO360, an object detection algorithm which can directly detect people in 360-degree videos with transfer learning, reducing the overhead of converting to Cubic projections and re-converting them back to Equirectangular frames to run the tracking algorithm. We've chosen YOLOv3 [17] and are improving upon the same to make it suitable for 360° videos. YOLOv3 is a CNN architecture, designed to detect multi-class multiple objects in normal videos. Owing to its speed and accuracy, it is most suitable for real-time object detection. We've employed online transfer learning [18] to retrain this architecture to make to suitable of 360° dataset. The architecture is originally trained of 80 classes of COCO dataset. We've retrained it for two classes, 'Person' and 'Car' classes of 360° images. There's a substantial increase in the efficiency with this retrained weights when compared to weights trained for normal images.

We use YOLOv3 architecture on which we perform transfer learning. Further we can increase the number of classes for training by changing the number of filters as per the equation 'filters'= (No of classes + 5) × 3. When we observe that average loss no longer decreases at many iterations, we stop the training, and This model can be used for inference. According to our experiments, YOLO360 overcomes the false detection made by YOLOv3 after 2000 iterations.

2.4 Object Tracking

Our main focus is to improve the detection algorithm in order to improve the tracking efficiency. We use SORT and DEEP SORT algorithms for tracking. SORT exploits both track by detection and kernel-based tracking methods. Multiple objects are first detected using a detection algorithm. These detection are assigned to their respective tracks based on IOU of detection and regions proposed by Kalman filter. The filter is then updated with the detection from each incoming frame, after assignment. DEEP SORT is an improvised version of SORT. In SORT only motion of the object is considered, whereas, in deep sort, both motion and appearance of the objects are considered. An association metrics is built for incoming detection and exiting tracklet, using formula

$$c_{i,j} = \lambda d^{(1)}(i,j) + (1 - \lambda)d^{(2)}(i,j) \qquad (1)$$

where $d^{(1)}$ is value given by motion model and $d^{(2)}$ by appearance model. λ decides the weight to be given to the two models. With repeated experiments, we conclude that both models combined, give better results than considering either one of them. A CNN network is used to obtain an appearance descriptor, which in combination with the position is used to assign objects to tracks.

We make use of both the motion model and the appearance model of the DEEP SORT and create a variant of DEEP SORT by changing the values of λ to demonstrate our results on 360° videos.

3   Results and Discussions

The 9 (1280 × 720) 360° videos selected for our dataset are selected from YouTube, captured and uploaded by users and from the Stanford University 360° videos dataset. We have chosen a total of 4000 frames and annotated 'Person' and 'Car' classes in each of these frames. We provide a pre-trained model along with dataset generated for training, ground truth and their results in our GitHub repository. The chosen videos represent moving

Fig. 3. Detection result for Person and Car for the frame using YOLO360.

objects as this would help us in building accuracy of detection, which is further fed to the tracking algorithm. We perform training on NVIDIA DGX-1 and also observe that it takes five hours for the weights to converge.

We demonstrate the detection and tracking results on 360° videos. Multi class detection results are shown in Figs. 3 and 4. We can observe that YOLO360 detects 15 objects from person class and 7 objects from car class out of total 20 objects from person class and 7 objects from car class respectively.

The results of tracking using MOTA and MOTP are shown in Table 1. We compare MOTA and

Fig. 4. Detection result for person class.

MOTP results using SORT and DEEP SORT tracking algorithm, with detection from both YOLOv3 and YOLO360. The results are calculated using CLEAR MOT benchmark. The cumulative average of the results obtained with YOLO360 when compared with those of YOLOv3 for SORT is 75.71 MOTA and 68.31 MOTP and 42.44 MOTA and 73.98 MOTP. For DEEP SORT, scores of 77.38 MOTA and 68.58 MOTP with YOLO360 and 47.86 MOTA and 68.52 MOTP with YOLOv3 are obtained.

The reason behind the high accuracy of 'rope walk' video is the nature of the video, with least motion and no occlusion. This video can be considered as an ideal condition for tracking algorithm.

Table 1. MOT results of tracking algorithms:

VIDEOS	FRAMES	Architecture	SORT		DEEP-SORT	
			MOTA	MOTP	MOTA	MOTP
Hawaii-Beach	500	YOLOv3	42.44	73.98	47.86	68.52
Hawaii-Beach		YOLO 360	64.45	60.10	67.86	69.66
Rope Walk	2000	YOLO 360	93.47	77.66	91.31	67.91
Person-Florida	300	YOLO 360	69.2	67.18	72.80	68.17

4   Conclusions

In this paper, we have proposed a robust algorithm for 'Person' and 'Car' tracking in 360° videos using efficient multiple object detection and a variant of DEEP SORT. The main challenges are limited dataset and geometric distortions in the 360° video frame. We have proposed an CNN based YOLO360 architecture to detect 'Person' and 'Car' objects in 360° videos using transfer learning approach on YOLOv3 to train YOLO360. We create dataset for 360° videos for objects of 'Person' and 'Car' class by annotating number of frames in Equirectangular frames. We also use the motion and appearance model of DEEP SORT for tracking. We demonstrated our proposed detection and tracking algorithms using a number of 360° videos and compared our results with SORT and DEEP SORT algorithms and achieved an average improvement of 5.42% increase in MOTA.

Acknowledgement. This work was supported and mentored by Samsung R&D Institute Bangalore, India.

References

1. Sujatha, C., Chivate, A.R., Ganihar, S.A., Mudenagudi, U.: Time driven video summarization using GMM. In: 2013 4th National Conference on Computer Vision, Pattern Recognition, Image Processing and Graphics, IIT Jodhapur, pp. 1–4 (2013)
2. Sujatha, C., Mudenagudi, U.: Gaussian mixture model for summarization of surveillance video. In: 2015 5th National Conference on Computer Vision, Pattern Recognition, Image Processing and Graphics, IIT Patna, pp. 1–4 (2015)
3. Tabib, R.A., Patil, U., Ganihar, S.A., Trivedi, N., Mudenagudi, U.: Decision fusion for robust horizon estimation using dempster shafer combination rule. In: 2013 4th National Conference on Computer Vision, Pattern Recognition, Image Processing and Graphics, NCVPRIPG 2013, IIT Jodhapur, pp. 1–4 (2013)

4. Lampert, C.H., Blaschko, M.B., Hofmann, T.: Beyond sliding windows Object localization by efficient subwindow search. In: 2008 IEEE Conference on Computer Vision and Pattern Recognition, pp. 1–8 (2008)
5. Harzallah, H., Jurie, F., Schmid, C.: Combining efficient object localization and image classification. In: 2009 IEEE 12th International Conference on Computer Vision (2009)
6. Wang, J., Yu, K., Lv, F., Gong, Y., Huang, T., Yang, J.: Locality-constrained linear coding for image classification. In: 2010 IEEE Conference on Computer Vision and Pattern Recognition (CVPR). https://doi.ieeecomputersociety.org/10.1109/CVPR.2010.5540018
7. Yang, J., Yu, K., Gong, Y., Huang, T.: Linear spatial pyramid matching using sparse coding for image classification. In: 2009 IEEE Computer Society Conference on Computer Vision and Pattern Recognition Workshops (CVPR Workshops) (2009). https://doi.ieeecomputersociety.org/10.1109/CVPR.2009.5206757
8. Reid, D.: An algorithm for tracking multiple targets. IEEE Trans. Autom. Control **24**, 843–854 (1979)
9. Yilmaz, A., Javed, O., Shah, M.: Object tracking: a survey. ACM Comput. Surv. **38**, 13 (2006). https://doi.org/10.1145/1177352.1177355
10. Peterfreund, N.: Robust tracking of position and velocity with Kalman Snakes. IEEE Trans. Pattern Anal. Mach. Intell. **21**, 564–569 (1999)
11. Bewle, A., Ge, Z., Ott, L., Ramos, F., Upcroft, B.: Simple online and realtime tracking. CoRR (2016) https://dblp.org/rec/bib/journals/corr/BewleyGORU16
12. Wojke, N., Bewley, A., Paulus, D.: Simple online and realtime tracking with a deep association metric. CoRR (2017). https://dblp.org/rec/bib/journals/corr/WojkeBP17
13. Bernardin, K., Stiefelhagen, R.: Evaluating multiple object tracking performance: the CLEAR MOT metrics. EURASIP J. Image Video Process. (2008). https://doi.org/10.1155/2008/246309
14. Lin, T.-Y., et al.: Microsoft COCO: common objects in context. CoRR (2014). https://dblp.org/rec/bib/journals/corr/LinMBHPRDZ14
15. Frey, B.J., Dueck, D.: Clustering by passing messages between data points (2007)
16. Gabriel, P., Verly, J., Piater, J., Genon, A.: Proceedings of Advanced Concepts for Intelligent Vision Systems (2014)
17. Redmon, J., Farhadi, A.: YOLOv3: an incremental improvement. CoRR (2018). https://dblp.org/rec/bib/journals/corr/abs-1804-02767
18. Pan, S.J., Yang, Q.: A survey on transfer learning. IEEE Trans. Knowl. Data Eng. **22**, 1345–1359 (2010)

Improving Change Detection Using Centre-Symmetric Local Binary Patterns

Rimjhim Padam Singh[✉] and Poonam Sharma

Visvesvaraya National Institute of Technology, Nagpur, India
rimjhimsingh1012@gmail.com, dr.poonamasharma@gmail.com

Abstract. Efficient change detection in real-time applications is a major research goal in computer vision. Several researchers have put efforts in this direction and have achieved notable performances in varied challenging situations. But handling all the challenges posed in real-time environments with a single change detection method is almost impracticable. On the other hand, ensemble based background modelling techniques have obtained improved results but they also suffer from trade-off between efficiency and hardware or time requirements, thereby hindering their real-time applicability. This paper proposes an effective hybrid change detection algorithm, light and simple enough to have an effective real-time applicability. The proposed hybrid change detection algorithm employs per-channel RGB colour features with centre-symmetric local binary patterns for pixel-modelling and feeds it to a sample-consensus classification technique for foreground segmentation. Finally, performance of the proposed technique has been tested on widely accepted change detection dataset namely, 2014 Change detection dataset (2014 CDnet dataset).

Keywords: Motion detection · Texture representation · Background model

1 Introduction

Motion detection also known as background segmentation is employed as a fundamental step in most of the computer-vision applications. These applications range from activities involving humans like traffic surveillance, human machine interaction, pedestrians tracking etc. to object and animal based applications like wild animal monitoring, object tracking etc. Detecting and tracking motion occurrences in real-time pose variety of environment-based challenges like (rain, winds, improper lights etc.) and foreground object-based challenges like (bootstrapping issues, camouflaged objects, longterm immobile objects etc.). Hence, employing simple traditional background subtraction approaches that create backgrounds from single frame or mean, median, variance etc. of a set of frames has ill-suitability towards such real-time challenging applications. Several researchers have put immense efforts to efficiently segment the foreground objects occurring in the video from rest of the background by employing different feature spaces (color and texture features) with varied machine learning

© Springer Nature Switzerland AG 2019
B. Deka et al. (Eds.): PReMI 2019, LNCS 11942, pp. 507–514, 2019.
https://doi.org/10.1007/978-3-030-34872-4_56

approaches (like statistical modelling, non-parametric approaches, fuzzy modelling, clustering techniques etc.).

Initially, the researchers used only colour features for pixel representation and fed corresponding matrices to background learning methods. Stauffer and Grimson [13] proposed creating background model using Gaussian mixture models (GMM) that employed a predefined set of frames for training pixel models using multiple Gaussians. The parameters of these models are learnt throughout the background subtraction procedure to adapt the background model to the changing environment and dynamic background scenery. The model performed fairly well as compared to other traditional methods but required more frames and time for training purpose and also could not handle improper illumination, shadows etc. efficiently. Later, several modifications [19] to classical GMM background model were proposed to enhance its over all performance but mostly failed to handle scenarios where pixel models did not follow Gaussian distribution. Later, more sophisticated machine learning approaches like principal component analysis (PCA) [4,11], clustering techniques [7], support vector machines (SVM) were employed that exploited color feature spaces for background model generation and segmentation procedures. These background models performed well against certain challenges but none of them is known to have handled all the background subtraction challenges. Eigen-based background models mostly required off-line processing and could work with occlusions, scaling cameras, dynamic scenes etc. while codebook based clustering models failed to differentiate camouflaged objects from the background. Models taking benefit from SVR and SVM techniques could handle varying illumination conditions and dynamic scenes to some extent but are time intensive and extremely sensitive to threshold mechanisms.

Appropriate parameter tuning, being a major drawback of statistical modelling motivated the development of several non-parametric approaches. Kernel density estimator (KDE) [3], a distinguished non-parametric method, instead of assuming the pixel distribution, generated the statistical representation of background pixels by estimating the distribution directly from a set of pixel values. Hence, it could generate probability functions of arbitrary shape but incurred more storage cost and used first-in-first-out update policy. SACON [15] and Vibe [1], two powerful non-parametric sample-consensus background models populated background models with colour features of observed pixel values and carried out sample voting for pixel classification. But Vibe proved efficient enough to be widely accepted for model building due to its fast and immediate initialization and random time-subsampled update.

Researchers also projected pixels in texture feature space with Local binary pattern (LBP) [5] proposed by Heikkila et al. to be the earliest texture descriptor. LBP code for each pixel value is generated by comparing the centre pixel's intensity with its entire 8-connected neighbourhood. Such feature spaces could easily model the backgrounds but failed when intensity changed in larger areas. Several modifications to LBP texture descriptor have been proposed that have proved to be powerful features for background modelling. Centre-symmetric LBP (CS-LBP) [6] and spatially extended centre-symmetric LBP (SCS-LBP) [17] features generated shorter histograms as they modelled only centre-symmetric pairs of neighbouring pixels but mostly failed to handle illumination variations. In 2010, authors in [9] proposed Scale invariant local ternary features (SILTP) that precisely handled noisy images but, it along with its improvised version proposed in [10] failed majorly at handling illumination variations. Spatial-color binary pattern (SCBP) [16] and Opponent color binary pattern (OCBP) [8] generated features by fusing both colour and texture information. They could outperform other texture features but led to longer histograms that reduced the practicability of the features in real-time scenarios.

Local binary similarity pattern (LBSP) [2] and modified LBSP [12] both generated inter-image texture descriptor and intra-image texture descriptor that captured both colour and intensity changes for better foreground segmentation. Due to improved feature modelling, these features were later combined with Vibe's classification procedure to produce outstanding results but required more storage spaces and time. Hence, the methodology proposed in this paper employs a novel combination of RGB colour features and compact CS-LBP features for efficient background modelling at lesser storage costs.

2   Proposed Methodology

This paper proposes a light-weight hybrid pixel model for background modelling that uses per-channel classification strategy in order to make the algorithm applicable in real-time scenarios.

2.1   Background Model Generation

Background model B_p for pixel p at location (x, y) contains N sample background values that are extracted from the initial frame of the video by using spatial 8-connected neighbourhood of pixel $p(x, y)$ as presented in Eq. 1. Spatial neighbourhood of a pixel shares similar information and hence, aids in better background modelling [1].

$$B_p(x, y) = \{SV_1, SV_2, \ldots, SV_N\} \tag{1}$$

where SV_i are sample background values obtained by randomly selecting the pixel values from $p's$ neighbourhood. It also has to be noted that sample value SV_i represents a combination of pixel values in two feature domains namely, RGB color space and CS-LBP texture descriptors. CS-LBP texture descriptors have been extracted from all three independent R, G and B colour channels. CS-LBP texture features for channel 'ch' of centre pixel $c(x, y)$ having 'M' neighbouring pixels lying within radius 'R' is given by:

$$CS - LBP_{M,R}(c_{ch}) = \sum_{i=0}^{\frac{M}{2}-1} f(v_{i,ch} - v_{(i+\frac{M}{2}),ch}) \cdot 2^i \tag{2}$$

where, $v_{i,ch}$ is the intensity value of channel 'ch' of neighbouring pixels, and 'f' defines a thresholding function presented in following equation.

$$f(y - z) = \begin{cases} 1 & (y - z) \geq T \\ 0 & otherwise. \end{cases} \quad (3)$$

CS-LBP has an advantage of producing shorter histograms of length $2^{\frac{M}{2}}$ over LBP and LBSP texture features that produce histograms of length 2^M. Hence, the memory required for storing the texture features of the background model B_p is reduced by 50%.

2.2   Classification Procedure

For pixel labeling, the proposed methodology follows the standard sample-consensus classification strategy. Pixel $p(x, y)$ is labelled background bg based on the number of samples voting for close similarity with the background (V_{bg}). Pixel $p(x, y)$ is labelled background as soon as it is found similar to minimum λ_{min} samples.

$$LB(p(x, y)) = \begin{cases} bg & V_{bg}(p(x, y)) < \lambda_{min} \\ fg & otherwise. \end{cases} \quad (4)$$

Here, $LB(p(x, y))$ represents the label outcome of pixel $p(x, y)$. It has to be noted that similarity of pixel $p(x, y)$ with the background sample SV_i is obtained both in RGB colour space and CS-LBP texture feature space respectively. Similarity in RGB color space has been obtained by computing $L1 - distance$ or city-block distance metric. While $Hamming\ Distance$ has been used for checking the similarity between CS-LBP feature strings instead of traditionally thresholding the difference between the decimal values of the obtained binary strings.

Due to employment of a hybrid feature space and to reduce the time and computation requirements, the proposed methodology sticks to the per-channel similarity decision of the background pixels. Sample SV_i can vote for a pixel to be $foreground$ as soon as any one of the channels from colour or texture domain meets the thresholding criteria. But, for a sample, in order to vote $background$ for pixel $p(x, y)$, both the over-all RGB color space and the over-all CS-LBP texture space must satisfy the thresholding criteria. The detailed pixel classification procedure can be studied by referring to Algorithm 1.

2.3   Model Update

In order to keep the methodology simple and applicable in real-time scenarios, we avoided employing complex feedback mechanisms for model update as they are too time intensive and delay pixel labelling process. The methodology follows a selective and random update mechanism where foreground pixels are discarded from labelling the background. Doing so avoids permanent absorption of foreground objects into the model. Also the model update is done in a time sub-sampled manner controlled by parameter ϕ. Such a random model update performed under time subsampling condition allows the model to retain very old to very new pixel values thereby, maintaining a larger time window for pixel values observed.

Considering the spatio-temporal information shared across whole neighbourhood, it is wise to update the neighbouring pixels also. Hence, a random neighbouring pixel is also updated conditionally in similar time sub-sampled manner. This allows to produce less noisy and more stable foreground segmentation results.

Input:
  P_{color} color map for pixel P.
  P_{CS-LBP} CS-LBP map for pixel P.
Output:
  $Vote(SV(i))$ Background sample's vote for pixel P.
begin
  1: $K \leftarrow 1/3$
  2: **for** $c \leftarrow 1$:noChannels **do**
  3:    $sim_{color} \leftarrow L1 - dist(P_{color}(ch), SV_{color}(p, i, ch))$
  4:    **If** $sim_{color} > Th_{color} \cdot K$
  5:        **return** $(Vote(SV(i)) =' fg')$; **foreground**
  6:    **else**
  7:        $TotSim_{color} \leftarrow TotSim_{color} + sim_{color}$
  8: **for** $c \leftarrow 1$:noChannels **do**
  9:    $sim_{CS-LBP} \leftarrow HD(P_{CS-LBP}(ch), SV_{CS-LBP}(p, i, ch))$
  10:   **if** $sim_{CS-LBP} > K \cdot Th_{CS-LBP}$
  11:       **return** $(Vote(SV(i)) =' fg')$; **foreground**
  12: **end for**
  13: **if** $(TotSim_{CS-LBP} < Th_{CS-LBP}) \wedge (TotSim_{color} < Th_{color})$
  14:   **return** $(Vote(SV(i)) =' bg')$; **background**
  15: **else**
  16:   **return** $(Vote(SV(i)) =' fg')$; **foreground**
  17: **end for**
  18: **end algo**
    Algorithm 1: Vote assignment for pixel $p^t(x, y)$.

3   Experiments and Results

In order to comprehensively evaluate the proposed background subtraction methodology, widely accepted standard benchmark 2014 Change Detection (2014 Cdnet) dataset has been used that contains total 54 videos. The comparisons with other change detection algorithms is made on the basis of three statistical measures namely, Recall (Re), Precision (Pr) and F-score (Fs) which are defined as follows:

$$Re = tp/(tp + fn) \tag{5}$$

$$Pr = tp/(tp + fp) \tag{6}$$

$$Fs = (2 * Pr * Re)/(Pr + Re) \tag{7}$$

where, tp: number of foreground pixels classified correctly
fp: number of foreground pixels classified incorrectly

tn: number of background pixels classified correctly
fb: number of background pixels classified incorrectly.

Table 1. Comparative results on 2014 CDnet dataset.

Video	Metric (%)	GMM	KDE	Vibe	AMBER	Ivibe	Ours
Dining	Recall	75.74	75.74	87.06	70.23	81.67	93.03
	Precision	88.41	88.41	91.39	97.62	84.45	91.07
	F-measure	81.59	81.59	89.17	81.69	83.03	92.04
Copymachine	Recall	53.91	88.55	89.52	84.8	55.93	97.47
	Precision	79.41	83.05	85.58	90.77	79.98	86.58
	F-measure	64.23	85.71	87.51	87.68	65.83	91.75
Library	Recall	27.99	92.20	90.63	93.94	93.25	94.96
	Precision	84.75	97.14	95.38	99.82	96.54	99.33
	F-measure	42.09	94.60	92.94	96.79	94.87	97.10
Turbulence2	Recall	74.79	77.40	76.04	78.67	65.30	80.90
	Precision	68.66	64.62	70.98	91.59	96.79	99.92
	F-measure	71.59	70.44	73.42	84.64	77.98	89.41

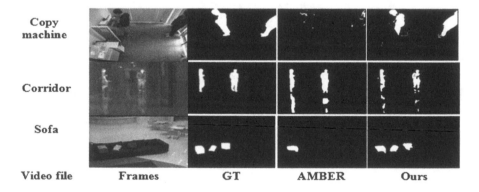

Fig. 1. Foreground masks obtained for sample videos.

Table 1 presents the comparisons of the proposed methodology with other change detection algorithms viz, GMM [13], KDE [3], Vibe [1], AMBER [14] and Ivibe [18]. Here, 'library' and 'dining' video sequences have been shot using thermal infrared cameras in closed environment. Both the videos have camouflaged objects, while 'library' video also posed long term immovable object problem. The outperforming performance metrics for these categories prove that light CS-LBP combined with sample-consensus technique is capable enough to handle camouflage problems in a better way. Also, it has to be noted that AMBER

[14] and Ivibe [18] ate up the long-term immovable person due to their feedback mechanism, but the proposed technique retained it completely with minimal amount of noise. The continuous and adaptive feedback mechanisms employed by other algorithms are responsible for eaten up objects and slower classification results.

On the other hand, 'copy machine' video belongs to shadow category. It contains shadows and long term immovable persons with improper illumination issues. Highest recall achieved in this category depicts that efficient suppression of the false negatives in the videos, while other change detection methods absorbed the foreground objects into the background. Highest precision and F-score in 'turbulence2' video sequences depicts better false positive suppression as compared to other methods. Figure 1 presents the sample videos where the proposed technique obtained more accurate foreground results.

Performance Evaluation and Memory Requirements: The unoptimized implementation of the proposed technique has been carried out on Intel-i5 CPU (2nd generation) with 4 GB DDR3 RAM using MATLAB 2014. For a 320×240 video sequence the proposed methodology requires only 7.062 MBs (approx) of memory to store the background model, where 3 bytes per pixel are required to store each colour and texture feature representations.

4 Conclusion

The hybrid pixel-modelling employing RGB colour space and light-weight CS-LBP texture features, proposed in this paper is able to effectively uncover the camouflaged objects, areas having improper illumination and long-term immovable objects. The sample consensus technique, when fed with the proposed pixel-model could efficiently suppress noise and shadow regions also as depicted in the results. The pixel-level modelling employed also allows parallel computation to make the algorithm further faster. The time-subsampled random update applied in the methodology failed to handle PTZ videos and bootstrapping issues efficiently. Further improving the efficiency of the proposed pixel model by employing adaptive learning feedback mechanisms can be treated as a future goal.

References

1. Barnich, O., Van Droogenbroeck, M.: ViBe: a universal background subtraction algorithm for video sequences. IEEE Trans. Image Process. **20**(6), 1709–1724 (2011)
2. Bilodeau, G.A., Jodoin, J.P., Saunier, N.: Change detection in feature space using local binary similarity patterns. In: International Conference on Computer and Robot Vision, CRV 2013, pp. 106–112. IEEE (2013)
3. Elgammal, A., Harwood, D., Davis, L.: Non-parametric model for background subtraction. In: Vernon, D. (ed.) ECCV 2000. LNCS, vol. 1843, pp. 751–767. Springer, Heidelberg (2000). https://doi.org/10.1007/3-540-45053-X_48

4. Gao, Z., Cheong, L.F., Wang, Y.X.: Block-sparse RPCA for salient motion detection. IEEE Trans. Pattern Anal. Mach. Intell. **36**(10), 1975–1987 (2014)
5. Heikkila, M., Pietikainen, M.: A texture-based method for modeling the background and detecting moving objects. IEEE Trans. Pattern Anal. Mach. Intell. **28**(4), 657–662 (2006)
6. Heikkilä, M., Pietikäinen, M., Schmid, C.: Description of interest regions with local binary patterns. Pattern Recogn. **42**(3), 425–436 (2009)
7. Ilyas, A., Scuturici, M., Miguet, S.: Real time foreground-background segmentation using a modified codebook model. In: Sixth IEEE International Conference on Advanced Video and Signal Based Surveillance, AVSS 2009, pp. 454–459. IEEE (2009)
8. Lee, Y., Jung, J., Kweon, I.S.: Hierarchical on-line boosting based background subtraction. In: 2011 17th Korea-Japan Joint Workshop on Frontiers of Computer Vision (FCV), pp. 1–5. IEEE (2011)
9. Liao, S., Zhao, G., Kellokumpu, V., Pietikäinen, M., Li, S.Z.: Modeling pixel process with scale invariant local patterns for background subtraction in complex scenes. In: 2010 IEEE Computer Society Conference on Computer Vision and Pattern Recognition, pp. 1301–1306. IEEE (2010)
10. Noh, S.J., Jeon, M.: A new framework for background subtraction using multiple cues. In: Lee, K.M., Matsushita, Y., Rehg, J.M., Hu, Z. (eds.) ACCV 2012. LNCS, vol. 7726, pp. 493–506. Springer, Heidelberg (2013). https://doi.org/10.1007/978-3-642-37431-9_38
11. Oliver, N.M., Rosario, B., Pentland, A.P.: A bayesian computer vision system for modeling human interactions. IEEE Trans. Pattern Anal. Mach. Intell. **22**(8), 831–843 (2000)
12. St-Charles, P.L., Bilodeau, G.A.: Improving background subtraction using local binary similarity patterns. In: 2014 IEEE Winter Conference on Applications of Computer Vision (WACV), pp. 509–515. IEEE (2014)
13. Stauffer, C., Grimson, W.E.L.: Adaptive background mixture models for real-time tracking. In: CVPR, p. 2246. IEEE (1999)
14. Wang, B., Dudek, P.: A fast self-tuning background subtraction algorithm. In: Proceedings of the IEEE Conference on Computer Vision and Pattern Recognition Workshops, pp. 395–398 (2014)
15. Wang, J., Bebis, G., Miller, R.: Robust video-based surveillance by integrating target detection with tracking. In: IEEE Conference on Computer Vision and Pattern Recognition Workshop, CVPRW 2006, p. 137. IEEE (2006)
16. Xue, G., Song, L., Sun, J., Wu, M.: Hybrid center-symmetric local pattern for dynamic background subtraction. In: 2011 IEEE International Conference on Multimedia and Expo, pp. 1–6. IEEE (2011)
17. Xue, G., Sun, J., Song, L.: Dynamic background subtraction based on spatial extended center-symmetric local binary pattern. In: 2010 IEEE International Conference on Multimedia and Expo, pp. 1050–1054. IEEE (2010)
18. Yang, S., Hao, K., Ding, Y., Liu, J.: Improved visual background extractor with adaptive range change. Memetic Comput. **10**(1), 53–61 (2018)
19. Zivkovic, Z.: Improved adaptive Gaussian mixture model for background subtraction. In: Proceedings of the 17th International Conference on Pattern Recognition, ICPR 2004, vol. 2, pp. 28–31. IEEE (2004)

A Multi-tier Fusion Strategy for Event Classification in Unconstrained Videos

Prithwish Jana[1], Swarnabja Bhaumik[2],
and Partha Pratim Mohanta[3(✉)]

[1] Department of Computer Science and Engineering, Jadavpur University,
Kolkata, India
jprithwish@gmail.com
[2] Department of Computer Science and Engineering,
Meghnad Saha Institute of Technology, Kolkata, India
swarnabjazq22@gmail.com
[3] Electronics and Communication Sciences Unit, Indian Statistical Institute,
Kolkata, India
partha.p.mohanta@gmail.com

Abstract. In this paper, we propose a novel fusion strategy of prediction vectors obtained from two different deep neural networks designed for the task of event recognition in unconstrained videos. Videos are suitably represented by a set of key-frames. Two types of features, namely spatial and temporal, are computed from a hybrid pre-trained CNN-RNN (Convolutional Neural Networks - Recurrent Neural Networks) framework for each video. These features are able to capture both the transient and long-term dependencies for understanding of the events. Frame-level and video-level prediction vectors are generated from two separate CNN-RNN (ResNet50-LSTM) frameworks exploiting spatial and temporal features respectively. The fusion is performed on these prediction vectors at different levels. The entire fusion framework relies on a concept of consolidation of probability distributions. This consolidation is implemented using conflation and a biasing technique. A multi-level fusion is achieved and at each level, a significant amount of classification accuracy is observed as improving. The experiment is performed on four benchmark datasets, namely, Columbia Consumer Video (CCV), Kodak's Consumer Video, UCF-101 and Human Motion Database (HMDB). The increment in *mAP* values achieved by the proposed fusion strategy is much higher than the conventional fusion strategies in use. Also, the classification accuracies of all the four datasets are comparable to other state-of-the-art methods for event classification in unconstrained videos

Keywords: Motion and video analysis · Event classification · Deep neural networks · Spatio-temporal features · Fusion · Conflation

1 Introduction

Due to easy availability and low cost of the video capturing devices, everyday moments are effortlessly captured by non-professionals. The quality of such videos is generally not up to the mark. Automatic annotation of these videos and retrieval of multi-scale

© Springer Nature Switzerland AG 2019
B. Deka et al. (Eds.): PReMI 2019, LNCS 11942, pp. 515–524, 2019.
https://doi.org/10.1007/978-3-030-34872-4_57

information is useful for applications such as surveillance, sports, web search, etc. With generation of millions of videos on social events, it is nearly impossible to annotate them manually. Thus, there is a rising demand for automatic event classification.

The videos captured by the non-professionals are unconstrained in nature and often down-sampled. It is quite a challenge for researchers to overcome quality disparities such as unsteady background motion, jittery camera motion, clothing variation, occlusion, etc.

Initially the event recognition systems extract the visual features from short time spans of a video. Yeung et al. [1] apply dense labeling of each frame to exploit intra and inter class temporal relevance. Kumar et al. [2] use dense features and study the contextual behavioral patterns for event understanding in traffic videos. Laptev et al. [3] focus on single frame classification by memorizing past actions. The limitation of sparse approaches is pointed out by Ng et al., in [4].

Video features are extracted predominantly from the visual and acoustic channel [5]. Nowadays researchers prefer collaborative features from multiple channels. Wu et al. [6] use a multi-stream network of parallel CNNs to fuse multimodal features. Li et al. [7] add a fusion network that learns from 3 streams and form an end-to-end framework.

Fusion techniques are being used to consolidate results from multiple CNNs. Ideally, a feature fusion technique should retain the individual advantages of its constituents and also generate some unique characteristics of its own. Early fusion as in Ng et al. [4], integrates pixel level unimodal information. It allows the network to learn local concepts before training. Late fusion [8] consolidates data from multiple single frame networks and consolidates their score after classification.

In early fusion, an interference to others' training phase may create a harmful situation when both the CNNs deal with diverse data, for example, images and optical-flow. Late fusion generally provides better performance [9], however, may require extra effort to learn concepts. Borda Count [10] is a classical method of late-fusion that assigns lowest score to the last ranker and increases score with increase of rank. Highest Rank Method [11] compares highest rank from each classifier. Umer et al. [12] have analyzed the effect of late fusion policies in video event recognition.

While quite a few of the event recognition techniques employ late fusion, the increment in score after applying fusion is mostly meagre. Moreover, they are not always fruitful in the domain of video processing because frame-level prediction and video-level prediction must be taken care of separately. This gives us a motivation to come up with a fusion strategy which will operate on frame-level as well as video-level.

This paper is organized into four sections. The proposed methodology is described in Sect. 2. The datasets, experiments, results and related discussion is presented in Sect. 3. Finally, Conclusion and Future Scope are discussed in Sect. 4.

2 The Proposed Methodology

We have subdivided the proposed methodology into three major stages, namely *Key-Frame Selection*, *Classification Framework* and *Fusion of Multi-stream data*.

2.1   Key-Frame Selection

Each video is represented by a predefined set of representative frames or key-frames. We have relied on a graph-based algorithm, as elaborated in Jana et al. [13], to select the temporally significant frames as the representative frames of a video. The temporal feature for each frame is computed as the concatenated histogram of magnitude and slope of the optical flow. Based on the temporally significant frames, a graph is constructed considering the absolute difference of the concerned flow histograms as the edge-weights Finally, considering the trade-off between diversity of temporal content and time of appearance of the frames, the key-frames are selected from the graph.

2.2   Classification Framework

We propose a deep learning framework for classification of various social events, human activities, etc. in unconstrained videos. This framework consists of a ResNet50 [7] (a CNN) followed by LSTM [6] (an RNN). Two different kind of information, viz. spatial and temporal, are separately exploited through this deep learning framework. The spatial framework relies on the RGB images as the input, whereas the flow vectors' magnitude in optical flow matrices is used as input to the temporal framework. The RGB images and optical flow matrices corresponding to the key-frames are the reference point of a video. A prediction vector corresponding to each key-frame, is obtained in the last fully-connected layer of ResNet50. Further, all these *frame-wise* prediction vectors pertaining to each video, are fed into the LSTM network. Finally, in the topmost dense logistic layer of the LSTM network, a *video-wise* prediction vector is obtained. Each element of such vector indicates the probability of belongingness of a frame/video to a class. Thus, a prediction vector may be looked upon as a probability distribution. The final classification result is obtained by fusing the *frame-level* as well as *video-level* prediction vectors generated from both the spatial and temporal frameworks.

2.3   Fusion of Multi-stream Data

We have adopted a novel fusion strategy that fuses the *frame-level* and *video-level* prediction vectors envisaged by the ResNet50 and LSTM respectively for spatial and temporal framework. The proposed fusion strategy relies on *self-fusion* in the frame-level and *cross-fusion* in frame as well as video-level.

Let V be the total number of videos in a dataset and n_{KF} be the number of key-frames extracted from each video. Assume that the dataset contains C_{Lbl} events or classes. All the $V \times n_{KF}$ frames are fed into the ResNet50 architecture. From the last fully-connected layer of ResNet50, a prediction vector of size C_{Lbl} is generated corresponding to each of the $V \times n_{KF}$ frames. The outcome of the ResNet50 i.e., all the $V \times n_{KF}$ prediction vectors are fed to the LSTM architecture as a 3-dimensional matrix of size (V, n_{KF}, C_{Lbl}). Finally, a prediction vector of size C_{Lbl} is generated corresponding to each of the V videos.

The spatial and temporal framework follow exactly the same pipeline. Once the execution of ResNet50 is over, two *frame-level prediction* matrices of size

$(V \times n_{KF}, C_{Lbl})$, viz., FP_S and FP_T are generated from spatial and temporal framework respectively. Similarly, in the next phase, LSTM generates two *video-level prediction* matrices of size (V, C_{Lbl}), namely, VP_S and VP_T. Essentially, each row of all the four prediction matrices (FP_S, FP_T, VP_S and VP_T) represents a probability distribution. Hence, sum of the elements in each row is 1 and each element is a positive real number lying between 0 and 1. In the proposed fusion policy, these four prediction matrices are being fused to generate a single *video-level* prediction matrix. First, the *frame-level* prediction matrices FP_S and FP_T are fused and generate a *frame-level* prediction matrix FP_{ST}. A *video-level* prediction matrix $VP_{ST}^{(1)}$ is generated from FP_{ST}. On the other hand, by fusing *video-level* prediction matrices VP_S and VP_T, another *video-level* prediction matrix $VP_{ST}^{(2)}$ is obtained. The *final* classification result i.e. a video level prediction matrix VP_{ST}, is obtained by fusing $VP_{ST}^{(1)}$ and $VP_{ST}^{(2)}$. The pictorial demonstration of entire fusion strategy is depicted in the Fig. 1.

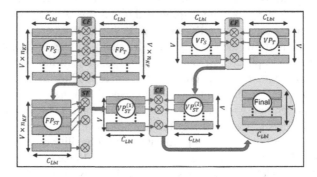

Fig. 1. Complete fusion strategy: cross-fusion and self-fusion on FP_S, FP_T, VP_S and VP_T

Here, two fusion strategies are adopted: (a) fusion between two prediction matrices and (b) fusion within a prediction matrix. The strategy (a) is termed as *cross-fusion* (CF) and (b) as *self-fusion* (SF). The fusion operation relies on the consolidation of probability distributions. The *cross-fusion* and *self-fusion* are defined as the consolidation of corresponding rows of two prediction matrices and number of rows in same prediction matrix respectively.

Consolidation of Probability Distributions

Let $P_1(A)$ and $P_2(A)$ be two *discrete probability distributions* defined on a set of class labels, A. The consolidation of $P_1(A)$ and $P_2(A)$ is also a discrete probability distribution, say $P(A)$. The objective is to find a consolidated distribution which is more confident about the accuracies of the class probabilities. Hill [14] has shown that the *conflation* of probability distributions provides reasonably good solution than the other straight forward technique such as averaging. Hill et al. [15] have also shown that, in case of $P(A) = \frac{P_1(A) + P_2(A)}{2}$ and P_1 and P_2 are nearly similar, it is not guaranteed that standard deviation of P is strictly less than the formers. So, the *average* is not a desirable solution. The *conflation* of P_1 and P_2 is defined as follows.

$$_{a \in A}P(X = a) = \frac{P_1(X = a) \times P_2(X = a)}{\sum_{b \in A}(P_1(X = b) \times P_2(X = b))} \tag{1}$$

Although conflation provides reasonably good solution, however there may cases where it fails to predict the true class. Thus, to enhance the chances to grasp the correct prediction, we have adopted a *biasing* technique. The consolidated distribution P would get biased towards P_1 or P_2. The biasing technique is described as follows.

Let $d_1 = bd(P, P_1)$ and $d_2 = bd(P, P_2)$, where $bd(p, q)$ is the Bhattacharyya distance [16] between two discrete probability distributions, p and q. The bias and the biased consolidated distribution are computed as $bias = \left(\frac{\min(d_1, d_2)}{4 \times \max(d_1, d_2)}\right)$ and

$$P_{biased} = (P + (P_x - P) \times bias)_{x = 1 \, or \, 2} \tag{2}$$

x attains value 1 when $d_1 < d_2$, otherwise takes value 2. The distance between P and P_1 (or P_2) get reduced by *bias*%. Likewise, P_{biased} gets closer towards its nearest one. *bias* attains 25% when P is equidistant from P_1 and P_2, whereas $\sim 0\%$ when its already biased. Algebraic sum of $(P_x - P)$ for all common elements is 0 and its addition to P will not change the sum. *bias* is applied only in case of *cross-fusion*. An example of consolidation of two prediction vectors is presented in Fig. 2 where consolidation of two wrong predictions gives rise to a correct prediction.

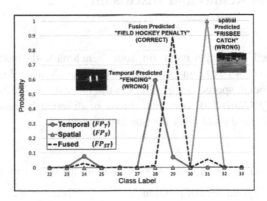

Fig. 2. Correct prediction in FP_{ST} as a consolidation of two wrong predictions in FP_S and FP_T

Cross-Fusion (CF)

The cross-fusion is applicable on *frame-level* as well as *video-level*. The *frame-level* cross-fusion is applied on FP_S and FP_T, whereas the *video-level* cross-fusion is applied first on VP_S and VP_T, and then on $VP_{ST}^{(1)}$ and $VP_{ST}^{(2)}$. Let M_1, M_2 be two prediction matrices and RM be the resultant prediction matrix of same dimension. RM is obtained by applying the fusion on M_1 and M_2. The fusion operation is defined on two corresponding rows of M_1 and M_2. The ith row of RM is obtained as follows:

$RM[i][1..C_{Lbl}] = f(M_1[i][1..C_{Lbl}], M_2[i][1..C_{Lbl}])$, where $f(.,.)$ is the fusion operation applied on M_1 and M_2.

Self-Fusion (SF)

The *self-fusion* is applicable only on *frame-level*. This fusion operation is applied to generate the video-level prediction matrix $VP_{ST}^{(1)}$ from the frame-level prediction matrix FP_{ST}. Essentially, it would generate a prediction vector for each of the V videos by fusing corresponding n_{KF} frame-level prediction vectors or rows of FP_{ST}. The ith row of $VP_{ST}^{(1)}$ is obtained as follows:

$$VP_{ST}^{(1)}\left[\frac{i}{n_{KF}}\right][1..C_{Lbl}] = f(FP_{ST}[i,..,i+n_{KF}-1][1..C_{Lbl}]) \qquad \text{for} \qquad i = k \times n_{KF}+1$$

$(k = 0,1,2,\ldots)$ where $f(.,.,\ldots)$ is the fusion operation on the n_{KF} rows of FP_{ST}.

The size of the input prediction matrix FP_{ST} is $(V \times n_{KF}, C_{Lbl})$ and that of output prediction matrix, $VP_{ST}^{(1)}$, is (V, C_{Lbl}).

The *confidence* of a prediction vector is defined as the difference between the highest and second-highest class probability values. A class is predicted by applying the majority voting among the all n_{KF} prediction vectors of a video. The *most-confident* prediction vector among the majority group is considered as *video-level* prediction. In case more than one group qualify in the majority voting, the *most-confident* prediction vector from each group are fused to generate consolidated prediction vector.

3   Experimental Results and Discussion

3.1   Dataset

We have performed our experiment on four benchmark datasets. The Columbia Consumer Videos (*CCV*) [17] and Kodak's Consumer Video (*KCV*) [18] contains videos of various social, sports, etc. events. On the other hand, *UCF-101* [19], Human Motion Database (*HMDB*) [20] contains videos of different human actions. Table 1 shows details of all the dataset used for our experiment.

Table 1. Overview of train-test splits used for the benchmark datasets

	Type/Source of partition	Number of available videos		Total classes
		Train	Test	
CCV [17]	Standard Split [17] (not all available)	3520	1224	20
KCV [18]	Random 70:30 Split	897	268	29
UCF-101 [19]	'Trainlist-1' and 'Testlist-1' [19]	9529	3780	101
HMDB [20]	'Split-1' [20]	3570	1530	51

3.2 Performance Evaluation

The performance of the proposed methodology is evaluated considering two aspects: *(a)* accuracy in event classification, and *(b)* efficacy of the fusion strategy. The classification accuracy is computed as the average of all the accuracies (mean Average Precision (*mAP*)) generated in different runs of the experiment. Table 2 summarizes the classification accuracy obtained in each stages of the proposed method. The performance of the conventional late fusion strategies is summarized in Table 3. The comparison with other state-of-the-art methods of event classification are presented in Table 4. It is evident that the event classification accuracy of the proposed methodology is comparable with state-of-the-art methods. Moreover, the performance of the proposed feature fusion strategy is much better than the prevailing late fusion strategies as well as the fusion strategies adopted by the state-of-the-art methods of event classification.

Table 2. *mAP* values of proposed method at successive stages of the fusion. Values in bracket denotes gain (+)/loss (−) from highest among FP_T, FP_S, VP_T and VP_S

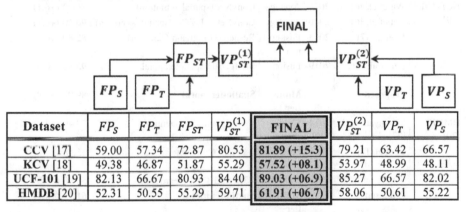

Dataset	FP_S	FP_T	FP_{ST}	$VP_{ST}^{(1)}$	FINAL	$VP_{ST}^{(2)}$	VP_T	VP_S
CCV [17]	59.00	57.34	72.87	80.53	**81.89 (+15.3)**	79.21	63.42	66.57
KCV [18]	49.38	46.87	51.87	55.29	**57.52 (+08.1)**	53.97	48.99	48.11
UCF-101 [19]	82.13	66.67	80.93	84.40	**89.03 (+06.9)**	85.27	66.57	82.02
HMDB [20]	52.31	50.55	55.29	59.71	**61.91 (+06.7)**	58.06	50.61	55.22

Table 3. Performance of late fusion techniques on benchmark datasets. Values in bracket denotes gain (+)/loss (−) from highest among VP_T and VP_S

Fusion methods	Inputs	*mAP (in %)* in dataset			
		CCV [17]	KCV [18]	UCF-101 [19]	HMDB [20]
Borda count [10]	Spatial (VP_S) &	61.84 (−4.7)	50.29 (+1.3)	68.53 (−13.5)	52.99 (−2.2)
Highest rank [11]	Temporal (VP_T)	67.63 (**+1.1**)	50.93 (+1.9)	74.64 (−7.4)	58.91 (**+3.7**)
Mean of prediction [14]		66.58 (+0.0)	52.87 (**+3.9**)	84.02 (**+2.0**)	55.86 (+0.6)

Table 4. Comparison of performance of state-of-the-art methods. Values in bracket denotes gain (+)/loss (−) from highest *mAP* value among respective fusion inputs

Dataset	Methods				mAP (in %) in dataset
	Publication	Year	Fusion type	Method description	
CCV [17]	Wu et al. [6]	2016	Average	ConvNet (spatial + motion)	75.80 (+0.8)
	Wu et al. [6]	2016	Average	ConvNet + LSTM (spatial + motion)	81.70 (+3.8)
	Umer et al. [12]	2017	Product	Multi-scale spatial	80.46 (+2.0)
	Jiang et al. [21]	2018	rDNN	(Spatial + Trajectory + Audio)	73.50 (+6.8)
	Zhang et al. [22]	2019	1-layer NN	VGG + ResNet + CNNM (Spatial + Temporal)	79.30 (+4.0)
	Proposed	–	**Multi-tier**	**Spatio-temporal**	**81.89 (+15.3)**
KCV [18]	Duan et al. [23]	2012	DSM	SIFT features	35.46
	Chen et al. [24]	2013	Average	Space-time features	49.61 (+4.8)
	Luo et al. [25]	2018	–	Semi-supervised analysis	47.70
	Proposed	–	**Multi-tier**	**Spatio-temporal**	**57.52 (+8.1)**
UCF-101 [19]	Wu et al. [6]	2016	Average	ConvNet (spatial + motion)	86.20 (+6.1)
	Wu et al. [6]	2016	Average	ConvNet + LSTM (spatial + motion)	90.10 (+6.8)
	Li et al. [7]	2018	Fusion n/w	Resnet50 (3-stream ConvNet)	82.80 (+6.6)
	Cai et al. [26]	2019	Fusion n/w	Multi-scale spatiotemporal	**93.50** (+6.0)
	Proposed	–	**Multi-tier**	**Spatio-temporal**	89.03 (**+6.9**)
HMDB [20]	Simonyan et al. [27]	2014	SVM	Two-stream spatiotemporal	59.40 (+4.8)
	Li et al. [7]	2018	Fusion n/w	3-stream ConvNet	55.58
	Song et al. [28]	2018	SVM	C3D (RGB + Foreground)	50.90 (+1.3)
	Proposed	–	**Multi-tier**	**Spatio-temporal**	**61.91 (+6.7)**

4   Conclusion and Future Scope

We have developed a novel multi-tier fusion strategy and its efficacy is tested on the classification accuracies of the event recognition in four benchmark unconstrained video datasets. Each video is represented by a preset number of key-frames. The classification is performed on CNN-RNN (ResNet50-LSTM) framework for spatial and temporal features. The *frame-level* and *video-level* prediction vectors are generated from spatial and temporal ResNet50-LSTM framework separately. A multi-tier late fusion strategy is thus adopted to arrive at a final prediction. A significant improvement in *mAP* values at every step of the proposed fusion strategy are reported.

The proposed fusion strategy is a late fusion and may be looked upon as a post-processing on prediction vectors. The fusion does not have any dependencies on the training phase of both spatial and temporal CNN-RNN framework. Hence, the efficacy of the parallelism of two frameworks is established. However, the fusion of these two frameworks not yet explored. Early fusion of the spatial and temporal feature may be explored in the near future.

References

1. Yeung, S., Russakovsky, O., Jin, N., Andriluka, M., Mori, G., Fei-Fei, L.: Every moment counts: Dense detailed labeling of actions in complex videos. Int. J. Comput. Vis. **126**(2–4), 375–389 (2018)
2. Kumar, P., Ranganath, S., Weimin, H., Sengupta, K.: Framework for real-time behavior interpretation from traffic video. IEEE Trans. Intell. Transp. Syst. **6**(1), 43–53 (2005)
3. Laptev, I., Pérez, P.: Retrieving actions in movies. In: 2007 IEEE 11th ICCV (2007)
4. Yue-Hei Ng, J., Hausknecht, M., Vijayanarasimhan, S., Vinyals, O., Monga, R., Toderici, G.: Beyond short snippets: Deep networks for video classification. In: IEEE Conference on CVPR, pp. 4694–4702 (2015)
5. Jiang, Y.G., Bhattacharya, S., Chang, S.F., Shah, M.: High-level event recognition in unconstrained videos. Int. J. Multimedia Inf. Retr. **2**(2), 73–101 (2013)
6. Wu, Z., Jiang, Y.G., Wang, X., Ye, H., Xue, X.: Multi-stream multi-class fusion of deep networks for video classification. In: 24th ACM Multimedia Conference, pp. 791–800. ACM (2016)
7. Li, C., Ming, Y.: Three-stream convolution networks after background subtraction for action recognition. In: Bai, X., et al. (eds.) FFER/DLPR -2018. LNCS, vol. 11264, pp. 12–24. Springer, Cham (2019). https://doi.org/10.1007/978-3-030-12177-8_2
8. Karpathy, A., Toderici, G., Shetty, S., Leung, T., Sukthankar, R., Fei-Fei, L.: Large-scale video classification with convolutional neural networks. In: IEEE Conference on CVPR, pp. 1725–1732 (2014)
9. Lee, J., Abu-El-Haija, S., Varadarajan, B., Natsev, A.P.: Collaborative deep metric learning for video understanding. In: 24th ACM SIGKDD International Conference on KDDM, pp. 481–490. ACM (2018)
10. Emerson, P.: The original Borda count and partial voting. Soc. Choice Welf. **40**(2), 353–358 (2013)
11. Ye, G., Liu, D., Jhuo, I.H., Chang, S.F.: Robust late fusion with rank minimization. In: 2012 IEEE Conference on CVPR, pp. 3021–3028. IEEE (2012)
12. Umer, S., Ghorai, M., Mohanta, P.P.: Event recognition in unconstrained video using multi-scale deep spatial features. In: 2017 9th ICAPR, pp. 1–6. IEEE (2017)
13. Jana, P., Bhaumik, S., Mohanta, P.P.: Key-frame based event recognition in unconstrained videos using temporal features. In: 2019 IEEE Region 10 Symposium (TENSYMP). IEEE (2019)
14. Hill, T.: Conflations of probability distributions. Trans. Am. Math. Soc. **363**(6), 3351–3372 (2011)
15. Hill, T.P., Miller, J.: How to combine independent data sets for the same quantity. Chaos: An Interdiscip. J. Nonlinear Sci. **21**(3), 033102 (2011)
16. Bhattacharyya, A.: On a measure of divergence between two multinomial populations. Sankhyā: Indian J. Stat. 401–406 (1946)
17. Columbia Consumer Video (CCV) Database. http://www.ee.columbia.edu/ln/dvmm/CCV/. Accessed May 2019
18. Kodak's consumer video benchmark data set. http://www.ee.columbia.edu/ln/dvmm/consumervideo/. Accessed May 2019

19. UCF101 - Action Recognition Data Set. http://crcv.ucf.edu/data/UCF101.php. Accessed May 2019
20. HMDB: A large human motion database. http://serre-lab.clps.brown.edu/resource/hmdb-a-large-human-motion-database/#Downloads. Accessed May 2019
21. Jiang, Y.G., Wu, Z., Wang, J., Xue, X., Chang, S.F.: Exploiting feature and class relationships in video categorization with regularized deep neural networks. IEEE Trans. Pattern Anal. Mach. Intell. **40**(2), 352–364 (2017)
22. Zhang, J., Mei, K., Zheng, Y., Fan, J.: Exploiting mid-level semantics for large-scale complex video classification. IEEE Trans. Multimedia **21**, 2518–2530 (2019)
23. Duan, L., Xu, D., Chang, S.F.: Exploiting web images for event recognition in consumer videos: a multiple source domain adaptation approach. In: 2012 IEEE Conference on CVPR, pp. 1338–1345. IEEE (2012)
24. Chen, L., Duan, L., Xu, D.: Event recognition in videos by learning from heterogeneous web sources. In: 2013 IEEE Conference on CVPR, pp. 2666–2673 (2013)
25. Luo, M., Chang, X., Nie, L., Yang, Y., Hauptmann, A.G., Zheng, Q.: An adaptive semisupervised feature analysis for video semantic recognition. IEEE Trans. Cybern. **48**, 648–660 (2017)
26. Cai, Y., Lin, W., See, J., Cheng, M.M., Liu, G., Xiong, H.: Multi-scale spatiotemporal information fusion network for video action recognition. In: IEEE VCIP, pp. 1–4 (2018)
27. Simonyan, K., Zisserman, A.: Two-stream convolutional networks for action recognition in videos. In: Advances in Neural Information Processing Systems, pp. 568–576 (2014)
28. Song, H., Tian, L., Li, C.: 3D convolutional network based foreground feature fusion. In: 2018 IEEE ISM, pp. 253–258. IEEE (2018)

Visual Object Tracking Using Perceptron Forests and Optical Flow

Gaurav Nakum$^{(\boxtimes)}$, Prithwijit Guha, and Rashmi Dutta Baruah

Indian Institute of Technology Guwahati, Guwahati, India
{n.gaurav,pguha,r.duttabaruah}@iitg.ac.in

Abstract. This work proposes a new approach for target representation and candidate proposal selection to track single objects in videos. Constructing an appearance model that is robust to visual appearance changes is a challenging problem in visual tracking. Typically, complicated models are used to achieve this robustness. However, such models have high computational cost and thus, are not suitable for real-time tracking. In this paper, we propose a visual tracking algorithm in which target objects are represented by compact low-dimensional codes, constructed using perceptron forests. We employ the highly successful tracking-by-detection framework to use these representations for training an online discriminative classifier to separate the object from the background. Moreover, the candidate proposals are generated from a two-phase strategy. First, we employ dense optical flow to estimate the geometric transformation undergone by the objects to generate one set of proposals. These proposals are used to initialize a particle swarm optimization framework which localizes the objects in the current frame. We integrated our algorithm with the VOT 2016 toolkit and report its performance compared to several state-of-the-art trackers.

Keywords: Perceptron forests · Object representation · Optical flow · Affine transformation · Particle swarm optimization · Passive-aggressive classifier

1 Introduction

Visual Object Tracking is the process of locating an object in all frames of a video given its location in the first frame. It has numerous applications in domains such as security and surveillance, perceptual user interfaces, augmented reality, traffic control and medical imaging. However, it is also one of the most challenging fields of computer vision, mainly due to the problems arising from illumination variations, occlusions, background clutter, object deformation and scale changes.

Several methods have been proposed which purport to tackle these problems using either sophisticated appearance models [3,17] or motion models [15,16]. Methods which model the object's appearance usually use multiple feature cues

© Springer Nature Switzerland AG 2019
B. Deka et al. (Eds.): PReMI 2019, LNCS 11942, pp. 525–532, 2019.
https://doi.org/10.1007/978-3-030-34872-4_58

to cope with object appearance changes. Information from these multiple feature types is typically concatenated to form a unified feature vector for representing the object. However, such a feature vector is usually high dimensional and contains redundant information, thereby decreasing their power to discriminate object from the background. In this work, we focus on learning compact and robust object representations for tracking objects. The discriminative approach to visual tracking treats it as a binary classification problem, where a discriminative classifier is trained to distinguish the object from its background. Since this method inculcates background information into the tracking system, it is more robust to occlusions and scene variations. For this work, we use the discriminative approach to object tracking.

Specifically, we construct a novel object representation for tracking by the use of perceptron forests. Perceptron trees [6] are a variant of decision trees in which each node is a single perceptron. They generally possess higher classification accuracy than traditional decision trees, since each perceptron considers multiple features of the data points to split the dataset at the node corresponding to it. For our purpose, we devise a hashing method following Li et al. [11] to reduce the dimensions of multiple image feature descriptors to a single vector. The codes are constructed by training the perceptron trees on randomly sampled patches of the image around the current target location followed by aggregating the posterior distributions of all leaf nodes reached in the perceptron forests. This generates a compact discriminative image representation on which a passive-aggressive classifier is trained to differentiate between the object and the background. The candidates for predicting the object location are selected using a combined optical flow and particle swarm optimization (PSO) based scheme. The optical flow method generates candidates by modeling the geometric transformation undergone by the object, while the PSO scheme is used to iteratively refine the candidates. In the experiments section, we demonstrate the results of our tracker on multiple sequences from the VOT 2016 dataset.

The rest of the paper is organized as follows: In Sect. 2, we provide a review of related works on object-tracking. In Sect. 3 we present our algorithm for object tracking. We show the results of evaluation of our algorithm on sample sequences in Sect. 4. Finally, in Sect. 5 we conclude the paper and offer directions for future research.

2   Related Works

Works on object tracking aim to model the appearance of the object or the motion of the object. Numerous techniques for object representation have been proposed [12].

Many methods exist to generate a candidate set of object locations for target location. Most of the well-known strategies use the tracking-by-detection framework. [2] uses a simple radius based method for constructing candidate location set.

However, a shortcoming of all these methods is that they do not take into account the transformations that an object can undergo, such as size changes, rotation, etc. Since we aim to construct a tracker which generalizes to these complex scenarios, we utilize an approach which generates a rich candidate proposal set where the test templates can be rotated and scaled with respect to the current bounding box.

To achieve this goal, we use optical flow method to generate candidate locations proposed by Hua et. al. [5]. In contrast to the works mentioned above, their approach considers the geometric transformations undergone by the target object. Although they consider only isotropic scaling of the object, we consider a more general object size change and thus our proposals are applicable even in cases when the object's size changes by different factors in the x and y direction.

There have also been a few works based on using decision trees for object tracking [11,18–20]. [18] (ORF) was the first work on the use of random forests for object tracking. It uses an out-of-bag error metric for deciding whether a tree should be used for classification or be discarded on the arrival on new data. At every node of the tree it uses two features to make a split. [20] uses a set of features selected using proximal-SVM to make a split at each node. In [19], target objects are modeled at three levels of granularity (bounding box level, and parts-based level, and pixel level). It uses a decision tree to dynamically select the minimum combination of features necessary to represent the target part at each frame. It also uses superpixels for learning new parts of the target.

3   Proposed Scheme

In this section, we present our entire tracking-by-detection framework for object tracking. We use the hashing method proposed by [11] to construct object representations using perceptron forests. Also, we use proposals generated by optical flow based affine transformation method with an online discriminative classifier and an object localization strategy. The classifier used is the Passive-Aggressive classifier which discriminates the object and the background. Subsequently, particle swarm optimization (PSO) algorithm is presented to demonstrate how the confidence scores obtained by the classifier can be refined by taking its decision function as the fitness function of the optimization algorithm and hence reliably track and localize the object.

3.1   Perceptron Forests for Compact Code Learning

We adopt the randomized forest method for constructing compact object codes following Li et. al.'s method [11]. Given a sample, their hash code generation algorithm essentially generates binary code based on the posterior distribution of the leaf nodes reached by the sample when passed through a random forest. The hash code function has the form

$$h(\mathbf{u}) = \mathcal{S}(\mathbf{u}, u_i^+) - \mathcal{S}(\mathbf{u}, u_i^-) \tag{1}$$

where $\mathbf{u}$ is a test sample. Now, to design an efficient and discriminative similarity function $\mathcal{S}$, we use perceptron forests. Let $\mathbb{C} \in \{-1, 1\}$ be the set of all classes, where -1 represents the background and 1 represents the object. Also, let $\mathbb{L}_t$ be the set of all leaves for a given perceptron tree $t \in \mathbb{T}$. We need to learn the posterior probability $Pr_{t,l}(c)$ for each class $c \in \mathbb{C}$ at each leaf node $l \in \mathbb{L}_t$ during the training stage. For this, we store the number of examples $\mathbb{Q}_{t,l,c}$ of class c reaching each leaf node and the total number of examples $\mathbb{Q}_{t,l}$ at the leaf node. Thereafter, $Pr_{t,l}(c)$ is calculated as the ratio $\frac{\mathbb{Q}_{t,l,c}}{\mathbb{Q}_{t,l}}$.

Based on the perceptron forest $\mathbb{T}$, the similarity function in 1 is defined as :

$$\mathcal{S}(\mathbf{u}, \{\mathbb{Q}_{t,l_u^t,c} | t \in \mathbb{T}\}) = \sum_{t \in \mathbb{T}} Pr_{t,l_u^t,c}(c). \tag{2}$$

There are two differences between their hash function generation algorithm and our method. First, we use perceptron forests instead of decision tree forests. Perceptron forests are a collection of perceptron trees [6]. We also employ the KMeans-SMOTE data balancing method [10] for balancing the trees [6]. Second, although Li et. al. use binary codes for object representation, we use code vectors having continuous values in $[0.0, 1.0]$. This is achieved by omitting the sign function in the equation for hash code value. Using this technique, we are able to achieve a greater flexibility in the values of code vectors while maintaining their compactness. Also, the designed hash function has an important property that each individual hash function is balanced such that

$$\int_{h(\mathbf{u}>0)} Pr(\mathbf{u})d\mathbf{u} = \int_{h(\mathbf{u}<0)} Pr(\mathbf{u})d\mathbf{u} = \frac{1}{2} \tag{3}$$

3.2 Optical Flow for Proposal Generation

We adopt the optical flow based method of Hua et. al. [5] for generating the candidate set of object locations. The authors aim to model the transformation undergone by the object using optical flow and subsequently use it to enrich the candidate set with geometry proposals. We employ Hua et. al.'s method with one small modification. The authors assumed isotropic scaling of the object and hence needed only one parameter corresponding to scale in the affine transformation. However, we assume a more general object size change in that the object size can change by different amounts in the x and y directions. Therefore, the object's transformation is represented with an affine matrix, unlike Hua et al.'s method which uses a similarity matrix. The affine transformation is defined by five parameters - one for rotation and two each for translation and scale changes. An affine transformation matrix for a general 2D point (x, y) is given as

$$\begin{bmatrix} s_x \cos\theta & -s_y \sin\theta & t_x s_x \cos\theta - t_y s_y \sin\theta \\ s_x \sin\theta & s_y \cos\theta & t_x s_x \sin\theta + t_y s_y \cos\theta \\ 0 & 0 & 1 \end{bmatrix} \tag{4}$$

Thus, it is defined by five parameters $(\theta, s_x, s_y, t_x, t_y)$. Thereafter, a general affine transformation is defined as $\mathbf{x'} = M\mathbf{x}$ where $\mathbf{x} = [x, y, 1]^T$ is a point in frame $(t-1)$, $\mathbf{x'} = [x', y', 1]^T$ is a point in frame t, and M is the transformation matrix (Eq. 4). Following this, the method proposed by Hua et al. [5] is used for candidate proposal generation.

3.3   Passive-Aggressive Classifier for Classification

To classify the features as object/background, we build an online discriminative appearance model based on a passive aggressive (PA) classifier [4]. The PA classifier is a margin maximization based learning algorithm which is easy to update on the receipt on new data. For training the PA classifiers we construct the training samples by extracting patches around the current target location and transforming them into compact features using the perceptron forests. For positive (object) samples, we crop out a set of patches within α distance from the current target location x_t^* and for negative (background) samples, we crop out a set of patches having distance in the range (α, β) from x_t^*. The classifier is updated after every p frames of the video. The classification score function of the classifier is used as the objective function in the PSO localization strategy explained next.

3.4   Object Localization Using PSO

The perceptron forest + PA classifier constitute our main decision making and score computing functions for evaluating candidate proposals. The Optical Flow + PSO scheme is used to generate and search for better candidates in the first place. The PSO [8] is a population based stochastic search algorithm used for optimization applications. Details of the algorithm can be found in [8]. The PSO and its variants have been earlier used in object localization [7].

We employ PSO for updating the target location and bounding box of the object in the next frame. For this purpose, the particles are defined as the patches cropped from the image around the current target location. The role of fitness function $f(.)$ is taken by the PA classifier's decision function, i.e., to evaluate the fitness of a particle (patch), we calculate the confidence score from the PA classifier. Thereafter, the PSO algorithm is applied to find the updated target state $\mathbf{X}_{i+1} = (x_{i+1}, y_{i+1}, w_{i+1}, h_{i+1}, \theta_{i+1})$ where (x_{i+1}, y_{i+1}) denotes the center of the new bounding box of the object, (w_{i+1}, h_{i+1}) are the bounding box dimensions, and θ is the orientation of the box with respect to the horizontal.

4   Experiments and Evaluation

The values for the parameters used in various algorithms can be found on the project homepage[1]. We tested our method on the Visual Object Tracking (VOT) Toolkit's 2016 version, which contains about 70 trackers and 55 video sequences. More details on the toolkit can be found at [9].

Figure 1 shows the performance of our tracker on the 4 sample video sequences. Table 1 compares the accuracy and robustness values of our tracker ("PForest") with 4 other trackers on 12 sequences picked from the VOT 2016 dataset. The four trackers used are MIL [2], MatFlow [13], FRT [1] and LoFT-Lite [14]. The accuracy of the tracker in VOT is defined as the average overlap between the

---

[1] https://github.com/nakumgaurav/Object-Tracking-Perceptron-Trees.

Fig. 1. Output for various VOT2016 sequences. Robustness to occlusion, illumination change, size change and motion change can be seen in these sequences. Green box shows the predicted locations while red shows ground truth. (Color figure online)

Table 1. Accuracy/Robustness values as obtained from the VOT2016 toolkit. Our method is labelled as 'PForest' (1st column). The best values on each sequence are marked in bold.

Sequence	PForest	MIL	MatFlow	FRT	LoFT-Lite
godfather	**0.44/0**	0.42/0.44	0.43/ **0**	0.29/0.13	0.12/1.31
gymnastics3	**0.25**/2.18	0.24/**1.35**	0.19/2.84	0.23/1.70	0.22/2.84
gymnastics4	**0.39/0.62**	0.37/0.64	0.37/0.97	0.29/1.85	0.29/2.07
pedestrian2	**0.46**/10.66	0.22/**0.43**	0.24/0.83	0.39/0.71	0.24/0.71
wiper	**0.56**/5.57	0.42/0.79	0.48/**0.79**	0.45/3.36	0.38/1.30
fish2	0.24/2.71	**0.28**/1.65	0.23/**1.37**	0.23/1.60	0.11/1.58
road	0.54/3.94	**0.58/0.50**	0.39/1.45	0.36/1.94	0.34/2.16
bmx	0.31/0	**0.34/0**	0.17/0.80	0.25/0.69	0.20/0.17
singer3	0.27/2.64	**0.29/0**	0.19/0	0.14/0.39	0.24/3.77
tunnel	0.40/9.52	**0.45**/1.4	0.33/0	0.41/0	0.27/1
car1	0.51/19.77	0.46/4.17	**0.53/0.61**	0.41/7.05	0.51/1
racing	0.38/0.47	0.38/0	**0.41/0**	0.37/0	0.38/0

ground truth bounding box and tracker-predicted bounding box during successful tracking, while robustness is defined as the average number of times a tracker drifts off the target, i.e., has zero overlap with the ground-truth bounding box. Since our tracker is stochastic, 15 runs are performed over each sequence to get average performance measures. Here, average accuracy is obtained by averaging per-frame accuracies which in turn are obtained by averaging overlaps over 15 runs, while average robustness is obtained by averaging failure rates over the 15 runs.

The average computation time over all sequences of our tracker was 9.7335 frames per second (FPS) for the baseline experiment and 10.7776 FPS for the unsupervised experiment.

5 Conclusion

In this work, we proposed a new method for target representation in visual object tracking. The representation was obtained using a perceptron forest hashing algorithm, whereby clusters of perceptron trees were employed to obtain compact low-dimensional object representation, which were shown to be robust to changes in object appearance. Moreover, we employed a modified optical flow-based algorithm to generate candidate object locations in succeeding frames by modeling the geometric transformation of objects in the form of affine representations. The candidate sets so generated were supplemented with an object localization strategy based on particle swarm optimization. Evaluation of the proposed scheme was performed using the VOT 2016 toolkit and noticeable results were obtained when compared to state-of-the-art techniques which use similar approaches.

For future work, we wish to focus on the use of deep neural decision forests and other hierarchical machine learning structures for constructing highly robust target representations. In the line of the current work, more optimized techniques can be found for the construction of perceptron forest codes, which can provide a significant boost to the speed of the algorithm.

References

1. Adam, A., Rivlin, E., Shimshoni, I.: Robust fragments-based tracking using the integral histogram. In: 2006 IEEE Computer Society Conference on Computer Vision and Pattern Recognition (CVPR 2006), vol. 1, pp. 798–805, June 2006
2. Babenko, B., Yang, M., Belongie, S.: Robust object tracking with online multiple instance learning. IEEE Trans. Pattern Anal. Mach. Intell. **33**(8), 1619–1632 (2011)
3. Comaniciu, D., Ramesh, V., Meer, P.: Kernel-based object tracking. IEEE Trans. Pattern Anal. Mach. Intell. **25**(5), 564–577 (2003)
4. Crammer, K., Dekel, O., Keshet, J., Shalev-Shwartz, S., Singer, Y.: Online passive-aggressive algorithms. J. Mach. Learn. Res. **7**, 551–585 (2006)
5. Hua, Y., Alahari, K., Schmid, C.: Online object tracking with proposal selection. In: 2015 IEEE International Conference on Computer Vision (ICCV), pp. 3092–3100, December 2015
6. Kannao, R., Guha, P.: Generic TV advertisement detection using progressively balanced perceptron trees. In: Proceedings of the Tenth Indian Conference on Computer Vision, Graphics and Image Processing, p. 8. ACM (2016)
7. Kate, P., Francis, M., Guha, P.: Visual tracking with breeding fireflies using brightness from background-foreground information. In: 2018 24th International Conference on Pattern Recognition (ICPR), pp. 2570–2575, August 2018. https://doi.org/10.1109/ICPR.2018.8546216

8. Kennedy, J., Eberhart, R.: Particle swarm optimization. In: Proceedings of ICNN 1995 - International Conference on Neural Networks. vol. 4, pp. 1942–1948, November 1995

9. Kristan, M., et al.: The visual object tracking VOT2016 challenge results. In: Hua, G., Jégou, H. (eds.) ECCV 2016. LNCS, vol. 9914, pp. 777–823. Springer, Cham (2016). https://doi.org/10.1007/978-3-319-48881-3_54

10. Last, F., Douzas, G., Bacao, F.: Oversampling for imbalanced learning based on k-means and smote (2017)

11. Li, X., Shen, C., Dick, A., van den Hengel, A.: Learning compact binary codes for visual tracking. In: 2013 IEEE Conference on Computer Vision and Pattern Recognition, pp. 2419–2426, June 2013

12. Li, X., Hu, W., Shen, C., Zhang, Z., Dick, A., Hengel, A.V.D.: A survey of appearance models in visual object tracking. ACM Trans. Intell. Syst. Technol. **4**(4), 58:1–58:48 (2013)

13. Maresca, M.E., Petrosino, A.: MATRIOSKA: a multi-level approach to fast tracking by learning. In: Petrosino, A. (ed.) ICIAP 2013. LNCS, vol. 8157, pp. 419–428. Springer, Heidelberg (2013). https://doi.org/10.1007/978-3-642-41184-7_43

14. Palaniappan, K., et al.: Efficient feature extraction and likelihood fusion for vehicle tracking in low frame rate airborne video. In: 2010 13th International Conference on Information Fusion, pp. 1–8, July 2010

15. Park, D.W., Kwon, J., Lee, K.M.: Robust visual tracking using autoregressive hidden Markov model. In: 2012 IEEE Conference on Computer Vision and Pattern Recognition, pp. 1964–1971, June 2012

16. Perez, P., Vermaak, J., Blake, A.: Data fusion for visual tracking with particles. Proc. IEEE **92**(3), 495–513 (2004)

17. Ross, D.A., Lim, J., Lin, R.S., Yang, M.H.: Incremental learning for robust visual tracking. Int. J. Comput. Vision **77**(1–3), 125–141 (2008)

18. Saffari, A., Leistner, C., Santner, J., Godec, M., Bischof, H.: On-line random forests. In: 2009 IEEE 12th International Conference on Computer Vision Workshops, ICCV Workshops, pp. 1393–1400, September 2009

19. Xiao, J., Stolkin, R., Leonardis, A.: Single target tracking using adaptive clustered decision trees and dynamic multi-level appearance models. In: 2015 IEEE Conference on Computer Vision and Pattern Recognition (CVPR), pp. 4978–4987, June 2015

20. Zhang, L., Varadarajan, J., Suganthan, P.N., Ahuja, N., Moulin, P.: Robust visual tracking using oblique random forests. In: 2017 IEEE Conference on Computer Vision and Pattern Recognition (CVPR), pp. 5825–5834, July 2017

Smart and Intelligent Sensors

Fabrication and Physical Characterization of Different Layers of CNT-BioFET for Creatinine Detection

Kshetrimayum Shalu Devi[✉], Gaurav Keshwani,
Hiranya Ranjan Thakur, and Jiten Chandra Dutta

Tezpur University, Tezpur 784028, Assam, India
{shalu, jitend}@tezu.ernet.in,
gauravkeshwani@gmail.com, hrtnerist@gmail.com

Abstract. In this work, two important layers namely transporting and bio-sensitive layer of CNT-BioFET have been fabricated for Creatinine detection using solution process. Creatinine is a waste product which is required for the detection of renal, muscle and thyroid function. For the fabrication of the BioFET, layers of CNT and ZrO_2 have been deposited one after another to serve as transporting layer and oxide layer respectively. The oxide layer depositions and CNT nano-composite layer have been done using PGSTAT128 N and programmable spin coating system (spinNXG-P2H) respectively. The immobilization of enzyme (Creatinine Deiminase) can be done using silicalite as adsorbent by drop casting. The physical characterization of deposited layers was done by SEM (Scanning electron Microscope) and X-ray diffraction (XRD). The XRD results of the deposited layers show the characteristic peaks of the corresponding layer. SEM analysis show physical structure and dimensions of the deposited films. The sizes of the particles were in nanometer range and show good structure and morphology.

Keywords: Creatinine · CNT-BioFET · Creatinine Deiminase

1 Introduction

Creatinine is a waste product of Creatine Phosphate and Creatine in normal body muscle function. Creatinine (2-amino-1 methyl-5H-imidazol-4 one) is an amino acid compound first named by Liebig in 1847 [1]. About 2% of the creatine in body is converted into creatinine and 0.2–8 gm/l is excreted in the urine [2] and is a marker for kidney function [3]. Constant amount of creatinine is formed and released in Glomalular filtration, in time span of 24 h. For blood serum/plasma the normal range of creatinine is 34–140 $\mu molL^{-1}$ and it can go up to concentration over 1000 $\mu molL^{-1}$ in case of kidney disorder [4]. Therefore, development of a reliable, accurate and cost-effective process for creatinine detection is extremely important and noteworthy.

Creatinine determination in clinical laboratories is mostly based on Jaffe reaction which is a spectrophotometric detection. This method has low selectivity due to the interference from other biomolecules [5]. Amperometric biosensors have been developed for detection of creatinine [6]; however, these devices have low selectivity, stability

B. Deka et al. (Eds.): PReMI 2019, LNCS 11942, pp. 535–542, 2019.
https://doi.org/10.1007/978-3-030-34872-4_59

and reproducibility [7, 8]. Among various concepts used for development of creatinine biosensors, the incorporation of enzyme with ion sensitive field effect transistor (ISFET) is one of the most captivating techniques. Using this concept, few creatinine sensitive ENFET devices (C-ENFET) have been developed in the recent past [9–11].

However, these devices have few disadvantages such as low sensitivity, high value of threshold voltage and small on-off current ratio. Using CNT in place of Silicon as channel material offer unique properties such as ballistic conduction, large surface area and high chemical stability [11]. Higher carrier mobility offered by CNT helps to improve sensitivity of the device as compared to silicon devices [12–17]. Moreover, compatibility of CNT with high-κ dielectric materials enhances device performance and also pave a way towards molecular electronics and devices miniaturization [14, 15]. For fabrication of such devices, electrochemical deposition (ECD) process has greater advantages as compared to the other microfabrication technologies presently used [18, 19].

For the fabrication of a CNT-BioFET different nanomaterials have been used. Zirconium dioxide (ZrO_2) is used as the gate oxide layer as it is one of the most versatile ceramic materials with good physic-chemical properties. Its high strength, high melting point, low thermal conductivity, toughness, abrasion and corrosion resistance make it desirable for broad range of applications in technology. Given its benefits, ZrO_2 is one of the most suitable candidates for gate dielectric material of in transistors [20–22].

The schematic figure and chemical bonding of ZrO_2 and PEI doped COOH-MWCNT nanocomposite are given in Fig. 1(a) and (b) respectively.

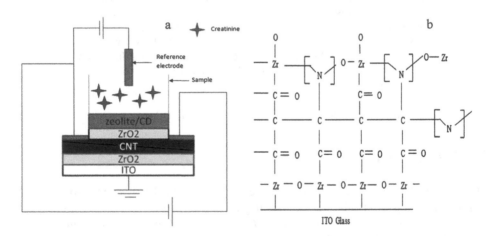

Fig. 1. (a) Schematic figure and (b) Chemical bonding of ZrO_2, PEI doped COOH-MWCNT and ZrO_2.

2   Experimental

2.1   Materials

Creatinine and Creatinine Deiminase enzyme were purchased from Sisco Research laboratory. COOH functionalized multi walled carbon nanotubes (MWCNTs) having carbon purity of $\sim 99\%$, were purchased from Alibaba. Indium tin oxide (ITO) glass was purchased from NANOCS. Polyethylene imine (PEI) was purchased from Sigma-Aldrich. Tetraethyl Orthosilicate (TEOS) (28-28.8% SiO2) and Tetrapropylammonium Hydroxide 20% were purchased from Sisco Research laboratory (SRL). All the materials and chemicals were of analytical grade.

2.2 Fabrication

ITO glass with dimension ~ 10 mm $\times$ 10 mm $\times$ 1 mm was used as substrate for fabrication of ENFET device. The ITO glass was rinsed using a solution composed of water, ammonium hydroxide and hydrogen peroxide in ratio (5:2:2) and then washed thoroughly with distill water [23]. For electrochemical deposition three electrodes system was used, ITO glass (cathode) as working electrode, platinum wire as counter electrode and Ag/AgCl electrode as reference electrode. A solution has been made by hydrolyzing $ZrCl_4$ in water (H_2O) (10 mg of $ZrCl_4$ in 10 ml of H_2O) and ultrasonicated for 20 min [24]. On the substrate ITO glass, the resulting solution of ZrO_2 was deposited by electrochemical deposition. The deposition was performed at room temperature, the resultant product after deposition was heated at 200 °C for 2 h.

Above ZrO_2 layer, PEI-doped COOH-MWCNT has been deposited using chemical solution process. Two different types of solvent are used for the preparation of the CNT solution. First, 10 mg of CNT has been added to 10 ml PEI/methanol and ultrasonicated for 30 min [12]. Secondly, by changing the solvent with ethanol, CNT has been added to 10 ml PEI/ethanol and ultrasonicated for 30 min [25]. The resulting solutions have been deposited by spin coating and dried at room temperature. For the oxide layer, ZrO_2 has been deposited above composite CNT layer using the same procedure.

A solution of silicalite was prepared in 5 mM PBS pH 7.4 using silicalite of 10% (v/v) and ultrasonicated for 10 min and then deposited over the gate surface by drop casting; after which, sample was heated for 15 min at 120 °C which will be used for the immobilization of enzyme. Tetraethyl Orthosilicate (TEOS) and tetrapropylammonium hydroxide (TPAOH) as a silica source and as a template respectively were used for silicalite preparation in molar ratio 1TPAOH:4TEOS:350H$_2$O [26, 27]. Before the deposition of the enzyme, silicalite-ZrO_2 layer was washed with distilled water to removed unbound silicalite from the ZrO_2 surface.

3 Results and Discussions

3.1 Structures of the Deposited Layers

Figures 2 and 3 shows the XRD pattern of ZrO_2 deposited on ITO glass and CNT composite. The two crystalline forms of ZrO_2 i.e., monoclinic, tetrahedral structures can be seen on the deposited layers. In Fig. 2, the peaks on the XRD shows the presence of characteristic peaks of both monoclinic 2θ = (111), (022), (122), (002), (200) reflections [JCPDS No. 37-1484] [28] and tetrahedral crystalline 2θ = 30.2, 34.5, 50.2, and 60.2 equivalent to the (101), (110), (200), and (211) reflections [JCPDS No. 70-1769] [29] form in the film. The bottom ZrO_2 layer acts as the insulating layer to avoid current leakage from transporting layer to ITO glass. In Fig. 3, ZrO_2 characteristic peaks can be seen but the intensity of the peak is lower as the crystal is poorly crystalline in structure. Poor crystallization of ZrO_2 deposited on CNT nanocomposite layers might be due to some effect of CNT layer beneath the deposited layer.

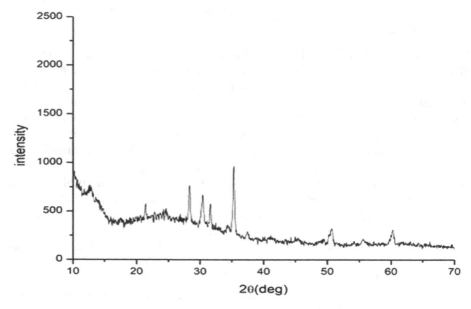

Fig. 2. XDR pattern of ZrO_2 on ITO glass

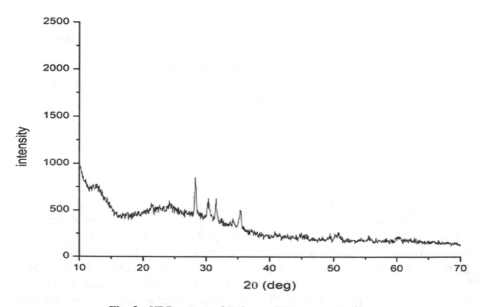

Fig. 3. XDR pattern of ZrO_2 on CNT composite film.

3.2   Morphology of the Deposited Layers

SEM analysis has been done after deposition of each layer for the study of surface morphology of the layers. SEM equipment (JEOL JSM 6390 LV, Singapore) was used for the investigation of the surface.

Figure 4(a) shows the SEM result of ZrO_2 deposited on ITO glass, most of the deposited ZrO_2 particles were in nanometer range.

Figures 5(a) and (b) show the SEM image of PEI doped COOH-MWCNT in methanol and ethanol solvent respectively. The PEI doped COOH-MWCNT composite deposited on the ZrO_2 surface is uniformly distributed in the second solution as compared to the first. The CNTs were dispersed homogeneously as the tubes of MWCNTs can be seen in the ethanol solvent much clearly than the methanol solvent, so the second sample is further used for ZrO_2 deposition. Because of the amorphous loading of PEI coating on the surface of CNTs [30] and high loading of CNT in PEI, the CNTs are tightly bound to one another which make the CNTs increase its electrical conductivity. The CNT layer act as source, drain and channel of fabricated Bio-FET.

Figure 4(b) shows surface morphology of ZrO_2 deposited on the CNT composite layer. The particles are formed in an amalgam so the surface is non-uniform. The shape and size of the particles in ZrO_2 deposited on CNT composite layer are different from the ZrO_2 deposited on ITO glass.

Immobilization of Creatinine Deiminase can be done using adsorption technique on silicalite adsorbent membrane. Excluding the gate area, polydimethylsilaxane can be used to seal the device for the passivation of the layers to the electrolyte.

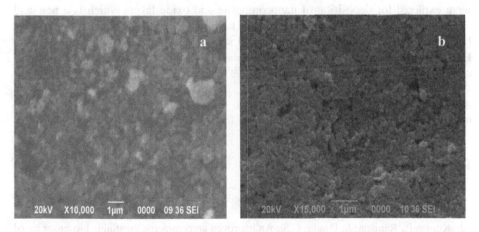

Fig. 4. (a) SEM image of ZrO_2 on ITO glass. (b) SEM image of ZrO_2 deposited on CNT composite film.

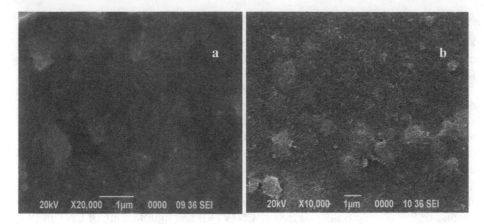

Fig. 5. (a) SEM image of PEI doped COOH-MWCNT using methanol as solvent and (b) SEM image of PEI doped COOH-MWCNT using ethanol as solvent

4 Conclusion

In this work, layers of CNT based Bio-FET has been deposited and physically characterized. High k-dielectric material ZrO_2 and polymer nanocomposite of functionalized CNT were deposited using solution process. The crystal structure and morphology have been studied using XRD and SEM analysis. Results show good crystalline structure and surface morphology. Electrochemical deposition and spin coating process were explored for deposition of nanocomposite and oxide layers which is a new and cost effective methodology for fabrication of ENFET devices. In this study, efforts have been made to fabricate the transporting and sensitive layers of nano-structured CNT-BioFET for Creatinine detection that may play important role in bioelectronics applications in near future.

Acknowledgements. The authors would like to extend their gratitude towards the Bioelectronics laboratory, ECE Dept., Tezpur University for allowing us to use the laboratory facilities.

References

1. Liebig, J.: Kreatin und Kreatinin, Bestandtheile des Harns der Menschen. J. für praktische Chemie **40**(1), 288–292 (1847)
2. Kumar, P., Ranjana, J., Pundir, C.S.: An improved amperometric creatinine biosensor based on nanoparticles of creatininase, creatinase and sarcosine oxidase. Anal. Biochem. **537**, 41–49 (2017)
3. Uchino, S.: Creatinine. Curr. Opin. Crit. Care **16**(6), 562–567 (2010)
4. Tietz, N.W.: Textbook of Clinical Chemistry, 1st edn. Saunders, Philadelphia (1968)
5. Jaffé, M.: Ueber den Niederschlag, welchen Pikrinsäure in normalem Harn erzeugt und über eine neue Reaction des Kreatinins. Zeitschrift für physiologische Chemie **10**(5), 391–400 (1886)

6. Weber, J.A., Van Zanten, A.P.: Interferences in current methods for measurements of creatinine. Clin. Chem. **37**(5), 695–700 (1991)
7. Kubo, I., Karube, I.: Immobilization of creatinine deiminase on a substituted poly (methylglutamate) membrane and its use in a creatinine sensor. Anal. Chim. Acta **187**, 31–37 (1986)
8. Khan, G.F., Wernet, W.: A highly sensitive amperometric creatinine sensor. Anal. Chim. Acta **351**(1–3), 151–158 (1997)
9. Soldatkin, A.P., Montoriol, J., Sant, W., Martelet, C., Jaffrezic-Renault, N.: Creatinine sensitive biosensor based on ISFETs and creatinine deiminase immobilised in BSA membrane. Talanta **58**(2), 351–357 (2002)
10. Soldatkin, A.P., Montoriol, J., Sant, W., Martelet, C., Jaffrezic-Renault, N.: Development of potentiometric creatinine-sensitive biosensor based on ISFET and creatinine deiminase immobilised in PVA/SbQ photopolymeric membrane. Mater. Sci. Eng.: C **21**(1–2), 75–79 (2002)
11. Sant, W., et al.: Development of a creatinine-sensitive sensor for medical analysis. Sens. Actuators B: Chem. **103**(1–2), 260–264 (2004)
12. Barik, M.A., Dutta, J.C.: Fabrication and characterization of junctionless carbon nanotube field effect transistor for cholesterol detection. Appl. Phys. Lett. **105**(5), 053509 (2014)
13. Peng, L.M., Zhang, Z., Wang, S.: Carbon nanotube electronics: recent advances. Mater. Today **17**(9), 433–442 (2014)
14. Javey, A., et al.: High-κ dielectrics for advanced carbon nanotube transistors and logic gates. Nat. Mater. **1**, 241–246 (2002)
15. Javey, A., et al.: Carbon nanotube field-effect transistors with integrated ohmic contacts and high-κ gate dielectrics. Nano Lett. **4**(3), 447–450 (2004)
16. Javey, A., et al.: High performance n-type carbon nanotube field-effect transistors with chemically doped contacts. Nano Lett. **5**(2), 345 (2005)
17. Gruner, G.: Carbon nanotube transistors for biosensing applications. Anal. Bioanal. Chem. **384**(2), 322–335 (2006)
18. Dutta, J.C., Thakur, H.R.: Sensitivity determination of CNT-based ISFETs for different high-dielectric materials. IEEE Sens. Lett. **1**(2), 1–4 (2017)
19. Barik, M., Deka, A.R., Dutta, J.C.: Carbon nanotube-based dual-gated junctionless field-effect transistor for acetylcholine detection. IEEE Sens. J. **16**(2), 280–286 (2016)
20. Robertson, J., Robert, M.W.: High-K materials and metal gates for CMOS applications. Mater. Sci. Eng.: R: Rep. **88**, 1–41 (2015)
21. Panda, D., Tseung-Yuen, T.: Growth, dielectric properties, and memory device applications of ZrO2 thin films. Thin Solid Films **531**, 1–20 (2013)
22. Fu, L., Yu, A.M.: Carbon nanotubes based thin films: fabrication, characterization and applications. Rev. Adv. Mater. Sci. **36**(1), 40–61 (2014)
23. Barik, M.A., et al.: Highly sensitive potassium-doped polypyrrole/carbon nanotube-based enzyme field effect transistor (ENFET) for cholesterol detection. Appl. Biochem. Biotechnol. **174**(3), 1104–1114 (2014)
24. Rahtu, A., Mikko, R.: Reaction mechanism studies on the zirconium chloride–water atomic layer deposition process. J. Mater. Chem. **12**(5), 1484–1489 (2002)
25. Lee, M.-S., Lee, S.-Y., Park, S.: Preparation and characterization of multi-walled carbon nanotubes impregnated with polyethyleneimine for carbon dioxide capture. Int. J. Hydrog. Energy **40**(8), 3415–3421 (2015)
26. Marchenko, S.V., et al.: Creatinine deiminase adsorption onto silicalite-modified pH-FET for creation of new creatinine-sensitive biosensor. Nanoscale Res. Lett. **11**(1), 173 (2016)
27. Kasap, B.O., et al.: Biosensors based on nano-gold/zeolite-modified Ion selective field-effect transistors for creatinine detection. Nanoscale Res. Lett. **12**(1), 162 (2017)

28. Kumar, S., et al.: Biofunctionalized nanostructured zirconia for biomedical application: a smart approach for oral cancer detection. Adv. Sci. 2(8), 1500048 (2015)
29. Singh, A.K., Nakate, U.T.: Microwave synthesis, characterization, and photoluminescence properties of nanocrystalline zirconia. Sci. World J. (2014)
30. Geng, X., et al.: In situ synthesis and characterization of polyethyleneimine-modified carbon nanotubes supported PtRu electrocatalyst for methanol oxidation. J. Nanomater. 16(1), 19 (2015)

Long Term Drift Observed in ISFET Due to the Penetration of H+ Ions into the Oxide Layer

Chinmayee Hazarika[1], Sujan Neroula[2], and Santanu Sharma[2(✉)]

[1] Department of E.C.E, G.I.M.T, Guwahati 781017, Assam, India
[2] Department of E.C.E, Tezpur University, Tezpur 784028, Assam, India
santanu.sharma@gmail.com

Abstract. Ion Sensitive Field Effect Transistor (ISFET) and related biosensors when undergo prolonged hours of operation, witnesses a temporal change in threshold voltage which is termed as drift. Drift is a secondary effect which leads to instability of the device resulting into inaccuracy in both in vivo and in vitro measurements. Various types of drift have been of great interest for researchers and long term drift is one of its kinds. Long term drift is observed in different sensing layer in ISFET devices of which silicon dioxide (SiO_2) gate ISFET witnesses the maximum. This paper presents the modeling and analysis of long term drift observed in ISFET due to diffusion of H+ ions into the oxide layer and the field caused by this penetration. The additional hydrogen ions left after the protonation of dangling bonds penetrate through the sensing layer which results into an electric field that influences the threshold voltage. A physical model has been designed explaining the effect of this penetration of positive ions into sensing layer and simulations of this model has been carried out. This model has been further experimentally validated using a Schottky based ISFET device with SiO_2 as the sensing layer. Both the theoretical and experimental data indicates toward the existence of this drift and its prominence can be observed in lower pH values.

Keywords: Biosensors · Diffusion · Ion Sensitive Field Effect Transistor · Long term drift · Threshold voltage

1 Introduction

In 1970, ion sensitive field effect transistor abbreviated as ISFET was introduced to the semiconductor industry by Bergveld [1–3], which had SiO_2 as the sensing layer. Around at the same time Matsuo and Wise develop similar device using silicon nitride as the sensing layer [4]. Right from its inception, work progressed mostly along two directions- one regarding its sensitivity to ions other than hydrogen ions and the other was explaining the mechanism involved in the operation of pH ISFET. However, limited work is reported on the secondary issues such as drift, hysteresis etc. Drift is an inherent problem related to ISFETs and related bioFETs as it results into instability of the device and degradation in accuracy of measurement with time. Drift can be defined as slow, monotonic, temporal change in the value of threshold voltage [5]. It is also

© Springer Nature Switzerland AG 2019
B. Deka et al. (Eds.): PReMI 2019, LNCS 11942, pp. 543–553, 2019.
https://doi.org/10.1007/978-3-030-34872-4_60

defined as shift in the rate of change in the gate to source voltage under constant current. Two types of drift are mentioned by Hein and Egger- storage drift and long term drift [6]. Few probable causes of drift stated in reported works are- Variation of surface state density D_{it} at the Si/SiO$_2$ interface, partially dehydrated surfaces which observe slow surface effect such as rehydration, electric field enhances the ion migration such as sodium ion within the gate insulator, negative space charges inside SiO$_2$ films due to injection of electrons from electrolyte at strong anodic polarization, buried layer beneath the surface [7]. Long term drift is significantly low in silicon nitrite sensing layer when compared with silicon dioxide but its effect cannot be neglected. This takes place in silicon nitride due to slow conversion of the nitride surface to hydrated SiO$_2$ or oxynitride layer and this hydration rate is due to dispersive mechanism [8]. The variation observed in threshold voltage with time for ISFET with different sensing membrane when immersed for a duration of 10^3 min was 3–30 mV for Ta$_2$O$_5$, 30–60 mV for SiO$_2$, 40 mV for silicon nitride, 50 mV for aluminum oxide etc. [9]. In a work by Jamasb et al. it was indicated that the pH dependent threshold voltage change is function of the effective charges induced in the semiconductor by the total charges in the insulator [8]. Further, a technique was presented in a work by Anita Topkar et al. to investigate the penetration of ions into the silicon dioxide when it is exposed in electrolyte [10]. Although works as mentioned above indicates about drift and causes of it; but no proper explanation is provided on the penetration of ions which leads to the creation of the electric field due to diffusion of hydrogen ions which causes long term drift in ISFET and other BioFET. In this work, a mathematical model has been proposed to formulate a relation of the voltage change as the function of time due to the field caused by the diffusion of hydrogen ions which leads to the long term drift in SiO$_2$ gate pH ISFET. This can lead to further refinement of the data acquisition process using ISFET based sensor and can enhance the accuracy.

2   Mathematical Model

Diffusion of protons into the sensing layer is a process by which species moves as a result of concentration gradient. The longer is the duration for which the ISFET is immersed in the electrolyte higher is the number of protons penetrating into the oxide layer. The electric field due to this diffusion is given by [11].

$$E_{diffusion} = \frac{qD_p}{\sigma} \frac{dP}{dx} \tag{1}$$

The electric field developed due to the diffusion itself act as a field which causes movement of the additional protons inside silicon dioxide layer. Therefore the electric field responsible for additional proton drift is the same field that occurs due to the diffusion of hydrogen ion at the edge of the oxide layer i.e. $E_{diffusion} = E_{drift}$. E_{drift} is termed here as the electric field that causes proton drift inside oxide layer as depicted in Fig. 1.

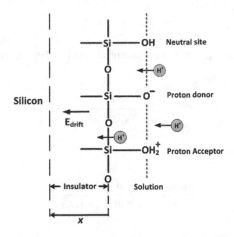

Fig. 1. Figure depicting diffusion of protons and the field caused by it

$$E_{drift} = \frac{qD_p}{\sigma}\frac{dP}{dx} \tag{2}$$

Here, D_p is the co-efficient of hydrogen ion near the oxide-solution interface, q is the electronic charge, σ is the conductivity, P is the concentration of the proton and dP/dx is the concentration gradient. Differentiating both sides of the Eq. (2), the equation obtained is

$$\frac{dE_{drift}}{dx} = \frac{qD_p}{\sigma}\frac{d^2P}{dx^2} \tag{3}$$

Further Fick's law of diffusion is given by

$$\frac{dP}{dt} = -D_p\frac{d^2P}{dx^2} \tag{4}$$

The Eqs. 3 and 4 can be simplified as,

$$\frac{dE_{drift}}{dx} = -\frac{q}{\sigma}\left(\frac{dP}{dt}\right) \tag{5}$$

Further, time variant concentration P is given by

$$P = \frac{Q}{\sqrt{\pi D_p t}}e^{-\frac{x^2}{4D_p t}} \tag{6}$$

Here Q is the surface concentration. Since the temporal variation of threshold voltage happen to be observed for comparatively longer duration of time and the minimum distance x is very small, we can approximate as

$$x^2 << 4D_p t \tag{7}$$

With the above assumption, the exponential term in Eq. (6) tends to unity. Hence Eq. (6) can be written as

$$P = \frac{Q}{\sqrt{\pi D_p t}} \tag{8}$$

Substituting Eq. (8) in (5)

$$\frac{dE_{drift}}{dx} = -\frac{q}{\sigma}\frac{d}{dt}\left(\frac{Q}{\sqrt{\pi D_p t}}\right) \tag{9}$$

Differentiating Eq. (9) can be written as

$$\frac{dE_{drift}}{dx} = \frac{q}{2\sigma}\left(\frac{Q}{\sqrt{\pi D_p}}\right)(t^{-1.5}) \tag{10}$$

Further integrating Eq. (10), the obtained equation is

$$E_{drift} = \frac{qQ}{2\sigma}\frac{1}{\sqrt{\pi D_p}}t^{-1.5}x \tag{11}$$

The constant of integration has been formulated using a boundary condition as E_{drift} $(x) = 0$, when $x_{minimum} \leq x < 0$. Here we have considered that the field is created only after the protons have moved to a certain distance which is explained in Eq. 14. Integrating Eq. (11) with necessary constraints we get

$$-\int_0^V dV = \frac{qQ}{2\sigma}\frac{1}{\sqrt{\pi D_p}}(t)^{-1.5}\int_{x_j}^{x_{ox}} xdx \tag{12}$$

Equation (12) can be simplified as

$$V = \frac{qQ}{4\sigma}\frac{1}{\sqrt{\pi D_p}}(t)^{-1.5}(x_j^2 - x_{ox}^2) \tag{13}$$

x_{ox} is the oxide thickness. Further, x_j is the distanced travelled by proton in silicon dioxide which is given as follows [11].

$$x_j = 2\sqrt{D_p t}\, erfc^{-1}\left(\frac{C_{sensing}}{C_s}\right) \tag{14}$$

Here $erfc^{-1}$ is the inverse complementary error function, C_S is the surface concentration and $C_{sensing}$ is the trapped charges in the oxide layer. Therefore, substituting Eq. (14) in (13),

$$V_{drift} = \frac{qQ}{4\sigma} \frac{1}{\sqrt{\pi D_p}} (t)^{-1.5} (4D_p terfc^{-1} \left(\frac{C_{sensing}}{C_s}\right)^2 - x_{ox}^2) \tag{15}$$

Equation (15) is the voltage developed because of the field created due the diffusion of additional protons into the oxide. The voltage developed due to diffusion of protons into the silicon dioxide is found as [12]

$$V_{th} = 2|\phi_f| + \frac{Q_B}{C_{ox}} - \frac{q}{\varepsilon_{ox}} (x_{ox} - 2\sqrt{Dt} erfc^{-1} \frac{C_{sensing}}{C_s}) \tag{16}$$

The later part of the Eq. 16 can be expressed as the voltage developed due to the diffusion of hydrogen ions. This voltage is further addressed here as $V_{diffusion}$ as shown in Eq. 17.

$$V_{diffusion} = \frac{q}{\varepsilon_{ox}} (x_{ox} - 2\sqrt{Dt} erfc^{-1} \frac{C_{sensing}}{C_s}); \tag{17}$$

$$V_{th} = 2|\phi_f| + \frac{Q_B}{C_{ox}} \tag{18}$$

The effect of the diffusion and field due to it on the threshold voltage can be expressed as

$$V_{th(final)} = V_{th} - (V_{diffusion} + V_{drift}) \tag{19}$$

Substituting the Eqs. 16, 17 and 18 in Eq. 19 the final expression of the threshold voltage can be obtained, as given in Eq. 20

$$V_{th(final)} = 2|\phi_f| + \frac{Q_B}{C_{ox}} - \left(\begin{array}{c} \left(\frac{q}{\varepsilon_{ox}} (x_{ox} - 2\sqrt{Dt} erfc^{-1}) \frac{C_{sensing}}{C_s} + \right) \\ \frac{qQ}{4\sigma} \frac{1}{\sqrt{\pi D_p}} (t)^{-1.5} \left(4D_p - terfc^{-1} \left(\frac{C_{sensing}}{C_s}\right)^2 - x_{ox}^2\right) \end{array} \right) \tag{20}$$

Here q is the charge per unit area, ϕ_f is the fermi-potential, Q_B is the bulk depletion charge, C_{ox} is the gate capacitance. The whole analysis has been carried out in isothermal condition and reference electrode voltage is considered zero. It has been observed from Eq. (20) that the temporal change of threshold voltage due to diffusion is proportional to $t^{1/2}$ and due to electric field resulting from diffusion is $t^{-3/2}$. Hence, it can be concluded that temporal change of threshold voltage due to diffusion is more dominant than that of the field created by diffusion with the passage of time.

3   Experimental Details

A Schottky based ISFET fabrication has been discussed in brief in this section, however, the detailed fabrication process flow of the same device has been reported in [2]. Silicon dioxide of thickness 480 nm has been deposited by thermal oxidation technique at a temperature of 1050 °C with a gas flow of 2 slpm (standard liquid per minute) of oxygen on a 3 inch double-sided polished p-type silicon wafer. Then first lithography was done to define the area where the metal has to be directly deposited on silicon substrate to form Schottky contact. The second lithography was done to partially etch out the active area to form the sensing area. Buffer oxide etchant was used for the etching purpose. Thus the sensing region obtained is about 120 nm thick measured using prism coupler (Metricon 2010M). Finally silver was deposited on the region exposed in the first step of lithography thus forming the Schottky contacts. The silicon substrate was then mounted on PCB board with adhesive. For the source and drain connections male connectors were soldered on the PCB board. The connectors were further bonded to the metal (Ag) layers using silver paste. A glass chamber was constructed using cover slip so that the electrolyte can be poured on the sensing layer. Silicone has been used for the sealing and passivation purpose. The schematic of the fabricated device and the picture of the actual device are illustrated in the Fig. 2(a) and (b) respectively.

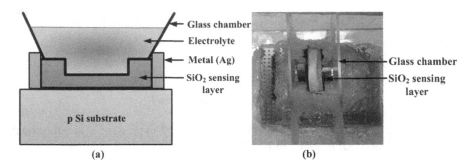

Fig. 2. (a) The schematic of the fabricated device (b) The picture of the actual device fabricated

3.1   Measurement Procedure

Measurement of the fabricated ISFET device was performed to witness the change in threshold voltage with the passage of time for different pH values. The experiment was carried out by maintaining a constant voltage of 0.27 V across the source and drain using a two electrode Electrometer (Keithley 6517B) as illustrated in the Fig. 3(a). The electrometer also measures the drain current. With passage of every 1200 s, the transfer characteristics of the device are obtained by varying the gate to source voltage and carefully noting the drain current at constant ambient temperature of 25 °C. The threshold voltage can be extracted by extrapolation of the linear region of transfer characteristics. The experiment was carried out for the three pH value (pH 4, 7 and 10). The schematic of the measurement has been illustrated in Fig. 3(b).

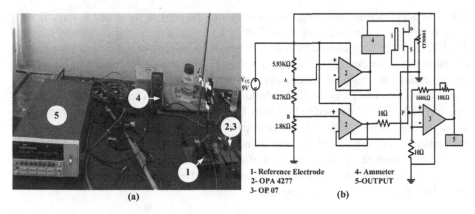

(a)   (b)

1- Reference Electrode   4- Ammeter
2- OPA 4277   5-OUTPUT
3- OP 07

Fig. 3. (a). The measurement set up (b) Schematic of measurement set up

4 Results and Discussion

The above explained analysis has been simulated considering the necessary boundary conditions and feasible values for the constant parameters as reported in [12]. The following parameters have been set prior to simulation of the derived mathematical expressions. The fixed oxide charges ($C_{sensing}$) present in carefully treated Si/SiO$_2$ interface system for <100> surface is found to be 1×10^{10} atoms/cm^3 [13]. Doping concentration N_a is taken as 1×10^{16}/cm^3. Other parameter values for simulation considered are as stated in Table 1.

Table 1. Following Table shows the values of volume concentration, surface concentration and surface charge density for pH values of 4, 7, 10

pH values	Volume concentration C_S (atom/cm^3)	Surface concentration (atom/cm^2)	Surface charge per unit area (C/cm^2)
4	6.022×10^{16}	1.53×10^{11}	2.448×10^{-8}
7	6.022×10^{13}	1.53×10^{9}	2.448×10^{-10}
10	6.022×10^{10}	1.53×10^{7}	2.448×10^{-12}

It is observed from Fig. 4(a) that for pH 4 the temporal variation in threshold voltage due to the electric field caused by diffusion becomes almost steady after 5000 s and the variation seems significantly higher before this point. This occurs because of the higher electric field caused by diffusion due to the higher concentration gradient at the oxide/electrolyte interface. The electric field decreases as it moves from the interface of the electrolyte/oxide into the depth of the oxide layer. When considering only the variation in threshold voltage due to the diffusion of hydrogen ions into the SiO$_2$ layer the change is observed to be 11.75 mV for a period of 10 h. In the first couple of hours significant change in the threshold voltage can be observed and this change gradually decreases in the later hours. Further, when solely the field caused by diffusion is considered, the contribution of this in the drift of threshold voltage is seen to be 0.969 mV for a period of 10 h. Further Fig. 4(a) depicts the combined effect of the diffusion and field caused by it on the threshold voltage of the device for pH 4. In the later hours the change in threshold voltage is dominated by the diffusion and the overall change in threshold voltage almost traces the plot of resultant threshold voltage due to diffusion. The variation in threshold voltage for pH 7 is found to be 0.12 mV for a period of 10 h as illustrated in Fig. 4(b) This can be explained as the concentration of hydrogen ions in the solution is lower in case of higher pH values. Further for pH 10, the variation in threshold voltage was found to be approximately 1.2 μV as depicted in Fig. 4(c). Therefore, it was observed from the simulated data that the change in threshold voltage for SiO_2 gate ISFET is prominent for lower pH values and for higher pH values these drift effect is quite low.

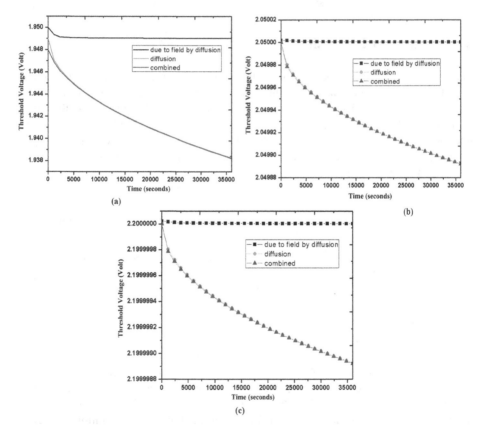

Fig. 4. (a, b & c) variation of threshold voltage with time, considering the diffusion of hydrogen ions into the sensing layer, the field caused by the diffusion and the combined effect of both for pH 4, 7 and 10 respectively.

The simulated data is compared with experimental data obtained from a fabricated SiO_2 gate pH ISFET. The experimentally obtained data along with the theoretical data for pH 4 is illustrated in Fig. 5(a). The fabricated ISFET showed a drift of 31.38 mV for a period of 10 h. The variation in the threshold voltage for both experimentally and theoretically obtained data shows a similar trend. However, a noticeable margin in the two values was observed as not all the parameters are considered in the MATLAB simulation environment. Further, in Fig. 5(b), the theoretical and the experimental values for the drift in threshold voltage for pH 7 are plotted. The fabricated ISFET showed a drift of approximately 0.4 mV for a period of 10 h. The variation in the threshold voltage exhibit similar trend for both experimentally and theoretically obtained results. In Fig. 5(c), the theoretical and the experimental plots for drift in threshold voltage for pH 10 are illustrated. Negligible variation between the results is observed as hydrogen ion concentration for pH 10 is significantly low.

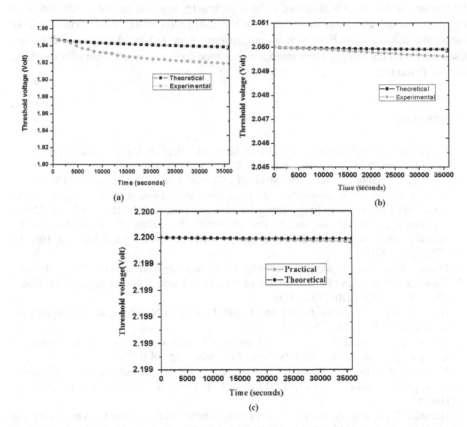

Fig. 5. (a, b, & c). Theoretical estimation of variation in threshold voltage with time for considering both diffusion and field caused by it along with the practical data obtained for SiO_2 gate pH ISFET when exposed for 36000 s (10 h) in pH 4, 7 and 10 respectively.

5 Conclusion

Long term drift, that brings instability to ISFET devices when kept immersed in electrolytes for long hours, cannot be overlooked as it results into inaccuracy in measurements. Most of the ISFETs and related BioFETs exhibit this effect. The diffusion of protons into the oxide layer results into this long term drift. The electric field resulting due to the diffusion of protons is more prominent in lower pH values and must be taken into account. The combined effect on the threshold voltage due to diffusion and due to electric field caused by diffusion of ions lowers the threshold voltage of the ISFET devices at lower pH values. The effect of electric field which brings about temporal change of threshold voltage in ISFET becomes less significant with the increase in pH values. The experiments carried out in this work using a fabricated Schottky based ISFET device the drift was observed to be high in the first few hour and attains a constantly decreasing trend after that. The decrease in threshold voltage after the passage of first hour is observed to be significantly low and can be concluded from the experimental and simulation data that the infiltration of H^+ ions occurs majorly during the initial hours. However, a comprehensive study considering all the factors associated with the sensing layer interface may produce better results and effective long term drift model.

References

1. Bergveld, P.: Development of an ion-sensitive solid-state device for neurophysiological measurements. IEEE Trans. Biomed. Eng. **17**(1), 70–71 (1970)
2. Hazarika, C., Sarma, D., Puzari, P., Medhi, T., Sharma, S.: Use of cytochrome P450 enzyme isolated from *Bacillus Stratosphericus sp.* as recognition element in designing schottky-based ISFET biosensor for hydrocarbon detection. IEEE Sens. J. **18**(15), 6059–6069 (2018)
3. Hazarika, C., Sarma, D., Neroula, S., Das, K., Medhi, T., Sharma, S.: Characterisation of a Schottky ISFET as Hg-MOSFET and as cytochrome P450-ENFET. Int. J. Electron. **105**(11), 1855–1865 (2018)
4. Matsuo, T., Esashi, M.: Methods of ISFET fabrication. Sens. Actuators **1**, 77–96 (1981)
5. Chou, J.C., Hsiao, C.N.: Drift behaviour of ISFETs with a-Si: H-SiO$_2$ gate insulator. Mater. Chem. Phys. **63**(3), 270–273 (2000)
6. Hein, P., Egger, P.: Drift behaviour of ISFETs with Si$_3$N$_4$-SiO$_2$ gate insulator. Sens. Actuators B: Chem. **14**, 655–656 (1993)
7. Bousse, L., Bergveld, P.: The role of buried OH sites in the response mechanism of inorganic-gate pH-sensitive ISFETs. Sens. Actuators **6**, 65–78 (1984)
8. Jamasb, S., Collins, S.D., Smith, R.L.: A physical model for threshold voltage instability in Si$_3$N$_4$-gate H^+-sensitive FET's (pH ISFET's). IEEE Trans. Electron Dev. **45**(6), 1239–1245 (1998)
9. Hazarika, C., Sharma, S.: Survey on ion sensitive field effect transistor from the view point of pH sensitivity and drift. Indian J. Sci. Technol. **10**(37), 1–18 (2017)
10. Topkar, A., Lal, R.: Effect of electrolyte exposure on silicon dioxide in electrolyte-oxide-semiconductor structures. Thin Solid Films **232**(2), 265–270 (1993)

11. Wolf, S., Tauber, R.N.: Silicon Processing for the VLSI Era. Process Technology, vol. 1. Lattice Press, Sunset Beach (1986)
12. Hazarika, C., Sharma, S.: A mathematical model describing drift in SiO_2 gate pH ISFET's due to hydrogen ion diffusion. Int. J. Appl. Eng. Res. **9**(23), 21099–21113 (2014)
13. May, G.S., Sze, S.M.: Fundamentals of Semiconductor Fabrication. Wiley, New York (2004)

Modelling and Simulation of a Patch Electrode Multilayered Capacitive Sensor

Mukut Senapati and Partha Pratim Sahu[✉]

Department of Electronics and Communication Engineering, Tezpur University,
Sonitpur, Tezpur, Assam, India
mukutsenapati@gmail.com, ppstezu@gmail.com

Abstract. In this paper, a MIS based patch electrode multilayered capacitive sensor has been modelled mathematically. The sensor consists of ZnO sensing layer having patch top electrode, SiO_2 layer and silicon substrate with bottom electrode. The overall capacitance of electrode is formulated by considering central, corner and fringe capacitances. The simulation was performed using mathematical mode to design the sensing layer thickness, patch electrode height and area and oxide layer thickness.

Keywords: Capacitive sensor · Multilayered sensor · Patch electrode

1 Introduction

Capacitive sensors have gained interest due to its simple structure, high reliability, low power consumption and low cost [1]. Capacitive sensors are used in many emerging fields such as automobile industry [2, 3], safety applications [4], industrial production [5] etc. However, effective design of the same is still a challenge for researchers when the design dimension goes beyond micrometric and nanometric scale. The most common capacitive sensor structure consists of two electrodes and a dielectric layer sandwiched between them. The geometries of the capacitive sensor electrode can be of (i) two parallel plate electrodes (ii) interdigitated electrodes (IDE) [6, 7]. Two parallel plate electrodes configuration of capacitive sensor is simple in design and its performance analysis can be easily done.

In a capacitive type sensor, change in the capacitance of the sensor occurs depending on the change in the dielectric properties of sensing layer [8, 9]. The variation of dielectric constant (ε_1), area of cross section (A_S) and thickness (h_1) of sensing layer would trigger change in capacitance. In gas sensing application, change in capacitance is observed due to change in the dielectric properties of the sensing layer on physisorption/chemisorption reaction between target gas and sensing surface.

In this paper a mathematical model of a multilayered capacitive sensor with two parallel plate electrodes configuration has been developed. Construction design details of the sensor exhibiting contribution from central, corner and fringe capacitance from the developed model has been obtained. Performance of the designed sensor has been analysed using MATLAB tool for high capacitive response with respect to change in dielectric properties.

B. Deka et al. (Eds.): PReMI 2019, LNCS 11942, pp. 554–560, 2019.
https://doi.org/10.1007/978-3-030-34872-4_61

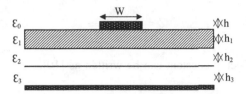

Fig. 1. Geometry of a metal patch top electrode multilayered capacitive sensor.

2   Patch Electrode Multilayered Capacitance Concept

The multilayered capacitive sensor with patch electrode is shown in the Fig. 1 where, the top squared shaped Au metal patch electrode is of height h. d width W. ε_0, is the permittivity in vacuum. The $\varepsilon_1, \varepsilon_2$ and ε_3 are relative permittivity of ZnO sensing layer, SiO$_2$ silicon oxide layer and silicon substrate. The h_1, h_2 and h_3 are the height of the sensing layer, oxide layer and substrate respectively. The bottom layer is the bottom metal electrode. The overall capacitance of the device is the contribution of central capacitance of the patch, corner capacitance under the multidielectric layer condition and fringe capacitance along the boundary. Thus, overall capacitance (C_T) of the multilayered device with squared shaped patch top electrode is written as [10],

$$C_T = C_0 + 2C_{f1} + 2C_{f2} + 4C_c \tag{1}$$

where, C_0 is the central capacitance, it depends upon permittivity and height of the dielectric layer. C_{f1} and C_{f2} are the fringe capacitances along the length L and width W, respectively. C_c is the corner capacitance. The central capacitance (C_0) can be calculated as,

$$C_0 = \frac{\varepsilon_0 A}{\left(\frac{h_1}{\varepsilon_1} + \frac{h_2}{\varepsilon_2} + \frac{h_3}{\varepsilon_3} \right)} \tag{2}$$

where,. A is the area of the squared shaped metal patch electrode.

The fringe capacitance (C_f) is given by,

$$C_{f1} = \frac{1}{2} \left(\frac{Z_0 W}{V_0 Z^2} - \frac{\varepsilon_1 \varepsilon_{eq} A}{h_1 + h_2 + h_3} \right) \tag{3}$$

here, Z_0 and Z are the characteristic impedances of the patch of width W on the air and the multilayered dielectric substrate respectively. V_0 is the velocity of light. Similarly, the fringe capacitance along the length L is obtained. Since, the patch is selected as squared shaped, therefore, $C_{f1} = C_{f2}$.

$$Z = \frac{Z_0}{\sqrt{\varepsilon_{eff}(W)}} \tag{4}$$

The effective dielectric constant can be written as [10],

$$\varepsilon_{eff} = \frac{C}{C_0} \tag{5}$$

$$Z_0 = \frac{1}{C_0 V_0} \tag{6}$$

$$Z = \frac{1}{V_0 \sqrt{CC_0}} \tag{7}$$

C and C_0 are capacitance per unit length of line with dielectrics and all dielectrics replaced by the air [11].

$$C \approx \pi \varepsilon_0 Y \tag{8}$$

where, Y is the admittance function and has been derived for a three layered structure as,

$$Y = 1 + \frac{\varepsilon_2 \varepsilon_3 (h_2 + h_3) coth\beta(h_2 + h_3) + \varepsilon_1 (h_2 \varepsilon_3 + h_3 \varepsilon_2) tanh\beta h_1}{\varepsilon_1 (h_2 \varepsilon_3 + h_3 \varepsilon_2) + \varepsilon_2 \varepsilon_3 (h_2 + h_3) coth(coth\beta(h_2 + h_3)) tanh\beta h_1} \tag{9}$$

The Corner capacitance (C_c) is obtained as,

$$C_c = \frac{1}{16} \left(\frac{W_{eq} - W}{W} \right) \times (CW - C_0) \tag{10}$$

where, W_{eq} is the equivalent width and is given as,

$$W_{eq} = \frac{120\pi h}{Z \sqrt{\varepsilon_{eff}(W)}} \tag{11}$$

Therefore, the Overall capacitance (C_T) of the device is obtained as,

$$C_T = \frac{1}{4} C(7W + W_{eq}) - \frac{C_0}{4} \left(3 + \frac{W_{eq}}{W} \right) \tag{12}$$

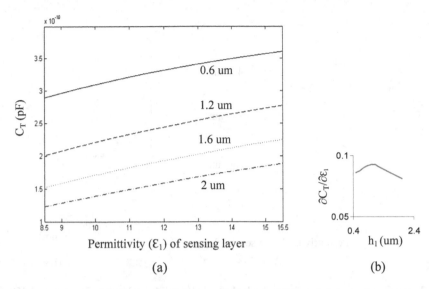

Fig. 2. (a) C_T versus ε_1 for different thickness of sensing layer, $h_1 = 0.6$ um, 1.2 um, 1.6 um and 2 um. (b) Gradient $\frac{\partial C_T}{\partial \varepsilon_1}$ vs h_1.

3   Simulation and Results

The formulated mathematical model was computed using MATLAB to obtain various design parameters for device construction. Various parameters were assessed using this model to check its effectiveness for better capacitive response. In Fig. 2(a) the overall capacitance (C_T) was obtained with respect to change in relative permittivity (ε_1) of the ZnO sensing layer for different thickness of the sensing layer ($h_1 = 0.6$ um, 1.2 um, 1.6 um and 2 um). Due to the adsorption reaction between the sensing surface and gas analytes, leading to change in the dielectric properties of sensing film, resulting in variation in capacitance. As shown in Fig. 2(b) the gradient $\frac{\partial C_T}{\partial \varepsilon_1}$ for different values of sensing layer thickness (h_1), is highest for 1.2 um. Thus, sensing layer thickness of 1.2 um was selected for device fabrication.

From Fig. 3(a) MATLAB plot between overall capacitance (C_T) and relative permittivity (ε_1) of the sensing layer for different height of patch electrode. It has been observed that the overall capacitance of the device depends upon the height of the patch electrode. As shown in Fig. 3(b) the $\frac{\partial C_T}{\partial \varepsilon_1}$ first increased and then it saturated as the patch electrode height increased beyond 350 nm.

Similarly, the plot between overall capacitance (C_T) and relative permittivity (ε_1) of the sensing layer for different patch electrode area (A) has been shown in Fig. 4(a). It is evident from Fig. 4(b), that the patch electrode area of 25 $(mm)^2$ is effective for device construction consideration since, $\frac{\partial C_T}{\partial \varepsilon_1}$ is maximum when the patch electrode area (A) was 25 $(mm)^2$.

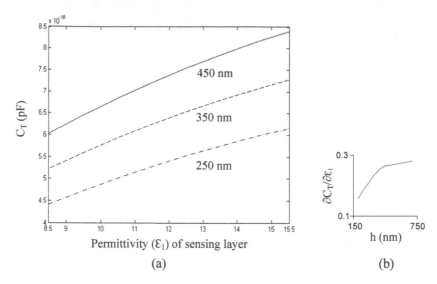

Fig. 3. (a) C_T versus ε_1 for different patch electrode height, h_1 = 250 nm, 350 nm, and 450 nm (b) Gradient $\frac{\partial C_T}{\partial \varepsilon_1}$ vs h.

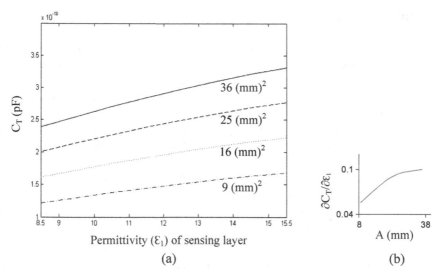

Fig. 4. (a) CT versus ε_1 for different patch electrode area, A = 9 $(mm)^2$, 16 $(mm)^2$, 25 $(mm)^2$ and 36 $(mm)^2$ (b) Gradient $\frac{\partial C_T}{\partial \varepsilon_1}$ vs A.

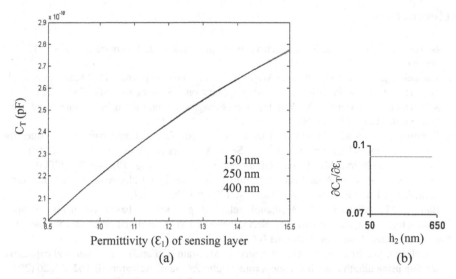

Fig. 5. (a) C_T versus ε_1 for different oxide layer height, h_2 = 150 nm, 250 nm and 400 nm. (b) Gradient $\frac{\partial C_T}{\partial \varepsilon_1}$ vs h_2.

Finally, the change in overall capacitance (C_T) with respect to relative permittivity (ε_1) of the sensing layer for different oxide thickness (h_2) has been shown in Fig. 5(a). It can be seen from the Fig. 5(b), as the height of oxide thickness increases, the gradient $\frac{\partial C_T}{\partial \varepsilon_1}$ remains constant. Therefore, for fabrication suitability, we have selected 250 nm of oxide layer thickness.

Table 1 mentions the design values of sensing layer thickness (h_1), patch electrode height (h), patch electrode area (A) and oxide layer thickness (h_2) obtained using mathematical model of capacitive sensor.

Table 1. Design parameters and its values.

Design parameters	Values
Sensing layer thickness (h_1)	1.2 um
Patch electrode height (h)	350 nm
Patch electrode area (A)	25 $(mm)^2$
Oxide layer thickness (h_2)	250 nm

4 Conclusion

In this paper, we have developed a mathematical model of a patch electrode multilayer capacitive sensor. A morphological design of capacitive sensor has been made to enhance the capacitive response of the sensor with variation in the dielectric constant of its sensing layer on exposure of gas samples. The mathematical model promises to design and develop a smart capacitive sensor for gas sensing application.

References

1. Ishihara, T., Matsubara, S.: Capacitive type gas sensors. J. Electroceramics **4**, 215–228 (1998)
2. Tamaekong, N., Liewhiran, C., Wisitsoraat, A., Phanichphant, S.: Flame-spray-made undoped zinc oxide films for gas sensing applications. Sensors **10**, 7863–7873 (2010)
3. Matsuzaki, R., Todoroki, A.: Wireless monitoring of automobile tires for intelligent tires. Sensors **8**, 8123–8138 (2008)
4. Carminati, M., et al.: Capacitive detection of micrometric airborne particulate matter for solid-state personal air quality monitors. Sens. Actuators A: Phys. **219**, 80–87 (2014)
5. Renard, S.: Industrial MEMS on SOI. J. Micromech. Microeng. **10**, 245–249 (2000)
6. Igreja, R., Dias, C.J.: Analytical evaluation of the interdigital electrodes capacitance for a multi-layered structure. Sens. Actuators, A **112**, 291–301 (2004)
7. Babaei, M., Alizadeh, N.: Methanol selective gas sensor based on nano-structured conducting polypyrole prepared by electrochemically on interdigital electrodes for biodiesel analysis. Sens. Actuators B **183**, 617–626 (2013)
8. Kavalenka, M., Striemer, C., DesOrmeaux, J., McGrath, J., Fauchet, P.: Chemical capacitive sensing using ultrathin flexible nanoporous electrodes. Sens. Actuators, B **162**, 22–26 (2012)
9. Patel, S., Mlsna, T., Fruhberger, B., Klaassen, E., Cemalovic, S., Baselt, D.: Chemicapacitive microsensors for volatile organic compound detection. Sens. Actuators **96**, 541–550 (2003)
10. Verma, A., Rostamy, Z.: Static capacitance of some multilayered microstrip Capacitors. IEEE Trans. Microw. Theory Tech. **43**, 1144–1152 (1995)
11. Yamashita, E., Mittra, R.: Variational method for the analysis of microstrip lines. IEEE Trans. Microw. Theory Tech. **4**, 251–256 (1968)

Systematic Design of Approximate Adder Using Significance Based Gate-Level Pruning (SGLP) for Image Processing Application

Sisir Kumar Jena$^{(\boxtimes)}$ (iD), Santosh Biswas, and Jatindra Kumar Deka

Department of CSE, Indian Institute of Technology Guwahati, Guwahati, India
sisir.jena@iitg.ac.in

Abstract. Approximate computing techniques emerged as a novel design paradigm that utilizes the error-resilience property of many applications and helps in reducing the power and area consumption with an expense of loss in accuracy of the result. In this paper, we introduce Significance-based gate-level pruning (SGLP) technique to design an approximate adder circuit whose accuracy can be controlled using an Error-Threshold provided by the application user. All previous method are nonsystematic and conceptually different from each other. Those methods can either apply to a chain-based adder (adders made up of a chain of full adders, e.g., Ripple Carry Adder) or unchain-based adder (e.g., Kogge-Stone Adder) but not both. SGLP follows a systematic approach to generate an approximate version of a Full Adder which is later used to produce multi-bit adder. By using the SGLP method, we can also realize an approximate variant of an unchained adder. This characteristic makes the SGLP more superior than the previous methods. To check the quality and reliability, we have tested our approach using a DCT architecture for image processing particularly image compression and found that our result is acceptable to human perception-behavior on image clarity.

Keywords: Approximate computing · Approximate circuit design · Approximate adder · Low power design · Gate-level pruning

1 Introduction

Approximate circuit design (ACD) paradigm is an emerging technique to realize future digital systems with less area and power consumption at a loss of a negligible amount of Quality of Result (QoR) or accuracy. Compared to contemporary IC design flow, ACD generates good enough result rather than an accurate result. There is a vast application area where these circuits can be used such as image processing, web search, machine learning, and many others those have some error resilience property. ACD can be used to design (i) Fundamental

© Springer Nature Switzerland AG 2019
B. Deka et al. (Eds.): PReMI 2019, LNCS 11942, pp. 561–570, 2019.
https://doi.org/10.1007/978-3-030-34872-4_62

arithmetic circuits like an adder, multiplier, and a divider, (ii) CPUs and GPUs, and (iii) Approximate accelerators. This paper mainly focuses on the design of an approximate adder circuit for image processing application. There are several ACD schemes proposed in the literature for designing approximate adder circuits [1–11]. We categorize them into three groups, shown in Fig. 1(a) based on either reducing the carry chain or redesigning a fundamental block. (1) *Block adder* [1–5]: This approach is also called full-adder approximation (AFA). Each FA is considered as a *block*, and the FAs present at the lower part of the multi-bit adder, are replaced with an approximate version of it. Figure 1(b) shows the overall idea of block adder technique. An approximate variant of an FA is built through redesigning either by substituting a gate with another (e.g., XOR is replaced with OR gate) or removing one or more component (gates or transistors). (2) *Segment Adder* [6–10]: This approach divides a given adder into equal/variable sized sub-adders called *segments*, shown in Fig. 1(c). One or more segments present in the lower part are executed inaccurately with no carry propagation. Hence, reducing the carry chain. The remaining segments being in the most significant portion will execute accurately and requite in producing good enough output. The two categories explained above, are often implemented on adders built from a chain of full adder cells (e.g., Ripple Carry Adder (RCA)), refer Fig. 1(a). So we introduce another category of ACD technique known as uncut adder that are applied on unchained adders to generate its approximate version. (3) *Uncut Adder* [11]: This approach neither divides an adder into blocks nor segments rather it directly prunes some components, and realizes an approximate variant of the adder (refer Fig. 1(d)). This technique mostly applied to adders like Kogge Stone Adder (KSA), Brent Kung Adder (BKA), etc. The pruning process, in this case, generally starts from the Least Significant Bits (LSBs) and moves towards the Most Significant Bits (MSBs). Unfortunately, Literature [11] is the only work published in this category, which is known as Gate-Level Pruning (GLP) technique. According to GLP, a circuit netlist is represented as a directed acyclic graph where nodes and edges represent gates and wires, respectively. The decision to prune a node depends on Significance-Activity Product (SAP) calculation. The nodes with the lowest SAP are pruned first. This process uses Error-Rate as the measure of approximation. GLP treats chained adders as a bad candidate for pruning process. In this paper, we propose an ACD technique known as Significance-based Gate-Level Pruning (SGLP) for designing adder circuits. If the above three categories are concerned, SGLP method comes under *uncut* as well as *block* adder category. i.e., we can apply this method to obtain an approximate version of chained (adders made up of a chain of full-adders, e.g., RCA), Unchained (e.g., KSA) and full adder block. As mentioned in [11], chained adders are not suitable for GLP approach but, with SGLP this is possible. In summary, SGLP approach can do the following that defines our contribution in this paper. (1)It is quite easy to get an FA approximation which can later be used in the lower part of a multi-adder to get a multi-bit approximate adder (Block adder category). (2) We can apply SGLP directly on the chain of FA (gate-level netlist) to realize its approximate variant.

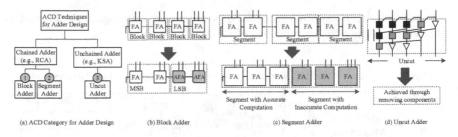

Fig. 1. ACD scheme for approximate adder

(3) Unlike GLP [11], SGLP can also be used to obtain the approximate version of uncut adders (e.g., KSA).

The other contribution of this paper includes: (1) We introduces a way of categorizing the ACD techniques for approximate adder design. To the best of our knowledge, this is the first paper that categorizes the ACD techniques. (2) We propose a systemic approach that removes gates and reduces the logic complexity at gate-level. (3) We demonstrate the benefits (in terms of power, area, and accuracy) of SGLP over previous approximate procedures and conventional adder design. (4) We built a DCT architecture using the approximate adders generated through SGLP for image compression application, and the result is outrightly acceptable.

2   SGLP Technique and Implementation

2.1   SGLP for FA Approximation

In this section, we describe the detailed procedure to generate an approximate FA block (AFA) and later part of this section describes how this AFA block is used to form a multi-bit adder circuit. We use RCA as the primary architecture upon which the AFA is implemented to build the required approximate multi-bit adder. SGLP follows a systematic approach and prune gates one by one and on every removal, we get one approximate version of the FA. Figure 2 shows the overall process. There are mainly two processes: (1) Significance Assignment (SA) and (2) Prune and Truth Table Analysis (PTA). The objective of SA is to assign an integer numeral to each gate of the given FA netlist. The purpose of PTA is to prune the gate having the lowest significance and analyze the effect in the truth table for all exhaustive set of inputs. One gate removal originates one approximate FA (AFA) which is again going through the SA and PTA method to produce another AFA. By connecting AFAs, we can generate several multi-bit approximate adders. Figure 3 shows the detail of the entire AFA generation started with the SA shown in Fig. 3(a). There are two output lines y and c_{out}, which is assigned with an initial value 2^0 and 2^1, respectively. Hence, the significance of the gates connected to y and c_{out} will be 1 and 2, respectively. The significance of the remaining gates follows a reverse topological order and is calculated by adding the significance of its descendants.

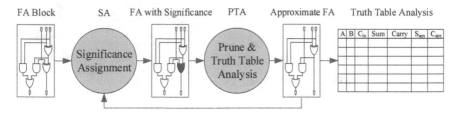

Fig. 2. Systematic process of SGLP for AFA

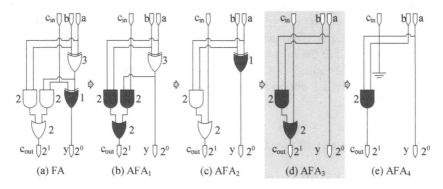

Fig. 3. Systematic generation of AFAs using SGLP

Figure 3(a) shows the significance of each gate calculated through the above approach. We can now prune the gates one by one to obtain AFAs starting with the gate having the lowest significance. Figure 3(b) shows the first AFA obtained by removing one gate from the original FA. After the removal, we recalculate the significance of the newly obtained AFA. The similar process continues to obtain the remaining AFAs. Figure 3(c)–(e) shows the entire AFAs obtained through the above approach. Table 1 shows the corresponding truth table analysis of each AFAs. The first five columns specify the input combinations (a, b and C_{in}) and the output of a conventional FA (Sum and C_{out}). The remaining columns show the outputs of all the AFAs (AFA_1 through AFA_4) where a tick mark ($\checkmark$) is used to indicate the correct output and cross ($\times$) to indicate incorrect output. Now, we can construct multi-bit adders using the AFAs depending on the application where it is used and accuracy of output the application needs.

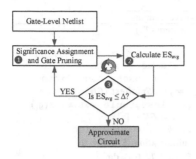

Fig. 4. Significance-based Gate-Level Pruning process

Table 1. Truth table analysis of approximate FAs

a	b	C_{in}	FA		AFA_1		AFA_2		AFA_3		AFA_4	
			S	C_o	S	C_o	S	C_o	S	C_o	S	C_o
0	0	0	0	0	0✓	0✓	0✓	0✓	0✓	0✓	0✓	0✓
0	0	1	1	0	0✗	0✓	0✗	1✗	0✗	1✗	0✗	0✓
0	1	0	1	0	1✓	0✓	1✓	0✓	0✗	0✓	0✗	0✓
0	1	1	0	1	1✗	1✓	1✗	1✓	0✓	1✓	0✓	0✗
1	0	0	1	0	1✓	0✓	1✓	0✓	1✓	0✓	1✓	0✓
1	0	1	0	1	1✗	1✓	1✗	1✓	1✗	1✓	1✗	0✗
1	1	0	0	1	0✓	1✓	0✓	1✓	1✗	1✓	1✗	1✓
1	1	1	1	1	0✗	1✓	0✗	1✓	1✓	1✓	1✓	1✓

2.2 SGLP for Uncut Adder

This section describes the detailed procedure and algorithm of Significance based Gate-Level Pruning (SGLP) method for Uncut adder. Figure 4 shows the overall process flow of SGLP. The SGLP method comprehensively defined as the process of removing (pruning) the netlist component (in our case, gate) such that, on execution, the netlist may produce an error but within the threshold defined by the designer. It is an iterative process (refer Fig. 4), begins with (1) significance assignment and pruning the gate with lower significance followed by (2) average Error-Significance (ES_{avg}) calculation and finally, (3) checking whether ES_{avg} is less than or equal to the error-threshold (Δ). The process repeats until $ES_{avg} \leq \Delta$, and on completion, we get a pruned version of the netlist with less number of gates, which produce a result within the error-threshold. Algorithm 1 in Fig. 5 shows the detailed procedure of SGLP method. GN denote the given Gate-Level netlist with x_i and y_i as the input and output lines, respectively. We represent the circuit under consideration (GN) as a tuple set $GN = \left\{ \langle g_i, s(g_i) \rangle \right\}$ Where: g_i represents gate(s) in GN and $s(g_i)$ represents significance of each gate Our algorithm generates a pruned version of GN (PGN) such that, the following conditions hold:

1. $|PGN(g_i)| \leq |GN(g_i)|$
2. $|PGN(x_i)| \equiv |GN(x_i)|$ and $|PGN(y_i)| \equiv |GN(y_i)|$
3. $ES_{avg}(PGN) \leq \Delta$

i.e., PGN has less number of gate and an equal number of input/output lines compared to GN. Lastly, the result produced is less than or equal to the error-threshold (Δ). We use the following notations in our proposed algorithm: T_e: Number of exhaustive test patterns of a circuit which is 2^n, where n is the number of input lines. T_k: Set of sample test patterns. T_k is much less than T_e. T_k^j: Represent a test pattern in T_k. i.e., $T_k^j \in T_k$. Υ: Output produced by the netlist (GN or PGN). R and $R^\dagger$: Set of all output produced by GN and PGN, respectively.

Our algorithm starts with the calculation of the golden result produced by the given netlist (GN). For smaller circuits, we took exhaustive test patterns (T_e),

---

Algorithm 1: Significance based Gate-Level Pruning

Data: Gate-Level Netlist (GN), Error Threshold (Δ), Sample test pattern (T_k)

Result: Pruned Gate-Level Netlist (PGN)

// Calculate the Golden result of GN and store it in a set **R**	// Gate Pruning: Removing the gate from the netlist having lowest significance.
1 **foreach** $T_k^j \in T_k$ **do**	13 $\tau = findSmallest\ (GN)$ // Find smallest tuple w.r.t. $s(g_i)$
2 $\Upsilon = y_{n-1}2^{n-1} + y_{n-2}2^{n-2} + \cdots + y_0 2^0$	14 $PGN = GN - \{\tau\}$
3 $R = R \cup \{\Upsilon\}$	// ES_{avg} calculation
4 **end**	15 **Initialization:** $ES_{sum} = 0$
5 **do**	16 **foreach** $T_k^j \in T_k$ **do**
//Assign Significance to each gate	17 $\Upsilon = y_{n-1}2^{n-1} + y_{n-2}2^{n-2} + \cdots + y_0 2^0$
6 **foreach** $g_i \in GN$ **do**	18 $R^\dagger = R^\dagger \cup \{\Upsilon\}$
7 **if** g_i *has no successor* **then**	19 **end**
8 $s(g_i) = 2^m, m = 0,1,2,\ldots$	20 $ES_{sum} = \sum_{i=1}^{|T_k|} (|R_i - R_i^\dagger|)$
9 **else**	21 $ES_{avg} = \dfrac{ES_{sum}}{|T_k|}$
10 $s(g_i) = \sum s(g_i^{descendant})$	22 $GN = PGN \backslash GN$
11 **end**	23 **while** $ES_{avg} \leq \Delta$
12 **end**	24 **return** PGN

Fig. 5. SGLP algorithm

and for a larger one, we randomly chose a subset (T_k). After applying them to GN, the output (Υ) is calculated using $\Upsilon = y_{n-1}2^{n-1} + y_{n-2}2^{n-2} + \ldots + y_0 2^0$ and stored in a set R (refer line number 1 to 3 in Algorithm 1).

The next task of our algorithm is to assign significance to each gate, which helps in identifying the first one to be removed. The process starts with the lowest level gates connected to output line and having no successor. It takes the form 2^m (assigns from the Least significant bit) where $m = 0, 1, 2\ldots$, represents the number of lines present in the output. Then a reverse topological traversal is performed to assign significance to the remaining gates present in the circuit. Significance calculation is carried out using $s(g_i) = \sum s\left(g_i^{descendant}\right)$ (refer line number 6 to 12 in Algorithm 1). Where: $s\left(g_i^{descendant}\right)$ is the significance of the immediate descendant of gate i.

After successfully assigning the significance to each gate of the circuit the pruning process is carried out. Line number 13 and 14 of our algorithm is doing this job. The function findSmallest(GN) is used to search the entire set and finds the gate having the lowest significance. This gate is pruned from the GN in line number 14 to get the PGN. Again, the same set of test-pattern (T_k) is applied on PGN to calculate the result. The output produced is stored in the set $R^\dagger$ (refer line number 16 to 19 in Algorithm 1). We have two sets of result R and $R^\dagger$, obtained from GN and PGN, respectively. We subtract them element-by-element using $ES_{sum} = \sum\limits_{i=1}^{|T_k|} \left(|R_i - R_i^\dagger| \right)$ to get our error-significance sum (ES_{sum}) in line number 20. Line number 21 calculates the average error-significance, and in line number 23 it is compared with the Error-Threshold (Δ). The process repeats only if $ES_{avg} \leq \Delta$, else our algorithm returns the PGN shown in line number 24. In case it reiterates then, the PGN obtained in the last iteration, is treated as the GN for the current iteration (refer line number 22).

3 Experimental Evaluation

In this section, we describe the detail implementation of the 16-bit adder circuit for FA approximation. Due to size limitation, we are not showing the detail implementation of the approximate uncut adder. But we can follow the same procedure as FA approximation to get our desired result. A simple Ripple Carry Adder (RCA) is considered in our case which is implemented using Verilog HDL in Xilinx Vivado 2018.1 environment (Vivado System Generator for DSP 2018.1). Compatible version Matlab 2017b is used to run our design. A system with Core i5 processor and 8 GB RAM is used to execute our experiments.

For FA approximation, we divided the entire circuit into two segments not necessarily equal. We replace the LSBs with the proposed AFAs, and the remaining MSBs uses the regular FA. We use the Design Compiler (DC) EDA tool from Synopsys with 45-nm open cell library to transform the RTL design into Gate-Level netlist. The detail of gain in terms of area, power, and delay is shown in Table 2. Here l_b represents the number of FA replaced with AFAs from LSB. For instance, $l_b = 4$ means we divide the entire circuit into two segments one segment contains 4 FAs from LSB, and other contains 12 FAs from MSB. We replace the 4 FAs from LSB with 4 AFAs.

After obtaining all these approximate 16-bit Adders (AFA-16), we generate the black box for each of them using Vivado System Generator Tool. These black boxes will be further used to implement the DSP application for image processing. We have generated black boxes for four AFA-16 with $l_b = 8$ which replaces AFA_1 through AFA_4 in each of these AFA-16.

DCT Application: DCT is a computationally unblemished component for image processing application. For our experiment, we took 8×8 pixel blocks DCT. Several DCT architecture has been proposed in the literature [12–14]. The conventional method requires 64 multiplication and 56 addition operations which is a substantial number and hence cannot solve our goal. The scope of our paper needs a multiplier-less DCT architecture. This paper does not present any new DCT rather we are using an existing multiplier-less state-of-the-art architecture to test our proposed method. One such architecture is presented in [14] which is a multiplication-free transform suitable for image compression commonly referred BAS-2011 in literature. The proposed hardware architecture of BAS-2011 has total 18 addition operations represent a 1D DCT. Using two 1D DCT block along with a transpose buffer, we can realize a complete 2D-DCT transform.

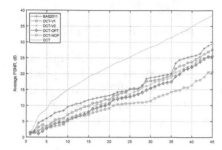

Fig. 6. Average PSNR for several compression ratios

Table 2. Area, Power, and delay characteristics

Adder	Matrix	$l_b = 2$	$l_b = 4$	$l_b = 8$	$l_b = 16$
RCA	Area	68			
	Power	0.54			
	Delay	0.8			
AFA_1	Area	66	64	61	54
	Power	0.52	0.51	0.48	0.43
	Delay	0.78	0.76	0.72	0.64
AFA_2	Area	64	62	55	41
	Power	0.50	0.46	0.42	0.32
	Delay	0.77	0.72	0.63	0.48
AFA_3	Area	59	55	46	27
	Power	0.49	0.45	0.37	0.21
	Delay	0.74	0.68	0.56	0.32
AFA_4	Area	61	54	40	14
	Power	0.47	0.42	0.30	0.10
	Delay	0.72	0.64	0.48	0.16

Area→$[nm^2]$, Power→$[mW]$, Delay→$[ns]$

FPGA Implementation: Initially, each 1D DCT is modeled by replacing the conventional adders with the proposed adder circuit (implemented as black box using system generator tool) and then linked to form a comprehensive 2D transform. The entire design is realized using Vivado System Generator tool. By this process, we get four different 2D DCT model named as DCT_{v1}, DCT_{v2}, DCT_{opt}, and DCT_{nop}. The realized models are physically built using Xilinx Virtex-6 XC6VLX240T field programmable gate array (FPGA) and connected to the host computer running Matlab Simulink version 2017b. Image processing activity is carried out by estimating the DCT of sample images acquired from each model. The transformed image is then fed into Inverse DCT function to obtain the compressed image. Finally, we calculate the quality measure PSNR (peak signal-to-noise ratio) of the original and compressed image for image degradation using $PSNR = 10 \log_{10} \left(\frac{MAX^2}{MSE} \right)$. Where MAX represents the maximum possible pixel value and MSE is the mean square error: the cumulative squared error between the original image I and obtained compressed image $\hat{I}$ using $MSE = \frac{1}{MN} \sum_{i=1}^{M} \sum_{j=0}^{N} [I(i,j) - \hat{I}(i,j)]^2$.

Image Compression and Result Analysis: To show that our proposed approach does not provide any unreasonable output, we conducted the image compression experiment described in [15] and carried by [16–18]. We considered 45 512 × 512 grayscale images obtained from a public image library [19]. Each image is divided into 8 × 8 blocks and submitted to the 2D transformation similar to [17]. For a particular transformation, each block furnished 64 coefficients in the approximate transform domain. Following the standard zigzag sequence [20] reconstruction of the image is done by employing $r (1 \leq r \leq 45)$ initial coefficient to each block and zero to the remaining coefficient. Finally, we obtain the compressed image by applying the actual inverse transformation. We then compare the original image with the compressed one for image degradation using PSNR as the quality measure. Figure 6 shows the average PSNR plot obtained by the experiment. Analyzing the result, we conclude that the proposed testing method does not produce any unreasonable output and quite competent for image compression.

Our objective is not to compare our result with other existing DCT algorithm. The main purpose of the experiment is to determine that approximate testing of circuit generates a tolerable result. Hence we also present a visual quality evaluation of our experimental result applied to the standard Lena image for r = 25. Figure 7 shows the effects of the experiment and supports our claim acquainted in introduction section.

Fig. 7. Lena image produced with (a) DCT, (b) BAS-2011, (c-f) Proposed method DCT_{v1}, DCT_{v2}, DCT_{opt}, DCT_{nop}

4   Conclusion

Approximate Computing technique is a novel design paradigm provides several benefits in terms of area, power consumption, and delay. In this paper, we have presented an ACD technique named as SGLP which can be used to reproduce adder circuits for chained and unchained adders. With the previously developed technique, this is not possible that shows the novelty of our work. We have tested our approach using a DCT architecture for image processing particularly image compression and found that our result is acceptable to human perception-behavior on image clarity.

References

1. Shin, D., Gupta, S.K.: A re-design technique for datapath modules in error tolerant applications. In: 2008 17th Asian Test Symposium, pp. 431–437. IEEE (2008)
2. Xu, S., Schafer, B.C.: Exposing approximate computing optimizations at different levels: from behavioral to gate-level. IEEE Trans. Very Large Scale Integr. (VLSI) Syst. **25**(11), 3077–3088 (2017)
3. Almurib, H.A.F., Kumar, T.N., Lombardi, F.: Inexact designs for approximate low power addition by cell replacement. In: 2016 Design, Automation & Test in Europe Conference & Exhibition (DATE), pp. 660–665. IEEE (2016)

4. Gupta, V., Mohapatra, D., Park, S.P., Raghunathan, A., Roy, K.: Impact: imprecise adders for low-power approximate computing. In: Proceedings of the 17th IEEE/ACM International Symposium on Low-Power Electronics and Design, pp. 409–414. IEEE Press (2011)
5. Gupta, V., Mohapatra, D., Raghunathan, A., Roy, K.: Low-power digital signal processing using approximate adders. IEEE Trans. Comput.-Aided Design Integr. Circuits Syst. **32**(1), 124–137 (2012)
6. Verma, A.K., Brisk, P., Ienne, P.: Variable latency speculative addition: a new paradigm for arithmetic circuit design. In: Proceedings of the Conference on Design, Automation and Test in Europe, pp. 1250–1255. ACM (2008)
7. Zhu, N., Goh, W.L., Wang, G., Yeo, K.S.: Enhanced low-power high-speed adder for error-tolerant application. In: 2010 International SoC Design Conference, pp. 323–327. IEEE (2010)
8. Zhu, N., Goh, W.L., Zhang, W., Yeo, K.S., Kong, Z.H.: Design of low-power high-speed truncation-error-tolerant adder and its application in digital signal processing. IEEE Trans. Very Large Scale Integr. (VLSI) Syst. **18**(8), 1225–1229 (2009)
9. Mahdiani, H.R., Ahmadi, A., Fakhraie, S.M., Lucas, C.: Bio-inspired imprecise computational blocks for efficient VLSI implementation of soft-computing applications. IEEE Trans. Circuits Syst. I: Regul. Pap. **57**(4), 850–862 (2009)
10. Dalloo, A., Najafi, A., Garcia-Ortiz, A.: Systematic design of an approximate adder: the optimized lower part constant-or adder. IEEE Trans. Very Large Scale Integr. (VLSI) Syst. **26**(8), 1595–1599 (2018)
11. Schlachter, J., Camus, V., Palem, K.V., Enz, C.: Design and applications of approximate circuits by gate-level pruning. IEEE Trans. Very Large Scale Integr. (VLSI) Syst. **25**(5), 1694–1702 (2017)
12. Karakonstantis, G., Banerjee, N., Roy, K.: Process-variation resilient and voltage-scalable DCT architecture for robust low-power computing. IEEE Trans. Very Large Scale Integr. (VLSI) Syst. **18**(10), 1461–1470 (2009)
13. Cintra, R.J., Bayer, F.M.: A DCT approximation for image compression. IEEE Signal Process. Lett. **18**(10), 579–582 (2011)
14. Bouguezel, S., Ahmad, M.O., Swamy, M.N.S.: A low-complexity parametric transform for image compression. In: 2011 IEEE International Symposium of Circuits and Systems (ISCAS), pp. 2145–2148. IEEE (2011)
15. Haweel, T.I.: A new square wave transform based on the DCT. Signal Process. **81**(11), 2309–2319 (2001)
16. Lengwehasatit, K., Ortega, A.: Scalable variable complexity approximate forward DCT. IEEE Trans. Circuits Syst. Video Technol. **14**(11), 1236–1248 (2004)
17. Bouguezel, S., Ahmad, M.O., Swamy, M.N.S.: Low-complexity 8×8 transform for image compression. Electron. Lett. **44**(21), 1249–1250 (2008)
18. Bouguezel, S., Ahmad, M.O., Swamy, M.N.S.: A fast 8×8 transform for image compression. In: 2009 International Conference on Microelectronics-ICM, pp. 74–77. IEEE (2009)
19. The USC-SIPI image database, University of Southern California, Signal and Image Processing Institute. http://sipi.usc.edu/database/
20. Wallace, G.K.: The JPEG still picture compression standard. IEEE Trans. Consum. Electron. **38**(1), xviii–xxxiv (1992)

Syringe Based Automated Fluid Infusion System for Surface Plasmon Resonance Microfluidic Application

Surojit Nath, Kristina Doley, Ritayan Kashyap,
and Biplob Mondal$^{(\boxtimes)}$

Department of Electronics and Communication Engineering, Tezpur University,
Tezpur, Assam, India
biplobm@tezu.ernet.in

Abstract. This paper presents the design, implementation and working of a syringe based automated fluid infusion system integrated with microchannel platforms for lab-on-a-chip application. Polymer based microchannels are fabricated and adhered to the silver coated glass substrates to direct the flow of the liquid at rates down to the value of 60 µL/min. The fluid infusion system integrated with the microfluidic platforms accounts for the precise and accurate monitoring of fluids at the micro-level presenting a scope for study of surface plasmon resonance based bio-sensing application.

Keywords: Lab-on-a-chip · Microfluidic · Surface plasmon resonance

1 Introduction

The field of microfluidics deals with the manipulation of the fluid flow to achieve steady flow, mixing of liquids etc. at micro, nano level [1, 2]. Recently Microfluidics devices are being used in diverse field or areas like bio-chemistry, food safety, pharmaceutics etc. for various bio-sensing, immune-sensing applications [3–5]. In such applications for many reasons detection of biomolecules like bacteria, proteins etc. are to be done at extremely small level and many a times presence of such target biomolecules in the sample is also very low [6]. Also point-of-care application demands for miniaturized devices that can integrate sensing and transducing element on same platform. The sensing devices in Lab-on-a-chip are to be exposed to the target sample at micro to nano levels. This demands the development of micfluidic devices that can allow precise manipulation of liquids at such extreme low volumes.

One of the specific applications of microfluidics is surface plasmon resonance (SPR) based biosensing which uses an optical transduction method for detection [7]. The measurement process requires the passage of the samples extremely precisely and accurately over the surface of the SPR sensor chip in order to produce extraordinary detection limits [8].The integration of microfluidics with SPR sensing provides the advantages of automation leading to precise control over reactions, fast processing with

© Springer Nature Switzerland AG 2019
B. Deka et al. (Eds.): PReMI 2019, LNCS 11942, pp. 571–578, 2019.
https://doi.org/10.1007/978-3-030-34872-4_63

small sample volumes and better sensing efficiency [9–12]. The use of microfluidic integrated platforms allows for the utilization of SPR lab-on-a-chip application for point-of-care testing. The increasing popularity of the SPR technique is observed in fundamental biological studies, health care research, drug discovery and clinical diagnosis, environmental and agricultural research etc. [13]. In this direction, a miniaturized pumping system with a flow rate of 250 µL/min is reported by Liu et al. for injection test samples to a SPR sensor for the detection of lectin concanavalin A (Con A) and glycoprotein ribonuclease B (RNase B) [14]. Tee et al. designed an automated syringe based fluidic system for microfluidic application that can provide flow rate as low as 100 µL/min using a 5 ml syringe [15]. The authors also reported a syringe based system integrated with polyimide fabricated microfluidic device to allow flow of liquids at relatively high flow rate in the range 1000–5000 µL/min [16].

This paper presents the design and implementation of a syringe based automated fluid infusion system to pump liquid through microchannels. Polymer based microchannels were fabricated and later adhered to silver coated glass substrate for the construction of microfluidic platform. This allows directing liquid samples and reagents to the sensor surface at flow rates down to 60 µL/min. The developed system integrated with the microfluidic platforms thereby presents a scope for use in surface plasmon resonance (SPR) based biosensing and multichannel outlet systems for lab-on-a-chip applications.

2   Experimental Methods

2.1   Design of the Syringe Based Fluid Infusion System

The working of the infusion system is based on the displacement of a syringe piston through a program controlled bipolar stepper motor. The shaft of the stepper motor (4.2 kg cm) is coupled to a linear slider of length 300 mm which is connected to the plunger of the syringe with the help of a sliding support to convert the rotational motion of motor into linear movement of the piston. The microcontroller controlled program signals the motor driver to rotate the stepper motor which exerts a force on the slider thereby pushing the plunger of the syringe forward causing the infusion of fluids. The infusion system has a provision for mounting of two syringes of different volumes that allows fluid flow in different ranges. The 10 ml syringe provides flow rates as high as 3000 µL/min while the 1 ml syringe provides flow rates as low as 60 µL/min. A soft flexible tube of inner diameter and outer diameter 0.2 mm and 0.4 mm respectively is used to direct the flow of liquids to a microfluidic device made of epoxy resin.

A power supply (12 V, 2A) is used to feed the power to the microcontroller (Arduino) and the stepper motor driver (L293 N). A 4 × 4 keypad panel and a 20 × 4 Liquid Crystal Display are assigned to the input and output ports of the microcontroller respectively from which they are powered for user interface purpose. Figure 1 shows the schematic representation of the syringe based fluid infusion system.

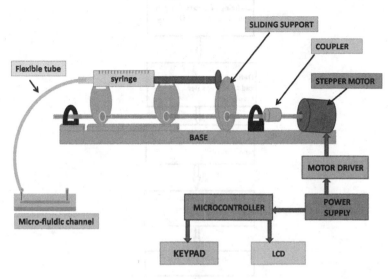

Fig. 1. Schematic representation of the syringe based fluid infusion system

The programming of the syringe based fluid infusion system is governed by a microcontroller to determine the step delay of the motor based on an algorithm. The library functions and the initialization of the input and output ports are defined in the program followed by the defining of step size of the bipolar motor (1.8°) and the total number of steps (200) required for one revolution. The linear displacement of the slider (6 mm) due to one revolution of the motor is thereafter defined in the program. The volume of the syringe is calibrated with its length (length of 10 ml syringe = 5.25 cm) and defined in the program to determine the volume of fluid infused (1.14285714 ml for 10 ml syringe) per revolution of the motor. Thereafter the volume of fluid infused due to a unit step of the motor is determined (5.7143 µl). Finally, the step delay, which is the time required to move unit step of motor is defined by dividing the infused volume per revolution of the motor (5.7143 µL for the 10 ml syringe) with the desired flow rate. The developed low cost infusion system amounting to a sum of nearly rupees two rupees thus enables the variation of different flow rates using syringes of varied dimensions and volumes. Figure 2 shows the program flowchart.

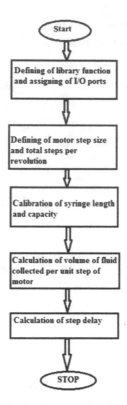

Fig. 2. Program flowchart

2.2   Fabrication of the Fluidic Microchannels

Polymer based fluidic microchannels were fabricated to direct the flow of the fluid infusion system through the microfluidic chambers. Figure 3 shows the schematic representation of the polymer based fluidic microchannels. High gloss epoxy resin clear coat was procured from Haksons containing the resin and the hardener. An epoxy resin flow cell was then fabricated by mixing the resin and the hardener in a solution of ratio 2:1. The solution was degassed for ten minutes in a water bath at a temperature of 50°c before being stored for a duration of twelve hours at room temperature to undergo solidification. Multiple cylindrical shaped channels with radius measuring 0.5 mm and length measuring 30 mm (volume ~23.55 µl) were designed on the fabricated flow cell with separate inlets and outlets for the flow of fluids. The fabricated epoxy resin flow cell was then adhered to a silver film coated BK 7 glass substrate of thickness around 50 nm with the help of a suitable adhesive to prevent any leakage of liquid.

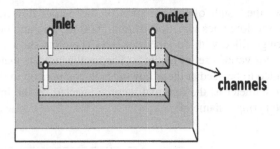

Fig. 3. Schematic representation of the polymer based fluidic microchannels

3 Results and Discussion

The syringe based fluid infusion system is experimentally tested by flowing fluids at different flow rates using two types of syringes of volume 1 ml and 10 ml.

Figure 4 shows the graph of the variation of the output flow rate with time at different values of set flow rates viz., 300 µL/min, 600 µL/min, 900 µL/min, 1000 µL/min, 2000 µL/min and 3000 µL/min using a 10 ml syringe filled with de-ionized water. The plots indicate a fluctuation in the output flow rate with time for the lower values of set flow rates viz., 300 µL/min–900 µL/min and a constant output flow rate with time for higher values of set flow rates viz., 1000 µL/min–3000 µL/min for various set of readings. Thus, the experimental results suggest that the 10 ml syringe based fluid infusion system provides a laminar flow of liquid only at the higher flow rates ranging from 1000 µL/min–3000 µL/min.

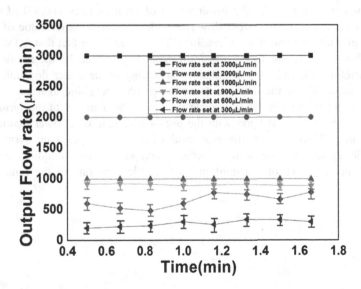

Fig. 4. Variation of the output flow rate of the infusion system with time at different values of set flow rates using a 10 ml syringe of DI water

Figure 5 shows the graph of the variation of the output flow rate with time at different values of set flow rates viz., 300 µL/min, 600 µL/min and 900 µL/min using a 1 ml insulin syringe filled with de-ionized water. The plots show a constant output flow rate with time for various set of readings. The results indicate that the 1 ml insulin syringe is more suited to be used at these flow rates in comparison to the 10 ml syringe owing to its small diameter as the minimal volume injected by the infusion system is proportional to the syringe diameter.

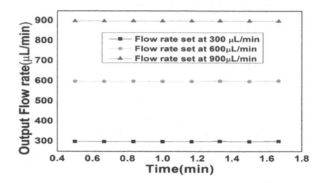

Fig. 5. Variation of the output flow rate of the infusion system with time at different values of set flow rates using a 1 ml syringe of DI water

Figure 6 shows the graph of the variation of the output flow rate with time at different values of set flow rates viz; 40 µL/min, 50 µL/min and 60 µL/min using a 1 ml insulin syringe filled with de-ionized water. The plots show a fluctuation in the output flow rate with time for the lower values of set flow rates viz., 40 µL/min and 50 µL/min while a constant output flow rate with time for a higher value of set flow rate of 60 µL/min for various set of readings. This is due to the fact that the stability in the fluid flow of a syringe based infusion system is governed by the rotation of the motor. Therefore, the pulses or oscillations appearing at the lower flow rates, represented by fluctuation in the data points on the graph, occur due to the mismatch of movement of the stepper motor in discrete steps with the diameter of the syringe. This mismatch can be improved upon with the use of a syringe of a lower diameter than a 1 ml syringe. Thus the experimental results indicate that the 1 ml syringe based fluid infusion system provides a stable, accurate and precise flow of liquid at a flow rate of 60 µL/min indicative of the minimum achievable flow rate of the fluid infusion system.

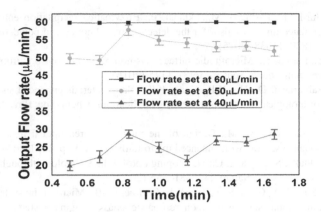

Fig. 6. Variation of the output flow rate of the infusion system with time (at low flow rate) using 1 ml syringe of DI water

4 Conclusion

This paper presents the design and working of a custom made syringe based automated fluid infusion system for precise monitoring of fluid flow at micro level. The system allows for the experimental variation of the flow rates down to the value of 60 µL/min using a 1 ml insulin syringe for use in microfluidics based application. The paper also presents the fabrication of polymer based microfluidic platforms to direct the fluidic flow at micro level without any leakage. The fluid infusion system integrated with the fabricated microfluidic platforms presents a future scope for use in surface plasmon resonance (SPR) based bio-sensing applications.

Acknowledgments. The authors are thankful to the Department of Science and Technology, Government of India and Science and Engineering Research Board, Government of India for providing the financial assistance. They are also thankful to Noman Hanif Barbhuiya, M.Sc., Department of Physics, Tezpur University.

References

1. Whitesides, G.M.: Nature **442**(7101), 368–373 (2006)
2. Hung, L., Lee, A.: Microfluidic devices for the synthesis of nanoparticles and biomaterials. J. Med. Biol. Eng. **27**(1), 1–6 (2007)
3. Jiang, K., Thomas, P.C., Forry, S.P., DeVoe, D.L., Raghavan, S.R.: Microfluidic synthesis of monodisperse PDMS microbeads as discrete oxygen sensors. Soft Matter **8**(4), 923–928 (2012)
4. Jayaraj, N., Cherian, C.M., Vaidyanathan, S.G.: Intelligent insulin infuser. Paper presented at the Third UK Sim European Symposium Computer Modeling Simulation, pp. 1–9 (2009)
5. Pessi, J., Santos, H.A., Miroshnyk, I., Yliruusi, J., Weitz, D.A., Mirza, S.: Microfluidics-assisted engineering of polymeric microcapsules with high encapsulation efficiency for protein drug delivery. Int. J. Pharm. **472**, 82–87 (2014)

6. Farka, Z., Juřik, T., Pastucha, M., Skládal, P.: Enzymatic precipitation enhanced surface plasmon resonance immunosensor for the detection of Salmoneela in powdered milk. Anal. Chem. **88**, 11830–11838 (2016)
7. Wang, D.-S., Fan, S.-K.: Microfluidic surface plasmon resonance sensors: from principles to point-of-care applications. Sensors **16**(8), 1175 (2016)
8. Zeng, S., Baillargeat, D., Ho, H.P., Yong, K.T.: Nanomaterials enhanced surface plasmon resonance for biological and chemical sensing applications. Chem. Soc. Rev. **43**, 3426–3452 (2014)
9. Hassani, A., Skorobogatiy, M.: Design of the microstructured optical fiber-based surface plasmon resonance sensors with enhanced microfluidics. Opt. Exp. **14**, 11616–11621 (2006)
10. Psaltis, D., Quake, S.R., Yang, C.: Developing optofluidic technology through the fusion of microfluidics and optics. Nature **442**, 381–386 (2006)
11. Lee, K.-H., Su, Y.-D., Chen, S.-J., Tseng, F.-G., Lee, G.-B.: Microfluidic systems integrated with two-dimensional surface plasmon resonance phase imaging systems for microarray immunoassay. Biosens. Bioelectron. **23**, 466–472 (2007)
12. Demello, A.J.: Control and detection of chemical reactions in microfluidic systems. Nature **442**, 394–402 (2006)
13. Singh, P.: SPR biosensors: historical perspectives and current challenges. Sens. Actuators, B **229**, 110–130 (2016)
14. Liu, Q., Liu, Y., Chen, S., Wang, F., Peng, W.: A low-cost and portable dual-channel fiber optic surface plasmon resonance system. Sensors **17**(12), 2797 (2017)
15. Tee, K.S., Saripan, M.S., Yap, H.Y., Soon, C.F.: Development of a mechatronic syringe pump to control fluid flow in a microfluidic device based on polyimide film. In: IOP Conference Series: Material Science and Engineering, vol. 226 (2017)
16. Yap, H.Y., Soon, C.F., Tee, K.S., Zainal, N., Ahmad, M.K.: Customizing a high flow rate syringe pump for injection of fluid to a microfluidic device based on polyimide film. ARPN J. Eng. Appl. Sci. **11**(6), 3849–3855 (2016)

Author Index

Printed in the United States
By Bookmasters